Plant Cell Wall Patterning and Cell Shape

Plant Cell Wall Patterning and Cell Shape

Edited by
HIROO FUKUDA

WILEY Blackwell

Published by John Wiley & Sons, Inc., Hoboken, New Jersey
Published simultaneously in Canada

Library of Congress Cataloging-in-Publication Data

Plant cell wall patterning and cell shape / edited by Hiroo Fukuda.
pages cm
Includes bibliographical references and index.
ISBN 978-1-118-64737-0 (cloth)
1. Plant cell walls. I. Fukuda, H (Hiroo)
QK725.P4723 2014
572′.472–dc23

2014024315

Cover images: Foreground images © Clara Sánchez Rodríguez;
Background image © BeholdingEye (iStockphoto.com)

Printed and bound in Malaysia by Vivar Printing Sdn Bhd

1 2015

Contents

List of Contributors

Takashi Aoyama
Institute for Chemical Research
Kyoto University
Kyoto, Japan

Ruthie Arieti
Department of Biological Sciences and Purdue University Life Sciences Graduate Program
Purdue University
West Lafayette, IN, USA

Rin Asaoka
Laboratory of Plant Cell Wall Biology
Graduate School of Life Sciences
Tohoku University
Sendai, Japan

Dominique C. Bergmann
Department of Biology and Howard Hughes Medical Institute
Stanford University
Stanford, CA, USA

Ying Fu
State Key Laboratory of Plant Physiology and Biochemistry
College of Biological Sciences
China Agricultural University
Beijing, China

Hiroo Fukuda
Department of Biological Sciences
Graduate School of Science
University of Tokyo
Tokyo, Japan

Kaori Furuta-Miyashima
Institute of Biotechnology and Department of Biology and Environmental Sciences
University of Helsinki
Helsinki, Finland

Ykä Helariutta
Institute of Biotechnology and Department of Biology and Environmental Sciences
University of Helsinki
Helsinki, Finland

Eva Hellmann
Institute of Biotechnology and Department of Biology and Environmental Sciences
University of Helsinki
Helsinki, Finland

Mariko Kato
Institute for Chemical Research
Kyoto University
Kyoto, Japan

Chin-Min Kimmy Ho
Department of Biology
Stanford University
Stanford, CA, USA

Hiroaki Kusano
Department of Biological Science and Technology
Tokyo University of Science
Tokyo, Japan

Jiejie Li
Department of Biological Sciences
Purdue University
West Lafayette, IN, USA

Raffael Lichtenberger
Institute of Biotechnology and Department of Biology and Environmental Sciences
University of Helsinki
Helsinki, Finland

M. David Marks Department of Plant Biology
University of Minnesota
St Paul, MN, USA

David W. McCurdy School of Environmental and Life Sciences
University of Newcastle
Callaghan, Australia

Shingo Nagawa Center for Plant Stress Biology
Shanghai Institutes for Biological Sciences
Chinese Academy of Sciences
Shanghai, China

Hideki Narukawa Laboratory of Plant Cell Wall Biology
Graduate School of Life Sciences
Tohoku University
Sendai, Japan

Kazuhiko Nishitani Laboratory of Plant Cell Wall Biology
Graduate School of Life Sciences
Tohoku University
Sendai, Japan

Yoshihisa Oda Center for Frontier Research, National Institute of Genetics
Shizuoka, Japan
Department of Genetics, Graduate University for Advanced Studies (SOKENDAI)
Shizuoka, Japan
PRESTO, Japan Science and Technology Agency
Saitama, Japan

Staffan Persson Max-Planck-Institute for Molecular Plant Physiology
Potsdam, Germany

Clara Sánchez-Rodríguez ARC Centre of Excellence in Plant Cell Walls
School of Botany University of Melbourne,
Victoria, Australia

Sidney L. Shaw Department of Biology
Indiana University
Bloomington, IN, USA

Naoki Shinohara Laboratory of Plant Cell Wall Biology
Graduate School of Life Sciences
Tohoku University
Sendai, Japan

Chris R. Somerville Energy Biosciences Institute
University of California, Berkeley
Berkeley, CA, USA

Christopher J. Staiger Department of Biological Sciences and
Bindley Bioscience Center
Purdue University
West Lafayette, IN, USA

Rumi Tominaga Graduate School of Biosphere Science
Hiroshima University
Hiroshima, Japan

Laura Vineyard Department of Biology
Indiana University
Bloomington, IN, USA

Takuji Wada Graduate School of Biosphere Science
Hiroshima University
Hiroshima, Japan

Ian S. Wallace Energy Biosciences Institute
University of California, Berkeley
Berkeley, CA, USA

Zhenbiao Yang Center for Plant Stress Biology
Shanghai Institutes for Biological Sciences
Chinese Academy of Sciences
Shanghai, China
Center for Plant Cell Biology
Department of Botany and Plant Sciences
University of California, Riverside
Riverside, CA, USA

Zheng-Hua Ye Department of Plant Biology
University of Georgia
Athens, GA, USA

Ryusuke Yokoyama
Laboratory of Plant Cell Wall Biology
Graduate School of Life Sciences
Tohoku University
Sendai, Japan

Ruiqin Zhong
Department of Plant Biology
University of Georgia
Athens, GA, USA

Lei Zhu
State Key Laboratory of Plant Physiology and Biochemistry
College of Biological Sciences
China Agricultural University
Beijing, China

Preface

Hundreds of thousands of species of plants have evolved to thrive on Earth. One of the reasons for their prosperity must be the acquisition and clever use of cell walls. Plant cell walls dictate the plant life form by preventing cell migration, resulting in immobility of plants. On the other hand, cell walls produce the strength that allows plants to grow as huge trees and enable a unique strategy for the plant body plan by using cells hardened with cell walls as building blocks. Moreover, cell walls define cell shape, cell function, and sometimes cell fate through mechanisms such as asymmetric cell division and intercellular communication. An understanding of plant cell walls is therefore essential for an understanding of plant life.

A number of books about plant cell walls have already been published. Most books focus on the structure and biosynthesis of plant cell walls, providing invaluable knowledge of the biochemical and structural nature of plant cell walls. However, there are few books that provide detailed information about cell wall functions and their underlying mechanisms, despite recent conspicuous progress in these areas. I therefore planned a book entitled *Plant Cell Wall Patterning and Cell Shape*, describing current knowledge of the spatiotemporal regulation of plant cell organization in view of cell walls.

This book is grouped into three sections: (1) *Factors Controlling Plant Cell Wall Patterning*; (2) *Cellular Mechanisms Underlying Various Cell Shapes*; and (3) *Developmental Regulations of Cell Shape*.

Spatiotemporal regulation of cell wall formation appears as cell wall patterns. Section 1 therefore deals with crucial components for cell wall patterning. In Chapters 1 and 2, current knowledge of the biosynthesis of cell wall components, including cellulose microfibrils, is described as a basis for cell wall patterning. Membrane traffic, which is another key component for cell wall patterning, is described in Chapter 3. Mechanisms underlying cell wall patterning involve microtubule and actin filament dynamics beneath the plasma membrane. Chapters 4 and 5 therefore highlight regulation of microtubule and actin filament arrangement, respectively, in cell wall pattern formation.

A conspicuous function of cell walls is the formation of diverse cell shapes. Section 2 therefore describes new insights into cellular mechanisms leading to distinctive cell shapes. In plants, polarization of the plasma membrane leads to the formation of a locally specialized architecture of cell walls, resulting in various shapes of plant cells with specific functions. Recent progress in this field has revealed that Rho-like GTPases from plants (ROPs) play a crucial role in polarization of the plasma membrane to form distinct plant cell shapes. Chapters 6–8 therefore deal with the role of ROPs in the shape of three different cells such as pavement cells, xylem vessel cells, and pollen tubes. Chapters 9–11 describe current knowledge of the cell shape formation in root hair, trichome, and transfer cells, in which different cellular mechanisms such as lipid signaling and cytoskeleton are discussed.

Section 3 deals with the developmental regulation of specific types of plant cells such as guard cells, xylem cells, and phloem cells. Cell wall pattern and cell shape are both under developmental control. Cell-non-autonomous extracellular signals derived from neighboring cells or environmental cues regulate cell fate. Within the cell, transcriptional cascades finally determine a cell fate and then execute distinct cell-specific wall formation to lead to distinct cell shape. Signaling and transcriptional cascades leading to cell differentiation are discussed in Chapters 12 and 13, respectively. Finally, Chapter 14 describes our current knowledge of inter- and intra-cellular signaling that determines phloem cell differentiation.

I would like to thank the authors who provided authoritative reviews as experts of their areas. I also thank Justin Jeffryes, Stephanie Dollan, and their colleagues at Wiley-Blackwell publishing who provided me with this opportunity as well as encouragement during completion of this task.

Hiroo Fukuda

Section 1
Factors Controlling Plant Cell Wall Patterning

1 The Biosynthesis and Function of Polysaccharide Components of the Plant Cell Wall

Ryusuke Yokoyama, Naoki Shinohara, Rin Asaoka, Hideki Narukawa and Kazuhiko Nishitani

Laboratory of Plant Cell Wall Biology, Graduate School of Life Sciences, Tohoku University, Sendai, Japan

Introduction

The cell wall of land plants consists of three layers, namely the middle lamella, the primary cell wall, and the secondary cell wall. The middle lamella is directly derived from the cell plate generated during cytokinesis and the primary cell wall is deposited onto the middle lamella during the cell expansion process. The two cell wall layers are generally found in all cell types, whereas the secondary wall is deposited onto the primary cell wall in certain specific cell types after cell expansion has ceased (Albersheim *et al.*, 2011; Fig. 1.1).

The three layers differ from each other in terms of their chemical nature and physical properties, and they serve different biological functions. Although both the primary and secondary cell walls directly function as a mechanical housing capable of resisting both turgor pressure from the inside out and compression force from the outside in, only the primary cell wall can extend or deform in response to the force applied and thereby determine the direction and rate of cell expansion (Burgert and Frantzl, 2007; Wasterneys and Collings, 2007; Fig. 1.1). In addition to these mechanical roles, the primary cell wall functions as an information processing system. Typical functions include non-cell-autonomous regulation of cell differentiation via apoplastic signaling (Irving and Gehring, 2012; Wolf *et al.*, 2012a), particularly in meristems, defensive responses to pathogens and parasites (Bradley *et al.*, 1992; Vorwerk *et al.*, 2004), and interactions with symbionts. The dynamics of the primary cell wall therefore play a pivotal role in determining cell shape and function during development and in response to environmental stimuli. Accordingly, in this chapter we will focus on the primary cell wall and the dynamic aspects of its major components, namely cellulose and matrix polysaccharides, in relation to its function.

Plant Cell Wall Patterning and Cell Shape, First Edition. Edited by Hiroo Fukuda.

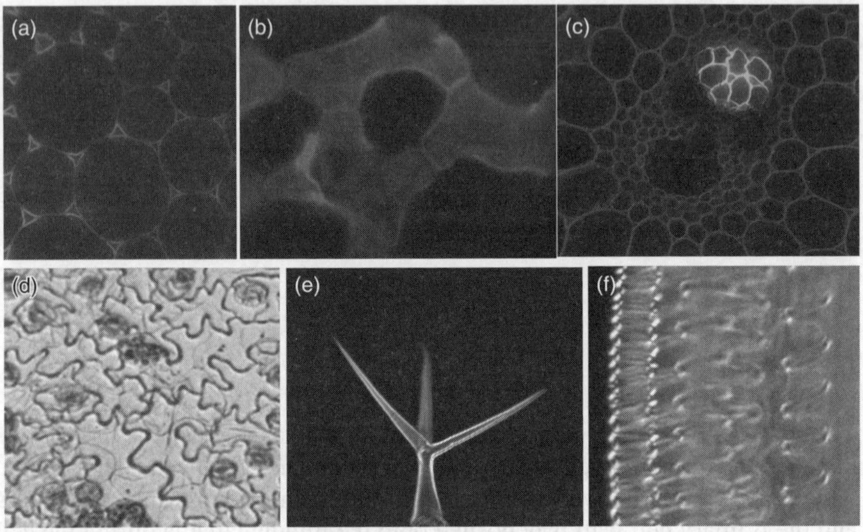

Figure 1.1 Various types of plant cells defined by the cell wall: (a–c) immunofluorescence labeling with monoclonal antibodies against cell wall polysaccharide epitopes; (a) JIM5, specific to homogalacturonan with a low degree of methylesterification; (b, c) CCRC-M1, specific to fucosylated xyloglucan; (d, e) bright field images of unstained specimens; (f) histochemical staining of lignin with phloroglucinol-HCl. A, parenchyma of *Oryza sativa*; B, spongy mesophyll of *Fagus crenata*; C, vascular of *O. sativa*; D, E and F, epidermis, trichome and xylem of *A. thaliana,* respectively.

Overview of the Plant Cell Wall

Plants devote a considerable amount of energy to constructing and maintaining the architecture of the plant cell wall, which is a biphasic composite consisting of crystalline microfibrils and an amorphous gel-like matrix; the former is embedded in the latter, which is intelligent enough to be able to self-organize and regulate cell shape and function during growth and, hence, the morphology of land plants.

For its assembly, remodeling, and disassembly, various types of structural and functional components must be secreted into the cell wall space. These include polysaccharides, structural proteins, enzymes, and small signaling molecules. Examination of the increasing number of currently available genome sequences of land plants tells us that each plant genome contains several thousand cell-wall-related genes which are implicated in biosynthesis, modification, and disassembly of the cell wall, and their regulation with respect to transcription, membrane trafficking, and enzyme actions (Henrissat *et al.*, 2001; Coutinho *et al.*, 2003; Somerville *et al.*, 2004; Yokoyama and Nishitani, 2004; Brown *et al.*, 2005). The presence of such a large number of genes and proteins committed to cell wall dynamics apparently reflects the fact that cell wall type is dependent upon cell type, of which there are estimated to be more than 40 in a land plant. Transcriptomic analysis has demonstrated that different cell types have different expression patterns of cell-wall-related genes (Zhu and Wang, 2000; Demura *et al.*, 2002; Birnbaum *et al.*, 2003; Imoto *et al.*, 2005; Demura and Fukuda, 2007).

In addition to cell-type-specific variations, the chemical and physical nature of the cell wall is also hugely dependent upon the stages of growth and differentiation of the cell. This is rather self-evident as we have seen that the rate and direction of cell growth, and thus the final shape of the cell, is ultimately determined by the nature of the cell wall. Continued reduction in the tensile strength of the cell wall, which is termed 'cell wall loosening', is the direct cause of cell wall expansion followed by cell expansion, the ubiquitous process by which cell expansion is regulated. Accordingly, an anisotropic or localized modification of the primary cell wall within a cell will cause anisotropic cell growth, such as cell elongation in stem cortical cells and polarized cell expansion in leaf trichomes and pavement cells. The chemical and physical nature of the primary cell wall can therefore precisely determine the size and shape of individual cells and play a vital role in determining the morphology of the plant as a whole (Fig. 1.1; Somerville *et al.*, 2004; Cosgrove, 2005).

By contrast, the secondary cell wall has a static structure consisting mainly of crystalline cellulose microfibrils impregnated with lignin and suberin, and is responsible for providing mechanical resistance as well as forming a diffusion barrier. In xylem and fiber cells, the secondary cell wall functions to resist compression force as well as tensile force, and it provides the cell with enough strength to support aerial parts of the plant body, or serves as a non-growing cellular pathway for the translocation of water and nutrients (Fig. 1.2; Demura and Fukuda, 2007). On the other hand, the diffusion resistance function of the secondary cell wall is most prominently found in the Casparian strip in the endodermis, in which lignin confers the hydrophobicity necessary for forming a diffusion barrier to the cell wall (Naseer *et al.*, 2012). These functions of the secondary wall are not directly related to the determination of cell shape and are therefore not discussed in this chapter.

Components of the Primary Cell Wall

The primary cell wall is composed of cellulose microfibrils, matrix polysaccharides, and structural proteins and can serve as an aqueous microenvironment harboring non-structural soluble components such as enzymes, signal ing molecules, and ions (Carpita and Gibeaut, 1993; Cosgrove, 1997). In this section, we first describe the structural features of the cellulose microfibrils and two major matrix polysaccharides – pectin and hemicellulose – before

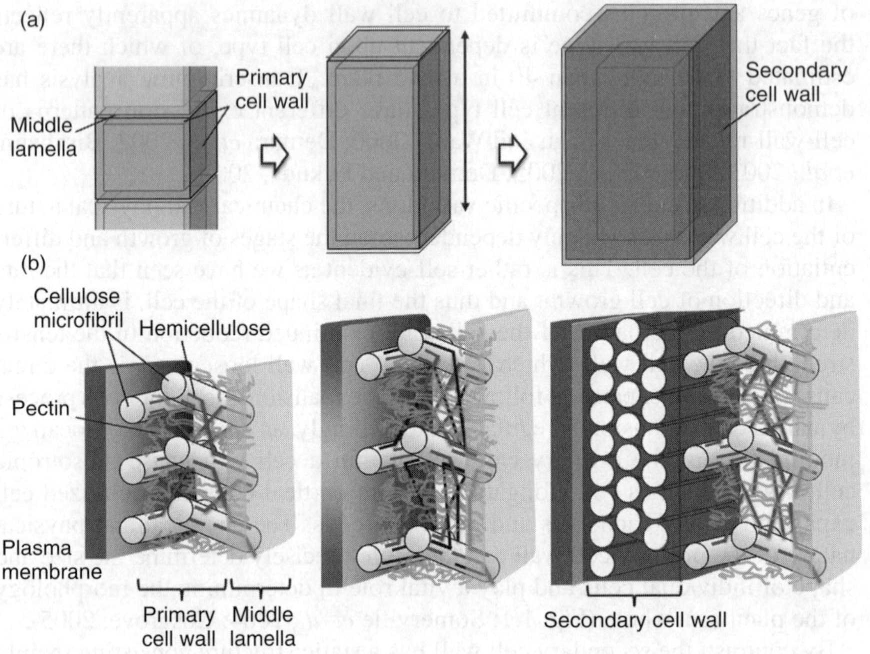

Figure 1.2 Cellulose/hemicellulose and pectin networks in the primary cell wall at successive stages of plant cell growth. (a) Processes of cell elongation and differentiation. (b) Major polymers and their likely arrangement in the cell wall. Newly secreted hemicelluloses (shown in black) and the other polymers (gray) are integrated into the cellulose/hemicellulose network.

describing how they are organized to form the dynamic architecture of the primary cell wall.

Basic Structure and Cellulose Microfibrils

A single microfibril in land plants is circular or square when observed in cross-section. The dimension of the cellulose microfibril in land plants has been estimated by transmission electron microscopy, X-ray scattering (Jakob *et al.*, 1995), and solid-state ^{13}C nuclear magnetic resonance (NMR) (Newman, 1999; Kennedy *et al.*, 2007). The diameters suggested by these analyses range from 2.5 nm to 3.6 nm, which corresponds to 15–32 chains of β-1,4-glucan molecules (Somerville, 2006; Fernandes *et al.*, 2011) if it is assumed that each chain occupies 0.317 nm^2 (Nishiyama *et al.*, 2002).

In cellulose microfibrils, there are two types of domains conforming to a triclinic (termed cellulose I-α) form and a monoclinic (termed cellulose I-β) form. In land plants, the I-β form predominates. In the crystalline domain, β-1,4-glucan chains are arranged in parallel and undergo self-association via several interactions, which include the formation of intramolecular hydrogen bonds at O3 … O5 and O2 … O5, an intermolecular hydrogen bond at O3 … O3, and hydrophobic intermolecular interactions. This structure renders the cellulose microfibrils insoluble in water, immune to enzymatic attack, and resistant to chemical agents.

Another important characteristic of cellulose is its high tensile strength and elastic modulus. The latter is estimated to be between 124 and 155 GPa for the cellulose I-β form, values that are comparable to that of gray cast iron (Nishino *et al.*, 1995). The crystallinity is frequently disrupted by dislocations, resulting in amorphous or *para*-crystalline regions in the microfibril. The cellulose microfibril therefore has a substructure consisting of highly organized crystalline domains linked together by less organized amorphous or para-crystalline regions (O'Sullivan, 1997; Nishiyama *et al.*, 2002).

In the primary cell wall, stable crystalline cellulose microfibrils are embedded in amorphous hydrophilic matrix polysaccharides through interaction with the less-organized *para*-crystalline domains. Land plants contain two major classes of matrix polysaccharides: hemicellulose and pectin. The former includes xyloglucans, glucomannans, and arabinoxylans (Scheller and Ulvskov, 2010), and the latter consists of homogalacturonan (HG) and the rhamnogalacturonan (RG) I and II domains (Mohnen, 2008).

Hemicellulosic Polysaccharides

The typical hemicellulosic polysaccharide is a linear polymer composed of a β-D-pyranosyl backbone substituted by short side chains with a single or a few glycosyl residues. Hemicellulosic polysaccharides and cellulose therefore share the structural feature of a β-D-pyranosyl backbone, which allows tight binding of hemicellulosic polysaccharides to the amorphous region of the para-crystalline cellulose microfibril by hydrogen bonding. The hemicellulose/cellulose interaction is of a chemical nature, such that concentrated alkaline solution is required to disrupt the interaction and liberate hemicellulose from the cell wall to aqueous solution (Cosgrove, 1997). Although the side chains on hemicellulosic polysaccharides modulate the efficiency of binding to the microfibrils, the interaction is still strong enough to resist tensile stress derived from turgor pressure.

Since the molecular lengths of certain hemicellulosic polysaccharides such as xyloglucans are longer than the distances between cellulose microfibrils, they can cross-link adjacent microfibrils to tether together and coat the surface of the cellulose microfibrils (Hayashi, 1989; McCann *et al.*, 1990), thereby forming a cellulose/hemicellulose network that functions as the major tension-bearing framework of the primary cell wall. It is worth noting that it is these hemicellulosic cross-links that confer extensibility to the network structure and act as a modulator of mechanical properties of the primary cell wall. Xyloglucans are typical hemicellulosic polysaccharides. These polysaccharides are ubiquitous in land plants and are particularly abundant in dicotyledonous plants (Talbott and Ray, 1992; Popper *et al.*, 2011). By contrast, commelinoid monocotyledons, which include cereals such as rice (*Oryza sativa*), have relatively few xyloglucans in most of the tissues, and a relatively large amount of xyloglucans are restricted to certain tissue types such as phloem (Fig. 1.1C; Brennan and Harris, 2011). The predominant glycans in these plant species are glucuronoarabinoxylan and β (1→3), (1→4)-mixed-linkage glucan, which are also hypothesized to cross-link the cellulose microfibrils in these plant species (Vogel, 2008).

In addition to xyloglucans, glucomannans and arabinoxylans may also bind to cellulose microfibrils in the primary cell walls of dicotyledonous and non-graminaceous monocotyledonous plants. The fact that the molecular lengths of these hemicelluloses are significantly shorter than those of xyloglucans, and that their extractability from the cell wall differs from that of xyloglucans, implies that these hemicelluloses may have different interactions with cellulose from those of xyloglucans.

Cellulose/xyloglucan Network

The primary cell wall, which determines cell shape, must be strong enough to withstand the mechanical stresses imposed upon it but flexible enough to allow deformation in response to developmental and environmental cues. Given the load-bearing function of cellulose/xyloglucan networks in the primary cell wall, the rearrangement of these networks is essential during cell growth. Two possible processes have been postulated to be involved in the rearrangement of cellulose/xyloglucan networks. One process envisages a remodeling process in which disruption of hydrogen bonding is followed by immediate reconnection at different positions, allowing remodeling of the cell wall network. The other process postulates remodeling by molecular grafting between xyloglucan cross-links by means of an endotransglycosylation reaction. Both of these processes can be achieved without the loss of cell wall integrity (Fig. 1.2). The protein families expansins (Cosgrove, 2005) and xyloglucan endotransglucosylase/hydrolases (XTHs) (Nishitani and Vissenberg, 2007) are implicated in these processes.

Expansins were originally discovered for their ability to cause acid-induced extension of isolated cell walls (McQueen-Mason *et al.*, 1992). Expansins form a family of small cell wall proteins characterized by a certain carbohydrate-binding domain, some of which dissociate interactions between hemicellulose and cellulose microfibrils (Cosgrove, 2000; Yennawar *et al.*, 2006). However, expansins do not exhibit hydrolytic or transglucosylation activity on any matrix polysaccharide examined, despite the fact that their amino acid sequences are similar to the catalytic domain of the family-45 endoglucanases. The expansin EXPB1 has been proposed to facilitate the local movement and stress relaxation of arabinoxylan–cellulose networks within the cell walls of maize by non-covalent rearrangement, but its molecular mechanism remains unclear (Yennawar *et al.*, 2006).

In growing cells, wall expansion must be coupled with the synthesis and integration of new wall components to maintain the thickness and mechanical properties of the cell wall. Since expansins cause no significant alterations in the chemical composition of the cellulose/xyloglucan network, their actions alone cannot explain long-term cell expansion in which incorporation of new wall materials is required to compensate for the thinning of wall thickness. Instead, it is likely that expansins are involved in rapid and transient cell expansion processes such as the 'acid growth' process observed during the first phase of auxin-induced cell expansion, which is based on rapid, localized changes in cell wall extensibility without additional polysaccharide synthesis (Rayle and Cleland, 1992).

XTHs form a subgroup in the Glycoside Hydrolase Family 16 (GH16). Several members of the XTH family specifically cleave a β (1→4) glucosidic linkage of an unsubstituted glucosyl residue in a xyloglucan main chain (donor substrate), and reconnect the reducing end generated by cleavage of the donor molecule to the non-reducing end of another xyloglucan molecule (acceptor substrate). This activity is termed xyloglucan endotransglucosylase (XET) activity. Certain XTHs only cleave xyloglucan without reconnecting the split end, an activity termed xyloglucan endohydrolase (XEH) activity (Nishitani, 1997; Nishitani and Vissenberg, 2007).

Both XET and XEH activities enable extension of xyloglucan chains tethering cellulose microfibrils and integrate new xyloglucans into the cellulose/xyloglucan networks (Rose *et al.*, 2002). This process provides an effective means of rearranging the cellulose/xyloglucan network and allows sustainable expansion of the cell wall. Although a complete description of the functions of the XTH family of proteins *in muro* is not yet available, the physiological role of XTH is becoming clearer. We will return to this topic in 'Function of Xyloglucan and XTH'.

Pectic Polysaccharides

Hemicellulose serves as a cross-linker between cellulose microfibrils in the primary cell wall, whereas pectin exists in a gel and serves as a space-filling or packing matrix in the primary cell wall.

Pectic polysaccharides are highly hydrophilic and are most soluble in the water of polysaccharides in the primary cell wall. Pectic polysaccharides are easily extracted by hot water, chelating agents, or dilute acidic solutions. These polysaccharides are composed of a complex, heterogeneous group of polysaccharide domains and characteristically contain galacturonic acid and rhamnose, arabinose, and galactose as major sugar components (Ridley *et al*., 2001). The main pectin domains include the homogalacturonan (HG), rhamnogalacturonan I (RG I), and rhamnogalacturonan II (RG II) domains. The HG domain has a relatively simple structure, forming a linear polymer consisting of 1,4-linked α-D-galacturonic acid residues. RG I has a backbone composed of alternating (1→2)α-L-rhamnose-(1→4) α-D-galacturonic acid residues decorated primarily with arabinan and galactan side chains. The RG II domain is a low molecular mass but highly complex carbohydrate domain composed of 11 different glycosyl residues.

Some structural models have been proposed for the pectic matrix in the primary cell wall. One conventional model envisages HG to be a long main chain connected in series to two branching domains, RG I and RG II. In this model, the pectin backbone consists of three connected domains. In a more recent model, however, RG I is postulated to function as a scaffold or the main chain to which long HG domains and RG II domains are covalently attached as side chains (Vincken *et al*., 2003). Pectic polysaccharides are therefore covalently joined *in vivo* (Willats *et al*., 2006).

Pectin Network

The pectic polysaccharides are subject to a number of modifications of conformation and covalent linkage, crucial processes that alter the chemical and physical nature of the matrix in the primary cell wall.

Borate binds to two apiose residues in RG II domains and forms an apiose-borate-apiose diester bridge. Most RG II molecules are spontaneously dimerized through the borate ester bridge upon secretion into the cell wall space (O'Neill *et al*., 2004). Borate-mediated cross-linking contributes to the strengthening of the primary cell wall as well as the control of wall porosity and intercellular adherence (Caffall and Mohnen, 2009).

Another modification of pectic polysaccharides is de-esterification of methylesterified galacturonic acid residues in the HG domain. Nascent HG domains localized in the Golgi are normally fully methylesterified. Methylester groups on the HG domain are removed upon secretion into the cell wall by pectin methylesterases (PMEs) present in the cell wall space. Demethylesterification of pectin is followed by two alternative pectin modification processes.

In one process, free carboxyl groups generated on the HG domain are cross-linked via Ca^{2+} bridges to assemble the pectin into a gel-like network. The resulting HG-Ca^{2+} complex acts as a space-filling hydrophilic filter to prevent aggregation and collapse of the cellulose/hemicellulose network, and is considered to make the network less sensitive to the actions of cell wall enzymes (Cosgrove, 1997) and thereby renders the cell wall more resistant to compression stress. This process is observed in the regulation of cell wall stiffening in basal parts of inflorescence stems (Hongo *et al.*, 2012).

In the other process, random demethylesterified HG domains are not fully cross-linked via Ca^{2+} bridges. These domains become more susceptible to hydrolytic degradation and are disassembled. This process is observed during primordial development in the shoot apical meristem (Peaucelle *et al.*, 2011a).

In addition to the boron and calcium bridges, pectic polysaccharides are linked to each other by various covalent bonds including ester linkages through phenolic dimers such as diferulic acid (Wallace and Fry, 1994). Furthermore, the pectin networks may be linked to the cellulose/xyloglucan networks and structural proteins via phenolics, such as *p*-coumaryl and feruloyl acids (Caffall and Mohnen, 2009). More recently, pectin was demonstrated to link covalently to arabinoxylan via a rhamnosyl residue in the arabinogalactan (AG) domain of an arabinogalactan protein (AGP). This wall structure, consisting of arabinoxylan, pectin, and AGP, is referred to as Arabinoxylan Pectin Arabinogalactan Protein1 (APAP1; Tan *et al.*, 2013). A large macromolecular network, which can interact functionally with the other components, is also thought to be responsible for various physical properties of the primary cell wall (Fig. 1.2).

The precise functions of many of the pectin networks are yet to be determined; however, recent work shows that modifying the pectic polysaccharides is a key process in elucidating functional network formation with respect to plant growth and development. We will return to this topic in 'Function of Pectin and PME'.

Biosynthesis and Assembly of the Cell Wall

Cellulose is synthesized at the plasma membrane by large complexes called rosette terminal complexes (TCs; Delmer and Amor, 1995), whereas matrix polysaccharides are polymerized exclusively in the Golgi lumen and secreted into the apoplast or cell wall space via a membrane trafficking system. This section focuses on the general mechanisms of the synthesis of the two types of cell wall polysaccharides and how these polysaccharides are assembled into the dynamic architecture of the primary cell wall.

Cellulose Synthesis

A rosette terminal complex consists of six subunits, with each subunit containing six catalytic units of cellulose synthase (CESA) proteins. Each catalytic unit has been proposed to mediate polymerization of one (1→4)-linked β-D-glucan chain using UDP (uridine diphosphate) -glucose as the substrate, which is supplied by a membrane-associated form of sucrose synthase localized in the vicinity of the catalytic units of cellulose synthase (Doblin *et al.*, 2002; Carpita, 2011). According to this model, 36 molecules of (1→4)-linked β-D-glucans are synthesized at the same time on a single rosette. This predicted number of molecules is not however supported by the actual observed numbers, which range from 15 to 32 as discussed in 'Components of the Primary Cell Wall'. Therefore, either two catalytic units are involved in the synthesis of a single glucan chain, or fewer than five of the six catalytic units are actually functional in the rosette. The mechanism underlying cellulose synthesis therefore remains controversial.

Nascent (1→4)-linked β-D-glucan chains are extruded as a self-assembling microfibril, forming a crystalline microfibril. Evidence from genetic experiments indicates that three different *CESA* genes are normally required to produce a functional complex, and that different sets of genes are involved in the formation of the primary and secondary walls. In Arabidopsis, for example, *CESA1*, *CESA3*, and *CESA6* are required for the synthesis of the primary cell wall, whereas *CESA4*, *CESA7*, and *CESA8* are required to form secondary cell walls (Burn *et al.*, 2002; Taylor *et al.*, 2003). Moreover, CESA2 and CESA5 are partially redundant with CESA6 (Desprez *et al.*, 2007).

The master regulatory transcription factors that specifically govern the synthesis of the secondary cell wall have been successfully identified using suspension-culture cell lines, which can be forced to undergo highly synchronized differentiation to tracheary elements (Yamaguchi and Demura, 2010). On the other hand, cellulose synthesis, especially in the primary cell wall, has been proposed to be controlled post-transcriptionally rather than by transcription factors (Somerville *et al.*, 2004).

In addition, a variety of correlative evidence shows that the oriented deposition of cellulose microfibrils seems to be guided by microtubules adjacent to the plasma membrane (Somerville, 2006). A microtubule-associated protein termed CESA interactive protein 1 (CSI1) functions as a bridge between CESA complexes and cortical microtubules. CSI1 plays a crucial role in regulating microtubule-directed cellulose synthesis (Li *et al.*, 2012; Mei *et al.*, 2012).

Chitinase-like (CTL) proteins, including CTL1/POM1 and CTL2, are another class of regulators of cellulose synthesis. These two proteins are secreted to the apoplast and interact with CESA. In *ctl1/ctl2* double mutants the crystalline cellulose content is reduced. This suggests that these two CTLs affect assembly of the glucan chains, thereby modulating the interactions between xyloglucan and cellulose (Sanchez-Rodriguez *et al.*, 2012).

Synthesis of Matrix Polysaccharides

Matrix polysaccharides, including hemicellulosic and pectic polysaccharides, are synthesized in the Golgi and secreted into the wall via an uncharacterized vesicle-mediated trafficking pathway. Given the complexity of the structural features of the matrix polysaccharides in terms of glycosidic linkages and sugar residues, it is obvious that a large number of enzymes are required for their synthesis.

Biosynthesis of xyloglucan (which has a relatively simple structure) is thought to require β (1→4)-glucan synthase, encoded by *CELLULOSE SYNTHASE-LIKE C* (*CslC*), to form the glucan backbone as well as at least three other types of glycosyltransferases – including α-fucosyltransferases, β-galactosyltransferases, and α-xylosyltransferase – to decorate the glucan main chain with side chains (Zabotina, 2012).

For the synthesis of pectic polysaccharides, which are more complex than xyloglucan, at least 67 transferases are thought to be required. These enzymes, which include glycosyl-, methyl-, and acetyltransferases (Mohnen, 2008), are typically encoded by large multigene families classified as glycosyl transferases in the CAZy (carbohydrate-active enzymes) database (Yokoyama and Nishitani, 2004). Most of these synthetic enzymes for matrix polysaccharides are integral membrane proteins and are considered to exist as complexes anchored to the Golgi (Atmodjo *et al.*, 2013).

The newly synthesized and retained polysaccharides in the Golgi lumen are secreted as soluble polymers into the cell wall space, where they diffuse within the aqueous extracellular environment to their final destination by an as-yet-unknown mechanism.

Cell Wall Assembly

Upon secretion into the cell wall space, the matrix polysaccharide precursors become associated with the pre-existing cell wall polymers or other newly secreted precursors *in muro*. Some of the polysaccharides are also assembled into larger polysaccharides via poorly understood mechanisms. Thus, the newly synthesized polysaccharides are integrated into the pre-existing framework of the primary cell wall to alter or maintain its chemical and physical nature.

Network formation involves both spontaneous interactions between the polysaccharides and, perhaps, enzymatic cross-linking. Although the precise molecular processes have not yet been fully identified, there are a few examples in which specific wall enzymes are involved in the assembly of newly synthesized polysaccharides into the pre-existing network. XTH may be the only potential candidate for this function. XTH catalyzes the molecular grafting or disassembly of xyloglucan cross-links within the cellulose/xyloglucan network as well as the integration of newly synthesized xyloglucans into the cellulose/xyloglucan network (Ito and Nishitani, 1999; Rose *et al.*, 2002; Eklöf and Brumer, 2010). PME-mediated demethylesterification of the HG domain of pectin is another example, which leads to the formation of Ca^{2+} bridges between the carboxyl groups of HG domains, thereby forming pectin gel (Micheli, 2001). These *in muro* network formations generally require a regulatory system for transporting the enzymes and their substrates to the proper location at the proper time and to perform reactions in a synergistic manner. PMEs are no doubt the key factor involved in the regulatory system for such *in muro* network formation of pectin.

To date, two types of key factors (XTH and PME) have been specifically implicated in the important processes of network formation which directly affects the physical properties of the primary cell wall and defines cell morphology. The functions of these two classes of proteins are discussed in the following sections.

Function of Xyloglucan and XTH

The first indication of the role for xyloglucan in cell wall expansion was obtained when its metabolism was studied using pulse-chase experiments employing ^{14}C-labeled glucose in pea stem tissues. This classical experiment clearly showed that xyloglucan metabolism is enhanced during auxin-induced cell expansion (Labavitch and Ray, 1974a, b). This study was followed by the finding that changes in the molecular weight of cell wall xyloglucans were generally induced by auxin and acidic pH in various land plants, including monocotyledonous plants and gymnosperms (Nishitani, 1997).

These observations provided strong evidence for the hypothesis that hydrolytic cleavage of xyloglucan cross-links between cellulose microfibrils is the key step controlling the mechanical properties of the cell wall. However, it was also observed that cleavage of load-bearing linkages alone cannot account for prolonged cell expansion, in which remodeling of the cell wall is required to integrate new wall components into the pre-existing framework (Nishitani, 1997). To explain this paradox, hypothetical endotransglycosylation, or molecular grafting between cross-linking molecules, was postulated (Albersheim, 1976).

The existence of this hypothetical enzyme was demonstrated when the enzyme capable of mediating molecular grafting between xyloglucans was isolated from the cell wall space of Azuki bean (Fry *et al.*, 1992; Nishitani and Tominaga, 1992). This enzyme is currently termed xyloglucan endotransglucosylase/hydrolase (XTH) (Rose *et al.*, 2002). In this section, we discuss the roles of XTH family enzymes and their specific substrates, xyloglucans, in determining cell shape in plants.

Xyloglucan Structure and Function

Xyloglucan (Fig. 1.3) is the most abundant cross-linking polymer in land plants (Fry, 1989; Hayashi, 1989; Popper *et al.*, 2011), consisting of a β (1→4)-glucan backbone frequently decorated with side chains of α (1→6)-xylosyl residues. The α (1→6)-xylosyl side chains are often further substituted with several different glycosyl residues depending on phylogeny and cell type. In non-commelinoid monocotyledons and non-solanaceous eudicotyledons, xylosyl residues are substituted with β (1→2)-galactosyl residues with or without further substitution with α (1→2)-fucosyl residues; in solanaceous plants, these residues are substituted with β (1→2)-galactosyl or β (1→2)-arabinofuranosyl residues. In Poales (and Commelinales), the β (1→4)-glucan backbone is less frequently substituted by xylosyl residues, and α (1→2)-fucosyl-β (1→2)-galactosyl substitution is only restricted to a specific cell wall type such as that in phloem (Hsieh and Harris, 2009; Brennan and Harris, 2011).

As for other polysaccharides in plant cell walls, hydroxyl groups in xyloglucans are O-acetylated at various positions. In Arabidopsis, for example, the galactosyl moiety of the side chain is intensively O-acetylated. This acetylation affects the hydrolytic degradation properties of xyloglucan side chains. An Arabidopsis mutant with reduced acetylation in xyloglucan (*rwa2*) is more resistant to the pathogen *Botrytis cinerea* than the wild type, but it exhibits no obvious morphological or growth differences (Manabe *et al.*, 2011). The role of O-acetylation in the cell wall extension process therefore remains unclear.

As we saw in ‘Components of the Primary Cell Wall’, xyloglucans can non-covalently associate with β (1→4)-glucan chains of the cellulose microfibrils to form the cellulose/xyloglucan network (Pauly *et al.*, 1999) which is found in the primary cell walls of flowering plants (Carpita and Gibeaut, 1993). *In vitro* experiments reveal that the side chains of xyloglucans can prevent self-association of xyloglucan molecules and that the terminal fucosyl residue helps to stabilize the planar configuration of the xyloglucan backbone, which is essential for optimal cellulose binding. Arabidopsis mutants lacking a fucosyl (*mur1*) or galactosyl (*mur3*) residue exhibit tissues with slightly reduced mechanical strength, but do not exhibit drastic phenotypic changes in vegetative growth. Moreover, the Arabidopsis double mutant *xxt1/xxt2*, which lacks detectable xyloglucan in its cell wall, exhibits

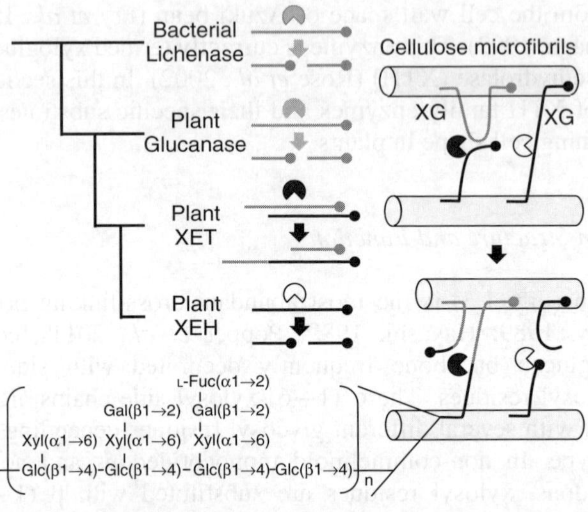

Figure 1.3 Phylogeny of the XTH family of enzymes and their roles in construction, modification, and disassembly of the cellulose/xyloglucan network in the primary cell wall. Black or gray lines indicate xyloglucans (XG), and the circle ends indicate the reducing termini. XET, xyloglucan endotransglucosylase; XEH xyloglucan endohydrolase.

significant changes in the mechanical properties of its tissue, but gross morphological phenotypes are not observed in this mutant (Cavalier *et al.*, 2008). These observations suggest that xyloglucans, and hence cellulose/xyloglucan networks, are not essential for basic plant architecture or that they may be compensated for by other components of the cell wall; however, they may be required for modulating the mechanical properties of the cell wall.

XTH Family

A molecular basis for a dynamic aspect of the cellulose/xyloglucan network was first demonstrated in the 1990s, when XET activity was isolated from growing plant tissues (Fry *et al.*, 1992; Nishitani and Tominaga, 1992).

The enzymes responsible for such activity (EC 2.4.1.207) are encoded by a multigene family. The genomes of three angiosperm species, Arabidopsis (*Arabidopsis thaliana* (L.) Heynh.), rice (*Oryza sativa* L.), and poplar (*Populus trichocarpa*) contain 33, 29, and 41 members (Yokoyama and Nishitani, 2001; Yokoyama *et al.*, 2004; Geisler-Lee *et al.*, 2006), respectively, while the basal land plant moss (specifically *Physcomitrella patens*) possesses 32 members (Yokoyama *et al.*, 2010). This family of proteins is classified into three groups – I/II, IIIA, and IIIB – based on amino acid sequence similarity. Biochemical and structural studies have predicted that group I/II XTH exhibits XET activity, whereas group IIIA XTH exclusively exhibits endohydrolase (XEH) activity to xyloglucan molecules (Rose *et al*., 2002; Chanliaud *et al*., 2004; Baumann *et al*., 2007; Eklöf and Brumer, 2010; Kaewthai *et al*., 2013). Accordingly, this family of enzymes was renamed the xyloglucan endotransglucosylase/hydrolase (XTH) family (Rose *et al*., 2002). Each member of this family can mediate either cleavage and reconnection or simple cleavage of xyloglucan molecules. It is worth noting that construction, remodeling, and disassembly of xyloglucan cross-linking in the cellulose/xyloglucan network can be explained by the collaborative actions of XTH family enzymes (Nishitani, 1997).

Phylogenetic analysis indicates that the XTH gene family has diversified during the evolution of land plants (Yokoyama and Nishitani, 2001; Yokoyama *et al*., 2010). In addition, both biochemical and crystallographic data indicate that the functional differences between XET and XEH can be ascribed to differences in the structural features around the acceptor binding site, namely whether a water molecule or the xyloglucan molecule has access to the acceptor substrate site (Baumann *et al*., 2007).

As mentioned in 'Biosynthesis and Assembly of the Cell Wall', the XTH enzymes belong to Glycoside Hydrolase Family 16 (GH16) (Barbeyron *et al*., 1998; Cantarel *et al*., 2009), which comprises a broad range of microbial endoglucanases and endogalactanases, as well as algal β-agarase and κ-carrageenase. Since the closest relatives of XTH in GH16 are bacterial licheninases (EC 3.2.1.73) which specifically hydrolyze β (1→3), β (1→4)-glucans, it has been hypothesized that the XTH genes evolved from these bacterial glucanases (Planas, 2000; Eklöf and Brumer, 2010). This notion was supported by the recent finding of a novel endoglucanase from poplar (*Populus trichocarpa*), which revealed a newly recognized small clade of GH16 genes that represent intermediates between the bacterial licheninases and plant XTHs (Eklöf *et al*., 2013).

Xyloglucan Dynamics as Mediated by XTH

XET activity: Before the discovery of XTH, growth-promoting and growth-inhibitory effects of exogenously applied xyloglucan oligosaccharides were documented, and the 'oligosaccharin' hypothesis was advanced in the late 1980s (McDougall and Fry, 1989; Vargas-Rechia *et al*., 1998). This hypothesis proposes that oligosaccharides derived from cell wall xyloglucans act as signaling molecules (Ryan, 1987).

Currently however, the term 'oligosaccharins' is seldom used for xyloglucan oligosaccharides, which are instead considered to function simply as acceptor substrates for XTH, which mediate incorporation of xyloglucan oligomers into the cellulose/xyloglucan network to facilitate disassembly of xyloglucan cross-links (Takeda *et al.*, 2002). This point of view is also supported by the observation that the growth of tobacco suspension-culture cells is accelerated by XTH-mediated incorporation of xyloglucan oligosaccharide into the cell wall (Ito and Nishitani, 1999; Kaida *et al.*, 2010).

Since the 1990s, the roles of XTHs in plant growth and differentiation have been extensively investigated (Nishitani and Vissenberg, 2007). These studies show a positive correlation between XET activity and the growth process in specific tissues or organs, such as fruit ripening (Redgwell and Fry, 1993) and root elongation (Ryan, 1987; Pritchard *et al.*, 1993; Vissenberg *et al.*, 2000). Furthermore, the effects of environmental factors (such as mechanical stimulus, light conditions, anoxia, and developmental signals) on XTH gene expression, as mediated by plant hormones, have been extensively studied (Lee *et al.*, 2005; Nishitani and Vissenberg, 2007).

Comprehensive expression analysis of all XTH genes of Arabidopsis reveals that virtually all members of the XTH family have specific expression patterns, many of which overlap. This result indicates that a specific set of XTH family genes are expressed in individual tissues and in response to individual environmental stimuli (Yokoyama and Nishitani, 2001; Lee *et al.*, 2005). This ubiquitous nature suggests the functional importance of xyloglucan dynamics throughout the life cycles of land plants.

The overwhelming redundancy in XTH genes has however hampered functional studies employing phenotypic analysis of XTH mutants. Only a few of the 33 Arabidopsis XTH genes have been shown to exhibit morphological phenotypes in their loss-of-function mutants. Nonetheless, some mutants of XTH genes exhibit tissue-specific phenotypes.

For example, *atxth27-1* is a transposon-tagged knockout line of *AtXTH27*. This mutant differentiates morphologically altered tracheary elements, fails to develop tertiary veins in leaves, and exhibits lesion-mimic yellow spots on leaves (Matsui *et al.*, 2005), indicating that this gene is involved in the formation of cell walls of tracheary elements. Another Arabidopsis mutant, *atxth28-1* (in which *AtXTH28* is disrupted by a T-DNA insertion), is defective in efficient self-pollination due to its shorter stamens, suggesting that *atxth28* plays a role in the elongation growth of filament cells in the stamen (Kurasawa *et al.*, 2009). AtXTH27 and AtXTH28 are similar to each other in terms of both expression profile and amino acid sequence. These proteins belong to the group IIIB subclass of the XTH family and are predicted to exhibit XET activity. However, the phenotype of *atxth28* is not enhanced by *atxth27*, indicating that AtXTH28 is not functionally redundant with AtXTH27 in planta. XTH is involved in a plant's responses to light. In general, plants are highly sensitive to the quality, quantity, and direction of light, and they usually escape from the shade of other plants by rapidly elongating their shoot. This phenomenon, termed shade avoidance, is triggered under low red/far red light and high green light conditions. These types of light upregulate the expression of *AtXTH9*, *AtXTH15*, *AtXTH16*, *AtXTH17*, *AtXTH19*, and *AtXTH22* (Sasidharan *et al.*, 2010). Interestingly, disrupting only one of these genes can retard accelerated shoot growth under such light conditions, indicating the non-redundant action of these genes on shoot growth.

XTH is also involved in plant defenses. Infestation of celery (*Apium graveolens*) by aphids (*Myzus persicae*) causes up-regulation of certain XTH genes, including celery *XTH1* specifically in the phloem, a tissue from which aphids suck plant sap (Divol *et al.*, 2007). Arabidopsis *AtXTH33*, which is the closest homolog to celery *XTH1*, is also up-regulated when Arabidopsis is infested by aphids. Since aphids preferentially settle on loss-of-function mutant of *AtXTH33* (*atxth33*) plants compared with wild-type plants, AtXTH33 and its celery counterpart may play a role in protecting plants from aphids. Since over-expression of *AtXTH33* in Arabidopsis does not increase the resistance of plants to aphids, factors other than AtXTH33 are likely to be involved in this defense mechanism.

XTHs are therefore involved in growth responses to environmental stimuli as well as defense to biotic stresses. For other functions of XTH in the hormonal regulation of plant growth and responses to biotic and abiotic environmental stimuli, the reader is referred to previous reviews (Lee *et al.*, 2005; Nishitani and Vissenberg, 2007).

XEH Activity: Storage xyloglucan, which occurs widely in plant seeds, was noted as early as the 1830s by Matthias J. Schleiden and was named amyloide because it was stained blue with iodine (Edwards *et al.*, 1985). The enzyme, which is capable of specifically hydrolyzing storage xylgolucans, was isolated and characterized from germinated cotyledons of nasturtium (*Tropaeolum majus* L.) and was named xyloglucan-specific endo-(1→4)-β-D-glucanase before the discovery of the XTH family of proteins (Edwards *et al.*, 1986). Currently, this hydrolase is classified as a group IIIA XTH.

During germination, storage xyloglucans in cotyledonary cell walls are degraded into monosaccharides by the sequential actions of hydrolytic enzymes. These enzymes include glycosidases acting on xyloglucan side chains as well as XTH members with XEH activity (Dos Santos *et al.*, 2004). XEH activity is therefore responsible for supplying carbohydrates used for both respiration and body construction during seed germination.

Interestingly, genes encoding group IIIA XTH are found even in plant species whose cotyledons do not contain storage xyloglucans (Eklöf and Brumer, 2010). Arabidopsis does not contain storage xyloglucans in its seeds, but it does contain two group IIIA XTHs, AtXTH31 and AtXTH32, which are expressed in various organs other than seeds. This observation suggests that group IIIA XTHs may play other roles than in degradation of storage xyloglucan. In fact, a recent study shows that neither a double mutant of *atxth31* and *atxth32* nor transgenic plants impaired in these genes exhibit prominent phenotypic changes in morphology (Kaewthai *et al.*, 2013). These results, together with those obtained using the xyloglucan-less double mutant *xxt1/xx2* described in 'Xyloglucan Structure and Function', raise questions about the essential role of xyloglucan in the primary cell wall in determining cell shape.

Another point worth noting is that rice (*Oryza sativa* L.), which contains little xyloglucan in most tissues, still possesses group IIIA XTH with hydrolase activity; the physiological role of this XTH remains elusive (Hara *et al.*, 2014).

One possible explanation for these discrepancies is that there is an as-yet-unknown essential factor (or factors) that is in charge of the regulation of the basic properties of the primary cell wall, and xyloglucans serve as modulators of this essential factor. This notion is consistent with the newest structural model of the primary cell wall which has been advanced based on solid-state, high-resolution, carbon-13 cross-polarization/magic angle spinning nuclear magnetic resonance studies. This model envisages that pectin, not xyloglucan, intensively interacts with cellulose microfibrils in the primary cell wall (Peaucelle *et al.*, 2012; Wang *et al.*, 2012). The possible role of pectin in determining cell shape is discussed in the next section.

Function of Pectin and PME

As for xyloglucans, pectins are broadly conserved among land plants and they act as the major component that fills spaces within the cellulose/xyloglucan network in the primary cell wall. Through the modification of pectin, plants regulate the physical properties of primary cell walls, thereby directly regulating developmental processes including the control of cell shape in plants. There have also been recent breakthroughs in understanding pectin biosynthesis and functionality, and a set of enzymes responsible for pectin biosynthesis has been identified (Mohnen, 2008; Peaucelle *et al.*, 2012; Wang *et al.*, 2012; Atmodjo *et al.*, 2013). In addition, new roles for pectin in plant growth and development are being elucidated (Palin and Geitmann, 2012; Peaucelle *et al.*, 2012). Since recent progress in this field has been reviewed by other authors (e.g., Atmodjo *et al.*, 2013), this section will focus on PME-mediated modification of pectin and its biological significance in the determination of cell shape in plants.

Basic Mechanisms of PME Action

Pectins as Regulators of the Physical Properties of the Primary Cell Wall: As stated in 'Components of the Primary Cell Wall', pectin consists of three major domains, namely HG (HG), RG I (RG-1), and RG II (RG-II). HG is the most abundant domain, accounting for more than 60% of the pectin in the primary cell wall. HG consists of an unbranched homopolymer of α-D-galacturonic acid residues, while RG I and RG II have complex side chains. Biosynthesis, methylesterification, and the addition of side chains to pectin main chains occur in the Golgi apparatus.

An immunolabeling study using sycamore maple (*Acer pseudoplatanus* L.) suspension-culture cells has revealed that pectins are polymerized at the cis-Golgi, methylesterified at the medial-Golgi, and modified with side chains at the trans-Golgi (Zhang and Staehelin, 1992).

Two galacturonosyltransferases (GalAT), GAUT1 and GAUT7, were recently demonstrated to form a catalytic core of GalAT in Arabidopsis (Atmodjo *et al.*, 2011). Interestingly, during the formation of a functional catalytic core, GAUT1 is processed to cleave the N-terminal transmembrane domain and is no longer retained in the Golgi, whereas GAUT7 remains unprocessed. GAUT1 is therefore anchored to the Golgi via association with GAUT7 to form the GAUT1: GAUT7 complex.

As identified in 'Components of the Primary Cell Wall', HG is secreted into the cell wall space in a highly methylesterified state and is then demethylesterified by pectin methylesterases (PMEs). The degree and pattern of methylesterification are critical factors in determining subsequent reactions, i.e., stiffening by cross-linking with Ca^{2+} or degradation by polygalacturonases (PGs). Unraveling the mode of action of PMEs on HG and its regulatory mechanism would therefore constitute an important advance in understanding how polysaccharide modification is controlled in the wall space and how closely it is related to cell wall integrity.

Two Modes of Demethylesterification of Pectin: PME acts on HG in two different manners, i.e., 'random' and 'linear' (Fig. 1.4; Micheli, 2001). In the random mode, PME randomly demethylesterifies galacturonic acid residues in HG and produces discontinuously demethylesterified regions in HG, which is susceptible to degradation by PGs, resulting in cell wall loosening or cell wall disassembly. In the linear mode on the other hand, PME acts on continuous blocks of galacturonic acid residues along HG, giving rise to blocks of demethylesterified residues in HG. This facilitates association among the residues via the highly frequent formation of Ca^{2+} bridges to form a gel structure, which contributes to the stiffening of cell walls. Multiple factors such as pH, cations, and the initial degree of methylesterification (DE) are thought to affect the mode of action and catalytic properties of PME (Catoire *et al.*,

1998; Denes *et al.*, 2000; Schmohl *et al.*, 2000; Goldberg *et al.*, 2001; Kim *et al.*, 2005).

Diversification of the PME Family: Pectin is conserved ubiquitously in the Streptophyta, from charophytes to angiosperms (Popper *et al.*, 2011). The PME family of enzymes is thought to have arisen along with pectins and undergone diversification. In Arabidopsis, 66 members comprise the PME family. Monocotyledons, especially grasses (whose cell walls contain fewer pectic polysaccharides than dicotyledonous plants), have relatively few PMEs (*c.* 40 members in rice and *Sorghum bicolor* L. Moench). In addition, 15 PME members are found in *Coleochaete orbicularis* (Wang *et al.*, 2013). The diversification of PMEs is likely to have contributed to both the functional and morphological diversification of land plants.

PME Inhibitors: The activity of PME is regulated by specific proteinaceous inhibitors termed pectin methylesterase inhibitors (PMEIs). PMEI was first isolated from kiwi fruit (*Actinidia chinensis* Planch) (Balestrieri *et al.*, 1990). Kiwi PMEI is a small protein composed of 152 amino acid residues. Its sequence shows significant similarity to the N-terminal region (termed 'PRO region') of plant PMEs and it binds to PMEs, forming a heterodimeric complex.

PMEIs also comprise a multigene family in angiosperms, whereas ferns, mosses, and charophytes have few or no PMEI members (Wang *et al.*, 2013). The genome of *A. thaliana* encodes 71 PMEI members. PMEI and PME form a stoichiometric 1:1 complex. Crystallographic analysis of the PME-PMEI complex has revealed that plant PMEI is structurally distinct from bacterial PMEI (Di Matteo *et al.*, 2005). This observation suggests that PMEIs specifically inhibit plant endogenous PMEs, unlike other inhibitor proteins such as polygalacturonase inhibitor proteins (PGIPs) which function equally against microbial polygalacturonases. Clustering analysis of PME and PMEI mRNA data in Arabidopsis has revealed that PMEs can be classified into five clusters with respect to their expression sites (seed coat, shoot apex, micropylar endosperm, chalazal endosperm, and pollen), and PMEIs are classified within the same categories as PMEs (Wolf *et al.*, 2009a).

The balanced regulation of PME activity via endogenous inhibitors may be beneficial to angiosperms, which may be related to the evolution of floral organs, fruits, and seeds. To date, functional analyses of PMEI reveal that PMEI is involved in various phenomena (which we discuss below), but little is known about the individual functions of PME–PMEI pairs, except for a few such pairs (Peaucelle *et al.*, 2008; Rockel *et al.*, 2008; Reca *et al.*, 2012).

PRO Region of PME: The PME family enzymes belong to Carbohydrate Esterase Family 8 (CE8) and are classified into two groups based on their

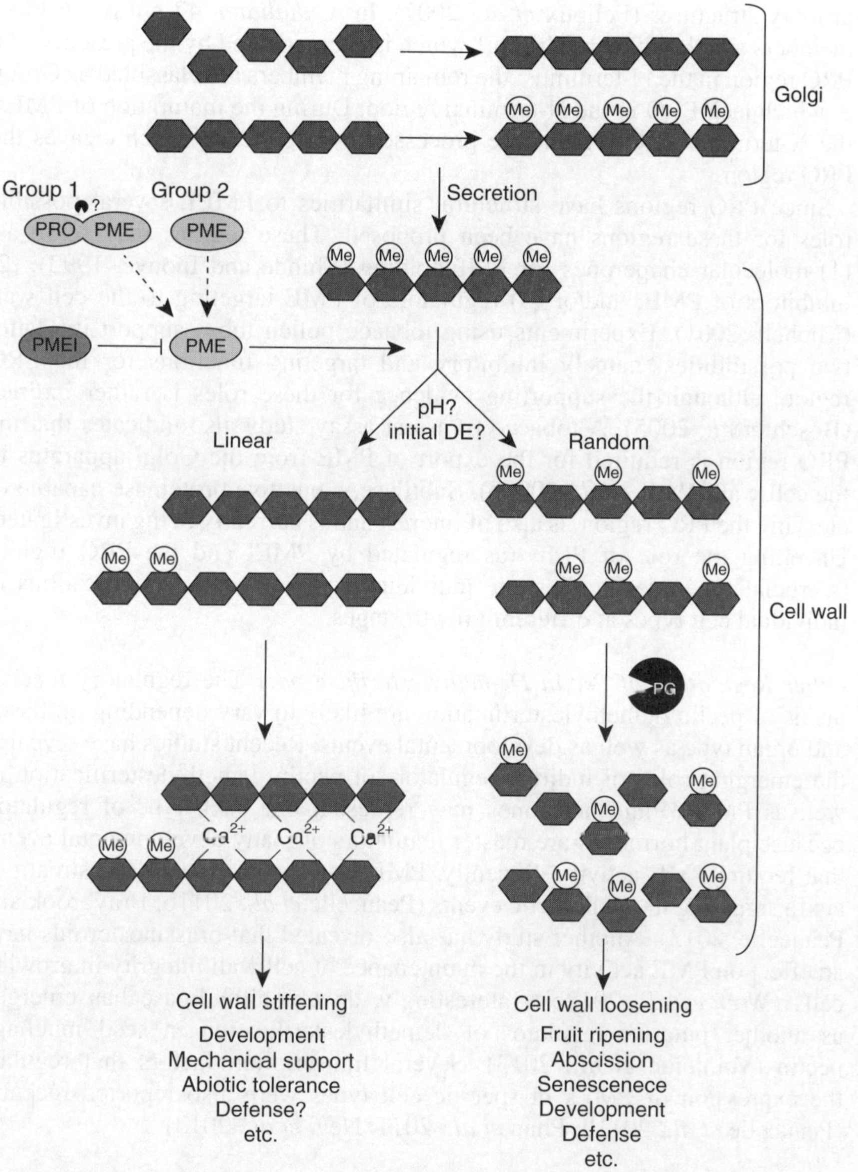

Figure 1.4 Two processes of HG modification. Two types of processes by PME ('linear' and 'random') have different effects on the physical properties of the primary cell wall. Gray hexagons represent galacturonic acid; white circles represent methyl group esterified to galacturonic acid.

primary structures (Pelloux *et al.*, 2007). In *A. thaliana*, 47 out of 66 PME members are classified as Group 1 which is characterized by the presence of a PRO region in the N-terminus; the remaining members are classified as Group 2, which lack PRO in the N-terminal region. During the maturation of PMEs, the N-terminal PRO regions are processed by a protease which cleaves the PRO region.

Since PRO regions have structural similarities to PMEI, several possible roles for these regions have been proposed. These regions may serve as: (1) molecular chaperones for PME folding (Shinde and Inouye, 1993); (2) inhibitors of PME; and/or (3) regulators of PME targeting to the cell wall (Micheli, 2001). Experiments using tobacco pollen tubes support the latter two possibilities, namely inhibitory and targeting functions for the PRO region, although the supporting evidence for these roles is rather indirect (Bosch *et al.*, 2005). A tobacco transient assay study also indicates that the PRO region is required for the export of PME from the Golgi apparatus to the cell wall (Wolf *et al.*, 2009b). Subtilase, a putative proteinase capable of cleaving the PRO region, is also of interest and is currently being investigated. Unveiling the role of PME (as regulated by PMEI and the PRO region) is crucial for understanding the individual roles of pectin modifications in individual cell types at different growth stages.

Other Regulators of Pectin Demethylesterification: The regulatory mechanisms of pectin demethylesterification are likely to vary depending on tissue and organ type as well as developmental events. Recent studies have revealed the emerging roles of indirect regulators of pectin demethylesterification as well as PMEI. Plant hormones may represent one such type of regulator, because plant hormones are master regulators of many developmental events that require PME activity. Recently, PME was shown to act downstream of auxin signaling in phyllotactic events (Peaucelle *et al.*, 2011b; Braybrook and Peaucelle, 2013). Another study has also revealed that brassinosteroids have an effect on PME activity in the maintenance of cell wall integrity in growing cells (Wolf *et al.*, 2012b). Interestingly, an ubiquitin ligase has emerged as another putative regulator of demethylesterification in seed mucilage pectin (Voiniciuc *et al.*, 2013). Several transcription factors that regulate the expression of PMEs in specific cell types were also reported recently (Peaucelle *et al.*, 2011b; Phan *et al.*, 2011; Negi *et al.*, 2013).

Physiological Roles of Pectin Modification

Pectin Modification in Plant Development and Adaptation to Environmental Changes: As mentioned in 'Components of the Primary Cell Wall' and at the beginning of this section, the modification of pectins by PMEs and their regulators plays a crucial role in multiple steps of development by affecting cell wall properties. PME-mediated random demethylesterification, followed by PG-mediated degradation of pectin, is thought to lead to fruit softening in some species such as watermelon, tomato, and peach (Tieman and Handa, 1994; Kagan-Zur *et al.*, 1995; Brummell *et al.*, 2004; Karakurt and Huber, 2004). PMEI plays an important role in fruit ripening. PMEI purified from ripe fruit is a well-known proteinaceous inhibitor (Balestrieri *et al.*, 1990), although the physiological function of this compound has not been fully clarified. PME-mediated enhancement of pectin degradation by PG is also involved in cell adhesion, which contributes to tissue abscission and senescence (Willats *et al.*, 2001; Arancibia and Motsenbocker, 2006).

On the other hand, PME and PMEI are involved in multiple steps of plant development because they regulate cell wall stiffening (or suppress stiffening in the case of PMEI). Many studies have suggested the involvement of mutual regulation of PME and PMEI in the growth of hypocotyls, flower stems, roots, and many other organs. Recently, mucilage extrusion in Arabidopsis seed coats was successfully used as a model system to study pectin degradation. Seed mucilage is chiefly composed of unbranched RG-I and small amounts of HG, cellulose, and xyloglucan, and contributes to protecting germinating seeds against drought. Recent work demonstrates that the suppression of pectin demethylesterification by PMEI6 promotes mucilage release, and a subtilase is also involved in this regulatory process (Saez-Aguayo *et al.*, 2013).

Another PME member, PME5, plays a role in generating proper shoot phyllotaxis. This PME acts in Arabidopsis shoot meristems and, unexpectedly, increases the elasticity of meristematic cell walls (Peaucelle *et al.*, 2008, 2011a, 2011b). By contrast, primary cell wall stiffening mediated by demethylesterification of pectin contributes to the mechanical support to the basal part of the stem (Hongo *et al.*, 2012). Pectin modification is therefore important, even in non-growing tissues. These results imply that the mode of action of PME varies depending on its site of action, even in the stem.

Cell wall modification also plays a critical role in the plant's interaction with the environment. PME activity affects a plant's susceptibility to fungi, bacteria, viruses, herbivores, and nematodes (reviewed in Lionetti *et al.*, 2012). Highly methylesterified pectin is likely to be most susceptible to degradation by microbial pathogens (Lionetti *et al.*, 2007; Volpi *et al.*, 2011), whereas the action of fruit-specific PME and the consequent generation of oligogalacturonides by PG reinforce plant defenses against pathogens in strawberries (Osorio *et al.*, 2008, 2011). Moreover, the involvement of PME in metal tolerance was recently revealed (Weber *et al.*, 2013). Since demethylesterification of HG increases its capacity to interact with apoplastic cations, HG may contribute to the maintenance of metal homeostasis in the cell wall space.

There are other studies showing that polarized or anisotropic demethylesterification occurs within a cell wall through the action of PME, and its inhibition by PMEI is coordinated with the formation of individual cell shape. This phenomenon is well-characterized during pollen tube growth, and the PME–PMEI interaction was shown to be involved in this process. In tobacco (*Nicotiana tabacum* L.) the apical region of the pollen tube is highly methylesterified, and demethylesterification by exogenously applied PME results in thickening of the apical cell wall and inhibition of pollen tube growth (Bosch *et al.*, 2005).

Using a transient expression system in tobacco pollen tubes, the growth-promoting activity of Arabidopsis AtPMEI2 and the growth-inhibiting effect of AtPPME1 have been demonstrated. In Arabidopsis, AtPMEI2 accumulates in the apical region of the growing pollen tube, whereas AtPPME1 exhibits a non-polarized localization (Rockel *et al.*, 2008).

In addition, PME activity may influence trichome formation and stomatal development although, to date, there is no direct evidence that demethylesterification of pectin is required for polarized growth of the cells (Francis *et al.*, 2006; Bischoff *et al.*, 2010; Negi *et al.*, 2013).

The PME family of enzymes, together with their specific inhibitors in the PMEI family proteins, plays many diverse roles in various developmental processes in almost every tissue. Their individual roles still remain elusive however, and the whole picture of PME/PMEI function has not yet been described. Elucidating these complicated mechanisms is highly important for unveiling the role of the cell wall in determining cell shape, and thus the growth patterns and function of the plant itself.

References

Albersheim, P. (1976) The primary cell wall, in *Plant Biochemistry*, 3rd edition (eds J. Bonner and J.E. Varner), Academic Press, New York, 225–273.

Albersheim, P., Darvill, A., Roberts, K., Sederoff, R., and Staehelin, A. (2011) Cell walls and plant anatomy, in *Plant Cell Walls: From Chemistry to Biology*, Garland Science, Taylor & Francis Group, LLC, New York, 1–42.

Arancibia, R.A. and Motsenbocker, C. E. (2006) Pectin methylesterase activity *in vivo* differs from activity *in vitro* and enhances polygalacturonase-mediated pectin degradation in tabasco pepper. *Journal of Plant Physiology*, **163**, 488–496.

Atmodjo, M.A., Sakuragi, Y., Zhu, X., Burrell, A.J., Mohanty, S.S., Atwood, J.A., Orlando, R., Scheller, H.V., and Mohnen, D. (2011) Galacturonosyltransferase (GAUT)1 and GAUT7 are the core of a plant cell wall pectin biosynthetic homogalacturonan: galacturonosyltransferase complex. *Proceedings of the National Academy of Sciences of the United States of America*, **108**, 20225–20230.

Atmodjo, M.A., Hao, Z., and Mohnen, D. (2013) Evolving views of pectin biosynthesis. *Annual Review of Plant Biology*, **66**, 747–779.

Balestrieri, C., Castaldo, D., Giovane, A., Quagliuolo, L., and Servillo, L. (1990) A glycoprotein inhibitor of pectin methylesterase in kiwi fruit (*Actinidia chinensis*). *European Journal of Biochemistry*, **193**, 183–187.

Barbeyron, T., Gerard, A., Potin, P., Henrissat, B., and Kloareg, B. (1998) The kappa-carrageenase of the marine bacterium *Cytophaga drobachiensis*. Structural and phylogenetic relationships within family-16 glycoside hydrolases. *Molecular Biology and Evolution*, **15**, 528–537.

Baumann, M.J., Eklof, J.M., Michel, G., Kallas, A.M., Teeri, T.T., Czjzek, M. and Brumer, H., 3rd, (2007) Structural evidence for the evolution of xyloglucanase activity from xyloglucan endo-transglycosylases: biological implications for cell wall metabolism. *Plant Cell*, **19**, 1947–1963.

Birnbaum, K., Shasha, D.E., Wang, J.Y., Jung, J.W., Lambert, G.M., Galbraith, D.W., and Benfey, P.N. (2003) A gene expression map of the Arabidopsis root. *Science*, **302**, 1956–1960.

Bischoff, V., Nita, S., Neumetzler, L., Schindelasch, D., Urbain, A., Eshed, R., Persson, S., Delmer, D., and Scheible, W. R. (2010) TRICHOME BIREFRINGENCE and its homolog AT5G01360 encode plant-specific DUF231 proteins required for cellulose biosynthesis in Arabidopsis. *Plant Physiology*, **153**, 590–602.

Bosch, M., Cheung, A.Y., and Hepler, P.K. (2005) Pectin methylesterase, a regulator of pollen tube growth. *Plant Physiology*, **138**, 1334–1346.

Bradley, D.J., Kjellbom, P., and Lamb, C.J. (1992) Elicitor- and wound-induced oxidative cross-linking of a proline-rich plant cell wall protein: A novel, rapid defense response. *Cell*, **70**, 21–30.

Braybrook, S.A. and Peaucelle, A. (2013) Mechano-chemical aspects of organ formation in *Arabidopsis thaliana*: the relationship between auxin and pectin. *PLoS One*, **8**, e57813.

Brennan, M. and Harris, P. J. (2011) Distribution of fucosylated xyloglucans among the walls of different cell types in monocotyledons determined by immunofluorescence microscopy. *Molecular Plant*, **4**, 144–156.

Brown, D.M., Zeef, L.A.H., Ellis, J., Goodacre, R., and Turner, S.R. (2005) Identification of novel genes in Arabidopsis involved in secondary cell wall formation using expression profiling and reverse genetics. *Plant Cell*, **17**, 2281–2295.

Brummell, D.A., Dal Cin, V., Crisosto, C.H., and Labavitch, J.M. (2004) Cell wall metabolism during maturation, ripening and senescence of peach fruit. *Journal of Experimental Botany*, **55**, 2029–2039.

Burgert, I. and Frantzl, P. (2007) Mechanics of the expanding cell wall, in *The Expanding Cell, Plant Cell Monographs,* vol. 5 (eds J.-P. Verbelen and V. Vissenberg), Springer-Verlag, Berlin, Heidelberg, 191–215.

Burn, J.E., Hocart, C.H., Birch, R.J., Cork, A.C., and Williamson, R.E. (2002) Functional analysis of the cellulose synthase genes CesA1, CesA2, and CesA3 in Arabidopsis. *Plant Physiology*, **129**, 797–807.

Caffall, K.H. and Mohnen, D. (2009) The structure, function, and biosynthesis of plant cell wall pectic polysaccharides. *Carbohydrate Research*, **344**, 1879–1900.

Cantarel, B.L., Coutinho, P.M., Rancurel, C., Bernard, T., Lombard, V. and Henrissat, B. (2009) The Carbohydrate-Active EnZymes database (CAZy): an expert resource for Glycogenomics. *Nucleic Acids Research*, **37**, D233–238.

Carpita, N.C. (2011) Update on mechanisms of plant cell wall biosynthesis: how plants make cellulose and other (1→4)-beta-D-glycans. *Plant Physiology*, **155**, 171–184.

Carpita, N.C. and Gibeaut, D.M. (1993) Structural models of primary cell walls in flowering plants: consistency of molecular structure with the physical properties of the walls during growth. *Plant Journal*, **3**, 1–30.

Catoire, L., Pierron, M., Morvan, C., Du Penhoat, C.H., and Goldberg, R. (1998) Investigation of the action patterns of pectinmethylesterase isoforms through kinetic analyses and NMR spectroscopy. Implications In cell wall expansion. *Journal of Biological Chemistry*, **273**, 33150–33156.

Cavalier, D.M., Lerouxel, O., Neumetzler, L., Yamauchi, K., Reinecke, A., Freshour, G., Zabotina, O.A., Hahn, M.G., Burgert, I., Pauly, M., Raikhel, N.V., and Keegstra, K. (2008) Disrupting two *Arabidopsis thaliana* xylosyltransferase genes results in plants deficient in xyloglucan, a major primary cell wall component. *Plant Cell*, **20**, 1519–1537.

Chanliaud, E., De Silva, J., Strongitharm, B., Jeronimidis, G., and Gidley, M.J. (2004) Mechanical effects of plant cell wall enzymes on cellulose/xyloglucan composites. *Plant Journal*, **38**, 27–37.

Cosgrove, D.J. (1997) Assembly and enlargement of the primary cell wall in plants. *Annual Review of Cell and Developmental Biology*, **13**, 171–201.

Cosgrove, D.J. (2000) Loosening of plant cell walls by expansins. *Nature*, **407**, 321–326.

Cosgrove, D.J. (2005) Growth of the plant cell wall. *Nature Reviews Molecular Cell Biology*, **6**, 850–861.

Coutinho, P.M., Stam, M., Blanc, E., and Henrissat, B. (2003) Why are there so many carbohydrate-active enzyme-related genes in plants? *Trends in Plant Sciences*, **8**, 563–565.

Delmer, D.P. and Amor, Y. (1995) Cellulose biosynthesis. *Plant Cell*, **7**, 987–1000.

Demura, T. and Fukuda, H. (2007) Transcriptional regulation in wood formation. *Trends in Plant Science*, **12**, 64–70.

Demura, T., Tashiro, G., Horiguchi, G., Kishimoto, N., Kubo, M., Matsuoka, N., Minami, A., Nagata-Hiwatashi, M., Nakamura, K., Okamura, Y., Sassa, N., Suzuki, S., Yazaki, J., Kikuchi, S. and Fukuda, H. (2002) Visualization by comprehensive microarray analysis of gene expression programs during transdifferentiation of mesophyll cells into xylem cells. *Proceedings of the National Academy of Sciences of the United States of America,* **99**, 15794–15799.

Denes, J.M., Baron, A., Renard, C.M., Pean, C. and Drilleau, J.F. (2000) Different action patterns for apple pectin methylesterase at pH 7.0 and 4.5. *Carbohydrate Research*, **327**, 385–393.

Desprez, T., Juraniec, M., Crowell, E.F., Jouy, H., Pochylova, Z., Parcy, F., Höfte, H., Gonneau, M., and Vernhettes, S. (2007) Organization of cellulose synthase complexes involved in primary cell wall synthesis in *Arabidopsis thaliana*. *Proceedings of the National Academy of Sciences of the United States of America*, **104**, 15572–15577.

Di Matteo, A., Giovane, A., Raiola, A., Camardella, L., Bonivento, D., De Lorenzo, G., Cervone, F., Bellincampi, D., and Tsernoglou, D. (2005) Structural basis for the interaction between pectin methylesterase and a specific inhibitor protein. *Plant Cell*, **17**, 849–858.

Divol, F., Vilaine, F., Thibivilliers, S., Kusiak, C., Sauge, M.H., and Dinant, S. (2007) Involvement of the xyloglucan endotransglycosylase/hydrolases encoded by celery *XTH1* and Arabidopsis *XTH33* in the phloem response to aphids. *Plant Cell and Environment*, **30**, 187–201.

Doblin, M.S., Kurek, I., Jacob-Wilk, D., and Delmer, D.P. (2002) Cellulose biosynthesis in plants: from genes to rosettes. *Plant and Cell Physiology*, **43**, 1407–1420.

Dos Santos, H.P., Purgatto, E., Mercier, H., and Buckeridge, M.S. (2004) The control of storage xyloglucan mobilization in cotyledons of *Hymenaea courbaril*. *Plant Physiology*, **135**, 287–299.

Edwards, M., Dea, I.C.M., Bulpin, P.V., and Reid, J.S.G. (1985) Xyloglucan (amyloid) mobilization in the cotyledons of *Tropaeolum majus* L. seeds following germination. *Planta*, **163**, 133–140.

Edwards, M., Dea, I.C.M., Bulpin, P.V. and Reid, J.S.G. (1986) Purification and properties of a novel xyloglucan-specific endo-(1→4)-beta-D-glucanase from germinated Nasturtium Seeds (*Tropaeolum majus* L.). *Journal of Biological Chemistry*, **261**, 9489–9494.

Eklöf, J.M. and Brumer, H. (2010) The XTH gene family: an update on enzyme structure, function, and phylogeny in xyloglucan remodeling. *Plant Physiology*, **153**, 456–466.

Eklöf, J.M., Shojania, S., Okon, M., Mcintosh, L.P. and Brumer, H. (2013) Structure-function analysis of a broad specificity *Populus trichocarpa* endo-beta-glucanase reveals an evolutionary link between bacterial licheninases and plant XTH gene products. *Journal of Biological Chemistry*, **288**, 15786–15799.

Fernandes, A.N., Thomas, L.H., Altaner, C.M., Callow, P., Forsyth, V.T., Apperley, D.C., Kennedy, C.J., and Jarvis, M.C. (2011) Nanostructure of cellulose microfibrils in spruce wood. *Proceedings of the National Academy of Sciences of the United States of America*, **108**, E1195–E1203.

Francis, K.E., Lam, S.Y., and Copenhaver, G.P. (2006) Separation of Arabidopsis pollen tetrads is regulated by QUARTET1, a pectin methylesterase gene. *Plant Physiology*, **142**, 1004–1013.

Fry, S.C. (1989) The Structure and functions of xyloglucan. *Journal of Experimental Botany*, **40**, 1–11.

Fry, S.C., Smith, R.C., Renwick, K.F., Martin, D.J., Hodge, S.K., and Matthews, K.J. (1992) Xyloglucan endotransglycosylase, a new wall-loosening enzyme activity from plants. *Biochemical Journal*, **282**, 821–828.

Geisler-Lee, J., Geisler, M., Coutinho, P.M., Segerman, B., Nishikubo, N., Takahashi, J., Aspeborg, H., Djerbi, S., Master, E., Andersson-Gunneras, S., Sundberg, B., Karpinski, S., Teeri, T.T., Kleczkowski, L.A., Henrissat, B., and Mellerowicz, E.J. (2006) Poplar carbohydrate-active enzymes. Gene identification and expression analyses. *Plant Physiology*, **140**, 946–962.

Goldberg, R., Pierron, M., Bordenave, M., Breton, C., Morvan, C., and Du Penhoat, C.H. (2001) Control of Mung bean pectinmethylesterase isoform activities. Influence of pH and carboxyl group distribution along the pectic chains. *Journal of Biological Chemistry*, **276**, 8841–8847.

Hara, Y., Yokoyama, R., Osakabe, K., Toki, S., and Nishitani, K. (2014) Function of xyloglucan endotransglucosylase/hydrolases in rice. *Annals of Botany,* published online 19 December 2013, doi: 10.1093/aob/mct292.

Henrissat, B., Coutinho, P.M., and Davies, G.J. (2001) A census of carbohydrate-active enzymes in the genome of Arabidopsis thaliana. *Plant Molecular Biology*, **47**, 55–72.

Hayashi, T. (1989) Xyloglucans in the primary-cell wall. *Annual Review of Plant Physiology and Plant Molecular Biology*, **40**, 139–168.

Hongo, S., Sato, K., Yokoyama, R., and Nishitani, K. (2012) Demethylesterification of the primary wall by PECTIN METHYLESTERASE35 provides mechanical support to the Arabidopsis stem. *Plant Cell*, **24**, 2624–2634.

Hsieh, Y.S.Y. and Harris, P.J. (2009) Xyloglucans of monocotyledons have diverse structures. *Molecular Plant*, **2**, 943–965.

Imoto, K., Yokoyama, R. and Nishitani, K. (2005) Comprehensive approach to genes involved in cell wall modifications in *Arabidopsis thaliana*. *Plant Molecular Biology*, **58**, 177–192.

Irving, H.R. and Gehring, C. (eds) (2012) *Plant Signaling Peptides, Signaling and Communication in Plants (16)*. Springer-Verlag, Berlin, Heidelberg.

Ito, H. and Nishitani, K. (1999) Visualization of EXGT-mediated molecular grafting activity by means of a fluorescent-labeled xyloglucan oligomer. *Plant and Cell Physiology*, **40**, 1172–1176.

Jakob, H.F., Fengel, D., Tschegg, S.E., and Fratzl, P. (1995) The elementary cellulose fibril in Picea abies: Comparison of transmission electron microscopy, small-angle X-ray scattering, and wide-angle X-ray scattering results. *Macromolecules*, **28**, 8782–8787.

Kaewthai, N., Gendre, D., Eklof, J.M., Ibatullin, F.M., Ezcurra, I., Bhalerao, R.P., and Brumer, H. (2013) Group III-A XTH genes of Arabidopsis encode predominant xyloglucan endohydrolases that are dispensable for normal growth. *Plant Physiology*, **161**, 440–454.

Kagan-Zur, V., Tieman, D.M., Marlow, S.J., and Handa, A.K. (1995) Differential regulation of polygalacturonase and pectin methylesterase gene expression during and after heat stress in ripening tomato (*Lycopersicon esculentum* Mill.) fruits. *Plant Molecular Biology*, **29**, 1101–1110.

Kaida, R., Sugawara, S., Negoro, K., Maki, H., Hayashi, T., and Kaneko, T.S. (2010) Acceleration of cell growth by xyloglucan oligosaccharides in suspension-cultured tobacco cells. *Molecular Plant* **3**, 549–554.

Karakurt, Y. and Huber, D.J. (2004) Ethylene-induced gene expression, enzyme activities, and water soaking in immature and ripe watermelon (*Citrullus lanatus*) fruit. *Journal of Plant Physiology*, **161**, 381–388.

Kennedy, C.J., Cameron, G.J., Sturcova, A., Apperley, D.C., Altaner, C., Wess, T.J., and Jarvis, M.C. (2007) Microfibril diameter in celery collenchyma cellulose: X-ray scattering and NMR evidence. *Cellulose*, **14**, 235–246.

Kim, Y., Teng, Q., and Wicker, L. (2005) Action pattern of Valencia orange PME de-esterification of high methoxyl pectin and characterization of modified pectins. *Carbohydrate Research*, **340**, 2620–2629.

Kurasawa, K., Matsui, A., Yokoyama, R., Kuriyama, T., Yoshizumi, T., Matsui, M., Suwabe, K., Watanabe, M., and Nishitani, K. (2009) The *AtXTH28* gene, a xyloglucan endotransglucosylase/hydrolase, is involved in automatic self-pollination in *Arabidopsis thaliana. Plant and Cell Physiology*, **50**, 413–422.

Labavitch, J.M. and Ray, P.M. (1974a) Turnover of cell-wall polysaccharides in elongating pea stem segments. *Plant Physiology*, **53**, 669–673.
Labavitch, J.M. and Ray, P.M. (1974b) Relationship between promotion of xyloglucan metabolism and induction of elongation by indoleacetic-acid. *Plant Physiology*, **54**, 499–502.
Lee, D., Polisensky, D.H., and Braam, J. (2005) Genome-wide identification of touch- and darkness-regulated Arabidopsis genes: a focus on calmodulin-like and XTH genes. *New Phytologist*, **165**, 429–444.
Li, S.D., Lei, L., Somerville, C.R., and Gu, Y. (2012) Cellulose synthase interactive protein 1 (CSI1) links microtubules and cellulose synthase complexes. *Proceedings of the National Academy of Sciences of the United States of America*, **109**, 185–190.
Lionetti, V., Raiola, A., Camardella, L., Giovane, A., Obel, N., Pauly, M., Favaron, F., Cervone, F., and Bellincampi, D. (2007) Overexpression of pectin methylesterase inhibitors in Arabidopsis restricts fungal infection by *Botrytis cinerea*. *Plant Physiology*, **143**, 1871–1880.
Lionetti, V., Cervone, F., and Bellincampi, D. (2012) Methyl esterification of pectin plays a role during plant-pathogen interactions and affects plant resistance to diseases. *Journal of Plant Physiology*, **169**, 1623–1630.
Manabe, Y., Nafisi, M., Verhertbruggen, Y., Orfila, C., Gille, S., Rautengarten, C., Cherk, C., Marcus, S.E., Somerville, S., Pauly, M., Knox, J.P., Sakuragi, Y., and Scheller, H.V. (2011) Loss-of-function mutation of REDUCED WALL ACETYLATION2 in Arabidopsis leads to reduced cell wall acetylation and increased resistance to *Botrytis cinerea*. *Plant Physiology*, **155**, 1068–1078.
Matsui, A., Yokoyama, R., Seki, M., Ito, T., Shinozaki, K., Takahashi, T., Komeda, Y., and Nishitani, K. (2005) AtXTH27 plays an essential role in cell wall modification during the development of tracheary elements. *Plant Journal*, **42**, 525–534.
McCann, M.C., Wells, B., and Roberts, K. (1990) Direct visualization of cross-links in the primary plant-cell wall. *Journal of Cell Science*, **96**, 323–334.
McDougall, G.J. and Fry, S.C. (1989) Structure-activity relationships for xyloglucan oligosaccharides with antiauxin activity. *Plant Physiology*, **89**, 883–887.
McQueen-Mason, S., Durachko, D.M., and Cosgrove, D.J. (1992) Two endogenous proteins that induce cell wall extension in plants. *Plant Cell*, **4**, 1425–1433.
Mei, Y., Gao, H.B., Yuan, M., and Xue, H.W. (2012) The Arabidopsis ARCP protein, CSI1, which is required for microtubule stability, is necessary for root and anther development. *Plant Cell*, **24**, 1066–1080.
Micheli, F. (2001) Pectin methylesterases: cell wall enzymes with important roles in plant physiology. *Trends in Plant Science*, **6**, 414–419.
Mohnen, D. (2008) Pectin structure and biosynthesis. *Current Opinion in Plant Biology*, **11**, 266–277.
Naseer, S., Lee, Y., Lapierre, C., Franke, R., Nawrath, C., and Geldner, N. (2012) Casparian strip diffusion barrier in Arabidopsis is made of a lignin polymer without suberin. *Proceedings of the National Academy of Sciences of the United States of America*, **109**, 10101–10106.
Negi, J., Moriwaki, K., Konishi, M., Yokoyama, R., Nakano, T., Kusumi, K., Hashimoto-Sugimoto, M., Schroeder, J.I., Nishitani, K., Yanagisawa, S., and Iba, K. (2013) A Dof transcription factor, SCAP1, is essential for the development of functional stomata in Arabidopsis. *Current Biology*, **23**, 479–484.
Newman, R.H. (1999) Estimation of the lateral dimensions of cellulose crystallites using C-13 NMR signal strengths. *Solid State Nuclear Magnetic Resonance*, **15**, 21–29.
Nishino, T., Takano, K., and Nakamae, K. (1995) Elastic-modulus of the crystalline regions of cellulose polymorphs. *Journal of Polymer Science Part B-Polymer Physics*, **33**, 1647–1651.
Nishitani, K. (1997) The role of endoxyloglucan transferase in the organization of plant cell walls. *International Review of Cytology*, **173**, 157–206.
Nishitani, K. and Tominaga, R. (1992) Endo-xyloglucan transferase, a novel class of glycosyltransferase that catalyzes transfer of a segment of xyloglucan molecule to another xyloglucan molecule. *Journal of Biological Chemistry*, **267**, 21058–21064.

Nishitani, K. and Vissenberg, K. (2007) Roes of the XTH protein family in the expanding cell. *Plant Cell Monographs*, **5**, 89–116.

Nishiyama, Y., Langan, P., and Chanzy, H. (2002) Crystal structure and hydrogen-bonding system in cellulose 1 beta from synchrotron X-ray and neutron fiber diffraction. *Journal of the American Chemical Society*, **124**, 9074–9082.

O'Neill, M.A., Ishii, T., Albersheim, P., and Darvill, A.G. (2004) Rhamnogalacturonan II: structure and function of a borate cross-linked cell wall pectic polysaccharide. *Annual Review of Plant Biology*, **55**, 109–139.

O'Sullivan, A.C. (1997) Cellulose: the structure slowly unravels. *Cellulose*, **4**, 173–207.

Osorio, S., Castillejo, C., Quesada, M.A., Medina-Escobar, N., Brownsey, G.J., Suau, R., Heredia, A., Botella, M.A., and Valpuesta, V. (2008) Partial demethylation of oligogalacturonides by pectin methyl esterase 1 is required for eliciting defence responses in wild strawberry (*Fragaria vesca*). *Plant Journal*, **54**, 43–55.

Osorio, S., Bombarely, A., Giavalisco, P., Usadel, B., Stephens, C., Araguez, I., Medina-Escobar, N., Botella, M.A., Fernie, A.R., and Valpuesta, V. (2011) Demethylation of oligogalacturonides by FaPE1 in the fruits of the wild strawberry *Fragaria vesca* triggers metabolic and transcriptional changes associated with defence and development of the fruit. *Journal of Experimental Botany*, **62**, 2855–2873.

Palin, R. and Geitmann, A. (2012) The role of pectin in plant morphogenesis. *Biosystems*, **109**, 397–402.

Pauly, M., Albersheim, P., Darvill, A., and York, W.S. (1999) Molecular domains of the cellulose/xyloglucan network in the cell walls of higher plants. *Plant Journal*, **20**, 629–639.

Peaucelle, A., Louvet, R., Johansen, J.N., Höfte, H., Laufs, P., Pelloux, J., and Mouille, G. (2008) Arabidopsis phyllotaxis is controlled by the methyl-esterification status of cell-wall pectins. *Current Biology*, **18**, 1943–1948.

Peaucelle, A., Braybrook, S.A., Le Guillou, L., Bron, E., Kuhlemeier, C., and Höfte, H. (2011a) Pectin-induced changes in cell wall mechanics underlie organ initiation in Arabidopsis. *Current Biology*, **21**, 1720–1726.

Peaucelle, A., Louvet, R., Johansen, J.N., Salsac, F., Morin, H., Fournet, F., Belcram, K., Gillet, F., Höfte, H., Laufs, P., Mouille, G., and Pelloux, J. (2011b) The transcription factor BELLRINGER modulates phyllotaxis by regulating the expression of a pectin methylesterase in Arabidopsis. *Development*, **138**, 4733–4741.

Peaucelle, A., Braybrook , S., and Höfte, H. (2012) Cell wall mechanics and growth control in plants: the role of pectins revisited. *Flontiers in Plant Science*, **3**, 1–6.

Pelloux, J., Rusterucci, C., and Mellerowicz, E. J. (2007) New insights into pectin methylesterase structure and function. *Trends in Plant Science*, **12**, 267–277.

Phan, H.A., Iacuone, S., Li, S.F., and Parish, R.W. (2011) The MYB80 transcription factor is required for pollen development and the regulation of tapetal programmed cell death in *Arabidopsis thaliana*. *Plant Cell*, **23**, 2209–2224.

Planas, A. (2000) Bacterial 1,3-1,4-beta-glucanases: structure, function and protein engineering. *Biochimica et Biophysica Acta*, **1543**, 361–382.

Popper, Z.A., Michel, G., Herve, C., Domozych, D.S., Willats, W.G.T., Tuohy, M.G., Kloareg, B., and Stengel, D.B. (2011) Evolution and diversity of plant cell walls: from algae to flowering plants. *Annual Review of Plant Biology*, **62**, 567–588.

Pritchard, J., Hetherington, P.R., Fry, S.C., and Tomos, A.D. (1993) Xyloglucan endotransglycosylase activity, microfibril orientation and the profiles of cell-wall properties along growing regions of maize roots. *Journal of Experimental Botany*, **44**, 1281–1289.

Rayle, D.L. and Cleland, R.E. (1992) The acid growth theory of auxin-induced cell elongation is alive and well. *Plant Physiology*, **99**, 1271–1274.

Reca, I.B., Lionetti, V., Camardella, L., D'avino, R., Giardina, T., Cervone, F., and Bellincampi, D. (2012) A functional pectin methylesterase inhibitor protein (SolyPMEI) is expressed during tomato fruit ripening and interacts with PME-1. *Plant Molecular Biology*, **79**, 429–442.

Redgwell, R.J. and Fry, S.C. (1993) Xyloglucan Endotransglycosylase Activity Increases during Kiwifruit (*Actinidia deliciosa*) Ripening (Implications for Fruit Softening). *Plant Physiology*, **103**, 1399–1406.

Ridley, B.L., O'Neill, M.A., and Mohnen, D. (2001) Pectins: structure, biosynthesis, and oligogalacturonide-related signaling. *Phytochemistry*, **57**, 929–967.

Rockel, N., Wolf, S., Kost, B., Rausch, T., and Greiner, S. (2008) Elaborate spatial patterning of cell-wall PME and PMEI at the pollen tube tip involves PMEI endocytosis, and reflects the distribution of esterified and de-esterified pectins. *Plant Journal*, **53**, 133–143.

Rose, J.K., Braam, J., Fry, S.C., Nishitani, K. (2002) The XTH family of enzymes involved in xyloglucan endotransglucosylation and endohydrolysis: current perspectives and a new unifying nomenclature. *Plant and Cell Physiology*, **43**, 1421–1435.

Ryan, C.A. (1987) Oligosaccharide Signaling in Plants. *Annual Review of Cell Biology*, **3**, 295–317.

Saez-Aguayo, S., Ralet, M.C., Berger, A., Botran, L., Ropartz, D., Marion-Poll, A., and North, H.M. (2013) PECTIN METHYLESTERASE INHIBITOR6 promotes Arabidopsis mucilage release by limiting methylesterification of homogalacturonan in seed coat epidermal cells. *Plant Cell*, **25**, 308–323.

Sanchez-Rodriguez, C., Bauer, S., Hematy, K., Saxe, F., Ibanez, A.B., Vodermaier, V., Konlechner, C., Sampathkumar, A., Ruggeberg, M., Aichinger, E., Neumetzler, L., Burgert, I., Somerville, C., Hauser, M.T., and Persson, S. (2012) CHITINASE-LIKE1/POM-POM1 and its homolog CTL2 are glucan-interacting proteins important for cellulose biosynthesis in Arabidopsis. *Plant Cell*, **24**, 589–607.

Sasidharan, R., Chinnappa, C.C., Staal, M., Elzenga, J.T., Yokoyama, R., Nishitani, K., Voesenek, L.A., and Pierik, R. (2010) Light quality-mediated petiole elongation in Arabidopsis during shade avoidance involves cell wall modification by xyloglucan endotransglucosylase/hydrolases. *Plant Physiology*, **154**, 978–990.

Scheller, H.V. and Ulvskov, P. (2010) Hemicelluloses. *Annual Review of Plant Biology*, **61**, 263–289.

Schmohl, N., Pilling, J., Fisahn, J., and Horst, W.J. (2000) Pectin methylesterase modulates aluminium sensitivity in *Zea mays* and *Solanum tuberosum*. *Physiologia Plantarum*, **109**, 419–427.

Shinde, U. and Inouye, M. (1993) Intramolecular chaperones and protein folding. *Trends in Biochemical Sciences*, **18,** 442–446.

Somerville, C. (2006) Cellulose synthesis in higher plants. *Annual Review of Cell and Developmental Biology*, **22**, 53–78.

Somerville, C., Bauer, S., Brininstool, G., Facette, M., Hamann, T., Milne, J., Osborne, E., Paredez, A., Persson, S., Raab, T., Vorwerk, S., and Youngs, H. (2004) Toward a systems approach to understanding plant cell walls. *Science*, **306**, 2206–2211.

Takeda, T., Furuta, Y., Awano, T., Mizuno, K., Mitsuishi, Y., and Hayashi, T. (2002) Suppression and acceleration of cell elongation by integration of xyloglucans in pea stem segments. *Proceedings of the National Academy of Sciences of the United States of America*, **99**, 9055–9060.

Talbott, L.D. and Ray, P.M. (1992) Molecular size and separability features of pea cell wall polysaccharides : implications for models of primary wall structure. *Plant Physiology*, **98**, 357–368.

Tan, L., Eberhard, S., Pattathil, S., Warder, C., Glushka, J., Yuan, C.H., Hao, Z.Y., Zhu, X., Avci, U., Miller, J.S., Baldwin, D., Pham, C., Orlando, R., Darvill, A., Hahn, M.G., Kieliszewski, M.J., and Mohnen, D. (2013) An Arabidopsis cell wall proteoglycan consists of pectin and arabinoxylan covalently linked to an arabinogalactan protein. *Plant Cell*, **25**, 270–287.

Taylor, N.G., Howells, R.M., Huttly, A.K., Vickers, K., and Turner, S.R. (2003) Interactions among three distinct CesA proteins essential for cellulose synthesis. *Proceedings of the National Academy of Sciences of the United States of America*, **100**, 1450–1455.

Tieman, D.M. and Handa, A.K. (1994) Reduction in pectin methylesterase activity modifies tissue integrity and cation levels in ripening tomato (*Lycopersicon esculentum* Mill.) fruits. *Plant Physiology*, **106**, 429–436.

Vargas-Rechia, C., Reicher, F., Rita Sierakowski, M., Heyraud, A., Driguez, H., and Linart, Y. (1998) Xyloglucan octasaccharide XXLGol derived from the seeds of hymenaea courbaril acts as a signaling molecule. *Plant Physiology*, **116**, 1013–1021.

Vincken, J.P., Schols, H.A., Oomen, R.J., Mccann, M.C., Ulvskov, P., Voragen, A.G., and Visser, R.G. (2003) If homogalacturonan were a side chain of rhamnogalacturonan I. Implications for cell wall architecture. *Plant Physiology*, **132**, 1781–1789.

Vissenberg, K., Martinez-Vilchez, I.M., Verbelen, J.P., Miller, J.G., and Fry, S.C. (2000) In vivo colocalization of xyloglucan endotransglycosylase activity and its donor substrate in the elongation zone of Arabidopsis roots. *Plant Cell*, **12**, 1229–1237.

Vogel, J. (2008) Unique aspects of the grass cell wall. *Current Opinion in Plant Biology*, **11**, 301–307.

Voiniciuc, C., Dean, G.H., Griffiths, J.S., Kirchsteiger, K., Hwang, Y.T., Gillett, A., Dow, G., Western, T.L., Estelle, M., and Haughn, G.W. (2013) Flying saucer1 is a transmembrane RING E3 ubiquitin ligase that regulates the degree of pectin methylesterification in Arabidopsis seed mucilage. *Plant Cell*, **25**, 944–959.

Volpi, C., Janni, M., Lionetti, V., Bellincampi, D., Favaron, F., and D'ovidio, R. (2011) The ectopic expression of a pectin methyl esterase inhibitor increases pectin methyl esterification and limits fungal diseases in wheat. *Molecular Plant-Microbe Interactions*, **24**, 1012–1019.

Vorwerk, S., Somerville, S., and Somerville, C. (2004) The role of plant cell wall polysaccharide composition in disease resistance. *Trends in Plant Sciences*, **9**, 2003–2009.

Wallace, G. and Fry, S.C. (1994) Phenolic components of the plant cell wall. *International Review of Cytology*, **151**, 229–267.

Wang, M., Yuan, D., Gao, W., Li, Y., Tan, J., and Zhang, X. (2013) A comparative genome analysis of PME and PMEI families reveals the evolution of pectin metabolism in plant cell walls. *PLoS One*, **8**, e72082.

Wang, T., Zabotina, O., and Hong, M. (2012) Pectin-cellulose interactions in the Arabidopsis primary cell wall from two dimensional magic-angle-spinning solid-state nuclear magnetic resonance. *Biochemistry*, **51**, 9846–9856.

Wasterneys, G. and Collings, D. (2007) The cytoskeleton and co-ordination of directional expansion in a multicellular context, in *The Expanding Cell, Plant Cell Monographs,* vol. 5 (eds J.-P. Verbelen and V. Vissenberg), Springer-Verlag, Berlin, Heidelberg, 217–248.

Weber, M., Deinlein, U., Fischer, S., Rogowski, M., Geimer, S., Tenhaken, R., and Clemens, S. (2013) A mutation in the *Arabidopsis thaliana* cell wall biosynthesis gene pectin methylesterase 3 as well as its aberrant expression cause hypersensitivity specifically to Zn. *Plant Journal*, **76,** 151–164.

Willats, W.G., Orfila, C., Limberg, G., Buchholt, H.C., Van Alebeek, G.J., Voragen, A.G., Marcus, S.E., Christensen, T.M., Mikkelsen, J.D., Murray, B.S., and Knox, J.P. (2001) Modulation of the degree and pattern of methyl-esterification of pectic homogalacturonan in plant cell walls. Implications for pectin methyl esterase action, matrix properties, and cell adhesion. *Journal of Biological Chemistry*, **276**, 19404–19413.

Willats, W.G.T., Knox, P., and Mikkelsen, J.D. (2006) Pectin: new insights into an old polymer are starting to gel. *Trends in Food Science & Technology*, **17**, 97–104.

Wolf, S., Mouille, G., and Pelloux, J. (2009a) Homogalacturonan methyl-esterification and plant development. *Molecular Plant*, **2**, 851–860.

Wolf, S., Rausch, T., and Greiner, S. (2009b) The N-terminal pro region mediates retention of unprocessed type-I PME in the Golgi apparatus. *Plant Journal*, **58**, 361–375.

Wolf, S., Hématy, K., and Höfte, H. (2012a) Growth control and cell wall signaling in plants. *Annual Review of Plant Biology*, **63**, 381–407.

Wolf, S., Mravec, J., Greiner, S., Mouille, G., and Höfte, H. (2012b) Plant cell wall homeostasis is mediated by brassinosteroid feedback signaling. *Current Biology*, **22**, 1732–1737.

Yamaguchi, M. and Demura, T. (2010) Transcriptional regulation of secondary wall formation controlled by NAC domain proteins. *Plant Biotechnology*, **27**, 237–242.

Yennawar, N.H., Li, L.C., Dudzinski, D.M., Tabuchi, A., and Cosgrove, D.J. (2006) Crystal structure and activities of EXPB1 (Zea m 1), alpha, beta-expansin and group-1 pollen allergen from maize. *Proceedings of the National Academy of Sciences of the United States of America*, **103**, 14664–14671.

Yokoyama, R. and Nishitani, K. (2001) A comprehensive expression analysis of all members of a gene family encoding cell-wall enzymes allowed us to predict cis-regulatory regions involved in cell-wall construction in specific organs of Arabidopsis. *Plant and Cell Physiology*, **42**, 1025–1033.

Yokoyama, R. and Nishitani, K. (2004) Genomic basis for cell-wall diversity in plants. A comparative approach to gene families in rice and Arabidopsis. *Plant and Cell Physiology*, **45**, 1111–1121.

Yokoyama, R., Rose, J.K., and Nishitani, K. (2004) A surprising diversity and abundance of xyloglucan endotransglucosylase/hydrolases in rice. Classification and expression analysis. *Plant Physiology*, **134**, 1088–1099.

Yokoyama, R., Uwagaki, Y., Sasaki, H., Harada, T., Hiwatashi, Y., Hasebe, M., and Nishitani, K. (2010) Biological implications of the occurrence of 32 members of the XTH (xyloglucan endotransglucosylase/hydrolase) family of proteins in the bryophyte *Physcomitrella patens*. *Plant Journal*, **64**, 645–656.

Zabotina, O.A. (2012) Xyloglucan and its biosynthesis. *Frontiers in Plant Science*, **3**, 134.

Zhang, G.F. and Staehelin, L.A. (1992) Functional compartmentation of the Golgi apparatus of plant cells : immunocytochemical analysis of high-pressure frozen- and freeze-substituted sycamore maple suspension culture cells. *Plant Physiology*, **99**, 1070–1083.

Zhu, T. and Wang, X. (2000) Large-scale profiling of the Arabidopsis transcriptome. *Plant Physiology*, **124**, 1477–1476.

2 Regulation of Cell Wall Formation by Membrane Traffic

Clara Sánchez-Rodríguez[1,2] and Staffan Persson[1]

[1]*Max-Planck-Institute for Molecular Plant Physiology, Potsdam, Germany*
[2]*ARC Centre of Excellence in Plant Cell Walls School of Botany, University of Melbourne, Victoria, Australia*

Introduction

Plant cell walls have been explored and described since the end of the 19th century, most notably concerning their biochemical and structural characteristics (see Chapter 1). This work has provided the foundation for where certain cell wall components are produced, i.e., most of the cell wall polysaccharides and glycoproteins are synthesized inside the cell and secreted to the apoplast, with the exception of cellulose and callose that are made at the plasma membrane. This means that many cell wall components must be secreted to the apoplast; the cell wall architecture and remodeling of this structure therefore depends on trafficking efficiency within the plant cell. However, the mechanisms by which the cell wall elements are trafficked to and from the apoplast have not received much attention, perhaps due to the difficulty of polysaccharide visualization. New tools and techniques, such as immuno-based electron microscopy of specific polysaccharide antibodies, fluorescently tagged-protein localization by high-resolution microscopy, click chemistry, chemical genomics, and vesicle proteomics have however been shedding some light on this topic. In this chapter we review the current status of how cell wall components are trafficked in the cell. We begin with an overview of the plant membrane trafficking system to familiarize the reader with its different paths and elements, highlighting those tracks necessary for cell wall formation. We then focus on the cell wall components of typical plant cells, and examine how they use the trafficking system to reach their final destinations.

Membrane Trafficking in Plant Cells

Eukaryotic cells are distinguished from prokaryotes by having a nucleus and internal membrane-bound organelles/compartments. Most of these organelles/

Plant Cell Wall Patterning and Cell Shape, First Edition. Edited by Hiroo Fukuda.

compartments are surrounded by a single membrane envelope and are connected to each other, and to the plasma membrane (PM), through a complex and highly regulated trafficking system. These transport pathways are essential for the secretion and internalization of proteins, carbohydrates, and lipids, and may also maintain distinct groups of resident macromolecules at each organelle/compartment. The membrane traffic is vital for the cell to maintain its homeostasis and to respond to stimuli via signal transduction. Moreover, in plants, the endomembrane transport system plays a key role in cell wall synthesis and remodeling by secretion and endocytosis of oligosaccharides, structural proteins, and cell wall enzymes.

Endomembrane Compartments and their Mutual Relationships

Among the various organelles/compartments associated with vesicle trafficking, we find the endoplasmic reticulum (ER), Golgi apparatus, trans-Golgi network/early endosome (TGN/EE), multivesicular body/prevacuolar compartment/late endosome (MVB/PVC/LE) and the vacuole. All of these organelles/compartments exchange proteins and lipids through membrane-enclosed trafficking. To maintain directed trafficking paths, the pH-homeostasis in the endomembrane system must be precisely regulated. In plants, the proton pumping system involved in this regulation consists of three main V-ATPases (two vacuolar and one TGN-based H^+-ATPases) and a vacuolar H^+-pyrophosphatase (V-PPase) (reviewed in Schumacher 2006).

Three endomembrane compartments act as hubs for trafficked compounds.

1. The first is the ER, which is a continuous membrane system that emanates from the nuclear envelope and ramifies the cytoplasm. The main role of the ER is largely attributed to protein folding, and subsequent post-transcriptional modifications and oligomerization. The ER is also necessary for lipid biosynthesis and works as an initiator of the secretory system.
2. The next compartment is represented by the Golgi apparatus, which is formed by 5–20 individual cisternae stacked in a polarized cis-to-trans orientation. Golgi bodies can typically be found in interphase cells as rapidly moving donut-like shapes that are loosely connected to the ER, and are moved via the actin cytoskeleton. The cis-Golgi is linked to the ER from which it receives certain cargo. These cargo compounds travel sequentially from cis- to medial- and trans-Golgi (Staehelin and Kang, 2008). In addition to its role as a hub of endomembrane transport, the Golgi apparatus is the major biosynthetic source of polysaccharides and also participates in protein modifications.

3. The TGN is a tubular-vesicular structure derived, but independent, from the trans-Golgi cisternae. The TGN is the starting point of diverse pathways which deliver cargo to the PM and vacuole via the MVB/PVC/LE. In addition, the TGN gets proteins from the MVB/PVC/LE and acts as an early endosome receiving internalized material from the PM (Dettmer *et al.*, 2006; Viotti *et al.*, 2010). The TGN is therefore a hub for many post-Golgi trafficking pathways and has a key role in controlling the distribution of proteins belonging to each trafficking path.

Membrane traffic involves two main pathways (Fig. 2.1): secretory and endocytic. The secretory path, also named anterograde, starts at the ER and flows through the Golgi and TGN/EE to the PM. The endocytic route, on the other hand, recruits cargo from the PM and typically transports it to the vacuole through the endosomes (i.e., TGN/EE and MVB/PVC/LE). In addition, various alternative pathways (so-called retrograde paths) exist that redirect cargo to various compartments, such as trafficking from Golgi to ER, or recycling of proteins from endosomes back to the PM. Another function of the retrograde pathways is to move newly synthesized cargo directly to the vacuole.

The best-characterized membrane-associated trafficking pathway is Golgi-dependent (Fig. 2.1, black arrows), which is mediated by vesicle formation and plays an essential role in cell wall formation. Golgi-mediated trafficking requires the recruitment of a plethora of cytosolic proteins to form vesicles and the action of organelle-specific complexes for correct targeting and fusion of the transport vesicles. The vesicle-dependent path through Golgi is essential for the connection among organelles inside the cell, and also for endocytosis, transcytosis, and so-called 'classical/conventional secretion'.

Recent data confirm the existence of transport paths that by-pass the Golgi apparatus in plants, which has also been described in animals and yeast (Fig. 2.1, white arrows; for recent reviews see Ding *et al.*, 2012; Drakakaki and Dandekar 2013). These so-called 'unconventional protein secretion' paths are used by the cell to deliver proteins lacking N-terminal signal peptides.

Regulators of Vesicle-mediated Membrane Trafficking in Plants

The Golgi-dependent vesicle membrane trafficking consists of several sequential steps: selection of the cargo and vesicle formation; vesicle shuttling through the cytosol; and targeting and tethering of the vesicle to the destination membrane. These steps are followed by membrane fusion between the vesicle and the sink membranes. While these different steps have been recently reviewed (Bassham *et al.*, 2008; Fujimoto and Ueda 2012), we attempt to amalgamate these views with new findings, paying special attention to those related to cell wall regulation.

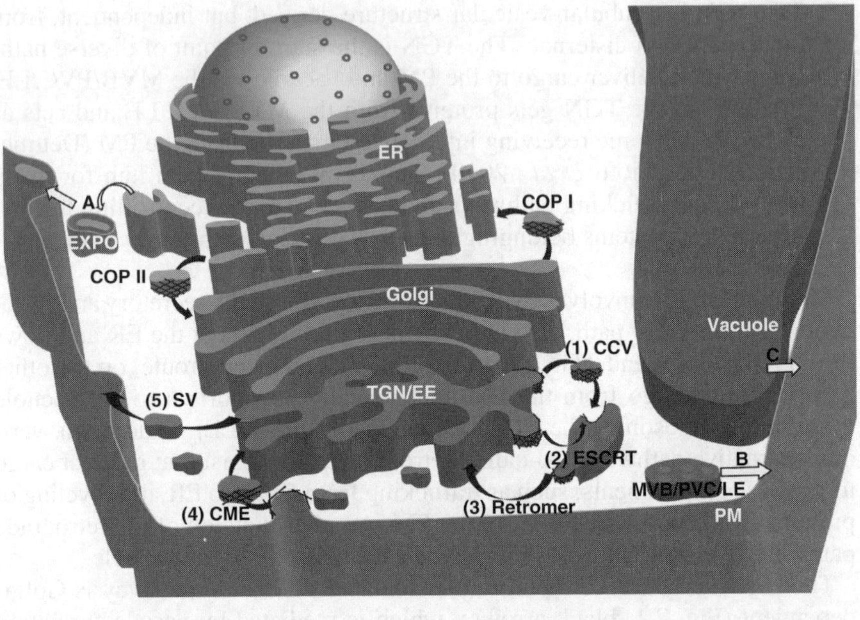

Figure 2.1 Plant endosomal trafficking. The endomembrane system of eukaryotes consists of a network of single-membrane organelles that exchange material with each other and with the plasma membrane. In plants, the traditional vesicle pathway (black arrows) starts off by the exchange of vesicles between the endoplasmic reticulum (ER) and the Golgi via the coat protein complex II (COP-II; ER to Golgi) or the coat protein complex I (COP-I; Golgi to ER). From the Golgi, cargo may be transferred to the trans-Golgi network, which in plants also acts as early endosome (TGN/EE). The TGN/EE is a hub for many post-Golgi pathways. 1, Clathrin-coat-vesicle (CCV) mediated trafficking of proteins to the vacuole via the multivesicular bodies/prevacuolar compartement/late endosome (MVB/PVC/LE); 2, Endocytosis of proteins into MVB/PVC/LE by the ESCRT system, which acts at the MVC/PVC/LE and at the TGN/EE; 3, Retromer-mediated recycling of proteins from the MVB/PVC/LE to the TGN; 4, Clathrin-mediated endocytosis (CME) of plasma membrane (PM) and apoplastic cargos; 5, Secretion of proteins to the PM via coat-independent secretory vesicles (SV). The cages of the coat complexes are represented by black lines. The secretion of certain cell wall components is also mediated by the so-called 'unconventional' trafficking pathways (white arrows). A, Exocytosis via the double membrane organelle exocyst-positive-organelle (EXPO); B, Deposition of callose synthase to the PM from MVB/PVC/LE; C, Fusion to the vacuole with the PM. The different organelles/compartments are indicated in the figure.

Vesicle Formation: The first phase of vesicle trafficking consists of selective recruitment of the cargo into a membrane structure, which is then pinched off the donor organelle/compartment. Most commonly, the cargo selection and membrane deformation relies on coat protein complexes (CPCs). Only the formation of secretory vesicles at the TGN, which transport cargo to the PM and apoplast, seems to be coat-independent. After sufficient membrane deformation, i.e. vesicle formation, dynamin-related proteins (DRPs) are involved in membrane tubulation and/or scission from the donor membrane (Fig. 2.2a).

The first information about how this process works was obtained during the 1960s when electron microscopy studies in animal cells identified small vesicles encircled by electron-dense 'coats' (Roth and Porter 1964). Different types of CPCs can be identified within a typical eukaryotic cell, each responsible for vesicle formation at a certain organelle/compartment. Interestingly, although the protein configuration of the coat varies, the coating mechanism is largely conserved among the different CPCs and among eukaryotes (Bonifacino and Lippincott-Schwartz 2003; Field *et al.*, 2011).

The first step in this process, i.e. CPC nucleation, is coordinated by a GTP-cycle of coat-GTPases. The coat-GTPases constitute one subfamily of plant small-GTPases, which also include RAB (RAS-related in brain), ROP (Rho-like proteins in plants), and RAN (RAS-related nuclear protein) (Vernoud *et al.*, 2003). All small GTPases typically cycle between an inactive soluble GDP-bound form and an active membrane GTP-bound state. The GTPases are recruited to the donor membrane and activated by a GTP-exchange factor (GEF). Subsequently, a GTPase activation protein (GAP) converts the small GTPases to inactive forms. The active GTP-bound coat-GTPases may also stimulate the synthesis of certain phosphatidylinositols in the donor membranes that act as coat-protein-docking sites.

The recruitment of different coat complex adaptors or 'cargo-selective' sub units initiates the cargo recognition and the vesicle nucleation. Once the 'cage' subunits (i.e., proteins that coat the vesicle) of the CPC stabilize the vesicle, it is separated from the donor membrane by DRPs and/or via structural changes of the coat proteins themselves and the lipids which form the neck of the vesicle. After inactivation, the GDP-bound coat-GTPases trigger the CPC un-coating, which releases the coat scaffolds to engage in new recruitment and allows for fusion of the vesicle to the destination membrane. In the following, we outline the different CPCs identified in plants.

Coat Protein Complex II (COP-II): ER-to-Golgi Trafficking: The COP-II coats recruit secretory cargo from the ER and move them to the Golgi. The COP-II mediated sorting follows similar steps and is common to all eukaryotes. In most cases, the COP-II machinery concentrates at certain points along the ER membrane, the ER-exit sites (ERES), which associate with the Golgi cisternae in plant cells. The coat-GTPase that initiates the COP-II is SAR1 (secretion-associated and ras-superfamily related 1), which recruits a cargo-selective subunit that is composed of two proteins: SEC23 and SEC24. The two proteins SEC13 and SEC31 then constitute the cage subunit for COP-II. Due to the tight association between ER-Golgi in plants (Sparkes *et al.*, 2009), the COP-II vesicle disassembles rapidly. Probably for the same reason, the anterograde transport between these organelles does not need to

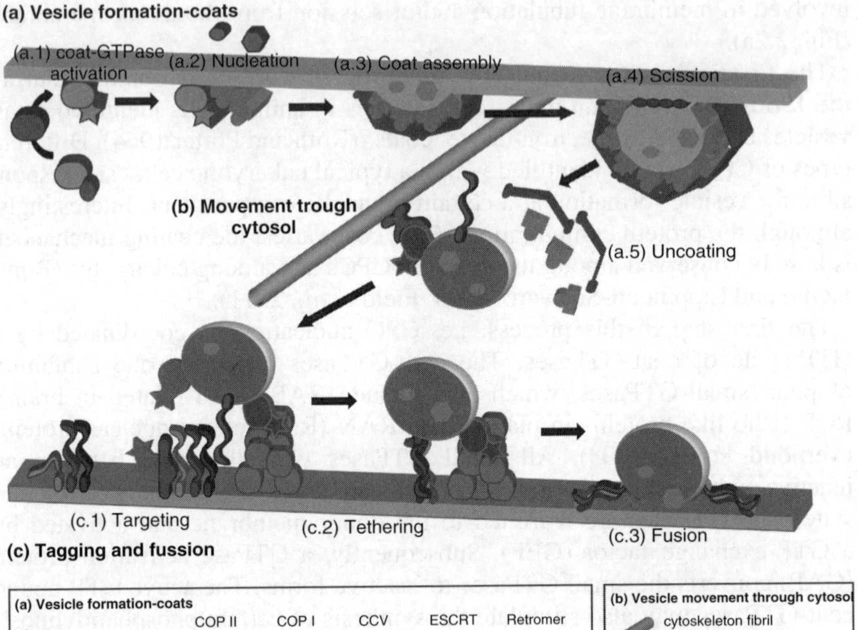

Figure 2.2 Vesicle mediated trafficking in plants. (a) Vesicle formation via coat-protein complexes. The inactive GDP-bounded coat-GTPase is activated by a GTP exchange factor (GEF) and recruited to the membrane (a.1). The active GTP-bound coat-GTPase recruits the coat adaptors that recognize the cargo and start the vesicle nucleation (a.2). The assembly of cage subunits stabilizes the vesicle (a.3), which is separated from the donor membrane by nTPase activity (a.4). Once in the cytosol, the vesicle is uncoated (a.5) and becomes free to travel to its destination. In plants, five different coat protein complexes have been identified (see Fig. 2.1): coat protein complex II and I (COP-II and COP-I), clathrin-coat vesicles (CCVs), endosomal sorting complex (ESCRT), and retromer. In the panel below the figure, the proteins involved in the different processes/complexes are indicated. (b) Vesicle movement through the cytosol. The uncoated vesicle travels via actin or microtubule fibrils, tethered by molecular motors and ROPs (a small GTPase). In the case of COP-I and II the donor and acceptor membranes are in close vicinity, which permits travel without cytoskeletal guidance. (c) Targeting and fusion. The vesicle is targeted to the sink membrane by the activity of small GTPases (ROP and RAB families), which recruit SM-protein, tethering factors (in the figure this is represented by the exocyst complex) and other effectors to attach the vesicle to the target membrane (c.1). A member of the SM family activates the syntaxin (one of the target-(t)-SNAREs), which promotes the formation of a three t-SNARE-helix that serves as a binding site for the vesicle-(v)-SNARE (c.2). Mutual twisting of multiple SNARE complexes leads to the fusion of the two bilayers and the cargo is delivered (c.3).

rely on the cytoskeleton, as is the case in other eukaryotes. Most secreted proteins, including those which will be part of the cell wall, are synthesized at the ER. The COP-II might therefore transport cell wall proteins to the Golgi, with the exception of some proteins that follow a non-conventional secretion pathway (see 'Membrane Traffic in Cell Wall Metabolism').

COP-I Coats: Golgi-to-ER Trafficking: COP-I is typically associated with retrograde transport from the Golgi. The coat GTPase for COP-I is ARF1 (ADP-ribosylation factor Ras GTPase 1). The active ARF1 recruits the cargo-selective subunit, F-COP, which is composed of four proteins (γ-COP, β-COP, δ-COP, and ζ-COP). The cage complex is named B-COP and consists of three proteins (α-COP, β'-COP and ε-COP; Pimpl *et al.*, 2000; Bassham *et al.*, 2008).

Clathrin Coat Vesicles (CCV): Post-Golgi Cargo Recruitment: The clathrin-coat vesicles (CCVs) are involved in sorting cargo into diverse endosomes from TGN (nicely visualized in Kang *et al.*, 2011), and endocytosis from the PM (reviewed in Chen *et al.*, 2011). In addition CCVs are implicated in a variety of other processes, some of them related to cell wall synthesis and remodeling, such as internalization of PM proteins and cell wall polysaccharides and the assembly of the cell plate (Baluska *et al.*, 2005; Miart *et al.*, 2013).

As in the case of COP-I coat formation, the CCV requires the activation of an ARF1. At least in the case of clathrin-mediated endocytosis (CME), synthesis of PIP_2 (phosphatidylinositol 4,5-biphosphate) at the PM is necessary for correct vesicle formation (Ischebeck *et al.*, 2013), suggesting that ARF1 may regulate PIP_2 production. Specific early adaptors are necessary for nucleation and cargo selection for each CCV-mediated pathway. For clathrin-mediated endocytosis (CME), an octameric protein complex (named TPLATE complex) has just been identified as the early adaptor module essential for plant PM endocytosis. Interestingly, this complex represents a unique evolutionary adaptation of the canonical eukaryotic pathway for CME (Gadeyne *et al.*, 2014). On the other hand, different plant-specific adaptor proteins have been shown to participate in CCV-mediated transport of vacuolar cargo selected by vacuolar sorting receptors, from the TGN to the PVC (Song *et al.*, 2006; Sauer *et al.*, 2013). Together with these early adaptors, the canonical adaptor protein (AP) complexes AP-1 to AP-4 play an essential role in cargo recognition and vesicle formation in post-Golgi trafficking pathways (reviewed in Robinson and Bonifacino, 2001; Bassham *et al.*, 2008). As for the early adaptors, each AP complex is assigned to a distinctive function: AP-1 seems to be part of the CCV trafficking from the TGN to the PVC and to the PM in interphase cells, and to the plane of cell division during cytokinesis (Song *et al.*, 2006; Park *et al.*, 2013; Teh *et al.*, 2013); AP-2 has recently been shown to participate in CME (Bashline *et al*., 2013; Di Rubbo *et al*., 2013; Fan *et al*., 2013; Kim *et al*., 2013; Yamaoka *et al*., 2013); and AP-3 is involved in Golgi-vacuolar trafficking (Niihama *et al*., 2009; Zwiewka *et al*., 2011; Wolfenstetter *et al*., 2012). Until now, the AP-4 function has not been well characterized, although it seems plausible that it participates in a subset of Golgi-to-endosomal transport (Bassham *et al*., 2008).

The cage complex for CCV formation consists of a three-legged structure named triskelion (Fotin *et al*., 2004). Each leg is made of one clathrin heavy chain (CHC) and one light chain (CLC). Interestingly, the plant triskelion has a higher molecular mass and longer arms than the mammalian counterpart, but the diameter of the CCV is smaller than in mammals. This may be due to the high internal turgor pressure and the presence of the cell wall, which might limit the size of the endocytic CCVs (McMahon and Boucrot, 2011).

Other Coats: ESCRT and Retromer: Other molecular cargo complexes are necessary for precise regulation of trafficking in endosomes, such as the retromer and the ESCRT (endosomal sorting complex required for transport). Both of these structures are implicated in protein degradation (Bassham *et al*., 2008; Reyes *et al*., 2011).

The ESCRT system, formed of three multi-subunit complexes ESCRT-I to -III, sorts ubiquitinated PM proteins into MVB/PVC internal vesicles. Interestingly, instead of forming a vesicle directed to the cytosol, ESCRT complex acts ‘in reverse’ by creating a vesicle that internalizes into the MVB/PVC/LE membrane and that engulfs cytosolic cargo. Interestingly, recent data showed that ESCRT proteins localize at the TGN/EE, indicating that the formation of intraluminal vesicles has already started at this point (Scheuring *et al*., 2011). The ESCRT network is conserved among eukaryotes, and plants contain orthologs for most of the ESCRT-I to -III components identified in other systems. However, only some of these plant ESCRTs have been characterized. Among them, ELCH is an ESCRT-I subunit that binds ubiquitin (Ub) and assembles with other ESCRT-I subunits. In addition to its role in the recognition of ubiquitinated proteins, ESCRT-I seems to bind PI_3P and clathrin at endosomal membranes, and can interact sequentially with ESCRT-II and -III to induce vesicle formation and scission. Finally, a large ATPase (atSKD1; Haas *et al*., 2007) disassembles and recycles the ESCRT coats to the cytosol. Different plant receptors have been identified to be ubiquitinated upon signal perception, probably to become internalized for degradation via ESCRT (reviewed in Reyes *et al*., 2011). However, only the auxin carrier PIN2 has been confirmed to be an ESCRT cargo (Spitzer *et al*., 2009). There is no precise information about ubiquitination of cell-wall-related proteins, although such a mechanism is appealing.

The plant retromer acts in the recycling of vacuolar sorting receptors (VSRs) from the PVC/MVB to the TGN. In addition, it modulates the trafficking of auxin carrier proteins (Jaillais *et al.*, 2007; Kleine-Vehn *et al.*, 2008). Different laboratories have localized the plant retromer complex at the TGN, MVB, or both. Consequently, there are still discrepancies regarding its exact subcellular localization, and the function of the retromer at these two endosomes is under debate (Niemes *et al.*, 2010). The cargo-selective subunit of the retromer is a heterotrimer composed of VPS35, VPS26, and VPS29. The sorting complex is formed of nexins (SNX), encoded by three genes in Arabidopsis (Jaillais *et al.*, 2007; Pourcher *et al.*, 2010).

Secretory Vesicle Clusters: The secretory vesicle cluster (SVC) was recently identified. The SVC is motile and originates from the TGN/EE, but is not associated with CCVs (Toyooka *et al.*, 2009). Hence, the TGN-like cisternal compartments seem to distinctly produce SVCs and CCVs. One group of proteins that has been identified as components of SVC are the SCAMPs (secretory carrier membrane proteins; Toyooka *et al.*, 2009). Both SCAMP1 and SCAMP2 co-localize with the TGN/EE markers syntaxin SYP41 (for a description of syntaxins see 'SNARE Proteins'), and the subunit VHA-a1 of the TGN V-ATPase (Dettmer *et al.*, 2006; Toyooka *et al.*, 2009). SVC seems to mature at the TGN and is subsequently involved in transport of secretory proteins and pectins from the Golgi to the PM and cell plate in plant cells (Toyooka *et al.*, 2009). In addition, both SCAMPs seem to be redirected from the PM to cell plates in dividing cells, probably via clathrin-mediated endocytosis (Toyooka *et al.*, 2009).

Recently, ECHIDNA (ECH; Gendre *et al.*, 2011) was identified as a new protein essential for secretory vesicles formation at the TGN (Boutté *et al.*, 2013). Together with two YIP4 proteins (YPT/RAB GTPase interacting protein), ECH is essential for secretion of polysaccharides and certain proteins from the TGN to the PM (Boutté *et al.*, 2013; Gendre *et al.*, 2013). Through different approaches, the authors elegantly confirmed that ECH/YIP complex mediates the trafficking of xyloglucan and pectins to the apoplast, although it does not partake in endocytosis and vacuolar targeting (Gendre *et al.*, 2013). The strong co-localization of ECH with the TGN markers SYP41, SYP61, and VHA-a1 suggests that they reside on a common subdomain of the TGN. These results strongly suggest that ECH/YIP complex is essential for the formation of SVCs, which secret SCAMPs and matrix polysaccharides.

Dynamins and Dynamin-related Proteins: The dynamins and DRPs are conserved throughout eukaryotic species. The first dynamin identified, dynamin I, was found to bind microtubules (Shpetner and Vallee, 1989). Since then, the dynamin superfamily has been implicated in diverse cellular activities such as endocytosis, actin nucleation and dynamics, cytokinesis, vacuolar sorting, mitochondrial and chloroplast biogenesis, and maintenance of ER morphology (Praefcke and McMahon, 2004). These diverse roles are accomplished due to the ability of the dynamin to bind and form helical structures around lipids, and either tubulate them or pinch them off of larger membrane sheets in a GTPase-dependent manner (Takei *et al.*, 2005).

Plants have six DRP families (DRP1–DRP6; Hong *et al.*, 2003a). Members of the DRP1 and DRP2 families function in clathrin-mediated trafficking, including CME and cell plate formation, and are therefore essential for cytokinesis and cell elongation (Kang *et al.*, 2003; Collings *et al.*, 2008; Konopka *et al.*, 2008; Fujimoto *et al.*, 2010; Taylor, 2011). DRP3A, DRP3B, and DRP5B have been involved in mitochondrial, chloroplast, and peroxisome biogenesis, and DRP5A is necessary for plant cytokinesis. The functions of the other plant DRPs are still unknown (Hong *et al.*, 2003a; Miyagishima *et al.*, 2008; Fujimoto *et al.*, 2009). The dynamins and DRPs are therefore associated with many different trafficking-related aspects in plant cells.

Vesicle Trafficking Through the Cytoskeleton: After scission from the donor membrane, and un-coating of the coat scaffolds, the vesicle has to reach the destination compartment. With the exception of when trafficking occurs between ER and Golgi, which are tightly connected (Sparkes *et al.*, 2009), the vesicle might track over cytoskeletal structures. Although little is known about cytoskeleton-mediated vesicle trafficking in plants, there is evidence for this process in plant cells (Fig. 2.2b; reviewed in Brandizzi and Wasteneys, 2013). For more information on the role of the cytoskeleton on plant development, see Chapters 4 and 5.

Actin-mediated Trafficking: Organelle movement and long-distance vesicle transport is dependent on the actin cytoskeleton. Accordingly, chemical disruption of either the actin cytoskeleton or myosin activity (the motor proteins driving movement along the actin cytoskeleton) interferes with Golgi and TGN movement (Nebenfuhr *et al.*, 1999; Akkerman *et al.*, 2011; Asaoka *et al.*, 2012). In addition, intact actin cables are essential for normal vesicle trafficking to the vacuole (Kim *et al.*, 2005), corroborating the role of the actin cytoskeleton in post-Golgi membrane trafficking in plants. Although direct proof of the interaction between actin filaments and organelle membranes has not yet been uncovered, co-localization data show that some myosins associate with secretory organelles (Yokota *et al.*, 2009; Ueda *et al.*, 2010). Furthermore, myosin mutant analyses confirmed that these proteins regulate TGN-, endosome-, and PVC-dynamics in plants (Avisar *et al.*, 2012).

A role for the actin cytoskeleton in clathrin-coat formation is suggested by other types of evidence. Firstly, DRP2 could modulate the actin organization during cytokinesis (Konopka *et al.*, 2006). Other studies showed that, due to the high turgor pressure in plant cells, the actin filaments are required for CME at the PM (Konopka *et al.*, 2008) perhaps through association to some

CME-early adaptors (Lam *et al.*, 2001). Moreover, actin filaments have been reported to be essential for relocation of pectins from the PM to the cell plate in dividing cells (Baluska *et al.*, 2002).

Members of the small GTPase subfamily of ROPs have recently been implicated in vesicle trafficking and regulation of cytoskeleton dynamics, probably acting as molecular motors that tether vesicles to the cytoskeleton. The close relationship between auxin distribution and ROP and RIC (ROP-interactive CRIB-motif-containing protein) activities in plant development is remarkable. Two recent studies show that auxin-mediated signaling activates both ROP6–RIC1 and ROP2–RIC4 pathways which antagonistically work on pavement cell expansion. RIC1 therefore appears to be activated by ROP6 and then becomes associated with microtubules, suppressing pavement cell lobing. At the same time, an auxin-activated ROP2–RIC4 pathway mediates cortical microtubule bundling, which promotes lobing by blocking CME of auxin carriers (Xu *et al.*, 2010; Nagawa *et al.*, 2012). Interestingly, the signaling modules of auxin ROP6–RIC1 and ROP2–RIC4 pathways that connect the cytoskeleton and CME are also active in roots (Yang *et al.*, 2007; Chen *et al.*, 2012). In both tissues, the asymmetric distribution of auxin carriers can lead to an irregular deposition of cell wall material and a concomitant disruption of both root development and interdigitation of leaf epidermal pavement cells.

Microtubule-mediated Trafficking: While long-distance vesicle trafficking is actin dependent, it has been proposed that microtubules and their motors might anchor and slow down organelles/compartments at their target membranes (Zhu and Dixit, 2011; Cai and Cresti, 2012). This stationary phase would be especially useful for a precise secretion of cell wall material.

Although the tethering mechanism of vesicles to the microtubules is not fully understood, it is plausible that certain ROPs may be involved in this process. ROP11 has been shown to alter the delivery of Cellulose Synthase (CesA) complexes (CSCs) to the PM during secondary cell wall formation (Oda and Fukuda, 2012; for further details, see Chapters 3, 4, and 7). This furthers the above-discussed role of ROP as molecular motors essential for coordination of cell wall deposition. Another group of proteins that might serve as putative microtubule–vesicle connectors during trafficking of cell wall material are the kinesins (motor proteins which move along microtubules; Brandizzi and Wasteneys, 2013). Interestingly, two cotton kinesins (KCH1 and 2) have been proposed to link microtubules and actin filaments. KCH1 therefore interacts with actin filaments in developing cotton fibers, and plays a putative role in dynamic microtubule–actin cross-linking (Preuss *et al.*, 2004). Similarly, the cotton KCH2 kinesin has been shown to bind actin filaments and to cross-link them with microtubules *in vitro* (Xu *et al.*, 2009). Apart from the canonical kinesins, a conceivable microtubule–vesicle linker is the microtubule-associated protein CLASP. CLASP mediates the motility of the TGN/EE retromer over microtubules, in which CLASP interacts with one of the sorting nexins. Interestingly, the authors of this study propose that microtubule-mediated traffic might be typical for all plant endosomes (Ambrose *et al.*, 2013).

Clathrin-mediated trafficking during cytokinesis and cell plate formation also relies on microtubules. One example of the evidence for this interaction is the association of the dynamins DRP1s with microtubules *in vivo* and *in vitro* and their essential role for secretory vesicle trafficking along phragmoplast microtubules (Hong *et al.*, 2003b). Moreover, electron microscopy micrographs demonstrated that CCVs clustered around cortical microtubules in vacuolated suspensor cells of white embryo cells (Fowke *et al.*, 1999).

Targeting and Fusion: Once vesicles have traversed the cytosol, they have to identify their target organelles/compartments and fuse to it. This process can be divided into three main steps – targeting, tethering, and fusion – and is regulated by three groups of proteins: Rab GTPases and SM-proteins, tethers, and SNARE proteins, respectively (Fig. 2.2c). Interestingly, the paralogous expansion of these three groups seen in different eukaryotic Kingdoms might explain the diversification of membrane-trafficking pathways (Dacks and Field, 2007).

Rab GTPases and SM-proteins: The Rab GTPases are one group of small G-proteins that regulate the vesicle targeting to the sink membrane. Their mode of action is similar to that previously explained for other small GTPases (see 'Vesicle Formation'); the inactive cytosolic GDP-bound-Rab is recruited to a membrane and activated by a Rab-GEF factor. The active GTP-Rab then further interacts with other specific vesicle-fusion effectors. All eukaryotes contain eight groups of Rab GTPases, named RABA to RABH in Arabidopsis (Vernoud *et al.*, 2003). Intriguingly, the TGN/EE localized RABA group has expanded dramatically in plants compared to other eukaryotes, suggesting that the RABA members have plant-specific functions. Indeed, various members of this RAB subclade have been implicated in cell wall metabolism (reviewed by Lycett, 2008). For example, RABA1a has recently been reported to participate in pectin deposition and fruit softening in tomato (Lunn *et al.*, 2013); RABA2 and A3 are involved in cell plate formation; and RABA1 and A4 are involved in pollen tip and root hair growth (reviewed by Fujimoto and Ueda, 2012). In addition, RABA4b seems to recruit phosphatidylinositol-4-kinase (PI-4Kβ1) to the TGN, and thereby regulates the formation of secretory vesicles by PI-4P synthesis (Preuss *et al.*, 2006). Furthermore, RAB4b co-localizes with the TGN markers SYP61 and VHA-a1, which also label ECHIDNA-containing SVCs (see 'Secretory Vesicle Clusters'; Boutté *et al.*, 2013). As in the case of *echidnia* mutant, the loss of PI-4Kβ1 activity results in defects in delivery of pectins and xyloglucans to the cell wall (Preuss *et al.*, 2006). While these data suggest some type of cooperation between ECHIDNA and RABA4b in secretory vesicle formation, such interactions remain to be confirmed.

The SM-family proteins (Sec1/Munc18) interact with inactive syntaxins at the target membrane (syntaxins are also named Qa-SNAREs; see 'SNARE Proteins'). This interaction seems to trigger a conformational change in the syntaxin, which 'opens' and binds to other SNARE proteins (Dulubova *et al.*, 1999). The SM-proteins can also associate with assembled SNARE complexes by binding to an N-terminal peptide of syntaxin (Yamaguchi *et al.*, 2002). Arabidopsis has six members of the SM family. One member KEULE, together with its interacting syntaxin KNOLLE, is involved in cytokinesis and cell wall deposition during cell plate formation in plants (Assaad *et al.*, 1996; Nickle and Meinke, 1998).

Tethers: Three tethering complexes can be distinguished in plants: the TRAPP (transport protein particle), the GARP (Golgi-associated retrograde protein), and the exocyst complexes. The TRAPP complex is required for cell plate formation (Thellmann *et al.*, 2010) and GARP is involved in pollen tube elongation (Guermonprez *et al.*, 2008). Both these processes (i.e., cell plate and pollen tube growth) rely on cell wall formation. Consistent with this notion, recent proteomic data identified one TRAPP subunit and proteins essential for primary wall cellulose synthesis (i.e., CesA1, CesA3, CesA6, CesA2, and KORRIGAN) in the same endosomes (Drakakaki *et al.*, 2012). However, a direct role for these tethering complexes in the synthesis and regulation of plant cell walls has to be consolidated.

The exocyst complex tethers secretory vesicle to the PM in all eukayotes. It typically consists of eight subunits: SEC3, SEC5, SEC6, SEC8, SEC10, SEC15, EXO70, and EXO84 (Munson and Novick, 2006). In plants, this complex is also involved in cytokinesis and cell plate maturation (Fendrych *et al.*, 2010). Consequently, mutations in various exocyst proteins exhibit dwarf phenotypes and defects in germination and pollen tube growth (Cole *et al.*, 2005; Hala *et al.*, 2008; Fendrych *et al.*, 2010). Vesicle tethering through the EXO70 family member, EXO70A1, has been implicated in tracheary element development (Li *et al.*, 2013) and, together with SEC8, in localized pectin deposition during seed coat formation (Kulich *et al.*, 2010). In addition, SEC3 may interact with ROP proteins, possibly providing a link between the exocyst and the cytoskeleton (Lavy *et al.*, 2007). However, many questions remain to be addressed regarding this complex. For example, a recent study localized EXO70E2 to a new endomembrane structure, the EXPO. This spherical double-membrane organelle seems to mediate unconventional protein secretion in plants (Wang *et al.*, 2010; see 'EXPO'), but the role of EXO70E2 at this organelle is unclear.

SNARE Proteins: The final step of the trafficking path is the vesicle fusion to the target membrane and the delivery of the luminal cargo. This job is executed by the SNARE family of proteins (Fig. 2.2c) which consists of four groups: Qa-SNAREs (also named syntaxins); Qb-; Qc-; and R-SNAREs (so-called VAMPS or vesicle-associated membrane proteins). In general, three Q-SNARES (Qa, Qb, and Qc) reside on the target membrane, which is why they are named t-SNAREs (target-SNAREs). Upon activation by an SM-family protein (described above in 'Rab GTPases and SM-proteins'), the three SNAREs bundle into a helix where the v-SNARE (vesicle-SNARE, usually a member of the R-SNAREs/VAMPs) can bind. Mutual twisting of multiple SNARE complexes leads to the fusion of the two bilayers. Afterwards, the four SNAREs are released and recycled (reviewed by Hong, 2005).

SNARE proteins and activity have been extensively reviewed elsewhere (Sanderfoot, 2007; Bassham *et al.*, 2008). Some of them have been implicated in cell wall formation, which is the case for the previously mentioned syntaxin KNOLLE (discussed in 'Rab GTPases and SM-proteins'), which has an essential role during cell plate assembly (Assaad *et al.*, 1996; Nickle and Meinke, 1998). The Qc-SNARE SYP61, which localizes at TGN/EE, co-localizes with different proteins/compartments involved in trafficking of cell wall components, such as CesA-containing compartments (MASCs/SmaCCs), ECHIDNA, and RABA4b (Preuss *et al.*, 2006; Crowell *et al.*, 2009; Gutierrez *et al.*, 2009; Gendre *et al.*, 2013). These data have been reinforced by a recent proteomic analysis of Arabidopsis endosomes containing SYP61, where various cell-wall-related proteins were identified, including primary wall CesAs (CesA1, CesA2, CesA3, and CesA6), the endoglucanase KORRIGAN, and the pectin-methyl esterase AtPME1 (Drakakaki *et al.*, 2012). Finally, plants impaired in the syntaxin SYP122 have been reported to show primary cell wall defects, based on FTIR (Fourier transform infrared analysis; Assaad *et al.*, 2004), although no direct role of this syntaxin on cell wall architecture has been reported.

Unconventional Protein Secretion in Plants

About half of the identified plant PM and apoplastic proteins lack a signal peptide (named leaderless secretory proteins or LSPs), reinforcing the importance of non-conventional secretion (Agrawal *et al.*, 2010). Interestingly, many LSPs identified in the plant secretome are related to stress response, mainly associated with pathogen attacks (Ding *et al.*, 2012). In the following, we describe the unconventional trafficking pathways identified in plants that may contribute to the deposition of certain cell wall components (Fig. 2.1, white arrows).

EXPO: A new organelle was recently discovered in Arabidopsis and tobacco BY2-cells (Wang *et al.*, 2010). This organelle was named EXPO (exocyst-positive-organelle) due to the presence of the exocyst protein homolog EXO70E2 at its membrane (see also 'Tethers'). EXPO seems to be common to plant cells, as it is found in several different cell types such as root tips and hairs and pollen tubes. This organelle is enclosed by a double membrane structure and can act as an exosome by releasing LSPs to the apoplast. The outer membrane of EXPO fuses to the plasma membrane, releasing a single membrane vesicle into the cell wall. This subsequently bursts and secrets the cargo proteins, such as the S-adenosylmethioninesynthetase SAMS2 (Wang *et al.*, 2010). Proteomic data also indicates that SAMS2 is present in the cell wall of Arabidopsis and might be a major substrate for lignin methylation in the cell wall (Shen *et al.*, 2002; Bayer *et al.*, 2006).

MVB-PM and Vacuole-PM: The cell wall is an essential defense layer for plant cells; one of the first steps after wounding and pathogen perception is therefore cell wall reinforcement. The multivesicular bodies (MVB) have a key role in this process by locally depositing callose synthases that produce callose-rich papillae around wounds or pathogen penetration areas (An *et al.*, 2006; Meyer *et al.*, 2009). Callose is a β-(1$\rightarrow$3)-glucan polymer synthesized directly at the PM and is also necessary for cell plate and plasmodesmata formation (Thiele *et al.*, 2009, Burch-Smith and Zambryski, 2012). Another local plant response against pathogens is the induction of localized programmed cell death, involving the fusion of the vacuole with the PM to release antibacterial vacuolar enzymes to the apoplast (Hatsugai and Hara-Nishimura, 2010).

Other Non-vesicle Mediated Paths: In addition to the non-conventional trafficking pathways discussed above there are other non-vesicle mediated paths in plants, although their roles in cell wall biology are less clear. Electron microscopy studies performed in the late 1980s showed ER-PM contact sites that suggested direct secretion from the ER to the apoplast. Recent data indicate that, under certain stress conditions, some lipids are secreted via this path (Samuels and McFarlane, 2012). Two LSPs have been shown to be secreted directly from the plant cytosol to the apoplast. One of them is the celery mannitol dehydrogenase (MTD), which mainly functions to control central metabolism and cell osmoregulation. MTD seems to be secreted in response to biotrophic pathogen perception (Chen *et al.*, 2009). The second example of cytosolic protein secreted through a vesicle-independent pathway is the bacterial enzyme hygromycin phosphotransferase (HYG^R; Zhang *et al.*, 2011).

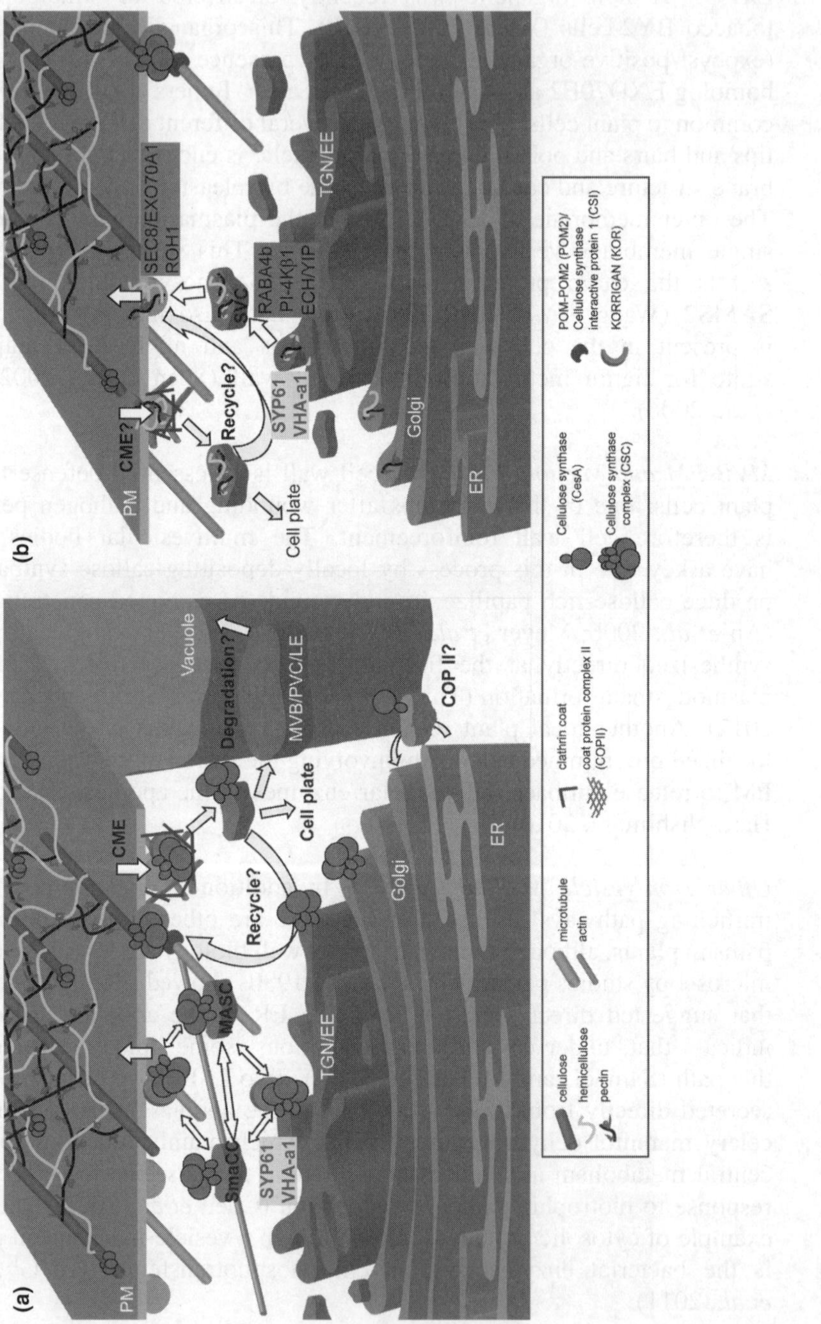
(a)
PM
CME
MASC
SmaCC
SYP61
VHA-a1
Recycle?
Cell plate
Degradation?
Vacuole
MVB/PVC/LE
COP II?
TGN/EE
Golgi
ER
(b)
PM
CME?
Recycle?
SYP61
VHA-a1
SVC
RABA4b
PI-4Kβ1
ECH/YIP
SEC8/EXO70A1
ROH1
Cell plate
ER
Golgi
TGN/EE
cellulose
hemicellulose
pectin
microtubule
actin
clathrin coat
coat protein complex II (COPII)
Cellulose synthase (CesA)
Cellulose synthase complex (CSC)
POM-POM2 (POM2)/ Cellulose synthase interactive protein 1 (CSI)
KORRIGAN (KOR)

Membrane Traffic in Cell Wall Metabolism

In the previous section ('Membrane Trafficking in Plant Cells') we described the plant endomembrane organelles/compartments and how they are connected. In this section, we focus on the cell wall components and describe in more detail the routes that they take to become secreted and internalized. The main components that contribute to primary walls are pectins, cellulose, and hemicelluloses. In contrast, the secondary walls largely lack pectins and also hold substantial amounts of the waterproof polymer lignin. For more detailed information about membrane trafficking during special cell wall formation, see Chapter 7 (tracheary elements) and Chapters 8 and 9 (tip growth of pollen tubes and root hair).

Cellulose

The biology behind cellulose synthesis and remodeling in plants has been accurately reviewed recently (McFarlane *et al.*, 2014) and is the topic of another chapter of this book (Chapter 3). Here we summarize the current view about the role of endomembrane trafficking on cellulose deposition (Fig. 2.3a).

←

Figure 2.3 Membrane trafficking in cell wall polysaccharide metabolism. (a) Membrane trafficking during cellulose synthesis. Cellulose synthase (CesA) subunits are assembled into CesA complexes (CSCs) at the endoplasmic reticulum (ER) or the Golgi apparatus (in the figure the last option is represented). Individual CesAs, or assembled CSCs, should reach the Golgi via coat protein complex II (COP-II). From the trans-Golgi network/early endosome (TGN/EE), the CSCs travel to the plasma membrane (PM) probably via SmaCCs/MASCs (small CESA compartments/microtubule-associated CSCs), which move over the actin and microtubule cytoskeleton. SmaCCs/MASCs partially co-localize with the TGN-markers SYP61 and VHA-a1. Once at the PM, the CSCs follow linear paths that are typically guided by cortical microtubules, due to the CSC connection to the microtubule-binding protein POM2/CSI1 (POM-POM2/cellulose synthase interactive protein 1). The CSCs are internalized, probably via clathrin-mediated endocytosis (CME). The CSCs might then be recycled back to the PM via TGN/EE or sent for degradation to the vacuole. In dividing cells, the CSCs are recruited to the cell plate through the CME machinery. The CSC interacting protein KORRIGAN (KOR) is also necessary for cellulose synthesis and traffics via similar paths as that of CSCs. (b) Membrane trafficking of matrix polysaccharides. Both pectins and hemicelluloses are synthesized in the Golgi apparatus and delivered to the PM via secretory vesicle clusters (SVCs). The SVCs are formed at the TGN/EE by the coordinated activity of various proteins such as the small-GTPase RABA4b, the phosphatidylinositol-4-kinase (PI-4Kβ1), and ECH/YIP complex (ECHIDNIA/YPT/RAB GTPase interacting proteins). The SVCs and some of the proteins required for their formation co-localize with the TGN/EE-markers SYP61 and VHA-a1. During seed coat formation, pectin-containing vesicles tether to the PM through the exocyst subunits SEC8 and EXO70A1 and their interactor ROH1. In dividing root cells xyloglucans and pectins are internalized and relocated to the cell plate, probably via CME. After endocytosis, part of the matrix polysaccharides might be recycled back to the PM.

Cellulose is the main load-bearing polymer of both primary and secondary walls. It is composed of β-(1→4)-D-glucan chains synthesized by CSCs localized at the PM. The CSCs are composed of CesAs that appear to be assembled into CSCs in the Golgi or ER. It is still not completely understood how the CSCs traffic through the endomembrane system. As for other secreted proteins, the CesAs should be synthesized at the ER and probably relocated to the Golgi via COP-II transport. Based on analyses of FP-CesA in Arabidopsis etiolated hypocotyl cells, the CSCs localize at the PM, in the Golgi (Paredez *et al.*, 2006), and in post-Golgi vesicles named MASCs (microtubule-associated CesA compartments; Crowell *et al.*, 2009) /SmaCCs (small CesA compartments; Gutierrez *et al.*, 2009). While a subpopulation of these CSC-containing vesicles is associated with microtubules, another subpopulation are clearly associated with the actin cytoskeleton. Interestingly, recent work indicates that there is an exchange of these vesicle subpopulations, i.e., it appears that there is an interchange of SmaCCs/MASCs between the microtubules and the actin cytoskeleton (Sampathkumar *et al.*, 2013).

MASC/SmaCC endomembranes partially co-localize with the TGN/EE markers SYP61, SYP41, and VHA-a1 (Crowell *et al.*, 2009; Gutierrez *et al.*, 2009). These data have been confirmed by a proteomic approach which identified primary wall CesAs in SYP61-enriched endosomes (Drakakaki *et al.*, 2012). As discussed above, the TGN/EE is extremely complex and dynamic and the markers used to define it do not allow the different paths that merge/split at this hub to be distinguished. The identity of the heterogeneous population of MASCs/SmaCCs is therefore not fully elucidated. Interestingly, the group of MASCs/SmaCCs which do not co-localize with TGN markers interacts with microtubules, and is likely to be involved in the microtubule-dependent positioning of CSC delivery to the PM (Gutierrez *et al.*, 2009). Moreover, MASCs/SmaCCs over-accumulate in the cell cortex upon osmotic stress and inhibition of cellulose synthesis activity, concomitantly with a depletion of CSCs from the PM (Crowell *et al.*, 2009; Gutierrez *et al.*, 2009). This relocation of CSCs can be a consequence of reduced secretion of CSCs and/or higher rate of internalization of CSCs, suggesting a putative role of MASCs/SmaCCs in either exo- or endocytosis of the CSCs.

Arabidopsis and rice plants impaired in CME showed aberrant cell plates, reduced cell elongation and decreased cellulose contents, suggesting a role for CME in cell wall formation (Kang *et al.*, 2003; Collings *et al.*, 2008; Xiong *et al.*, 2010). Recently, two primary wall CesAs have been shown to interact in yeast with one subunit of the AP-2 adaptor, AP2M (see 'Clathrin Coat Vesicles (CCV): Post-Golgi Cargo Recruitment'). Mutations in this AP2M subunit resulted in increased CSCs at the PM with no effect on CSC delivery and no apparent impact on overall plant growth (Bashline *et al.*, 2013). Surprisingly, the same *ap2m* mutant was more recently shown to affect different aspects of plant morphology, including cell elongation (Kim *et al.*, 2013). While these data are somewhat contradictory, they reinforce the role of CME in CSC endocytosis. The CSC internalization might be post-translationally regulated, probably via phosphorylation and/or ubiquitination of the CesAs. Both primary and secondary wall CesAs can be phosphorylated (Nuhse *et al.*, 2004; Benschop *et al.*, 2007; Taylor, 2007). However, the influence of this regulation on CSC internalization has not yet been investigated. Similarly, while there are indications that the CesAs may become ubiquitinated, these data were rendered not statistically significant (Manzano *et al.*, 2008). Further studies on CSC regulation at the PM are therefore necessary to understand the CSC function and dynamics.

The CSC activity is also regulated by other proteins such as: the glucan-binding protein chitinase-like 1 CTL1/POM1 (Sanchez-Rodriguez *et al.*, 2012); the CSCs-microtubule linker POM2/CSI1 (Bringmann *et al.*, 2012; Li *et al.*, 2012); the putative endo-(1→4)-β-glucanase (EGase), KORRIGAN (Nicol *et al.*, 1998); and the proteins KOBITO (Pagant *et al.*, 2002) and COBRA (Roudier *et al.*, 2005) of unknown function. Some of these proteins have been reported to influence CSC trafficking. Among them, POM2/CSI1 co-localizes with CSCs in MASCs/SmaCCs and at the PM, where it is required for correct microtubule-CSCs alignment (Bringmann *et al.*, 2012; Li *et al.*, 2012). KORRIGAN is a membrane-anchored endo-glucanase which may interact with the CesAs (Crowell *et al.*, 2010), and has been identified in the proteomic analysis of SYP61-enriched endomembranes (Drakakaki *et al.*, 2012). Although its precise function has not been fully elucidated, KORRIGAN is required for wild type levels of both primary and secondary cellulose synthesis (Lane *et al.*, 2001; Szyjanowicz *et al.*, 2004) and for cell plate formation (Zuo *et al.*, 2000). Based on the identification of an AP2-binding-motif in the KORRIGAN sequence, it has been suggested that this protein might be internalized via CME, although no conclusive data are available. Interestingly, in a recent paper it was shown that primary wall CesAs are recruited to the cell plate through the CME machinery (Miart *et al.*, 2013). These data may suggest that CSCs and KORRIGAN are trafficked through the same compartments, but further analyses are necessary to confirm this hypothesis. Similarly, KOBITO is essential for cellulose synthesis and, based on its localization to the PM and the endomembrane system (Pagant *et al.*, 2002), it is tempting to speculate that this protein is also involved in CSCs trafficking to/from the PM.

Matrix Polysaccharides: Pectin and Hemicelluloses

The cellulose microfibrils are para-crystalline in nature and cross-linked by hemicelluloses and pectins. Hemicelluloses contain a cellulose-like backbone of β-(1→4)-linked glucose or xylose, decorated by other sugar residues such as xylose, galactose, fucose, arabinose, and glucuronic acids. The most prominent hemicelluloses in dicot plants are xyloglucans, xylans, and mannans, which fill the gaps between cellulose microfibrils of primary and secondary cell walls, respectively. In addition, the so-called mixed-linked glucans composed of an unbranched β-(1→3, 1→4)-linked backbone are the hemicelluloses of type II cell walls of grasses (reviewed by Scheller and Ulvskov, 2010). Pectins are characterized by a D-galacturonic acid-containing backbone which, as in the case of hemicelluloses, can be adorned by diverse sugar-based side chains (reviewed by Atmodjo *et al.*, 2013). The pectins are classified as one of three types of polymers: the unbranched chain homogalacturonan (HG); the rhamnogalacturonan I (RGI), which is usually decorated by arabinan, galactan, and arabinogalactan side chains; and the extremely complex rhamnogalacturonan II (RGII).

Both groups of matrix polysaccharides are *de novo* synthesized and modified by acetylation in the Golgi apparatus. The delivery of certain pectins to the PM was shown to be mediated by secretory vesicle clusters (Toyooka *et al.*, 2009) which co-localize with TGN/EE markers. Notably, recent data indicate that pectins travel together with hemicelluloses from the TGN/EE to the PM in an ECH/YIP complex-dependent pathway (Boutté *et al.*, 2013; Gendre *et al.*, 2013). Moreover, this delivery depends on the activity of the phosphatidylinositol-4-kinase (PI-4Kβ1) at the TGN, which is regulated by RABA4b (see 'Rab GTPases and SM-proteins'; Preuss *et al.*, 2006). In addition, ECH is also necessary for pectin mucilage formation in Arabidopsis seeds (McFarlane *et al.*, 2013). Interestingly, the SEC8 and EXO70A1 exocyst subunits, together with their interactor ROH1, are also essential for the localized deposition of seed coat pectin in Arabidopsis (Kulich *et al.*, 2010). Both pectins and hemicellulose therefore seem to be secreted via SVC which are formed at the TGN by a correct regulation of RABA4b, PI-4Kβ1, and ECH/YIP activity. It is also plausible that the polysaccharides containing SVC are tethered to the PM by the exocyst (Fig. 2.3b). However, further analysis should be performed to confirm such a model.

The use of antibodies specific for certain pectins and xyloglucans, together with chemical inhibition of post-TGN/EE trafficking, led to the hypothesis that these polysaccharides are actively trafficked from the PM to the plant cell plate (Baluska *et al.*, 2002, 2005). However, over-accumulation of TGN/EE, which occurs after BrefeldinA treatment, can also be a consequence of defects in other vesicle trafficking pathways, such as secretion or vacuolar sorting. Nevertheless, the authors favored the endocytosis-based hypothesis largely due to the used antibodies, which were described as recognizing pectins modified *in muro* (Baluska *et al.*, 2002, 2005). In addition, these studies also indicated that the pectin relocation from the PM to the cell plate requires intact actin filaments (Baluska *et al.*, 2002). Considering the putative role of actin cytoskeleton in clathrin-mediated trafficking during cytokinesis, its influence over CSCs at the PM, and the higher levels of non-cellulosic polysaccharides observed in the walls of CME-deficient mutants (Collings *et al.*, 2008; Konopka *et al.*, 2008; Xiong *et al.*, 2010; Sampathkumar *et al.*, 2013), an internalization of matrix polysaccharides via clathrin-coats should not be discarded.

Cell Wall Proteins

Around 5–10% of the cell wall components are proteins, which are vital for cell wall assembly and modification and for cell stress response (Jamet *et al.*, 2008). The cell wall proteins (CWPs) are synthesized at the ER and then go through the Golgi, where they are glycosylated. Although not shown, the CWPs are likely to reach the Golgi via the COP-II pathway. There is evidence of localization of CWPs in compartments of the secretory pathways; however, these pathways have yet to be defined. For example, the use of an antibody against arabinogalactan proteins (AGPs) loosely indicated that these CWPs localize at the ER, Golgi, TGN/EE, MBV/PVC, and the tonoplast (Šamaj *et al.*, 2000). A recent study based on transcription and *in vitro* protein–protein interaction data supports trafficking of AGP6 and AGP11 through the TGN/EE, and that these proteins may reach the vacuole via MBV/PVC in pollen and pollen tubes (Costa *et al.*, 2013). Transient expression of a pectinmethylesterase inhibitor protein (PMEI1) and a polygalacturonase inhibitor protein (PGIP2) in tobacco protoplasts showed that they likely localize at the ER and Golgi during their relocation to the CW (De Caroli *et al.*, 2011). In addition, these studies suggested that the PGIP2 becomes internalized to endosomes when its ligand, a specific fungal endopolygalacturonase, is not present in the cell wall. Another report performed in Arabidopsis pollen tubes indicates that PMEI endocytosis regulates the PME activity and consequently the distribution of esterified and de-esterified pectins, which is essential for a correct pollen tube growth (Rockel *et al.*, 2008).

The protein secretion in plants is likely to be highly coordinated. Indeed, there is growing evidence of complex organization of protein delivery, such as the retention of some CWPs in the Golgi apparatus if their N-terminal domains are not cleaved. This is the case for endopolygalacturonases from clade B (Degan *et al.*, 2001), and the type-I pectinmethylesterase (PME; Wolf *et al.*, 2009). Interestingly, PME1 has been identified in SYP61-enriched endosomes, supporting that it is trafficked via TGN/EE (Drakakaki *et al.*, 2012).

Obviously, the exact mechanisms of CWP trafficking is largely unexplored. New insights in this field should clarify interesting questions regarding cell wall regulation and signal transduction mediated by CWPs.

Concluding Remarks

We hope that it is clear to the reader that while the trafficking system in plants is becoming clarified, it is not clear how certain cell wall polymers are secreted or internalized. Moreover, coordination between the synthesis/degradation of cell wall material and different trafficking pathways are a virtually untapped area. With the increase in cell-biology-based platforms and molecular tools, it appears likely that these questions will be resolved in the near future.

Acknowledgements

SP and CSR were funded by the Max-Planck Gesellschaft and by DFG grant PE1642/5-1.

References

Agrawal G.K., Jwa N.S., Lebrun M.H., Job D., and Rakwal R. (2010) Plant secretome: unlocking secrets of the secreted proteins. *Proteomics*, **10**, 799–827.

Akkerman M., Overdijk E.J., Schel J.H., Emons A.M., and Ketelaar T. (2011) Golgi body motility in the plant cell cortex correlates with actin cytoskeleton organization. *Plant Cell Physiology*, **52**(10), 1844–1855.

Ambrose C., Ruan Y., Gardiner J., Tamblyn L.M., Catching A., Kirik V., Marc J., Overall R., and Wasteneys G.O. (2013) CLASP interacts with sorting nexin 1 to link microtubules and auxin transport via PIN2 recycling in Arabidopsis thaliana. *Developmental Cell*, **24**(6), 649–659.

An Q., Huckelhoven R., Kogel K.H., and van Bel A.J. (2006) Multivesicular bodies participate in a cell wall-associated defence response in barley leaves attacked by the pathogenic powdery mildew fungus. *Cell Microbiology*, **8**(6), 1009–1019.

Asaoka R., Uemura T., Ito J., Fujimoto M., Ito E., Ueda T., and Nakano A. (2012) Arabidopsis RABA1 GTPases are involved in transport between the trans-Golgi network and the plasma membrane, and are required for salinity stress tolerance. *Plant Journal*, **73**(2), 240–249.

Assaad F.F., Mayer U., Wanner G., and Jurgens G. (1996) The KEULE gene is involved in cytokinesis in Arabidopsis. *Molecular Genetics and Genomics*, **253**(3), 267–277.

Assaad F.F., Qiu J.L., Youngs H., Ehrhardt D., Zimmerli L., Kalde M., Wanner G., Peck S.C., Edwards H., Ramonell K., Somerville C.R., and Thordal-Christensen H. (2004) The PEN1 syntaxin defines a novel cellular compartment upon fungal attack and is required for the timely assembly of papillae. *Molecular Biology of the Cell*, **15**(11), 5118–5129.

Atmodjo M.A., Hao Z., and Mohnen D. (2013) Evolving views of pectin biosynthesis. *Annual Review of Plant Biology*, **64**, 747–779.

Avisar D., Abu-Abied M., Belausov E., and Sadot E. (2012) Myosin XIK is a major player in cytoplasm dynamics and is regulated by two amino acids in its tail. *Journal of Experimental Botany*, **63**(1), 241–249.

Baluska F., Hlavacka A., Samaj J., Palme K., Robinson D.G., Matoh T., McCurdy D.W., Menzel D., and Volkmann D. (2002) F-actin-dependent endocytosis of cell wall pectins in meristematic root cells. Insights from brefeldin A-induced compartments. *Plant Physiology*, **130**(1), 422–431.

Baluska F., Liners F., Hlavacka A., Schlicht M., Van Cutsem P., McCurdy D.W., and Menzel D. (2005) Cell wall pectins and xyloglucans are internalized into dividing root cells and accumulate within cell plates during cytokinesis. *Protoplasma*, **225**(3-4), 141–155.

Bashline L., Li S., Anderson C.T., Lei L., and Gu Y. (2013) The Endocytosis of Cellulose Synthase in Arabidopsis Is Dependent on mu2, a Clathrin-Mediated Endocytosis Adaptin. *Plant Physiology*, **163**(1), 150–160.

Bassham D.C., Brandizzi F., Otegui M.S., and Sanderfoot A. A. (2008) The secretory system of Arabidopsis. *Arabidopsis Book*, **6**, e0116.

Bayer E.M., Bottrill A.R., Walshaw J., Vigouroux M., Naldrett M.J., Thomas C.L., and Maule A.J. (2006) Arabidopsis cell wall proteome defined using multidimensional protein identification technology. *Proteomics*, **6**(1), 301–311.

Benschop J.J., Mohammed S., O'Flaherty M., Heck A.J., Slijper M., and Menke F. L. (2007) Quantitative phosphoproteomics of early elicitor signaling in Arabidopsis. *Molecular Cell Proteomics*, **6**(7), 1198–1214.

Bonifacino J.S. and Lippincott-Schwartz J. (2003) Coat proteins: shaping membrane transport. *Nature Reviews Mololecular Cell Biology*, **4**(5), 409–414.

Boutté Y., Jonsson K., McFarlane H.E., Johnson E., Gendre D., Swarup R., Friml J., Samuels L., Robert S., and Bhalerao R. P. (2013) ECHIDNA-mediated post-Golgi trafficking of auxin carriers for differential cell elongation. *Proceedings of National Academy of Science, USA*, **110**(40), 16259–16264.

Brandizzi F. and Wasteneys G.O. (2013) Cytoskeleton-dependent endomembrane organization in plant cells: an emerging role for microtubules. *Plant Journal*, **75**(2), 339–349.

Bringmann M., Li E., Sampathkumar A., Kocabek T., Hauser M.T., and Persson S. (2012) POM-POM2/cellulose synthase interacting1 is essential for the functional association of cellulose synthase and microtubules in Arabidopsis. *Plant Cell*, **24**(1), 163–177.

Burch-Smith T.M. and Zambryski P.C. (2012) Plasmodesmata paradigm shift: regulation from without versus within. *Annual Review of Plant Biology*, **63**, 239–260.

Cai G. and Cresti M. (2012) Are kinesins required for organelle trafficking in plant cells? *Frontiers in Plant Science*, **3**, 170.

Chen X., Irani N.G., and Friml J. (2011) Clathrin-mediated endocytosis: the gateway into plant cells. *Current Opinion in Plant Biology*, **14**(6), 674–682.

Chen X., Naramoto S., Robert S., Tejos R., Lofke C., Lin D., Yang Z., and Friml J. (2012) ABP1 and ROP6 GTPase signaling regulate clathrin-mediated endocytosis in Arabidopsis roots. *Current Biology*, **22**(14), 1326–1332.

Cheng F.Y., Zamski E., Guo W.W., Pharr D.M., and Williamson J.D. (2009) Salicylic acid stimulates secretion of the normally symplastic enzyme mannitol dehydrogenase: a possible defense against mannitol-secreting fungal pathogens. *Planta*, **230**, 1093–1103.

Cole R.A., Synek L., Zarsky V., and Fowler J. E. (2005) SEC8, a subunit of the putative Arabidopsis exocyst complex, facilitates pollen germination and competitive pollen tube growth. *Plant Physiology*, **138**(4), 2005–2018.

Collings D.A., Gebbie L.K., Howles P.A., Hurley U.A., Birch R.J., Cork A.H., Hocart C.H., Arioli T., and Williamson R.E. (2008) Arabidopsis dynamin-like protein DRP1A: a null mutant with widespread defects in endocytosis, cellulose synthesis, cytokinesis, and cell expansion. *Journal of Experimental Botany*, **59**(2), 361–376.

Costa M., Nobre M.S., Becker J.D., Masiero S., Amorim M.I., Pereira L.G., and Coimbra S. (2013) Expression-based and co-localization detection of arabinogalactan protein 6 and arabinogalactan protein 11 interactors in Arabidopsis pollen and pollen tubes. *BMC Plant Biology*, **13**, 7.

Crowell E.F., Bischoff V., Desprez T., Rolland A., Stierhof Y.D., Schumacher K., Gonneau M., Hofte H., and Vernhettes S. (2009) Pausing of Golgi bodies on microtubules regulates secretion of cellulose synthase complexes in Arabidopsis. *Plant Cell*, **21**(4), 1141–1154.

Crowell E.F., Gonneau M., Stierhof Y.D., Hofte H., and Vernhettes S. (2010) Regulated trafficking of cellulose synthases. *Current Opinion in Plant Biology*, **13**(6), 700–705.

Dacks J.B. and Field M.C. (2007) Evolution of the eukaryotic membrane-trafficking system: origin, tempo and mode. *Journal of Cell Science*, **120**(17), 2977–2985.

De Caroli M., Lenucci M.S., Di Sansebastiano G.P., Dalessandro G., De Lorenzo G., and Piro G. (2011) Protein trafficking to the cell wall occurs through mechanisms distinguishable from default sorting in tobacco. *Plant Journal*, **65**(2), 295–308.

Degan F.D., Child R., Svendsen I., and Ulvskov P. (2001) The cleavable N-terminal domain of plant endopolygalacturonases from clade B may be involved in a regulated secretion mechanism. *Journal of Biological Chemistry*, **276**(38), 35297–35304.

Dettmer J., Hong-Hermesdorf A., Stierhof Y.D., and Schumacher K. (2006) Vacuolar H+-ATPase activity is required for endocytic and secretory trafficking in Arabidopsis. *Plant Cell*, **18**(3), 715–730.

Di Rubbo S., Irani N.G., Kim S.Y., Xu Z.Y., Gadeyne A., Dejonghe W., Vanhoutte I., Persiau G., Eeckhout D., Simon S., Song K., Kleine-Vehn J., Friml J., De Jaeger G., Van Damme D., Hwang I., and Russinova E. (2013) The Clathrin Adaptor Complex AP-2 Mediates Endocytosis of BRASSINOSTEROID INSENSITIVE1 in Arabidopsis. *Plant Cell*, **25**(8), 2986–2997.

Ding Y., Wang J., Wang J., Stierhof Y.D., Robinson D.G., and Jiang L. (2012) Unconventional protein secretion. *Trends in Plant Science*, **17**(10), 606–615.

Drakakaki G. and Dandekar A. (2013) Protein secretion: how many secretory routes does a plant cell have? *Plant Science*, **203–204**, 74–78.

Drakakaki G., van de Ven W., Pan S., Miao Y., Wang J., Keinath N.F., Weatherly B., Jiang L., Schumacher K., Hicks G., and Raikhel N. (2012) Isolation and proteomic analysis of the SYP61 compartment reveal its role in exocytic trafficking in Arabidopsis. *Cell Research*, **22**(2), 413–424.

Dulubova I., Sugita S., Hill S., Hosaka M., Fernandez I., Sudhof T.C., and Rizo J. (1999) A conformational switch in syntaxin during exocytosis: role of munc18. *EMBO Journal*, **18**(16), 4372–4382.

Fan L., Hao H., Xue Y., Zhang L., Song K., Ding Z., Botella M.A., Wang H., and Lin J. (2013) Dynamic analysis of Arabidopsis AP2 sigma subunit reveals a key role in clathrin-mediated endocytosis and plant development. *Development* **140**(18), 3826–3837.

Fendrych M., Synek L., Pecenkova T., Toupalova H., Cole R., Drdova E., Nebesarova J., Sedinova M., Hala M., Fowler J.E., and Zarsky V. (2010) The Arabidopsis exocyst complex is involved in cytokinesis and cell plate maturation. *Plant Cell*, **22**(9), 3053–3065.

Field M.C., Sali A., and Rout M.P. (2011) Evolution: On a bender--BARs, ESCRTs, COPs, and finally getting your coat. *Journal of Cell Biology*, **193**(6), 963–972.

Fotin A., Cheng Y., Sliz P., Grigorieff N., Harrison S.C., Kirchhausen T., and Walz T. (2004) Molecular model for a complete clathrin lattice from electron cryomicroscopy. *Nature*, **432**(7017), 573–579.

Fowke L., Dibbayawan T., Schwartz O., Harper J., and Overall R. (1999) Combined immunofluorescence and field emission scanning electron microscope study of plasma membrane-associated organelles in highly vacuolated suspensor cells of white spruce somatic embryos. *Cell Biology International*, **23**(6), 389–397.

Fujimoto M. and Ueda T. (2012) Conserved and plant-unique mechanisms regulating plant post-Golgi traffic. *Frontiers in Plant Science*, **3**, 197.

Fujimoto M., Arimura S., Mano S., Kondo M., Saito C., Ueda T., Nakazono M., Nakano A., Nishimura M., and Tsutsumi N. (2009) Arabidopsis dynamin-related proteins DRP3A and DRP3B are functionally redundant in mitochondrial fission, but have distinct roles in peroxisomal fission. *Plant Journal*, **58**(3), 388–400.

Fujimoto M., Arimura S., Ueda T., Takanashi H., Hayashi Y., Nakano A., and Tsutsumi N. (2010) Arabidopsis dynamin-related proteins DRP2B and DRP1A participate together in clathrin-coated vesicle formation during endocytosis. *Proceedings of National Academy of Science, USA*, **107**(13), 6094–6099.

Gadeyne A., Sanchez-Rodriguez C., Vanneste S., Di Rubbo S., Zauber H., Vanneste K., Van Leene J., De Winne N., Eeckhout D., Persiau G., Van De Slijke E., Vercruysse L., Adamowski M., Ehrlich M., Schweighofer A., Bednarek S., Ketelaar T., Maere S., Friml J., Gevaert K., Witters E., Russinova E., Persson S., De Jaeger G., and Van Damme D. (2014) The TPLATE adaptor complex drives clathrin-mediated endocytosis in plants. *Cell*, **156**(4), 691–704.

Gendre D., Oh J., Boutté Y., Best J. G., Samuels L., Nilsson R., Uemura T., Marchant A., Bennett M.J., Grebe M., and Bhalerao R.P. (2011) Conserved Arabidopsis ECHIDNA protein mediates trans-Golgi-network trafficking and cell elongation. *Proceedings of the National Academy of Science, USA*, **108**(19), 8048–8053.

Gendre D., McFarlane H.E., Johnson E., Mouille G., Sjodin A., Oh J., Levesque-Tremblay G., Watanabe Y., Samuels L., and Bhalerao R. P. (2013) Trans-Golgi Network Localized ECHIDNA/Ypt Interacting Protein Complex Is Required for the Secretion of Cell Wall Polysaccharides in Arabidopsis. *Plant Cell*, **25**(7), 2633–2646.

Guermonprez H., Smertenko A., Crosnier M.T., Durandet M., Vrielynck N., Guerche P., Hussey P.J., Satiat-Jeunemaitre B., and Bonhomme S. (2008) The POK/AtVPS52 protein localizes to several distinct post-Golgi compartments in sporophytic and gametophytic cells. *Journal of Experimental Botany*, **59**(11), 3087–3098.

Gutierrez R., Lindeboom J.J., Paredez A.R., Emons A.M., and Ehrhardt D.W. (2009) Arabidopsis cortical microtubules position cellulose synthase delivery to the plasma membrane and interact with cellulose synthase trafficking compartments. *Natural Cell Biology*, **11**(7), 797–806.

Haas T.J., Sliwinski M.K., Martinez D.E., Preuss M., Ebine K., Ueda T., Nielsen E., Odorizzi G., and Otegui M.S. (2007) The Arabidopsis AAA ATPase SKD1 is involved in multivesicular endosome function and interacts with its positive regulator LYST-INTERACTING PROTEIN5. *Plant Cell*, **19**(4), 1295–1312.

Hala M., Cole R., Synek L., Drdova E., Pecenkova T., Nordheim A., Lamkemeyer T., Madlung J., Hochholdinger F., Fowler J.E., and Zarsky V. (2008) An exocyst complex functions in plant cell growth in Arabidopsis and tobacco. *Plant Cell*, **20**(5), 1330–1345.

Hatsugai N. and Hara-Nishimura I. (2010) Two vacuole-mediated defense strategies in plants. *Plant Signal Behavior*, **5**(12), 1568–1570.

Hong W. (2005) SNAREs and traffic. *Biochimica et Biophysica Acta*, **1744**(3), 493–517.

Hong Z., Bednarek S.Y., Blumwald E., Hwang I., Jurgens G., Menzel D., Osteryoung K.W., Raikhel N.V., Shinozaki K., Tsutsumi N., and Verma D.P. (2003a) A unified nomenclature for Arabidopsis dynamin-related large GTPases based on homology and possible functions. *Plant Molecular Biology*, **53**(3), 261–265.

Hong Z., Geisler-Lee C.J., Zhang Z., and Verma D.P. (2003b) Phragmoplastin dynamics: multiple forms, microtubule association and their roles in cell plate formation in plants. *Plant Molecular Biology*, **53**(3), 297–312.

Ischebeck T., Werner S., Krishnamoorthy P., Lerche J., Meijón M., Stenzel I., Löfke C., Wiessner T., Im Y.J., Perera I.Y., Iven T., Feussner I., Busch W., Boss W.F., Teichmann T., Hause B., Persson S., and Heilmann I. (2013) Phosphatidylinositol 4,5-bisphosphate influences PIN polarization by controlling clathrin mediated membrane trafficking. *Plant Cell*, **25**(12), 4894–4911.

Jaillais Y., Santambrogio M., Rozier F., Fobis-Loisy I., Miege C., and Gaude T. (2007) The retromer protein VPS29 links cell polarity and organ initiation in plants. *Cell*, **130**(6), 1057–1070.

Jamet E., Boudart G., Borderies G., Charmont S., Lafitte C., Rossignol M., Canut H., and Pont-Lezica R. (2008) Isolation of plant cell wall proteins. *Methods in Molecular Biology*, **425**, 187–201.

Kang B.H., Busse J.S., and Bednarek S.Y. (2003) Members of the Arabidopsis dynamin-like gene family, ADL1, are essential for plant cytokinesis and polarized cell growth. *Plant Cell*, **15**(4), 899–913.

Kang B.H., Nielsen E., Preuss M.L., Mastronarde D., and Staehelin L.A. (2011) Electron tomography of RabA4b- and PI-4Kbeta1-labeled trans Golgi network compartments in Arabidopsis. *Traffic*, **12**(3), 313–329.

Kim H., Park M., Kim S.J., and Hwang I. (2005) Actin filaments play a critical role in vacuolar trafficking at the Golgi complex in plant cells. *Plant Cell*, **17**(3), 888–902.

Kim S.Y., Xu Z.Y., Song K., Kim D.H., Kang H., Reichardt I., Sohn E.J., Friml J., Juergens G., and Hwang I. (2013) Adaptor protein complex 2-mediated endocytosis is crucial for male reproductive organ development in Arabidopsis. *Plant Cell*, **25**(8), 2970–2985.

Kleine-Vehn J., Leitner J., Zwiewka M., Sauer M., Abas L., Luschnig C., and Friml J. (2008) Differential degradation of PIN2 auxin efflux carrier by retromer-dependent vacuolar targeting. *Proceedings of National Academy of Science, USA*, **105**(46), 17812–17817.

Konopka C.A., Schleede J.B., Skop A.R., and Bednarek S.Y. (2006) Dynamin and cytokinesis. *Traffic*, **7**(3), 239–247.

Konopka C.A., Backues S.K., and Bednarek S.Y. (2008) Dynamics of Arabidopsis dynamin-related protein 1C and a clathrin light chain at the plasma membrane. *Plant Cell*, **20**(5), 1363–1380.

Kulich I., Cole R., Drdova E., Cvrckova F., Soukup A., Fowler J., and Zarsky V. (2010) Arabidopsis exocyst subunits SEC8 and EXO70A1 and exocyst interactor ROH1 are involved in the localized deposition of seed coat pectin. *New Phytologist*, **188**(2), 615–625.

Lam B.C., Sage T.L., Bianchi F., and Blumwald E. (2001) Role of SH3 domain-containing proteins in clathrin-mediated vesicle trafficking in Arabidopsis. *Plant Cell*, **13**(11), 2499–2512.

Lane D.R., Wiedemeier A., Peng L., Hofte H., Vernhettes S., Desprez T., Hocart C.H., Birch R.J., Baskin T.I., Burn J.E., Arioli T., Betzner A.S., and Williamson R.E. (2001) Temperature-sensitive alleles of RSW2 link the KORRIGAN endo-1,4-beta-glucanase to cellulose synthesis and cytokinesis in Arabidopsis. *Plant Physiology*, **126**(1), 278–288.

Lavy M., Bloch D., Hazak O., Gutman I., Poraty L., Sorek N., Sternberg H., and Yalovsky S. (2007) A Novel ROP/RAC effector links cell polarity, root-meristem maintenance, and vesicle trafficking. *Current Biology*, **17**(11), 947–952.

Li S., Lei L., Somerville C.R., and Gu Y. (2012) Cellulose synthase interactive protein 1 (CSI1) links microtubules and cellulose synthase complexes. *Proceedings of the National Academy of Science, USA*, **109**(1), 185–190.

Li S., Chen M., Yu D., Ren S., Sun S., Liu L., Ketelaar T., Emons A.M., and Liu C. M. (2013) EXO70A1-mediated vesicle trafficking is critical for tracheary element development in Arabidopsis. *Plant Cell*, **25**(5), 1774–1786.

Lunn D., Phan T.D., Tucker G.A., and Lycett G.W. (2013) Cell wall composition of tomato fruit changes during development and inhibition of vesicle trafficking is associated with reduced pectin levels and reduced softening. *Plant Physiology and Biochemistry*, **66**, 91–97.

Lycett G. (2008) The role of Rab GTPases in cell wall metabolism. *Journal of Experimental Botany*, **59**(15), 4061–4074.

Manzano C., Abraham Z., Lopez-Torrejon G., and Del Pozo J. C. (2008) Identification of ubiquitinated proteins in Arabidopsis. *Plant Molecular Biology*, **68**(1–2), 145–158.

McFarlane H.E., Watanabe Y., Gendre D., Carruthers K., Levesque-Tremblay G., Haughn G.W., Bhalerao R.P., and Samuels L. (2013) Cell wall polysaccharides are mistargeted to the vacuole in echidna mutants. *Plant Cell Physiology*, **54**(11), 1867–1880.

McFarlane H.E., Döring A., and Persson S. (2014) The cell biology of cellulose synthesis. *Annual Reviews in Plant Biology*, **65**, 69–94.

McMahon H.T. and Boucrot E. (2011) Molecular mechanism and physiological functions of clathrin-mediated endocytosis. *Natural Reviews Molecular Cell Biology*, **12**(8), 517–533.

Meyer D., Pajonk S., Micali C., O'Connell R., and Schulze-Lefert P. (2009) Extracellular transport and integration of plant secretory proteins into pathogen-induced cell wall compartments. *Plant Journal*, **57**(6), 986–999.

Miart F., Desprez T., Biot E., Morin H., Belcram K., Höfte H., Gonneau M., and Vernhettes S. (2013) Spatio-temporal regulation of cellulose synthesis during cell plate formation in Arabidopsis. *Plant Journal*, **77**, 71–84.

Miyagishima S.Y., Kuwayama H., Urushihara H., and Nakanishi H. (2008) Evolutionary linkage between eukaryotic cytokinesis and chloroplast division by dynamin proteins. *Proeedings of the National Academy of Science, USA*, **105**(39), 15202–15207.

Munson M. and Novick P. (2006) The exocyst defrocked, a framework of rods revealed. *Nature Structural and Molecular Biology*, **13**(7), 577–581.

Nagawa S., Xu T., Lin D., Dhonukshe P., Zhang X., Friml J., Scheres B., Fu Y., and Yang Z. (2012) ROP GTPase-dependent actin microfilaments promote PIN1 polarization by localized inhibition of clathrin-dependent endocytosis. *PLoS Biology*, **10**(4), e1001299.

Nebenfuhr A., Gallagher L.A., Dunahay T.G., Frohlick J.A., Mazurkiewicz A.M., Meehl J.B., and Staehelin L.A. (1999) Stop-and-go movements of plant Golgi stacks are mediated by the acto-myosin system. *Plant Physiology*, **121**(4), 1127–1142.

Nickle T.C. and Meinke D.W. (1998) A cytokinesis-defective mutant of Arabidopsis (cyt1) characterized by embryonic lethality, incomplete cell walls, and excessive callose accumulation. *Plant Journal*, **15**(3), 321–332.

Nicol F., His I., Jauneau A., Vernhettes S., Canut H., and Hofte H. (1998) A plasma membrane-bound putative endo-1,4-beta-D-glucanase is required for normal wall assembly and cell elongation in Arabidopsis. *Embo Journal*, **17**(19), 5563–5576.

Niemes S., Langhans M., Viotti C., Scheuring D., San Wan Yan M., Jiang L., Hillmer S., Robinson D.G., and Pimpl P. (2010) Retromer recycles vacuolar sorting receptors from the trans-Golgi network. *Plant Journal*, **61**(1), 107–121.

Niihama M., Takemoto N., Hashiguchi Y., Tasaka M., and Morita M. T. (2009) ZIP genes encode proteins involved in membrane trafficking of the TGN-PVC/vacuoles. *Plant Cell Physiology*, **50**(12), 2057–2068.

Nuhse T.S., Stensballe A., Jensen O.N., and Peck S.C. (2004) Phosphoproteomics of the Arabidopsis plasma membrane and a new phosphorylation site database. *Plant Cell*, **16**(9), 2394–2405.

Oda Y. and Fukuda H. (2012) Initiation of cell wall pattern by a Rho- and microtubule-driven symmetry breaking. *Science*, **337**(6100), 1333–1336.

Pagant S., Bichet A., Sugimoto K., Lerouxel O., Desprez T., McCann M., Lerouge P., Vernhettes S., and Hofte H. (2002) KOBITO1 encodes a novel plasma membrane protein necessary for normal synthesis of cellulose during cell expansion in Arabidopsis. *Plant Cell*, **14**(9), 2001–2013.

Paredez A.R., Somerville C.R. and Ehrhardt D.W. (2006) Visualization of cellulose synthase demonstrates functional association with microtubules. *Science*, **312**(5779), 1491–1495.

Park M., Song K., Reichardt I., Kim H., Mayer U., Stierhof Y.D., Hwang I., and Jurgens G. (2013) Arabidopsis mu-adaptin subunit AP1M of adaptor protein complex 1 mediates late secretory and vacuolar traffic and is required for growth. *Proceedings of the National Academy of Science, USA*, **110**(25), 10318–10323.

Pimpl P., Movafeghi A., Coughlan S., Denecke J., Hillmer S., and Robinson D. G. (2000) In situ localization and in vitro induction of plant COPI-coated vesicles. *Plant Cell*, **12**(11), 2219–2236.

Pourcher M., Santambrogio M., Thazar N., Thierry A.M., Fobis-Loisy I., Miege C., Jaillais Y., and Gaude T. (2010) Analyses of sorting nexins reveal distinct retromer-subcomplex functions in development and protein sorting in Arabidopsis thaliana. *Plant Cell*, **22**(12), 3980–3991.

Praefcke G.J. and McMahon H.T. (2004) The dynamin superfamily: universal membrane tubulation and fission molecules? *Nature Reviews Molecular Cell Biology*, **5**(2), 133–147.

Preuss M.L., Kovar D.R., Lee Y.R., Staiger C.J., Delmer D.P., and Liu B. (2004) A plant-specific kinesin binds to actin microfilaments and interacts with cortical microtubules in cotton fibers. *Plant Physiology*, **136**(4), 3945–3955.

Preuss M.L., Schmitz A.J., Thole J.M., Bonner H.K., Otegui M.S., and Nielsen E. (2006) A role for the RabA4b effector protein PI-4Kbeta1 in polarized expansion of root hair cells in Arabidopsis thaliana. *Journal of Cell Biology*, **172**(7), 991–998.

Reyes F.C., Buono R., and Otegui M.S. (2011) Plant endosomal trafficking pathways. *Current Opinion in Plant Biology*, **14**(6), 666–673.

Robinson M.S. and Bonifacino J.S. (2001) Adaptor-related proteins. *Current Opinion in Cell Biology*, **13**(4), 444–453.

Rockel N., Wolf S., Kost B., Rausch T., and Greiner S. (2008) Elaborate spatial patterning of cell-wall PME and PMEI at the pollen tube tip involves PMEI endocytosis, and reflects the distribution of esterified and de-esterified pectins. *Plant Journal*, **53**(1), 133–143.

Roth T.F. and Porter K.R. (1964) Yolk protein uptake in the oocyte of the mosquito Aedes Aegypti. *L. Journal of Cell Biology*, **20**, 313–332.

Roudier F., Fernandez A.G., Fujita M., Himmelspach R., Borner G.H., Schindelman G., Song S., Baskin T.I., Dupree P., Wasteneys G.O., and Benfey P.N. (2005) COBRA, an Arabidopsis extracellular glycosyl-phosphatidyl inositol-anchored protein, specifically controls highly anisotropic expansion through its involvement in cellulose microfibril orientation. *Plant Cell*, **17**(6), 1749–1763.

Šamaj J., Šamajová O., Peters M., Baluška F., Lichtscheidl I., Knox J.P., and Volkmann D. (2000) Immunolocalization of LM2 arabinogalactan-protein epitope associated with endomembranes of plant cells. *Protoplasma*, **212**, 186–196.

Sampathkumar A., Gutierrez R., McFarlane H.E., Bringmann M., Lindeboom J., Emons A.M., Samuels L., Ketelaar T., Ehrhardt D.W., and Persson S. (2013) Patterning and lifetime of plasma membrane-localized cellulose synthase is dependent on actin organization in Arabidopsis interphase cells. *Plant Physiology*, **162**(2), 675–688.

Samuels L. and McFarlane H.E. (2012) Plant cell wall secretion and lipid traffic at membrane contact sites of the cell cortex. *Protoplasma*, **249**(1), 19–23.

Sanchez-Rodriguez C., Bauer S., Hematy K., Saxe F., Ibanez A.B., Vodermaier V., Konlechner C., Sampathkumar A., Ruggeberg M., Aichinger E., Neumetzler L., Burgert I., Somerville C., Hauser M.T., and Persson S. (2012) Chitinase-like1/pom-pom1 and its homolog CTL2 are glucan-interacting proteins important for cellulose biosynthesis in Arabidopsis. *Plant Cell*, **24**(2), 589–607.

Sanderfoot A. (2007) Increases in the number of SNARE genes parallels the rise of multicellularity among the green plants. *Plant Physiology*, **144**(1), 6–17.

Sauer M., Delgadillo M.O., Zouhar J., Reynolds G.D., Pennington J.G., Jiang L., Liljegren S.J., Stierhof Y.D., De Jaeger G., Otegui M.S., Bednarek S.Y., and Rojo E. (2013) MTV1 and MTV4 encode plant-specific ENTH and ARF GAP proteins that mediate clathrin-dependent trafficking of vacuolar cargo from the trans-Golgi network. *Plant Cell*, **25**(6), 2217–2235.

Scheller H.V. and Ulvskov P. (2010) Hemicelluloses. *Annual Review of Plant Biology*, **61**, 263–289.

Scheuring D., Viotti C., Kruger F., Kunzl F., Sturm S., Bubeck J., Hillmer S., Frigerio L., Robinson D.G., Pimpl P., and Schumacher K. (2011) Multivesicular bodies mature from the trans-Golgi network/early endosome in Arabidopsis. *Plant Cell*, **23**(9), 3463–3481.

Schumacher K. (2006) Endomembrane proton pumps: connecting membrane and vesicle transport. *Current Opinion in Plant Biology*, **9**(6), 595–600.

Shen B., Li C., and Tarczynski M.C. (2002) High free-methionine and decreased lignin content result from a mutation in the Arabidopsis S-adenosyl-L-methionine synthetase 3 gene. *Plant Journal*, **29**(3), 371–380.

Shpetner H.S. and Vallee R.B. (1989) Identification of dynamin, a novel mechanochemical enzyme that mediates interactions between microtubules. *Cell*, **59**(3), 421–432.

Song J., Lee M.H., Lee G.J., Yoo C.M., and Hwang I. (2006) Arabidopsis EPSIN1 plays an important role in vacuolar trafficking of soluble cargo proteins in plant cells via interactions with clathrin, AP-1, VTI11, and VSR1. *Plant Cell*, **18**(9), 2258–2274.

Sparkes I.A., Ketelaar T., de Ruijter N.C., and Hawes C. (2009) Grab a Golgi: laser trapping of Golgi bodies reveals in vivo interactions with the endoplasmic reticulum. *Traffic*, **10**(5), 567–571.

Spitzer C., Reyes F.C., Buono R., Sliwinski M.K., Haas T.J., and Otegui M.S. (2009) The ESCRT-related CHMP1A and B proteins mediate multivesicular body sorting of auxin carriers in Arabidopsis and are required for plant development. *Plant Cell*, **21**(3), 749–766.

Staehelin L.A. and Kang B.H. (2008) Nanoscale architecture of endoplasmic reticulum export sites and of Golgi membranes as determined by electron tomography. *Plant Physiology*, **147**(4), 1454–1468.

Szyjanowicz P.M., McKinnon I., Taylor N.G., Gardiner J., Jarvis M.C., and Turner S.R. (2004) The irregular xylem 2 mutant is an allele of korrigan that affects the secondary cell wall of Arabidopsis thaliana. *Plant Journal*, **37**(5), 730–740.

Takei K., Yoshida Y., and Yamada H. (2005) Regulatory mechanisms of dynamin-dependent endocytosis. *Journal of Biochemistry*, **137**(3), 243–247.

Taylor N.G. (2007) Identification of cellulose synthase AtCesA7 (IRX3) in vivo phosphorylation sites--a potential role in regulating protein degradation. *Plant Molecular Biology*, **64**(1–2), 161–171.

Taylor N.G. (2011) A role for Arabidopsis dynamin related proteins DRP2A/B in endocytosis; DRP2 function is essential for plant growth. *Plant Molecular Biology*, **76**(1–2), 117–129.

Teh O.K., Shimono Y., Shirakawa M., Fukao Y., Tamura K., Shimada T., and Hara-Nishimura I. (2013) The AP-1 mu adaptin is required for KNOLLE localization at the cell plate to mediate cytokinesis in Arabidopsis. *Plant Cell Physiology*, **54**(6), 838–847.

Thellmann M., Rybak K., Thiele K., Wanner G., and Assaad F.F. (2010) Tethering factors required for cytokinesis in Arabidopsis. *Plant Physiology*, **154**(2), 720–732.

Thiele K., Wanner G., Kindzierski V., Jurgens G., Mayer U., Pachl F., and Assaad F. F. (2009) The timely deposition of callose is essential for cytokinesis in Arabidopsis. *Plant Journal*, **58**(1), 13–26.

Toyooka K., Goto Y., Asatsuma S., Koizumi M., Mitsui T., and Matsuoka K. (2009) A mobile secretory vesicle cluster involved in mass transport from the Golgi to the plant cell exterior. *Plant Cell*, **21**(4), 1212–1229.

Ueda H., Yokota E., Kutsuna N., Shimada T., Tamura K., Shimmen T., Hasezawa S., Dolja V.V., and Hara-Nishimura I. (2010) Myosin-dependent endoplasmic reticulum motility and F-actin organization in plant cells. *Proceedings of the National Academy of Science, USA*, **107**(15), 6894–6899.

Vernoud V., Horton A.C., Yang Z., and Nielsen E. (2003) Analysis of the small GTPase gene superfamily of Arabidopsis. *Plant Physiology*, **131**(3), 1191–1208.

Viotti C., Bubeck J., Stierhof Y.D., Krebs M., Langhans M., van den Berg W., van Dongen W., Richter S., Geldner N., Takano J., Jurgens G., de Vries S.C., Robinson D.G., and Schumacher K. (2010) Endocytic and secretory traffic in Arabidopsis merge in the trans-Golgi network/early endosome, an independent and highly dynamic organelle. *Plant Cell*, **22**(4), 1344–1357.

Wang J., Ding Y., Hillmer S., Miao Y., Lo S.W., Wang X., Robinson D.G. and Jiang L. (2010) EXPO, an exocyst-positive organelle distinct from multivesicular endosomes and autophagosomes, mediates cytosol to cell wall exocytosis in Arabidopsis and tobacco cells. *Plant Cell*, **22**(12), 4009–4030.

Wolf S., Rausch T., and Greiner S. (2009) The N-terminal pro region mediates retention of unprocessed type-I PME in the Golgi apparatus. *Plant Journal*, **58**(3), 361–375.

Wolfenstetter S., Wirsching P., Dotzauer D., Schneider S., and Sauer N. (2012) Routes to the tonoplast: the sorting of tonoplast transporters in Arabidopsis mesophyll protoplasts. *Plant Cell*, **24**(1), 215–232.

Xiong G., Li R., Qian Q., Song X., Liu X., Yu Y., Zeng D., Wan J., Li J., and Zhou Y. (2010) The rice dynamin-related protein DRP2B mediates membrane trafficking, and thereby plays a critical role in secondary cell wall cellulose biosynthesis. *Plant Journal*, **64**(1), 56–70.

Xu T., Qu Z., Yang X., Qin X., Xiong J., Wang Y., Ren D. and Liu G. (2009) A cotton kinesin GhKCH2 interacts with both microtubules and microfilaments. *Biochem Journal*, **421**(2), 171–180.

Xu T., Wen M., Nagawa S., Fu Y., Chen J.G., Wu M.J., Perrot-Rechenmann C., Friml J., Jones A.M., and Yang Z. (2010) Cell surface- and rho GTPase-based auxin signaling controls cellular interdigitation in Arabidopsis. *Cell*, **143**(1), 99–110.

Yamaguchi T., Dulubova I., Min S.W., Chen X., Rizo J., and Sudhof T. C. (2002) Sly1 binds to Golgi and ER syntaxins via a conserved N-terminal peptide motif. *Developmental Cell*, **2**(3), 295–305.

Yamaoka S., Shimono Y., Shirakawa M., Fukao Y., Kawase T., Hatsugai N., Tamura K., Shimada T., and Hara-Nishimura I. (2013) Identification and dynamics of Arabidopsis adaptor protein-2 complex and its involvement in floral organ development. *Plant Cell*, **25**(8), 2958–2969.

Yang G., Gao P., Zhang H., Huang S., and Zheng Z.L. (2007) A mutation in MRH2 kinesin enhances the root hair tip growth defect caused by constitutively activated ROP2 small GTPase in Arabidopsis. *PLoS One*, **2**(10), e1074.

Yokota E., Ueda S., Tamura K., Orii H., Uchi S., Sonobe S., Hara-Nishimura I., and Shimmen T. (2009) An isoform of myosin XI is responsible for the translocation of endoplasmic reticulum in tobacco cultured BY-2 cells. *Journal of Experimental Botany*, **60**, 197–212.

Zhang H., Zhang L., Gao B., Fan H., Jin J., Botella M.A., Jiang L., and Lin J. (2011) Golgi apparatus-localized synaptotagmin 2 is required for unconventional secretion in Arabidopsis. *PLoS One*, **6**, e26477.

Zhu C. and Dixit R. (2011) Single molecule analysis of the Arabidopsis FRA1 kinesin shows that it is a functional motor protein with unusually high processivity. *Molecular Plant*, **4**(5), 879–885.

Zuo J., Niu Q.W., Nishizawa N., Wu Y., Kost B., and Chua N.H. (2000) KORRIGAN, an Arabidopsis endo-1,4-beta-glucanase, localizes to the cell plate by polarized targeting and is essential for cytokinesis. *Plant Cell*, **12**(7), 1137–1152.

Zwiewka M., Feraru E., Moller B., Hwang I., Feraru M.I., Kleine-Vehn J., Weijers D., and Friml J. (2011) The AP-3 adaptor complex is required for vacuolar function in Arabidopsis. *Cell Research*, **21**(12), 1711–1722.

3 A Blueprint for Cellulose Biosynthesis, Deposition, and Regulation in Plants

Ian S. Wallace[1,2] and Chris R. Somerville[1]

[1]*Energy Biosciences Institute, University of California, Berkeley, Berkeley, CA*

[2]*Present address: Department of Biochemistry and Molecular Biology, University of Nevada, Reno, Reno, NV*

Introduction

Terrestrial plants annually assimilate roughly 10^{11} metric tons of CO_2 through the process of photosynthesis (Beer *et al.*, 2010), and it is estimated that 70% of the resulting photosynthate is partitioned into the synthesis of plant cell walls (Pauly and Keegstra, 2008), the complex polysaccharide-rich extracellular matrices that surround all plant cells. Cell wall polysaccharides represent the most abundant component of plant biomass, which is a key renewable resource for the production of food, fiber, and textiles. Recently, plant biomass has also been proposed as a raw material for the production of renewable liquid transportation fuels and other commodity chemicals (Carroll and Somerville, 2009). In addition to the economic importance of plant biomass components, the synthesis, ordered deposition, and dynamic modification of cell wall polysaccharides critically influence the growth and development of plants. Cell walls also impart mechanical strength to plant tissues, providing them with the ability to grow upright and compete for light, which was a critical evolutionary advancement in the development of plants on land. Finally, cell walls serve as a physical barrier that must be circumvented by pathogens in order to gain access to the plant cytosol.

Plant cells execute a morphogenesis program that requires cell expansion, division, and differentiation, and all of these basic cellular processes must occur in the context of a rigid cell wall. Cell expansion is driven by turgor pressure generated inside the plant cell, and this pressure is resisted by the cell wall. Cell walls must therefore be strong enough to resist osmotic pressure, but must also remain sufficiently extensible to allow cell expansion to occur (Cosgrove, 2005). During cell division, new cell wall material is also deposited in

Plant Cell Wall Patterning and Cell Shape, First Edition. Edited by Hiroo Fukuda.

the phragmoplast to generate new cross walls (Moore *et al.*, 1991; Samuels *et al.*, 1995). Both of these processes influence the final acquisition of differentiated plant cell shape. It is also noteworthy to mention that the basic developmental processes of expansion, division, and differentiation can be modulated by a variety of developmental signals, such as hormones and light, or by environmental perturbations, such as abiotic stresses and pathogens, suggesting the existence of a complex regulatory network that modulates that amount and types of wall polysaccharides. Although the composition of cell walls can vary substantially in different plant species, or within cell types of an individual plant (Burton *et al.*, 2010), it is generally accepted that two developmentally distinct types of cell walls exist. Primary cell walls are produced in all plant cells, while secondary walls are deposited in specific cell types that require increased mechanical support (e.g. xylem fibers). Secondary wall deposition is also developmentally distinct in that this process occurs during cell differentiation after primary wall deposition and cell expansion are complete. Cell wall biosynthesis therefore plays an important role in cell expansion and division, as well as differentiation, and these processes can conceivably be modulated by developmental or environmental factors.

Cell wall polymers can be grouped into five distinct classes: cellulose, hemicelluloses, pectins, proteins, and lignin. Cellulose is the most abundant biomolecule on the planet, and typically constitutes 30–40% of the total cell wall polysaccharide composition, although this can increase in certain cell types (Burton *et al.*, 2010). Cellulose is composed of multiple long β-(1$\rightarrow$4)-linked glucan chains that are organized into a para-crystalline microfibril with a diameter of approximately 2.5 nm. This structural organization provides cellulose with a tensile strength similar to that of steel (Moon *et al.*, 2011), which supports the hypothesis that cellulose is the primary load-bearing polysaccharide that resists the force of osmotic pressure in plant cells. Cellulose microfibrils are embedded in a matrix of neutral hemicelluloses and acidic pectic polysaccharides that can cross-link the microfibrils to form a dynamic network (Cosgrove, 2005). The structure of these polysaccharides can also be modified by a plethora of enzymes to modulate the rigidity of the total cell wall polysaccharide network during cellular development and differentiation.

Due to the abundance of cellulose, the key role that this polysaccharide plays in normal plant development, and the vast utility of this polysaccharide as a raw material for everything from the clothes we wear to medical devices (Klemm *et al.*, 2005; Moon *et al.*, 2011), it is important to understand how cellulose is synthesized as well as how plants regulate the amount and quality of cellulose produced. While our mechanistic understanding of cellulose biosynthesis is not complete, recent advances in this field have allowed us to answer basic questions that have confounded plant biologists for decades and have led to the development of tools to study new problems in cellulose biosynthesis in unprecedented detail. In this chapter, we summarize current views of how plants synthesize cellulose, how the deposition of cellulose occurs, and the known factors that regulate cellulose biosynthesis, and discuss how these processes influence plant development and how they may be manipulated in useful ways. Some aspects of these topics have been reviewed elsewhere (Somerville *et al.*, 2004; Somerville, 2006; Mutwil *et al.*, 2008; Szymanski *et al.*, 2009; Endler and Persson, 2011).

Structure and Deposition of Cellulose

Cellulose is composed of multiple long β-(1→4)-linked glucan chains consisting of repeating cellobiose units. The hydroxyl groups as well as the pyranose ring oxygen of each repeating cellobiose unit hydrogen bond intramolecularly to increase the axial stability of the glucan chains, and multiple glucan chains associate with one another via hydrogen bonding and van der Waals interactions to form the crystalline regions of the cellulose microfibril. The hydrogen bonding pattern and the orientation of glucan chains with respect to one another determines the crystallographic form of cellulose. Most naturally occurring celluloses are in the cellulose I form, which can be further classified into cellulose I_α and I_β based on the precise hydrogen bonding pattern between glucan chains. Bacterial cellulose and cellulose isolated from some algal species exists in the I_α form, while plant cellulose primarily exists in the I_β form. It is interesting to note that cellulose I is converted to cellulose II upon sequential denaturation and regeneration (Moon *et al.*, 2011), suggesting that Cellulose I is metastable.

Celluloses from different sources have varied polymer lengths and associated glucan chains. The degree of polymerization of cellulose polymers varies from 300 to over 10,000 glucose units (Moon *et al.*, 2011), and the number of associated glucan chains as well as their average crystallinity is dependent upon the extraction method and source of isolation. It is commonly suggested that plant cellulose microfibrils contain 36 glucan chains due to the observation that microfibrils in plant samples have a diameter of 2.5–3 nm; the number of glucan chains is still a matter of debate however, and experimental evidence has suggested other numbers of associated glucan chains (Fernandes *et al.*, 2011).

In primary cell walls, cellulose microfibrils are spatially organized to resist cell expansion in certain directions. For example, cellulose is deposited in transverse arrays that are perpendicular to the expansion axis in rapidly elongating epidermal cells of root and dark-grown hypocotyl tissues (Sugimoto *et al.*, 2003; Refregier *et al.*, 2004; Anderson *et al.*, 2010). This organization prevents radial expansion by counteracting internally generated turgor pressure, while simultaneously facilitating expansion in the axial direction, resulting in the typical elongated cuboid cell shapes that are typical of these tissues. This localized deposition of cellulose can conceivably be utilized to produce more complex cell types, such as the jigsaw-piece-shaped pavement cells in leaf tissue, further supporting the notion that organized cellulose deposition plays a key role in the acquisition of differentiated plant cell shape.

Isolation of Genes Encoding Cellulose Synthase Catalytic Subunits

Although cellulose is the most abundant biomolecule in nature, efforts to understand the biosynthesis of this polysaccharide have met with considerable difficulty primarily due to the fact that it is challenging to produce crystalline cellulose from plant extracts *in vitro*. Cellulose is synthesized by the cellulose synthase complex, which utilizes cytosolic UDP-glucose as an activated sugar donor to extend elongating glucan chains into the extracellular matrix. However, plant extracts generally produce callose, a β-(1$\rightarrow$3)-glucan involved in responses to wounding and pathogens (Nishimura *et al.*, 2003), after the addition of UDP-glucose. In fact, at one time there was speculation that cellulose synthase and callose synthase activities might be catalyzed by the same enzyme; this notion was however ultimately discarded when distinct genes encoding both enzymes were isolated (Pear *et al.*, 1996; Arioli *et al.*, 1998; Nishimura *et al.*, 2003). While assay conditions have been developed in certain systems that favor cellulose synthase activity over callose synthase activity (Lai-Kee-Him *et al.*, 2002; Bessueille *et al.*, 2009), biochemical assays still require detailed product analysis of *in vitro* synthesized glucans to truly demonstrate cellulose biosynthesis. Biochemical analysis of cellulose biosynthesis has therefore remained largely intractable, and our current understanding of the process comes from a variety of other approaches.

Cellulose is synthesized at the plasma membrane by large multiprotein complexes that were first observed by freeze-fracture electron microscopy and were named ‘terminal complexes’ due to the fact that they associated with the ends of cellulose microfibrils (Montezinos and Brown, 1976; Mueller and Brown, 1980). While various species of algae display terminal complexes with an array of morphologies, all higher plant terminal complexes are roughly 25–30 nm in diameter and display a characteristic lobed structure with six-fold symmetry when viewed from the extracellular face of the plasma membrane. As a result, plant terminal complexes are typically referred to as ‘rosettes’ due to their apparent similarity in structure to floral organs. Further observations indicated that the cytosolic portions of the terminal complexes were 40–50 nm in diameter (Bowling and Brown, 2008). To put the size of this protein complex into perspective, the intramembrane dimensions are roughly equivalent to the size of the 80S eukaryotic ribosome and the intracellular portion is twice as large (Ben-Sham *et al.*, 2011).

Although the morphology is different from plant rosette structures, terminal complexes have also been observed in species of bacteria that produce cellulose, suggesting that these organisms posses the genetic requirements to organize these structures and produce cellulose. Indeed, some species of bacteria, such as *Gluconacetobacter xylinus*, produce crystalline cellulose with similar properties to plant cellulose (Moon *et al.*, 2011). Genetic screens of *Gluconacetobacter xylinus* (Wong *et al.*, 1990) and *Agrobacterium tumefaciens* (Matthyse *et al.*, 1995) revealed that four genes in the Bacterial Cellulose Synthase (Bcs) operon were absolutely required for cellulose biosynthesis in bacteria. BcsA encodes a polytopic membrane-bound glycosyltransferase with eight transmembrane domains and a central catalytic loop containing D,D,D,QxxRW motifs that are associated with processive glycosyltransferases. BcsB encodes a gene of unknown function that contains a predicted N-terminal signal peptide as well as a C-terminal transmembrane domain, suggesting that the majority of BcsB is exposed to the bacterial periplasm and that the protein is anchored in the bacterial plasma membrane. Genetic evidence indicates that BcsB is absolutely required for bacterial cellulose biosynthesis (Wong *et al.*, 1990), and the observation of BcsA–BcsB gene fusion events in some bacterial species (Saxena *et al.*, 1994) suggests that these proteins form a complex. BcsC is a predicted transmembrane β-barrel protein that is proposed to form a bacterial outer membrane pore which serves as a conduit for extrusion of glucan chains into the extracellular environment. Finally, BcsZ encodes a functional cellulase (Mazur and Zimmer, 2011), but the biochemical role of this protein in cellulose biosynthesis remains unclear. Certain bacterial species that produce crystalline cellulose also contain BcsD, a fifth gene in the Bcs operon that is necessary for cellulose crystallization (Wong *et al.*, 1990).

The identification of bacterial cellulose synthase genes led to the isolation of the first plant cellulose synthase cDNA from differentiating cotton fibers, which contain up to 95% pure cellulose. Random sequencing of cDNA clones during a period of increased cellulose biosynthesis resulted in the isolation of cDNA clones that shared amino acid sequence homology to the *Agrobacterium tumefaciens* BcsA homologue (Pear *et al.*, 1996). This gene, which was subsequently named CelA1, was also proposed to encode a polytopic membrane-bound glycosyltransferase containing an N-terminal zinc-finger domain, eight transmembrane domains, and a large central loop with the characteristic D,D,D,QxxRW motifs that are associated with processive glycosyltransferases. Additionally, a recombinant fragment of cotton fiber CelA1 containing these glycosyltransferase motifs was demonstrated to bind radiolabeled UDP-glucose, suggesting that this protein was capable of binding the substrate of the cellulose synthase reaction (Pear *et al.*, 1996). Further immuno-electron microscopy studies using an antibody raised against the central catalytic loop of cotton CelA1 revealed that this protein was a component of the higher plant rosette complex (Kimura *et al.*, 1999), demonstrating that CelA1-like proteins are also associated with cellulose-synthesizing terminal complexes.

Phylogenetic analysis of genes related to cotton CesA1 revealed that the fully sequenced Arabidopsis genome contains 10 highly similar genes organized into a small gene family (Richmond and Somerville, 2000). An additional 29 other sequences were found to have lower amino acid similarity but similar structural properties, which led to the cellulose synthase family members being renamed the Cellulose Synthase A (CesA) genes. CesA family members are present in all land plant genomes, as well as the genome of mosses, spikemosses, liverworts, and ferns (Yin *et al.*, 2009; Carroll and Specht, 2011), suggesting that these genes evolved early in land plant evolution; this is consistent with the important structural role of cellulose in the morphogenesis of land plants.

Structural Insights from the Bacterial Cellulose Synthase Complex

In addition to the importance of BcsA in the initial process of cloning plant CesA genes, structural studies of the bacterial cellulose synthase have provided the first structural detail of the cellulose synthase reaction. Recently, a heterodimer of BcsA and BcsB from *Rhodobacter sphaeroides* was crystallized at 3.25 Å resolution (Morgan *et al.*, 2013). The BcsA protein can be divided into three regions containing the glycosyltransferase (GT) domain, three amphipathic helices constituting an interfacial membrane domain, and the transmembrane spanning region (Fig. 3.1). The interfacial helices provide the major structural link between the GT domain and the transmembrane domains. BcsA and BcsB form a 1:1 heterodimer and the majority of BcsB is exposed to the periplasm, while the single BcsB transmembrane domain interacts with the BcsA transmembrane helices to anchor BcsB to the complex. Additionally, a single UDP molecule was co-crystallized in the BcsA GT domain, indicating that this domain is responsible for UDP-Glc substrate binding. A nascent glucan chain that begins in the GT domain, transits the transmembrane pore, and extends into the periplasm was also co-crystallized with the complex, suggesting that the BcsA-BcsB dimer is at least partially catalytically active.

The cytosolic GT domain of BcsA exhibits a typical glycosyltransferase A (GT-A) fold consisting of seven α-helices surrounding seven β-strands (Lairson *et al.*, 2008), and the single UDP substrate molecule forms numerous contacts with conserved residues in this domain. As observed in other glycosyltransferase structures (Tarbouiech *et al.*, 2001), the first two aspartate residues of the D,D,D,QxxRW motif coordinate the terminal phosphate moieties of UDP, possibly through a bridging magnesium ion. The third aspartate residue in this consensus sequence is located near the putative site of a glucose residue attached to UDP-Glc, and this residue is likely to serve as the catalytic base that mediates the glycosyltransferase reaction. The QxxRW motif interacts with the terminal glucose residue of the nascent glucan along with a second highly conserved FFCGS motif, potentially suggesting a role for these residues in glucan binding and chain elongation. The extended glucan also forms numerous contacts with residues in the transmembrane channel, suggesting that an amphipathic network of interactions is necessary for chain extrusion. Many of these important GT domain residues are conserved in plant CesAs, suggesting that CesAs catalyze the cellulose synthase reaction in a similar manner; this hypothesis is further supported by computational modeling of the CesA GT domain from cotton CesA1 (Sethaphong *et al.*, 2013).

The BcsA structure further addressed numerous questions regarding the mechanism of the cellulose synthase reaction (Fig. 3.1). Only one UDP molecule was co-crystallized with BcsA, suggesting that this protein is capable of synthesizing the cellobiose unit without two binding sites; this concept was suggested to account for the alternating glucose orientation of cellulose (Carpita, 2011). Additionally, the presence of a nascent glucan chain suggests that each BcsA monomer forms a single pore that extrudes one glucan chain. This observation implies that each plant CesA subunit may only synthesize a single glucan chain, and that the organization of CesAs within the

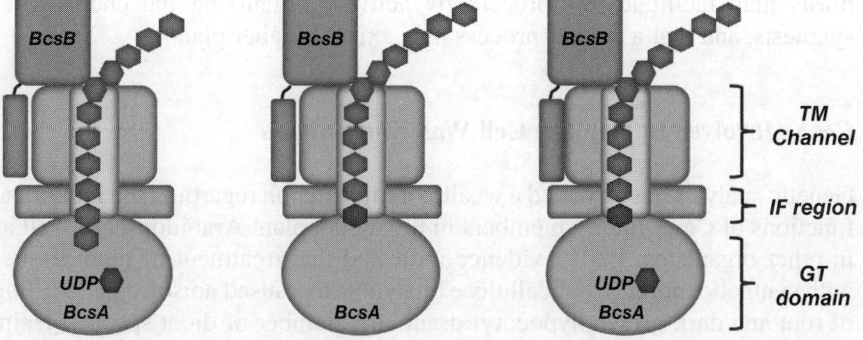

Figure 3.1 Features and proposed mechanism of the bacterial cellulose synthase complex. The BcsA-B dimer from Rhodobacter sphaeroides was crystallized as a 1:1 heterodimer (Morgan *et al.*, 2013). BcsA (yellow) consists of distinct transmembrane, interfacial, and glycosyltransferase domain regions. The single transmembrane domain of BcsB (blue) interacts with the transmembrane helices of BcsA to form a stable complex in the bacterial plasma membrane. BcsA was crystallized with a nascent glucan chain (green hexagons) that begins in the BcsA glycosyltransferase domain, transits a pore in the BcsA transmembrane region, and interacts with BcsB in the extracellular space. The glycosyltransferase domain binds a single UDP-glucose molecule (red hexagon), which serves as a sugar nucleotide donor that elongates the nascent glucan chain by a single glucose residue in each catalytic cycle. As a new glucose residue is added to the glucan chain (red hexagon), the polysaccharide is processively pushed further into the extracellular matrix. (See plate section for the color version of this figure.)

rosette complex governs the structure of the synthesized microfibril. Finally, the spatial separation of the UDP molecule in the GT domain compared to the end of the glucan chain in the periplasm seems to preclude the possibility that BcsA uses a glycolipid primer as a substrate (Peng *et al.*, 2002), a hypothesis that also seems unlikely based on genetic analysis of plants with defects in sterol glycoside synthesis (DeBolt *et al.*, 2009). Overall, the BcsA–B structure provides a basic mechanistic understanding of cellulose synthesis at atomic resolution (Fig. 3.1), and this structure is likely to provide key insights into the biochemistry of plant CesA proteins.

Structural analysis of other proteins associated with the bacterial cellulose synthase operon provides additional clues to the details of plant cellulose biosynthesis. For example, the structure of BcsD from *G. xylinus* suggests a potential mechanism for glucan chain crystallization (Hu *et al.*, 2010). BcsD was crystallized as a homo-octameric donut-shaped ring with four-fold symmetry surrounding a large central pore containing four cellapentaose molecules. While the structure must be considered in isolation, the organization of the cellapentaose chains within the pore of BcsD potentially suggests that BcsD facilitates glucan chain association after synthesis by the BcsA–B complex. Indeed, *G. xylinus* BcsD mutants are still capable of synthesizing glucan chains, but they are disorganized compared to the bacterial cellulose fibrils produced by wild-type strains (Saxena *et al.*, 1994; Hu *et al.*, 2010). These observations suggest that bacteria which produce organized cellulose fibrils may facilitate this process by actively organizing the chains after synthesis, and that a similar process may exist in higher plants.

CesAs Involved in Primary Cell Wall Biosynthesis

Genetic analysis has provided a wealth of information regarding the individual functions of CesA family members in the model plant Arabidopsis as well as in other organisms. Early evidence indicated that treatment of plant tissues with synthetic inhibitors of cellulose biosynthesis caused anisotropic swelling of root and dark-grown hypocotyl tissue in a number of dicot species (Heim *et al.*, 1989, 1990; Peng *et al.*, 2001; Vaughn and Turley, 2001). Subsequently, Arabidopsis forward genetic screens were designed to isolate mutant plants that recapitulated these phenotypes, leading to the identification of a number of important genes involved in cellulose biosynthesis.

A screen for temperature-sensitive root swelling (*rsw*) Arabidopsis mutants led to the first genetic evidence that CesAs were genetically required for cellulose biosynthesis (Arioli *et al.*, 1998). The *rsw* screen generated nine complementation groups; the first of these genes, $cesA1^{rsw1\text{-}1}$, mapped to an A549V mutation in Arabidopsis CesA1. Subsequent biochemical analysis revealed that the $cesA1^{rsw1\text{-}1}$ mutant displayed severely reduced cellulose content at the restrictive temperature, suggesting that this mutant was impaired in cellulose biosynthesis. Furthermore, cellulose-synthesizing rosette complexes lost their characteristic six-fold symmetry in the *cesA1*$^{rsw1\text{-}1}$ mutants at the restrictive temperature, linking this mutation to rosette structure (Arioli *et al.*, 1998). Additional forward genetic screens in Arabidopsis supported a role for CesA3 and CesA6 in primary cell wall cellulose biosynthesis. Mutations in CesA6 resulted in Arabidopsis seedlings that displayed severely reduced hypocotyl elongation and cellulose deficiency when grown in the dark (Desnos *et al.*, 1996; Fagard *et al.*, 2000). Interestingly, cesA6 null mutants are phenotypically similar to wild-type plants when grown in the light, suggesting that CesA6 plays a specific role in the elongating dark-grown hypocotyls (Fagard *et al.*, 2000). Further genetic analysis revealed that CesA2, 5, and 9 are partially functionally redundant with CesA6 in Arabidopsis (Desprez *et al.*, 2007; Persson *et al.*, 2007). However, CesA2 or CesA5 driven by the CesA6 promoter were not fully capable of rescuing cesA6 null mutant phenotypes, suggesting that these CesA family members are not functionally equivalent. Partial loss-of-function mutations in CesA3 display similar phenotypic defects to the *cesA1*$^{rsw1\text{-}1}$ mutant (Cano-Delgado *et al.*, 2000, 2003; Ellis and Turner, 2001; Daras *et al.*, 2009; Pysh *et al.*, 2012), including decreased cell elongation, decreased cellulose content, and extreme dwarfism. Null mutations of both CesA1 and CesA3 are pollen lethal and lead to defective pollen wall development (Persson *et al.*, 2007), suggesting that these CesA isoforms are absolutely required for primary cell wall cellulose biosynthesis.

Several lines of evidence suggest that CesAs implicated in primary cell wall biosynthesis physically associate with one another to form a complex that is dependent upon the unique activities of each protein (Fig. 3.2). The genetic requirement for three different types of CesA subunits in primary wall biosynthesis indicates that CesA1, CesA3, and at least one CesA6-like gene are necessary for primary cell wall biosynthesis (Desprez *et al.*, 2007; Persson *et al.*, 2007). In addition, CesA1, 3, and 6 have been demonstrated to physically interact by co-immunoprecipitation, bi-molecular fluorescence complementation, and split ubiquitin yeast two-hybrid assays (Desprez *et al.*, 2007; Carroll *et al.*, 2012). These observations suggest that rosettes contain multiple CesA isoforms that physically associate and have non-redundant functions.

CesAs Involved in Secondary Cell Wall Biosynthesis

Genetic analysis of Arabidopsis (Turner and Somerville, 1997; Taylor *et al.*, 1999), rice (Tanaka *et al.*, 2003; Yan *et al.*, 2007; Kotake *et al.*, 2011), and *Brachypodium distachyon* (Handakumbura *et al.*, 2013) mutants also implicate CesA proteins in the biosynthesis of secondary cell wall cellulose. Secondary cell walls contain much more cellulose than primary cell walls and are further structurally supported by xylan hemicelluloses as well as the polyphenolic polymer lignin (Turner *et al.*, 2007; Burton *et al.*, 2010). Defects in secondary cell wall biosynthesis are typically correlated with irregular xylem cell shape, decreased mechanical strength of the stem, and loss of upright growth (Turner and Somerville, 1997). Using these phenotypes as a proxy, a number of mutants specifically affecting secondary cell wall deposition have been isolated.

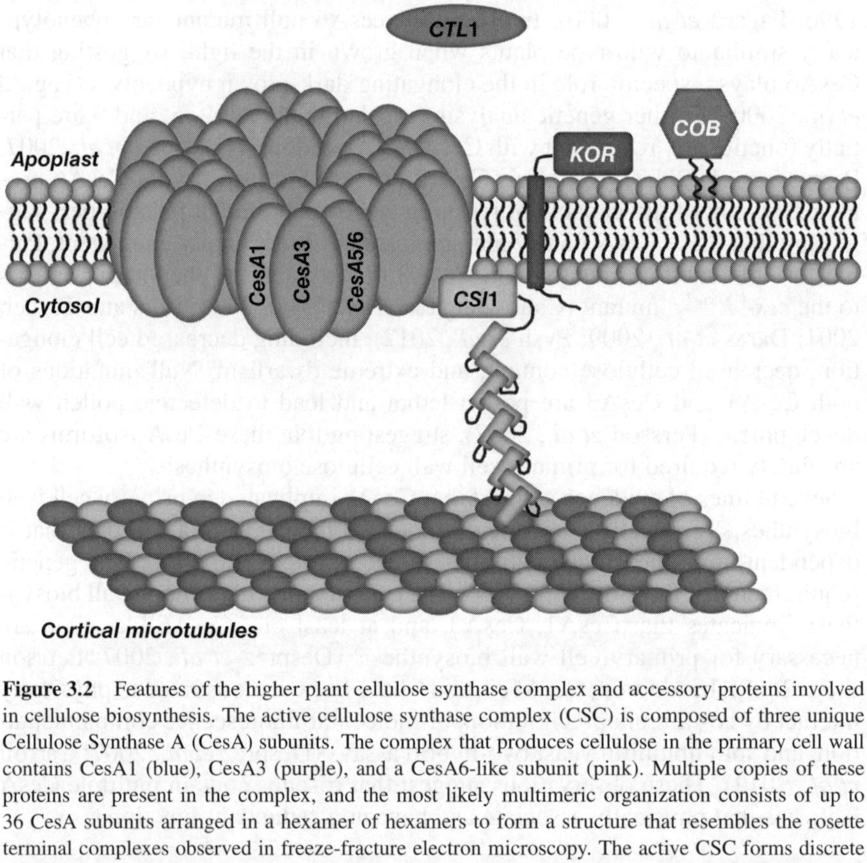

Figure 3.2 Features of the higher plant cellulose synthase complex and accessory proteins involved in cellulose biosynthesis. The active cellulose synthase complex (CSC) is composed of three unique Cellulose Synthase A (CesA) subunits. The complex that produces cellulose in the primary cell wall contains CesA1 (blue), CesA3 (purple), and a CesA6-like subunit (pink). Multiple copies of these proteins are present in the complex, and the most likely multimeric organization consists of up to 36 CesA subunits arranged in a hexamer of hexamers to form a structure that resembles the rosette terminal complexes observed in freeze-fracture electron microscopy. The active CSC forms discrete complexes in the plasma membrane, and these complexes are tethered to cortical microtubules through the influence of Cellulose Synthase Interacting protein 1 (CSI1). The complex moves along cortical microtubules as it is driven by synthesis of new crystalline cellulose in the apoplast. Other proteins have been implicated as positive regulators of cellulose biosynthesis. The KORRIGAN endoglucanase (red) localizes to the plasma membrane and contains a single membrane spanning region as well as an extracellularly localized endoglucanase domain. COBRA (green) is implicated in cellulose biosynthesis, is localized to the plasma membrane by a GPI anchor, and potentially facilitates the organization of nascent microfibrils. CTL1 (orange) is a secreted glucan binding protein that may also play a role in organizing glucan chains once they have been extruded into the apoplast. (See plate section for the color version of this figure.)

In Arabidopsis, three cellulose synthase catalytic subunits are required for secondary cell wall cellulose biosynthesis. Individual mutations in CesA4, CesA7, and CesA8 are all phenotypically similar and display short inflorescences with decreased mechanical strength as well as highly irregular xylem cell shape (Turner and Somerville, 1997; Taylor *et al.*, 1999). Similar phenotypic observations of the rice Brittle Culm (BC) mutant series revealed that three distinct CesA subunits are also required for secondary cell wall cellulose biosynthesis in this species (Tanaka *et al.*, 2003). Recently, artificial microRNA knockdown of two Brachypodium CesAs which are transcriptionally over-represented in vascular tissues also demonstrated that down-regulation of these genes causes similar phenotypes (Handakumbura *et al.*, 2013), suggesting that multiple functionally distinct CesA subunits are required for secondary cell wall cellulose biosynthesis in a variety of species.

CesAs involved in secondary cell wall biosynthesis have also been demonstrated to physically interact. Co-immunoprecipitation experiments using epitope-tagged CesA4 in transgenic Arabidopsis plants suggest that CesA4, 7, and 8 interact *in vivo* (Taylor *et al.*, 2003), and split-ubiquitin yeast two-hybrid experiments further support this conclusion (Timmers *et al.*, 2009; Carroll *et al.*, 2012). A subsequent analysis of immunoprecipitated CesA4 complexes by blue native PAGE (BN-PAGE) revealed that CesA4, 7, and 8 form an approximately 700 kDa complex, suggesting that multiple copies of these proteins are present (Atanassov *et al.*, 2009). While the predicted molecular weight of a CesA complex that could make 20–30 glucan chains is much larger than the complexes observed by BN-PAGE, these smaller complexes approximate the molecular weight of a CesA catalytic subunit hexamer and could represent a subunit of the rosette complex. Indeed, analysis of a computational model of cotton CelA1 in various symmetrical crystallographic orientations revealed that a hexameric orientation of this model approximates the size of a rosette lobe (Sethaphong *et al.*, 2013). It is also interesting to note that these complexes shifted to lower molecular weights when at least one CesA subunit was mutated (Atanassov *et al.*, 2009), suggesting that all three subunits are required to form stable high molecular weight complexes. Overall, these results suggest that CesA oligomerization is a conserved feature of the primary and secondary CesA complex, and that each CesA type fulfills a unique structural or functional role (Fig. 3.2).

Regulation of the Cellulose Synthase Complex (CSC) by Cytoskeletal Components and Vesicle Trafficking

Experiments as early as the 1960s suggested a relationship between cellulose biosynthesis and the microtubule cytoskeleton. Treatment of plant seedlings with the microtubule destabilizing drug colchicine led to disruption of cellulose biosynthesis and anisotropic cell expansion (Green, 1962), similar to the effects observed after treatment of plants with cellulose biosynthesis inhibitors or mutants impaired in cellulose biosynthesis. Further analysis demonstrated that cellulose microfibrils and underlying cortical microtubules were aligned, suggesting that microtubules might guide the deposition of newly synthesized cellulose (Ledbetter and Porter, 1963). These observations led to the proposal of the microtubule alignment hypothesis, which suggests that the cellulose synthase complex is either passively or directly guided by underlying cortical microtubules to facilitate the ordered deposition of cellulose (reviewed by Baskin and Gu, 2012; Bringmann *et al.*, 2012a). While this hypothesis remained controversial due to the observation that cellulose microfibrils do not align with cortical microtubules in some cell types (Sugimoto *et al.*, 2003), simultaneous live-cell imaging of fluorescently labeled CesA subunits and microtubules convincingly demonstrated that the CSC functionally associates with microtubules (Paredez *et al.*, 2006).

When examined in rapidly elongating epidermal cells of dark-grown Arabidopsis hypocotyls, fluorescently labeled CSCs involved in primary cell wall biosynthesis exhibit a complex localization pattern and are present in Golgi bodies, small vesicular compartments in the cell cortex, as well as small motile puncta at the plasma membrane (Paredez *et al.*, 2006; Desprez *et al.*, 2007; Persson *et al.*, 2007; Crowell *et al.*, 2009; Gutierrez *et al.*, 2009). The plasma membrane-localized puncta move with a constant velocity of approximately 250–300 nm/min along linear trajectories that coincide with underlying cortical microtubules (Paredez *et al.*, 2006). Treatment of Arabidopsis seedlings expressing fluorescently labeled CSCs with inhibitors of cellulose biosynthesis revealed that these inhibitors either influence the movement of CSCs at the plasma membrane (DeBolt *et al.*, 2007) or redistribute the complex into small vesicles in the cell cortex (Paredez *et al.*, 2006; Bischoff *et al.*, 2009; Gutierrez *et al.*, 2009; Harris *et al.*, 2012), suggesting that the active CSC is represented by motile particles at the plasma membrane and that the CSC velocity correlates with the rate of active cellulose biosynthesis.

The alignment of CSCs with microtubules potentially suggests that either the CesAs themselves or another scaffolding protein might mediate this association, and treatment of Arabidopsis seedlings with the microtubule depolymerizing herbicide oryzalin further supported this hypothesis. Oryzalin-induced microtubule depolymerization reduced CSC velocity of plasma-membrane-localized puncta, but did not alter their density at the plasma membrane (Paredez *et al.*, 2006). Instead, the complexes moved randomly around the cell rather than having predefined trajectories that are normally coincident with microtubules. These results suggested that the complex is capable of moving in the absence of microtubules and that microtubules are necessary for the organized deposition of cellulose. This observation also supports the hypothesis that the mobility of the CSCs is caused by polymer elongation (Paredez *et al.*, 2006; Diotallevi and Mulder, 2007), rather than the activity of a motor protein that actively pulls the complex along microtubules.

A very large protein called Cellulose Synthase Interacting protein 1 (CSI1) was identified in a yeast two-hybrid screen using the large central catalytic loop of CesA6 as a bait (Gu *et al.*, 2010), suggesting that CSI1 directly interacts with CesA proteins (Fig. 3.2). CSI1 was also found to be transcriptionally co-expressed with cellulose synthase complex components (Persson *et al.*, 2005; Gu *et al.*, 2010), and *csi1* loss-of-function Arabidopsis mutants were cellulose deficient and displayed defects in cell elongation (Gu *et al.*, 2010; Bringmann *et al.*, 2012b; Li *et al.*, 2012). CSI1 was subsequently demonstrated to link CSCs and cortical microtubules (Bringmann *et al.*, 2012b; Li *et al.*, 2012). Fluorescently-labeled CSI1 co-localizes with the CSC in the plasma membrane and moves with the complex at a similar velocity on the same microtubule-defined trajectories. Furthermore, imaging of CSCs in the *csi1* null background indicated that the complexes move with a reduced velocity in erratic trajectories, similar to the situation observed after oryzalin-induced microtubule depolymerization (Gu *et al.*, 2010; Li *et al.*, 2012).

CSI1 is a 2150 amino acid protein that contains multiple Armadillo (ARM) repeats as well as a C-terminal C2 domain (Gu *et al.*, 2010; Fig. 3.2). These domains are commonly involved in protein–protein interactions and calcium-dependent lipid binding, respectively (Duncan *et al.*, 2000; Tewari *et al.*, 2010), suggesting that CSI1 could serve as a scaffolding protein to link CesA subunits to other cytosolic components. Purified full-length CSI1 was demonstrated to bind microtubules *in vitro* with a dissociation constant of approximately 1 μM (Li *et al.*, 2012), which is similar to the binding constant for many other microtubule binding proteins. While the exact domains that contribute to microtubule and CesA binding are currently unknown, it was demonstrated that deletion of the C-terminal C2 domain abolished CSI1 localization at the plasma membrane (Bringmann *et al.*, 2012b), suggesting that this domain may be involved in CesA binding while other portions of CSI1 are involved in microtubule binding. It is also noteworthy that CSI1 does not co-localize with fluorescently-labeled CesAs in small compartments in the Golgi, suggesting that the CSI1–CesA interaction only occurs at the plasma membrane. Overall, these observations indicate that CSI1 is capable of interacting with both CesAs as well as microtubules, and it is very likely that this protein serves as the major link that guides the deposition of cellulose along cortical microtubule trajectories.

The microtubule cytoskeleton has also been demonstrated to play an important role in the delivery of new CSCs to the plasma membrane (Fig. 3.3). A fraction of fluorescently-labeled CSCs localize to small compartments that move rapidly through the cell cortex (Paredez *et al.*, 2006; Crowell *et al.*, 2009; Gutierrez *et al.*, 2009). Dual fluorescent imaging of these compartments,

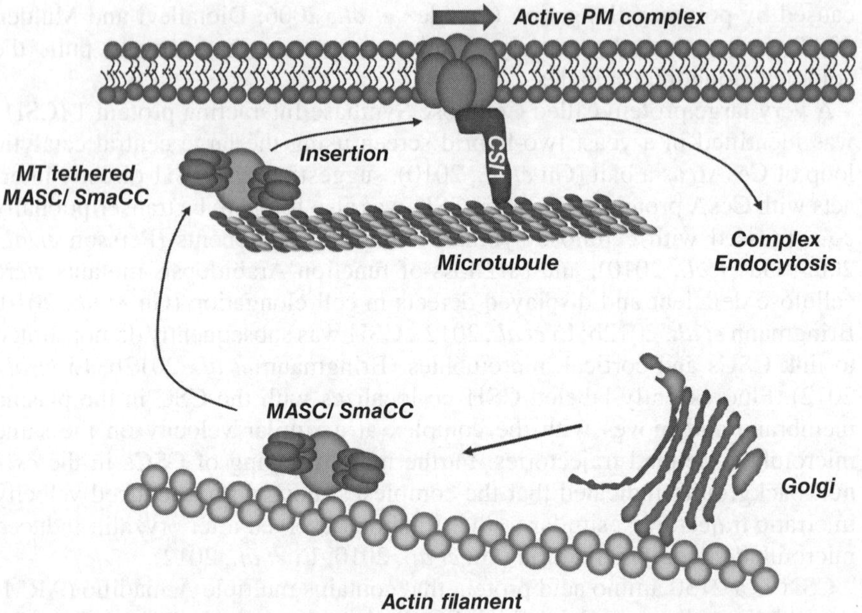

Figure 3.3 Trafficking of the cellulose synthase complex. Fluorescently labeled CesA subunits exhibit a complex subcellular localization pattern in living cells. CesAs are present in the Golgi apparatus, small microtubule associated compartments (MASCs/SmaCCs), and the plasma membrane. Live-cell imaging suggests that CesA-containing Golgi and MASC/SmaCCs move along actin cables throughout the cell to generate a homogenous distribution of CSC insertion into the plasma membrane. Actin-localized MASC/SmaCCs occasionally pause at, or become tethered to, cortical microtubules, and this event is correlated with the insertion of CSCs into the plasma membrane. Once CSCs arrive at the plasma membrane, they interact with CSI1 and move along cortical microtubule-defined tracks as they produce cellulose. Cortical microtubule arrays may reorient to guide the complex along different trajectories during environmental or developmental stimulus. The CSC is removed from the plasma membrane via incompletely defined endocytosis mechanisms, but at least one route of CSC endocytosis involves a clathrin-mediate process (Bashline *et al.*, 2013).

referred to as microtubule-associated small compartments (MASCs) or small microtubule-associated cytoplasmic compartments (SmaCCs), revealed that they exhibit an erratic 'stop–start' behavior in which they move along microtubules and pause for short amounts of time. Further analysis revealed that these pause events are associated with the insertion of new CSCs into the plasma membrane (Gutierrez *et al.*, 2009). Interestingly, the newly inserted CSCs were observed to remain stationary for approximately one minute before initiating their movement with constant velocity, suggesting the potential existence of an activation event at the plasma membrane.

The functional association of CSCs and microtubules further suggests that changing cortical microtubule orientations will influence cellulose deposition patterns. Indeed, cortical microtubule array orientations are highly dynamic and change in the context of normal cellular development as well as in response to environmental perturbations, such as hormones and light. In all cases where both CSCs and cortical microtubules have been observed simultaneously in living cells, CSC trajectories seem to follow altering cortical microtubule arrays. For example, in the epidermal cells of dark-grown Arabidopsis hypocotyls, blue light induces a rapid change in cortical microtubule array orientation from transverse to longitudinal. Exposure of transgenic YFP-CesA6 seedlings to blue light also induced corresponding changes in CSC trajectories over a similar time period (Paredez *et al.*, 2006), suggesting that this environmental perturbation also changes the pattern of cellulose distribution.

Cortical microtubule arrays in root epidermal tissue of light-grown Arabidopsis seedlings undergo a slow constant 360° rotation over a period of approximately 10 hours (Chan *et al.*, 2010), and CSC trajectories also follow these successive reorientations. The resulting pattern of cellulose deposition may explain the observed multi-lamellate cellulose microfibril network orientations which exist in this tissue (Sugimoto *et al.*, 2003; Anderson *et al.*, 2010). In the future, it will be useful to investigate CSC trajectories under other conditions that alter microtubule array orientations to further understand how these linked processes influence cell shape and differentiation.

The actin cytoskeleton also influences the trafficking and distribution of active CSCs at the plasma membrane (Fig. 3.3). Latrunculin B-induced disassembly of the actin cytoskeleton in GFP-CesA3-expressing transgenic Arabidopsis plants revealed that CSCs still arrive at the plasma membrane under these conditions, but that the global distribution of CSCs around the cell was markedly altered (Gutierrez *et al.*, 2009). Patches of plasma-membrane-localized CSCs correlated with underlying clumped Golgi bodies, suggesting that actin-dependent Golgi motility is required to produce an even distribution of CSCs around the cell. This situation seems to be due to the fact that the motility of the MASC/SmaCCs responsible for CSC delivery is actin-dependent (Sampathkumar *et al.*, 2013).

A model summarizing the facets of CSC regulation by cytoskeletal and vesicle-trafficking components is presented in Figure 3.3. CSC-containing Golgi compartments move along actin cables throughout the cell, and MASC/SmaCCs are conceivably derived from this organelle. MASC/SmaCCs also move along actin cables but pause occasionally on cortical microtubules to deliver CSCs to the plasma membrane. At the plasma membrane, the CSC interacts with CSI1 and begins to move along cortical microtubules as cellulose is deposited in the apoplast. Cortical microtubule dynamics can alter the pattern of cellulose deposition during reorientation under a variety of developmental or environmental conditions. The pathways to CSC endocytosis are less clear, but clathrin-mediated endocytosis via the

AP-2 complex seems to represent at least one route to CesA internalization (Bashline *et al.*, 2013).

Other Genes Involved in Cellulose Biosynthesis

In addition to CesA subunits, genetic analysis has revealed that a variety of other genes are required for normal cellulose biosynthesis in several plant species (Fig. 3.2). The COBRA (COB), KORRIGAN (KOR), KOBITO1 (KOB1), and chitinase-like 1 (CTL1) genes have all been genetically implicated in cellulose biosynthesis, but the precise function of these genes is still unclear. None of the corresponding protein products have been demonstrated to associate with the cellulose synthase complex, suggesting that their phenotypic defects may be indirect or that these proteins serve as 'accessory subunits' of the cellulose synthase complex.

The *kor1* mutant was originally described in an Arabidopsis forward genetic screen for mutants defective in hypocotyl elongation (Nicol *et al.*, 1998). Like mutations in cellulose synthase catalytic subunits, *kor1* mutant hypocotyls exhibit reduced cell elongation, concomitant radial expansion, and cellulose deficiency (Nicol *et al.*, 1998; Lane *et al.*, 2001; Sato *et al.*, 2001). The KOR1 gene encodes a membrane localized endo-(1→4)-glucanase (Fig. 3.2) that is part of a small gene family in Arabidopsis. Functional analysis indicates that an aspen homolog of KOR1 is an active cellulase (Rudsander *et al.*, 2008), and localization studies demonstrate that KOR1 localizes to the cell plate, the plasma membrane, and intracellular vesicles in Arabidopsis (Nicol *et al.*, 1998; Zuo *et al.*, 2000; Robert *et al.*, 2005). Additional analysis of *kor1* mutant cell walls revealed reduced cellulose crystallinity and disorganization of the cellulose deposition pattern (Sato *et al.*, 2001). Live-cell imaging of fluorescently tagged CesA subunits in *kor1* mutant backgrounds also indicate that CesA complex velocity is reduced in these mutants (Paredez *et al.*, 2008), suggesting that KOR1 is a positive regulator of cellulose biosynthesis.

The precise role of KOR in the cellulose synthase reaction remains unknown, but it has been suggested to regulate cellulose chain length by cleaving microfibrils during synthesis or to cleave putative lipid oligosaccharide primers of the cellulose synthase reaction (Peng *et al.*, 2002). However, neither of these hypotheses have been experimentally supported. In this context, it is noteworthy that bacterial cellulose synthases are also associated with an endoglucanase homologue (BcsZ); the structure of this protein is available (Mazur and Zimmer, 2011). Further biochemical analysis of KOR based on this structure may provide a more concise description of the functional role of this protein in cellulose biosynthesis.

The *cob* mutant was isolated in an Arabidopsis forward genetic screen for mutants with altered root morphology (Benfey *et al.*, 1993) and was found to encode a glycosylphosphatidylinositol (GPI) anchored protein of unknown function (Schindelman *et al.*, 2001; Fig. 3.2). The *cob* mutant alleles display similar phenotypic characteristics to other cellulose biosynthesis mutants, including cellulose deficiency, reduced cell elongation, and radial cell swelling; in contrast to other mutants, these defects are however dependent upon the sucrose concentration of the growth media. Phenotypes of the *cob* mutants are only evident on growth media containing high sucrose concentrations which stimulate high growth rates of rapidly elongating tissues (Benfey *et al.*, 1993). Immunolocalization analysis indicates that the COB protein localizes to plasma membranes as well as the cell wall, and this protein can be released from membranes by phospholipase C (Borner *et al.*, 2003; Roudier *et al.*, 2005); this suggests that the COB GPI anchor can be selectively cleaved to release the protein from the plasma membrane. While the function of COBRA is not yet fully understood, an analysis of cellulose deposition in *cob* mutants by Field Emission Electron Microscopy (FESEM) of the innermost layer of the cell wall revealed that cellulose is deposited in highly irregular orientations in this mutant, leading to the hypothesis that COBRA regulates the ordered deposition and perhaps the crystallinity of newly synthesized cellulose (Roudier *et al.*, 2005). This hypothesis is substantiated by the recent evidence that Brittle Culm 1 (BC1; Li *et al.*, 2003), a rice COBRA homolog, binds cellulose microfibrils through a putative cellulose binding domain and influences the crystallinity of the microfibrils (Liu *et al.*, 2013).

COBRA-like genes are conserved in plants and COBRA is part of a multi-gene family that includes 14 members in Arabidopsis (Roudier *et al.*, 2002; Brady *et al.*, 2007). COBRA gene family members display overlapping but distinct transcriptional profiles, and genetic analysis has revealed that other COBRA family members play distinct developmental roles in a variety of tissues. For example, COBRA-like family members have been implicated in directional tip growth processes, such as pollen tube (Li *et al.*, 2013) and root hair elongation (Hochholdinger *et al.*, 2008). Additionally, COBRA-like family members have been implicated in the synthesis of cellulose in secondary cell walls (Li *et al.*, 2003). These observations suggest that COBRA and COBRA-like proteins play a conserved role in the process of cell expansion. However, it is still unclear why plants encode many COBRA-like proteins that seem to be functionally distinct. It is possible that these proteins could serve to organize cellulose or other cell wall polymers in tissue-specific deposition patterns, but the true role of these proteins will only be revealed by further experiments.

Chitinase-like 1 (CTL1) mutants were also isolated as conditional root development mutants (Benfey *et al.*, 1993; Zhong *et al.*, 2002). CTL1 mutants exhibit radial swelling, reduced cell elongation, reduced cellulose content, and ectopic lignification as observed in other cellulose deficient mutants. The CTL1 gene encodes a chitinase-like protein, but it is unlikely that this protein encodes a functional chitinase since plant cell walls do not contain chitin and because CTL1 lacks key residues that are associated with chitinase activity, suggesting that CTL1 plays a distinct role in cellulose biosynthesis (Zhang *et al*., 2004). Recent work indicates that CTL1 localizes to the endomembrane system and the apoplast, suggesting that CTL1 is a secreted protein that operates on newly synthesized cellulose (Fig. 3.2). It was further demonstrated that CTL1, and its functionally redundant partner CTL2, are capable of binding glucan chains *in vitro*, suggesting that these proteins may play a role in organizing newly synthesized cellulose or could prevent other wall polymers from interdigitating into the nascent microfibril.

Mutations in the Arabidopsis glycosyltransferase-like protein KOBITO have been isolated in several forward genetic screens for genes impacting abscisic acid sensitivity (Brocard-Gifford *et al*., 2004), cell elongation (Cheng *et al*., 2000; Pagant *et al*., 2002; Lertpiriyapong and Sung, 2003), and stomatal differentiation (Kong *et al*., 2012). However, KOBITO loss-of-function mutants consistently display reduced cell elongation, cellulose deficiency, and other phenotypes that are typical of mutants impaired in cellulose biosynthesis. The KOBITO gene encodes a protein with low amino acid similarity to bacterial lipid glycosyltransferases (Kong *et al*., 2012), but it is unclear that this protein is directly involved in cellulose biosynthesis. KOBITO contains a predicted transmembrane domain, but the localization of this protein is still unknown. It will be useful to precisely characterize the localization and biochemical function of KOBITO in the future to determine the role of this protein in cellulose biosynthesis.

The variety of non-CesA proteins implicated in cellulose biosynthesis potentially suggests that these proteins could directly associate with the CSC and influence the biochemical activity of the complex. Further analysis is necessary to more precisely define whether these proteins and others are actually physically associated with the CSC. Transcriptional co-expression analysis (Brown *et al*., 2005; Persson *et al*., 2005), live-cell imaging (Paredez *et al*., 2006, 2008; Sanchez-Rodriguez *et al*., 2012), and proteomic analysis of immunoprecipitated cellulose synthases (Song *et al*., 2010) will likely provide more detailed information regarding whether these proteins and others directly interact with CesAs and how they influence the process of cellulose biosynthesis.

Regulation of the CSC by Phosphorylation

Protein phosphorylation is an extremely common regulatory mechanism in eukaryotes and is one of the most prevalent post-translational modifications in nature. Phosphorylation events have been demonstrated to regulate the activity, localization, stability, and interaction networks of modified proteins.

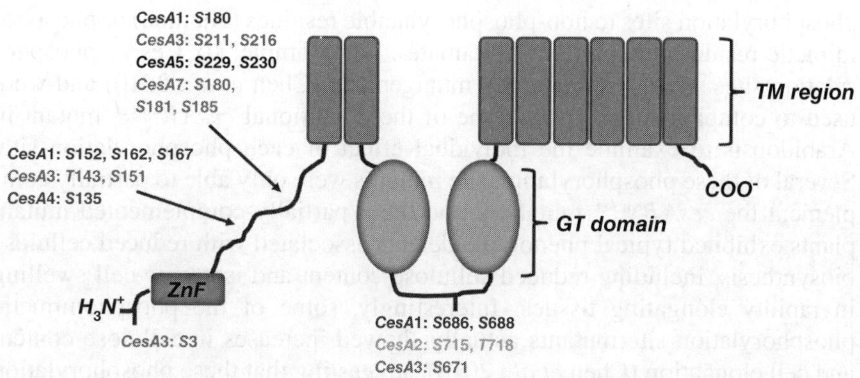

Figure 3.4 Summary of proteomically supported phosphorylation events on CesA subunits. CesAs exhibit a conserved domain architecture consisting of an N-terminal zinc finger (pink), eight transmembrane domains (blue), and a large central loop containing sequence motifs that are typical of processive glycosyltransferases (yellow). Phosphoproteomic surveys indicate that Arabidopsis CesA1, 2, 3, 4, 5, and 7 are phosphorylated at multiple positions within the cytosolic N-terminus and the large central catalytic loop. The relative position of each phosphorylation site is shown on a generalized model of CesA structure, with the residue indices and CesA isoform indicated. The color code for phosphorylation sites is: CesA1 (red), CesA2 (green), CesA3 (blue), CesA4 (purple), CesA5 (black), and CesA7 (orange). (See plate section for the color version of this figure.)

Large-scale phosphoproteomic surveys of Arabidopsis and rice phosphopeptides indicate that CesAs as well as other proteins implicated in cellulose biosynthesis are phosphorylated (Nuhse *et al.*, 2004, 2007; Nakagami *et al.*, 2010), and that many of these phosphorylation events are conserved in a variety of plant species (Carroll and Specht, 2011). In Arabidopsis, CesA1, 2, 3, 4, 5, and 7 are phosphorylated at multiple positions within hypervariable regions in their N-terminal domains and central catalytic loops (Nuhse *et al.*, 2004, 2007; Taylor, 2007; Nakagami *et al.*, 2010; Fig. 3.4). Additionally, the KOR endoglucanase is also phosphorylated at multiple positions within the cytosolic N-terminal domain.

Protein kinases generally recognize short characteristic peptide sequences surrounding a phosphorylated serine or threonine residue, and it is interesting to note that sequences surrounding CesA and KOR phosphorylation sites are relatively unique. The sequences surrounding each phosphorylation site are not conserved among phosphorylation sites but are conserved among similar CesA isoforms (Carroll and Specht, 2011), suggesting several protein kinases may be capable of phosphorylating components of the CSC. These observations suggest that the CSC may be regulated by multiple environmental and developmental processes through the action of specific protein kinases.

Several studies have attempted to address the role of protein phosphorylation in CSC regulation, primarily by mutating proteomically supported phosphorylation sites to non-phosphorylatable residues (Alanine) or phosphomimetic residues (Aspartate, Glutamate). For example, six CesA1 phosphorylation sites were systematically mutagenized (Chen *et al.*, 2010) and were used to complement the phenotype of the conditional *cesA1*$^{rsw1\text{-}1}$ mutant in Arabidopsis to examine the individual effect of each phosphorylation site. Several of these phosphorylation site mutants were only able to partially complement the *cesA1*$^{rsw1\text{-}1}$ mutation, and these partially complemented mutant plants exhibited typical phenotypic defects associated with reduced cellulose biosynthesis, including reduced cellulose content and isotropic cell swelling in rapidly elongating tissues. Interestingly, some of the phosphomimetic phosphorylation site mutants actually showed increases in cellulose content and cell elongation (Chen *et al.*, 2010), suggesting that these phosphorylation sites might serve as positive regulators of CSC catalytic activity.

In this study, CSC behavior was also analyzed in each phosphorylation site mutant using fluorescently tagged CesA subunits and confocal microscopy. Similar previous analyses indicated that CSCs moved along cortical microtubules with a similar velocity in either direction (Paredez *et al.*, 2006; Desprez *et al.*, 2007; Persson *et al.*, 2007; Crowell *et al.*, 2009; Gutierrez *et al.*, 2009), and this behavior was recapitulated in the *cesA1*$^{rsw1\text{-}1}$ plants complemented with a wild-type CesA1 construct. Some of these CesA1 phosphorylation site mutations resulted in complexes displaying differential velocities depending on their direction of microtubule migration, however. These results suggest that phosphorylation of CesA1 may differentially affect the interactions of the CSC with microtubules through an undefined mechanism, and this differential interaction may lead to preferentially reduced velocity in one direction versus another. Indeed, depolymerization of microtubules with the microtubule-destabilizing drug oryzalin was able to rescue the bi-directional velocity behavior of complexes in these mutants (Chen *et al.*, 2010), suggesting that the differential bi-directional velocity is microtubule-dependent. Due to the newly established role of CSI1 as a CesA guide protein (Gu *et al.*, 2010; Bringmann *et al.*, 2012b; Li *et al.*, 2012), it will be interesting to investigate whether these phosphorylation sites influence the interaction between CesAs and CSI1.

Phosphorylation events have also been suggested to regulate complex velocity (Bischoff *et al.*, 2011). Previous studies indicated CesA6 dark-grown hypocotyls exhibit severe dwarfism and contain much less cellulose than wild-type seedlings grown under the same conditions, but these effects are virtually absent in the same mutants grown in the light (Desnos *et al.*, 1996; Fagard *et al.*, 2000). Additionally, it was demonstrated that many of the phenotypic defects in *cesA6*$^{prc1\text{-}1}$ dark-grown hypocotyls could be rescued by exposure of the seedlings to red light (but not far-red light; Desnos *et al.*, 1996), suggesting a potential role for the red light photoreceptor phytochrome in cellulose biosynthesis. It was also demonstrated that CesA5, which is very similar in amino acid sequence to CesA6, could not completely rescue *cesA6*$^{prc1\text{-}1}$ mutant phenotypes in the dark mutants even when CesA5 was driven by the CesA6 promoter. This suggests that, despite their high sequence identity, CesA5 and CesA6 are not functionally equivalent.

Live-cell imaging of fluorescently labeled CesA5 in dark-grown seedlings revealed that CesA5 velocity is reduced in the *cesA6*$^{prc1\text{-}1}$ mutant background, but CesA5 velocity increased after red-light pretreatment (Bischoff *et al.*, 2011). This effect was not observed with a far-red light pretreatment, or with red-light treatment in the Phytochrome B (phyB) mutant, suggesting that red-light stimulation of phytochrome is responsible for the increase in CesA5 velocity. Furthermore, it was demonstrated that mutagenesis of four potential phosphorylation sites in the CesA5 N-terminus could recapitulate this increase in velocity in the absence of red-light pretreatment, linking the phosphorylation status of CesA5 to PhyB signaling and increased cellulose biosynthesis (Bischoff *et al.*, 2011). It is unclear whether PhyB or a protein kinase downstream of PhyB is responsible for directly phosphorylating CesA5 after red-light treatment. However, due to the kinetics of the CSC velocity increase and the fact that activated phytochrome quickly re-localizes to the nucleus after red-light stimulation (Sakamoto and Nagatani, 1996; Yamaguchi *et al.*, 1999; Huq *et al.*, 2003), we hypothesize that the involvement of PhyB is important, but indirect. These observations suggest that CesA activity can be differentially regulated by light quality and further demonstrate that CesA catalytic subunits exhibit functional specialization that can be modified by protein phosphorylation.

Secondary cell wall CesAs are also subject to phosphorylation (Taylor, 2007). Affinity purification of individual CesAs involved in Arabidopsis secondary wall biosynthesis and subsequent proteomic analysis indicated that CesA4 and CesA7 were phosphorylated in their N-terminal class-specific regions, similar to phosphorylation events observed in primary wall CesAs. The functional implications of these phosphorylation sites remain unclear, but further analysis of recombinant N-terminal fragments of CesA7 as well as full-length endogenous CesA7 revealed that phosphorylation of these fragments may affect the stability of the CesA7 protein (Taylor, 2007). Co-incubation with the proteasome inhibitor MG132 revealed that proteasome inhibition could stabilize both the N-terminal fragments as well as endogenous CesA7, potentially suggesting that these N-terminal phosphorylation events might be associated with ubiquitin-dependent protein degradation. This method of regulation could have potentially interesting implications if, for example, CesA7 phosphorylation status is regulated during cellular differentiation, and CesA7 is degraded after secondary cell wall biosynthesis is complete.

None of the protein kinases that directly phosphorylate cellulose synthases have been isolated and characterized, and this will be an important avenue of future research in cellulose biosynthesis. However, genetic analysis has revealed certain protein kinase candidates that mimic phenotypes of cellulose biosynthesis mutants and may be involved in the regulation of cellulose biosynthesis. For example, the FEI1 and FEI2 receptor-like protein kinases are implicated in cellulose biosynthesis (Xu *et al.*, 2008; Harpaz-Saad *et al.*, 2011), and double *fei1;fei2* mutants are defective in cell elongation as well as cellulose biosynthesis. While the precise role of these receptor-like kinases is unclear, it will be interesting to evaluate whether their kinase domains are capable of directly phosphorylating any proteomically supported phosphorylation sites in the CesA proteins. Overall, these observations suggest that post-translational modifications are likely to play an important role in CSC regulation, and the investigation of the functional implications of these regulatory events will be an important future goal in cellulose biosynthesis research.

Wall Integrity Sensing

In other walled organisms such as fungi, the integrity of the cell wall is monitored by a sophisticated signal transduction system which senses whether the wall is properly formed and inhibits growth if it is not (reviewed by Jendretzki *et al.*, 2011). Due to the importance of wall biosynthesis in plant development, a similar situation may exist in plants. When cellulose biosynthesis is inhibited, plants undergo a developmental suite of responses that include inhibition of cell elongation, radial cell expansion, ectopic lignification of tissues, and ectopic deposition of callose. Characterization of the response to an array of cellulose biosynthesis inhibitors, such as isoxaben, Dichlobenil (DCB), and Thaxtomin A, indicate that many genes are transcriptionally regulated in response to these chemicals, and that the response to these inhibitors resembles the wounding response in plants. A detailed characterization of the isoxaben response revealed that calcium influx and ROS production are linked to this process. Genetic analysis revealed that isoxaben-induced ROS production was specifically dependent on the Ca^{2+}-dependent NADPH oxidase AtRbohD, and that the upstream Ca^{2+} influx could be blocked by the calcium channel inhibitor $LaCl_3$ (Denness *et al.*, 2011). Furthermore, isoxaben treatment resulted in the production of the wounding response hormone jasmonic acid (JA). Accumulation of JA and downstream ectopic lignifications responses were dependent on OXI1 protein kinase, but were not dependent upon the JAR1 JA repector (Denness *et al.*, 2011).

Genetic analysis has also been instrumental in identifying genes involved in the response to cell wall damage; the THESEUS1 (THE1) receptor-like kinase is a good example of this approach. THE1 was isolated in an Arabidopsis forward genetic screen for suppressors of the *cesA6*$^{prc1\text{-}1}$ null mutant, using the rationale that these suppressors may be involved in the response to cellulose biosynthesis inhibition and the maintenance of cell wall integrity. Double *the1;cesA6*$^{prc1-}$ mutants partially rescue *cesA6*$^{prc1-1}$ developmental defects, including decreased hypocotyl elongation and ectopic lignification. However, the *the1* mutation does not rescue the cellulose biosynthesis defect imparted by *cesA6*$^{prc1-1}$ (Hematy *et al.*, 2007). These observations suggest that THE1 somehow perceives cell wall damage or cellulose biosynthesis inhibition and actively inhibits cell elongation when this defect is perceived. In the future, it will be important to fully characterize wall integrity sensing mechanisms in plants. Genetic suppressor screens of cell wall mutants represent one viable avenue toward isolating new cell wall sensing components, but further characterization of THE1 as well as other known factors that play a role in the response to wall biosynthesis inhibition will also be extremely important.

Conclusions and Future Perspectives

Factors Which Regulate the Amount of Cellulose Deposited in a Cell

It would be useful to understand how to genetically manipulate the amount and properties of cellulose produced in biomass feedstock crops. Based on our current understanding of the cellulose biosynthesis machinery, it seems that the two factors with the largest influence on total cellulose content are the number of CSCs at the plasma membrane, and the velocity of complex movement. The density of CSCs at the plasma membrane is likely regulated by the equilibrium of CSC membrane insertion from MASCs/ SmaCCs and removal of the complex from the plasma membrane. It will be important to completely characterize the factors regulating this process to potentially increase the average number of active CSCs at the plasma membrane, and the recent elucidation of a clathrin-dependent endocytosis pathway regulating CSC density is an exciting step in this direction (Bashline *et al.*, 2013).

It is also crucial to understand which factors influence complex velocity, since this parameter is often correlated with total cellulose content. It is clear that CSC accessory proteins, such as KORRIGAN and CTL1, can influence the rate of complex movement (Paredez *et al.*, 2008; Sanchez-Rodriguez *et al.*, 2012), suggesting that further analysis of CSC components or non-CesA proteins involved in cellulose biosynthesis will be a fruitful future endeavor. It will also be important to understand how the plethora of post-translational phosphorylation events on CesA proteins regulates the activity, localization, and interactions of these proteins.

Factors Which Influence the Properties of Cellulose

One aspect of cellulose quality may be the average length of the glucan chains that comprise the microfibrils or the length of assembled microfibrils. Based on current models, the lifetime of the CSC at the plasma membrane will most likely influence the length of the cellulose microfibrils. It will be interesting to manipulate this property to determine whether transgenic plants can be generated that produce cellulose of a defined length distribution for tailored engineering properties. Based on our current understanding of cellulose synthase, this property could also be manipulated by altering the velocity and plasma membrane distribution of the CSC at the plasma membrane (Sampathkumar *et al.*, 2013).

It will also be interesting to determine the mechanism for cellulose crystallization in the extracellular space, since relative crystallinity is correlated with cellulose digestibility and this property could conceivably be manipulated at the CesA level. Indeed, point mutations in the C-terminal region of CesA1 and CesA3 reduced the crystallinity of cellulose and enhanced cell wall digestibility, but the structural basis for these changes are unknown (Harris *et al.*, 2012). Based on the structure of cellulose and proposed models of the CSC, we suggest that the complex could be highly allosteric and that the addition of glucose residues to one new glucan chain is highly dependent on other catalytic subunits performing the reaction at the same time to generate the observed structure of cellulose. If this hypothesis is true, there could potentially be mechanisms to partially uncouple this allosterism and to produce cellulose with reduced crystallinity.

In summary, the field of cellulose biosynthesis has developed a fundamental platform for the understanding of this basic and important cellular process, but many questions remain unanswered. Addressing these questions will further our understanding of the role that the cell wall plays in plant development and has the potential to increase our capacity to manipulate this important renewable natural resource.

Acknowledgements

We would like to thank Heidi Szemenyei, Trevor Yeats, Nadav Sorek, and William Underwood for useful conversations during the preparation of this manuscript. This work was supported by the Energy Biosciences Institute and the Philomathia foundation.

References

Anderson, C. T., Carroll, A., Akhmetova, L., and Somerville, C. (2010) Real-time imaging of cellulose reorientation during cell wall expansion in Arabidopsis roots. *Plant Physiology*, **152**, 787–796.

Arioli, T., Peng, L., Betzner, A. S., Burn, J., Wittke, W., Herth, W., Camilleri, C., Hofte, H., Plazinski, J., Birch, R., Cork, A., Glover, J., Redmond, J., and Williamson, R. E. (1998) Molecular analysis of cellulose biosynthesis in Arabidopsis. *Science*, **279**, 717–720.

Atanassov, I. I., Pittman, J. K., and Turner, S. R. (2009) Elucidating the mechanisms of assembly and subunit interaction of the cellulose synthase complex of Arabidopsis secondary cell walls. *Journal of Biological Chemistry*, **284**, 3833–3841.

Bashline, L., Li, S., Anderson, C. T., Lei, L., and Gu, Y. (2013) The endocytosis of cellulose synthase in Arabidopsis is dependent on μ2, a clathrin-mediated endocytosis adaptin. *Plant Physiology*, **163**, 150–160.

Baskin, T. I. and Gu, Y. (2012) Making parallel lines meet: transferring information from microtubules to extracellular matrix. *Cell Adhesion and Migration*, **6**, 404–408.

Beer, C., Reichstein, M., Tomelleri, E., Ciais, P., Jung, M., Cavalhais, N., Rodenbeck, C., Arain, M. A., Baldocchi, D., Bonan, G. B., Bondeau, A., Cescatti, A., Lasslop, G., Lindroth, A., Lomas, M., Luyssaert, S., Margolis, H., Oleson, K. W., Roupsard, O., Veenendaal, E., Viovy, N., Williams, C., Woodward, F. I., and Papale, D. (2010) Terrestrial gross carbon dioxide uptake: global distribution and covariation with climate. *Science*, **329**, 843–848.

Benfey, P. N., Linstead, P. J., Roberts, K., Schiefelbein, J. W., Hauser, M. T., and Aeschbacher, R. A. (1993) Root development in Arabidopsis: four mutants with dramatically altered root morphogenesis. *Development*, **119**, 57–70.

Ben-Sham, A., Garreau de Loubresse, N., Melnikov, S., Jenner, L., Yusupova, G., and Yusupov, M. (2011) The structure of the eukaryotic ribosome at 3.0 Å resolution. *Science*, **334**, 1524–1529.

Bessueille, L., Sindt, N., Guichardant, M., Djerbi, S., Teeri, T. T., and Bulone, V. (2009) Plasma membrane microdomains from hybrid aspen cells are involved in cell wall polysaccharide biosynthesis. *Biochemistry Journal*, **420**, 93–103.

Bischoff V., Cookson, S. J., Wu, S., and Scheible, W. R. (2009) Thaxtomin A affects CesA-complex density, expression of cell wall genes, cell wall composition, and causes ectopic lignifications in Arabidopsis thaliana seedlings. *Journal of Experimental Botany*, **60**, 955–965.

Bischoff, V., Desprez, T., Mouille, G., Vernhettes, S., Gonneau, M., and Hofte, H. (2011) Phytochrome regulation of cellulose synthesis in Arabidopsis. *Current Biology*, **21**, 1822–1827.

Borner, G. H., Lilley, K. S., Stevens, T. J., and Dupree, P. (2003) Identification of glycosylphosphatidylinositol-anchored proteins in Arabidopsis: A proteomic and genomic analysis. *Plant Physiology*, **132**, 568–577.

Bowling, A. J. and Brown, R. M. Jr., (2008) The cytoplasmic domain of the cellulose-synthesizing complex in vascular plants. *Protoplasma*, **233**, 115–127.

Brady, S. M., Song, S., Dhugga, K. S., Rafalski, J. A., and Benfey, P. N. (2007) Combining expression and comparative evolutionary analysis: The COBRA gene family. *Plant Physiology*, **143**, 172–187.

Bringmann, M., Landrein, B., Schudoma, C., Hamant, O., Hauser, M. T., and Persson, S. (2012a) Cracking the elusive alignment hypothesis: the microtubule-cellulose synthase nexus unraveled. *Trends in Plant Science*, **17**, 666–674.

Bringmann, M., Li, E., Sampathkumar, A., Kocabek, T., Hauser, M. T., and Persson, S. (2012b) POM-POM2/ cellulose synthase interacting 1 is essential for the functional association of cellulose synthase and microtubules in Arabidopsis. *Plant Cell*, **24**, 163–177.

Brocard-Gifford, I., Lynch, T. J., Garcia, M. E., Malhotra, B., and Finkelstein, R. R. (2004) The Arabidopsis thaliana Abscisic Acid-Insensitive 8 encodes a novel protein mediating abscisic acid and sugar responses essential for growth. *Plant Cell*, **16**, 406–421.

Brown, D. M., Zeef, L. A., Ellis, J., Goodacre, R., and Turner, S. R. (2005) Identification of novel genes in Arabidopsis involved in secondary cell wall formation using expression profiling and reverse genetics. *Plant Cell*, **17**, 2281–2295.

Burton, R. A., Gidley, M. J., and Fincher, G. B. (2010) Heterogeneity in the chemistry, structure, and function of plant cell walls. *Nature Chemical Biology*, **6**, 724–732.

Cano-Delgado, A. I., Metzlaff, K., and Bevan, M. W. (2000) The eli1 mutation reveals a link between cell expansion and secondary cell wall formation in Arabidopsis thaliana. *Development*, **127**, 3395–3405.

Cano-Delgado, A., Penfield, S., Smith, C., Catley, M., and Bevan, M. (2003) Reduced cellulose synthesis invokes lignifications and defense responses in Arabidopsis thaliana. *Plant Journal*, **34**, 351–362.

Carpita, N. C. (2011) Update on mechanisms of plant cell wall biosynthesis: how plants make cellulose and other 1→4-β-D-glycans. *Plant Physiology*, **155**, 171–184.

Carroll, A. and Somerville, C. (2009) Cellulosic biofuels. *Annual Review of Plant Biology*, **60**, 165–182.

Carroll, A. and Specht, C. D. (2011) Understanding plant cellulose synthases through a comprehensive investigation of the cellulose synthase family sequences. *Frontiers in Plant Science*, **2**, 5.

Carroll, A., Mansoori, N., Li, S., Lei, L., Vernhettes, S., Visser, R. G., Somerville, C., Gu, Y., and Trindade, L. M. (2012) Complexes with mixed primary and secondary cellulose synthases are functional in Arabidopsis plants. *Plant Physiology*, **160**, 726–737.

Chan, J., Crowell, E., Eder, M., Calder, G., Bunnewell, S., Findlay, K., Vernhettes, S., Hofte, H., and Lloyd, C. (2010) The rotation of cellulose synthase trajectories is microtubule dependent and influences the texture of epidermal cell walls in Arabidopsis hypocotyls. *Journal of Cell Science*, **123**, 3490–3495.

Chen, S., Ehrhardt, D. W., and Somerville, C. R. (2010) Mutations of cellulose synthase (CesA1) phosphorylation sites modulate anisotropic cell expansion and bidirectional mobility of cellulose synthase. *PNAS*, **107**, 17188–17193.

Cheng, J. C., Lertpiriyapong, K., Wang, S., and Sung, Z. R. (2000) The role of the Arabidopsis ELD1 gene in cell development and photomorphogenesis in darkness. *Plant Physiology*, **123**, 509–520.

Cosgrove, D. J. (2005) Growth of the plant cell wall. *Nature Reviews Molecular Cell Biology*, **6**, 850–861.

Crowell, E. F., Bischoff, V., Desprez, T., Rolland, A., Stierhof, Y. D., Schumacher, K., Gonneau, M., Hofte, H., and Vernhettes, S. (2009) Pausing of Golgi bodies on microtubules regulates secretion of cellulose synthase complexes in Arabidopsis. *Plant Cell*, **21**, 1141–1154.

Daras, G., Rigas, S., Penning, B., Milioni, D., McCann, M. C., Carpita, N. C., Fasseas, C., and Hatzopoulos, P. (2009) The thanatos mutation in Arabidopsis thaliana cellulose synthase 3 (AtCesA3) has a dominant-negative effect on cellulose synthesis and plant growth. *New Phytologist*, **184**, 114–126.

DeBolt, S., Gutierrez, R., Ehrhardt, D. W., and Somerville, C. (2007) Nonmotile cellulose synthase subunits repeatedly accumulate with localized regions at the plasma membrane in Arabidopsis hypocotyl cells following 2,6-dichlorobenzonitrile treatment. *Plant Physiology*, **145**, 334–338.

DeBolt, S., Scheible, W. R., Schrick, K., Auer, M., Beisson, F., Bischoff, V., Bouvier-Nave, P., Carroll, A., Hematy, K., Li, Y., Milne, J., Nair, M., Schaller, H., Zemla, M., and Somerville, C. (2009) Mutations in UDP-glucose: sterol glucosyltransferase in Arabidopsis cause transparent testa phenotype and suberization defect in seeds. *Plant Physiology*, **151**, 78–87.

Denness, L., McKenna, J. F., Segonzac, C., Wormit, A., Madhou, P., Bennett, M., Mansfield, J., Zipfel, C., and Hamann, T. (2011) Cell wall damage-induced lignin biosynthesis is regulated by a reactive oxygen species- and jasmonic acid-dependent process in Arabidopsis. *Plant Physiology*, **156**, 1364–1374.

Desnos, T., Orbovic, V., Bellini, C., Kronenberger, J., Caboche, M., Traas, J., and Hofte, H. (1996) Procuste 1 mutants identify two distinct genetic pathways controlling hypocotyls cell elongation, respectively in dark- and light-grown Arabidopsis seedlings. *Development*, **122**, 683–693.

Desprez, T., Juraniec, M., Crowell, E. F., Jouv, H., Pochylova, Z., Parcy, F., Hofte, H., Gonneau, M., and Vernhettes, S. (2007) Organization of cellulose synthase complexes involved in primary cell wall synthesis in Arabidopsis thaliana. *PNAS*, **104**, 15572–15577.

Diotallevi, F. and Mulder, B. (2007) The cellulose synthase complex: a polymerization driven supramolecular motor. *Biophysics Journal*, **92**, 2666–2673.

Duncan, R. R., Shipston, M. J., and Chow, R. H. (2000) Double C2 protein: A review. *Biochemie*, **82**, 421–426.

Ellis, C. and Turner, J. G. (2001) The Arabidopsis mutant cev1 has constitutively active jasmonate and ethylene signal pathways and enhanced resistance to pathogens. *Plant Cell*, **13**, 1025–1033.

Endler, A. and Persson, S. (2011) Cellulose synthases and synthesis in Arabidopsis. *Molecular Plant*, **4**, 199–211.

Fagard, M., Desnos, T., Desprez, T., Goubet, F., Refregier, G., Mouille, G., McCann, M., Rayon, C., Vernhettes, S., and Hofte, H. (2000) Procuste 1 encodes a cellulose synthase required for normal cell elongation specifically in roots and dark-grown hypocotyls of Arabidopsis. *Plant Cell*, **12**, 2409–2424.

Fernandes, A. N., Thomas, L. H., Altaner, C. M., Callow, P., Forsyth, V. T., Apperley, D. C., Kennedy, C. J., and Jarvis, M. C. (2011) Nanostructure of cellulose microfibrils in spruce wood. *PNAS*, **108**, 1195–1203.

Green, P. B. (1962) Mechanism for plant cellular morphogenesis. *Science*, **138**, 1404–1405.

Gu, Y., Kaplinsky, N., Bringmann, M., Cobb, A., Carroll, A., Sampathkumar, A., Baskin, T. I., Persson, S., and Somerville, C. R. (2010) Identification of a cellulose synthase-associated protein required for cellulose biosynthesis. *PNAS*, **107**, 12866–12871.

Gutierrez, R., Lindeboom, J. J., Paredez, A. R., Emons, A. M., and Ehrhardt, D. W. (2009) Arabidopsis cortical microtubules position cellulose synthase delivery to the plasma membrane and interact with cellulose synthase trafficking compartments. *Nature Cell Biology*, **11**, 797–806.

Handakumbura, P. P., Matos, D. A., Osmont, K. S., Harrington, M. J., Heo, K., Kafle, K., Kim, S. H., Baskin, T. I., and Hazen, S. P. (2013) Perturbation of Brachypodium distachyon CELLULOSE SYNTHASE A4 or 7 results in abnormal cell walls. *BMC Plant Biology*, **13**, 131.

Harpaz-Saad, S., McFarlane, H. E., Xu, S., Divi, U. K., Forward, B., Western, T. L., and Kieber, J. J. (2011) Cellulose synthesis via the FEI2 RLK/ SOS5 pathway and cellulose synthase 5 is required for the structure of seed coat mucilage in Arabidopsis. *Plant Journal*, **68**, 941–953.

Harris, D. M., Corbin, K., Wang, T., Gutierrez, R., Bertolo, A. L., Petti, C., Smilgies, D. M., Estevez, J. M., Bonetta, D., Urbanowicz, B. R., Ehrhardt, D. W., Somerville, C. R., Rose, J. K., Hong, M., and DeBolt, S. (2012) Cellulose microfibril crystallinity is reduced by mutating C-terminal transmembrane region residues CesA1 A903V and CesA3 T942I of cellulose synthase. *PNAS*, **109**, 4098–4103.

Heim, D. R., Roberts, J. L., Pike, P. D., and Larrinua, I. M. (1989) Mutations of a locus of Arabidopsis thaliana confers resistance to the herbicide isoxaben. *Plant Physiology*, **90**, 146–150.

Heim, D. R., Roberts, J. L., Pike, P. D., and Larrinua, I. M. (1990) A second locus, IxrB1 in Arabidopsis thaliana, that confers resistance to the herbicide isoxaben. *Plant Physiology*, **92**, 858–861.

Hematy, K., Sado, P. E., Van Tuinen, A., Rochange, S., Desnos, T., Balzergue, S., Pelletier, S., Renou, J. P., and Hofte, H. (2007) A receptor-like kinase mediates the response of Arabidopsis cells to the inhibition of cellulose synthesis. *Current Biology*, **17**, 922–931.

Hochholdinger, F., Wen, T. J., Zimmerman, R., Chimot-Marolle, P., da Costa e Silva, O., Bruce, W., Lamkey, K. R., Wienland, U., and Schable, P. S. (2008) The maize (Zea Mays L.) roothairless 3 gene encodes a putative GPI-anchored, monocot-specific, COBRA-like protein that significantly affects grain yield. *Plant Journal*, **54**, 888–898.

Hu, S. Q., Gao, Y. G., Tajima, K., Sunagawa, N., Zhou, Y., Kawano, S., Fujiwara, T., Yoda, T., Shimura, D., Satoh, Y., Munekata, M., Tanaka, I., and Yao, M. (2010) Structure of bacterial cellulose synthase subunit D octamer with four inner passageways. *PNAS*, **107**, 17957–17961.

Huq, E., Al-Sady, B., and Quail, P. H. (2003) Nuclear translocation of the photoreceptor phytochrome B is necessary for its biological function in seedling morphogenesis. *Plant Journal*, **35**, 660–664.

Jendretzki, A., Wittland, J., Wilk, S., Straede, A., and Heinsch, J. J. (2011) How do I begin? Sensing extracellular stress to maintain yeast cell wall integrity. *European Journal of Cell Biology*, **90**, 740–744.

Kimura, S., Laosinchai, W., Itoh, T., Cui, X., Linder, C. R., and Brown, R. M. Jr., (1999) Immunogold labeling of rosette terminal cellulose-synthesizing complexes in the vascular plant Vigna angularis. *Plant Cell*, **11**, 2075–2086.

Klemm, D., Heublein, B., Fink, H. P., and Bohn, A. (2005) Cellulose: fascinating biopolymer and sustainable raw material. *Angewandte Chemie International Edition*, **44**, 3358–3393.

Kong, D., Karve, R., Willet, A., Chen, M. K., Oden, J., and Shpak, E. D. (2012) Regulation of plasmodesmatal permeability and stomatal patterning by the glycosyltransferase-like protein KOBITO1. *Plant Physiology*, **159**, 156–168.

Kotake, T., Aohara, T., Hirano, K., Sato, A., Kaneko, Y., Tsumuraya, Y., Takatsuji, H., and Kawasaki, S. (2011) Rice Brittle Culm 6 encodes a dominant-negative form of CesA protein that perturbs cellulose synthesis in secondary cell walls. *Journal of Experimental Botany*, **62**, 2053–2062.

Lai-Kee-Him, J., Chanzy, H., Muller, M., Putaux, J. L., Imai, T., and Bulone, V. (2002) In vitro versus in vivo cellulose microfibrils from plant primary wall synthases: structural differences. *Journal of Biological Chemistry*, **277**, 36931–36939.

Lairson, L. L., Henrissat, B., Davies, G. J., and Withers, S. G. (2008) Glycosyltranferases: structure, functions, and mechanisms. *Annual Review of Biochemistry*, **77**, 521–555.

Lane, D. R., Wiedemeier, A., Peng, L., Hofte, H., Vernhettes, S., Desprez, T., Hocart, C. H., Birch, R. J., Baskin, T. I., Burn, J. E., Arioli, T., Betzner, A. S., and Williamson, R. E. (2001) Temperature-sensitive alleles of RSW2 link the KORRIGAN endo-1,4-β-glucanase to cellulose synthesis and cytokinesis in Arabidopsis. *Plant Physiology*, **126**, 278–288.

Ledbetter, M. C. and Porter, K. R. (1963) A "microtubule" in plant cell fine structure. *Journal of Cell Biology*, **19**, 239–250.

Lertpiriyapong, K. and Sung, Z. R. (2003) The elongation defective 1 mutant of Arabidopsis is impaired in the gene encoding a serine-rich secreted protein. *Plant Molecular Biology*, **53**, 581–595.

Li, S., Lei, L., Somerville, C. R., and Gu, Y. (2012) Cellulose synthase interacting protein 1 (CSI1) links microtubules and cellulose synthase complexes. *PNAS*, **109**, 185–190.

Li, S., Ge, F. R., Xu, M., Zhao, X. Y., Huang, G. Q., Zhou, L. Z., Wang, J. G., Kombrink, A., McCormick, S., Zhang, X. S., and Zhang, Y. (2013) Arabidopsis COBRA-like 10, a GPI-anchored protein, mediates directional growth of pollen tubes. *Plant Journal*, **74**, 486–497.

Li, Y., Qian, Q., Zhou, Y., Yan, M., Sun, L., Zhang, M., Fu, Z., Wang, Y., Han, B., Pang, X., Chen, M., and Li, J. (2003) BRITTLE CULM1, which encodes a COBRA-like protein, affects the mechanical properties of rice plants. *Plant Cell*, **15**, 2020–2031.

Liu, L., Shang-Guan, K., Zhang, B., Liu, X., Yan, M., Zhang, L., Shi, Y., Zhang, M., Qian, Q., Li, J., and Zhou, Y. (2013) Brittle Culm1, a COBRA-Like protein, functions in cellulose assembly through binding cellulose microfibrils. *PLoS Genetics*, **9**, e1003704.

Matthyse, A. G., White, S., and Lightfoot, R. (1995) Genes required for cellulose synthesis in *Agrobacterium tumefaciens*. *Journal of Bacteriology*, **177**, 1069–1075.

Mazur, O., and Zimmer, J. (2011) Apo- and cellopentaose-bound structures of the bacterial cellulose synthase subunit BcsZ. *Journal of Biological Chemistry*, **286**, 17601–17606.

Montezinos, D., and Brown, M. Jr., (1976) Surface architecture of the plant cell: biogenesis of the cell wall, with special emphasis on the role of the plasma membrane in cellulose biosynthesis. *Journal of Supramolecular Structure*, **5**, 277–290.

Moon, R. J., Martini, A., Nairn, J., Simonsen, J., and Youngblood, J. (2011) Cellulose nanomaterials review: structure, properties, and nanocomposites. *Chemical Society Reviews*, **40**, 3941–3994.

Moore, P. J., Swords, K. M., Lynch, M. A., and Staehelin, L. A. (1991) Spatial organization of the assembly pathways of glycoproteins and complex polysaccharides in the Golgi apparatus of plants. *Journal of Cell Biology*, **112**, 589–602.

Morgan, J. L., Strumillo, J., and Zimmer, J. (2013) Crystallographic snapshot of cellulose synthesis and membrane translocation. *Nature*, **493**, 181–186.

Mueller, S. C. and Brown, R. M. Jr., (1980) Evidence for an intramembrane component associated with a cellulose microfibril-synthesizing complex in higher plants. *Journal of Cell Biology*, **84**, 315–326.

Mutwil, M., DeBolt, S., and Persson, S. (2008) Cellulose synthesis: a complex complex. *Current Opinion in Plant Biology*, **11**, 252–257.

Nakagami, H., Sugiyama, N., Mochida, K., Daudi, A., Yoshida, Y., Toyoda, T., Tomita, M., Ishihama, Y., and Shirasu, K. (2010) Large-scale comparative phosphoproteomics identifies conserved phosphorylation sites in plants. *Plant Physiology*, **153**, 1161–1174.

Nicol, F., His, I., Jauneau, A., Vernhettes, S., Canut, H., and Hofte, H. (1998) A plasma membrane-bound putative endo-1,4-β-D-glucanase is required for normal wall assembly and cell elongation in Arabidopsis. *EMBO Journal*, **17**, 5563–5576.

Nishimura, M. T., Stein, M., Hou, B. H., Vogel, J. P., Edwards, H., and Somerville, S. C. (2003) Loss of a callose synthase results in salicylic acid-dependent disease resistance. *Science*, **301**, 969–972.

Nuhse, T. S., Stensballe, A., Jensen, O. N., and Peck, S. C. (2004) Phosphoproteomics of the Arabidopsis plasma membrane and a new phosphorylation site database. *Plant Cell*, **16**, 2394–2405.

Nuhse, T. S., Bottrill, A. R., Jones, A. M., and Peck, S. C. (2007) Quantitative phosphoproteomic analysis of plasma membrane proteins reveals regulatory mechanisms of plant innate immune responses. *Plant Journal*, **51**, 931–940.

Pagant, S., Bichet, A., Sugimoto, K., Lerouxel, O., Desprez, T., McCann, M., Lerouge, P., Vernhettes, S., and Hofte, H. (2002) KOBITO1 encodes a novel plasma membrane protein necessary for normal synthesis of cellulose during cell expansion in Arabidopsis. *Plant Cell*, **14**, 2001–2013.

Paredez, A. R., Somerville, C. R., and Ehrhardt, D. W. (2006) Visualization of cellulose synthase demonstrates functional association with microtubules. *Science*, **312**, 1491–1495.

Paredez, A. R., Persson, S., Ehrhardt, D. W., and Somerville, C. R. (2008) Genetic evidence that cellulose synthase activity influences microtubule cortical array organization. *Plant Physiology*, **147**, 1723–1734.

Pauly, M., and Keegstra, K. (2008) Cell-wall carbohydrates and their modification as a resource for biofuels. *Plant Journal*, **54**, 559–568.

Pear, J. R., Kawagoe, Y., Schreckengost, W. E., Delmer, D. P., and Stalker, D. M. (1996) Higher plants contain homologs of the bacterial celA genes encoding the catalytic subunit of cellulose synthase. *PNAS*, **93**, 12637–12642.

Peng, L., Xiang, F., Roberts, E., Kawagoe, Y., Greve, L. C., Kreuz, K., and Delmer, D. P. (2001) The experimental herbicide CGA 325'615 inhibits synthesis of crystalline cellulose and causes accumulation of non-crystalline β-1,4-glucan associated with CesA protein. *Plant Physiology*, **126**, 981–992.

Peng, L., Kawagoe, Y., Hogan, P., and Delmer, D. (2002) Sitosterol-β-glucoside as primer for cellulose synthesis in plants. *Science*, **295**, 147–150.

Persson, S., Wei, H., Milne, J., Page, G. P., and Somerville, C. R. (2005) Identification of genes required for cellulose synthesis by regression analysis of public microarray data sets. *PNAS*, **102**, 8633–8638.

Persson, S., Paredez, A., Carroll, A., Palsdottir, H., Doblin, M., Poindexter, P., Khitrov, N., Auer, M., and Somerville, C. R. (2007) Genetic evidence for three unique components in primary cell-wall cellulose synthase complexes in Arabidopsis. *PNAS*, **104**, 15566–15571.

Pysh, L., Alexander, N., Swatzyna, L., and Harbert, R. (2012) Four alleles of AtCesA3 form an allelic series with respect to root phenotype in Arabidopsis thaliana. *Plant Physiology*, **144**, 369–381.

Refregier, G., Pelletier, S., Jaillard, D., and Hofte, H. (2004) Interaction between wall deposition and cell elongation in dark-grown hypocotyl cells in Arabidopsis. *Plant Physiology*, **135**, 959–68.

Richmond, T. A., and Somerville, C. R. (2000) The cellulose synthase superfamily. *Plant Physiology*, **124**, 495–498.

Robert, S., Bichet, A., Grandjean, O., Kierzkowski, D., Satiat-Jeunemaitre, B., Pelletier, S., Hauser, M. T., Hofte, H., and Vernhettes, S. (2005) An Arabidopsis endo-1,4-β-D-glucanase involved in cellulose synthesis undergoes regulated intracellular cycling. *Plant Cell*, **17**, 3378–3389.

Roudier, F., Schindelman, G., DeSalle, R., and Benfey, P. N. (2002) The COBRA family of putative GPI-anchored proteins in Arabidopsis: A new fellowship in expansion. *Plant Physiology*, **130**, 538–548.

Roudier, F., Fernandez, A. G., Fujita, M., Himmelspach, R., Borner, G. H., Schindelman, G., Song, S., Baskin, T. I., Dupree, P., Wasteneys, G. O., and Benfey, P. N. (2005) COBRA, an Arabidopsis extracellular glycosyl-phosphatidyl inositol-anchored protein, specifically controls highly anisotropic expansion through its involvement in cellulose microfibril orientation. *Plant Cell*, **17**, 1749–1763.

Rudsander, U. J., Sandstrom, C., Piens, K., Master, E. R., Wilson, D. B., Brumer, H., Kenne, L., and Teeri, T. T. (2008) Comparative NMR analysis of cellooligosaccharide hydrolysis by GH9 bacterial and plant endo-1,4-β-glucanases. *Biochemistry*, **47**, 5235–5241.

Sakamoto, K., and Nagatani, A. (1996) Nuclear localization activity of phytochrome B. *Plant Journal*, **10**, 859–868.

Sampathkumar, A., Gutierrez, R., McFarlane, H. E., Bringmann, M., Lindeboom, J., Emons, A. M., Samuels, L., Ketelaar, T., Ehrhardt, D. W., and Persson, S. (2013) Patterning and lifetime of plasma membrane-localized cellulose synthase is dependent on actin organization in Arabidopsis interphase cells. *Plant Physiology*, **162**, 675–688.

Samuels, A. L., Giddings, T. H. Jr.,, and Staehelin, L. A. (1995) Cytokinesis in tobacco BY-2 and root tip cells: a new model of cell plate formation in higher plants. *Journal of Cell Biology*, **130**, 1345–1357.

Sanchez-Rodriguez, C., Bauer, S., Hematy, K., Saxe, F., Ibanez, A. B., Vodermaier, V., Konlechner, C., Sampathkumar, A., Ruggeberg, M., Aichinger, E., Neumetzler, L., Burger, I., Somerville, C., Hauser, M. T., and Persson, S. (2012) Chitinase-like 1/ pom-pom1 and its homolog CTL2 are glucan-interacting proteins important for cellulose biosynthesis in Arabidopsis. *Plant Cell*, **24**, 589–607.

Sato, S., Kato, T., Kakegawa, K., Ishii, T., Liu, Y. G., Awano, T., Takabe, K., Nihiyama, Y., Kuga, S., Sato, S., Nakamura, Y., Tabata, S., and Shibata, D. (2001) Role of the putative membrane-bound endo-1,4-beta-glucanase KORRIGAN in cell elongation and cellulose synthesis in Arabidopsis thaliana. *Plant Cell Physiology*, **42**, 251–263.

Saxena, I. M., Kudlicka, K., Okuda, K., and Brown, R. M. Jr., (1994) Characterization of genes in the cellulose-synthesizing operon (acs operon) of *Acetobacter xylinum*: implications for cellulose crystallization. *Journal of Bacteriology*, **176**, 5735–5752.

Schindelman, G., Morikami, A., Jung, J., Baskin, T. I., Carpita, N. C., Derbyshire, P., McCann, M. C., and Benfey, P. N. (2001) COBRA encodes a putative GPI-anchored protein, which is polarly localized and necessary for oriented cell expansion in Arabidopsis. *Genes and Development*, **15**, 1115–1127.

Sethaphong, L., Haigler, C. H., Kubicki, J. D., Zimmer, J., Bonetta, D., DeBolt, S., and Yingling, Y. G. (2013) Tertiary model of a plant cellulose synthase. *PNAS*, **110**, 7512–7517.

Somerville, C. (2006) Cellulose synthesis in higher plants. *Annual Review of Cell and Developmental Biology*, **22**, 53–78.

Somerville, C., Bauer, S., Brininstool, G., Facette, M., Hamann, T., Milne, J., Osborne, E., Paredez, A., Persson, S., Raab, T., Vorwerk, S., and Young, H. (2004) Toward a systems approach to understanding plant cell walls. *Science*, **306**, 2206–2211.

Song, D., Shen, J., and Li, L. (2010) Characterization of cellulose synthase complexes in Populus xylem differentiation. *New Phytologist*, **187**, 777–790.

Sugimoto, K., Himmelspach, R., Williamson, R. E., and Wasteneys, G. O. (2003) Mutation or drug-dependent microtubule disruption causes radial swelling without altering parallel cellulose microfibril deposition in Arabidopsis root cells. *Plant Cell*, **15**, 1414–1429.

Szymanski, D. B. (2009) Plant cells taking shape: new insights into cytoplasmic control. *Current Opinion in Plant Biology*, **12**, 735–744.

Tanaka, K., Murata, K., Yamazaki, M., Onosato, K., Miyao, A., and Hirochika, H. (2003) Three distinct rice cellulose synthase catalytic subunit genes required for cellulose synthesis in the secondary wall. *Plant Physiology*, **133**, 73–83.

Tarbouiech, N., Charnock, S. J., and Davies, G. J. (2001) Three-dimensional structures of the Mn and Mg dTDP complexes of the family GT-2 glycosyltransferase SpsA: a comparison with related NDP-sugar glycosyltransferases. *Journal of Molecular Biology*, **314**, 655–661.

Taylor, N. G. (2007) Identification of cellulose synthase AtCesA7 (IRX3) in vivo phosphorylation sites—a potential role in regulation protein degradation. *Plant Molecular Biology*, **64**, 161–171.

Taylor, N. G., Scheible, W. R., Cutler, S., Somerville, C. R., and Turner, S. R. (1999) The irregular xylem 3 locus of Arabidopsis encodes a cellulose synthase required for secondary cell wall synthesis. *Plant Cell*, **11**, 769–780.

Taylor, N. G., Howells, R. M., Huttly, A. K., Vickers, K., and Turner, S. R. (2003) Interactions among three distinct CesA proteins essential for cellulose synthesis. *PNAS*, **100**, 1450–1455.

Tewari, R., Bailes, E., Bunting, K. A., and Coates, J. C. (2010) Armadillo-repeat protein functions: questions for little creatures. *Trends in Cell Biology*, **20**, 470–481.

Timmers, J., Vernhettes, S., Desprez, T., Vincken, J. P., Visser, R. G., and Trindade, L. M. (2009) Interactions between membrane-bound cellulose synthases involved in the synthesis of the secondary cell wall. *FEBS Letters*, **583**, 978–982.

Turner, S. R., and Somerville, C. R. (1997) Collapsed xylem phenotype of Arabidopsis identifies mutants deficient in cellulose deposition in the secondary cell wall. *Plant Cell*, **9**, 689–701.

Turner, S., Gallois, P., and Brown, D. (2007) Tracheary element differentiation. *Annual Review of Plant Biology*, **58**, 407–433.

Vaughn, K. C., and Turley, R. B. (2001) Ultrastructural effects of cellulose biosynthesis inhibitor herbicides on developing cotton fibers. *Protoplasma*, **216**, 80–93.

Wong, H. C., Fear, A. L., Calhoon, R. D., Eichinger, G. H., Mayer, R., Amikam, D., Benziman, M., Gelfand, D. H., Meade, J. H., Emerick, A. W., Bruner, R., Ben-Bassat, A., and Tal, R. (1990) Genetic organization of the cellulose synthase operon in *Acetobacter xylinum*. *PNAS*, **87**, 8130–8134.

Xu, S. L., Rahman, A., Baskin, T. I., and Kieber, J. J. (2008) Two leucine-rich repeat receptor kinases mediate signaling, linking cell wall biosynthesis and ACC synthase in Arabidopsis. *Plant Cell*, **20**(11), 3065–3079.

Yamaguchi, R., Nakamura, M., Mochizuki, N., Kay, S. A., and Nagatani, A. (1999) Light-dependent translocation of phytochrome B-GFP fusion protein to the nucleus in transgenic Arabidopsis. *Journal of Cell Biology*, **145**, 437–445.

Yan, C., Yan, S., Zeng, X., Zhang, Z., and Gu, M. (2007) Fine mapping and isolation of Bc7(t), allelic to OsCesA4. *Journal of Genetics and Genomics*, **34**, 1019–1027.

Yin, Y., Huang, J., and Xu, Y. (2009) The cellulose synthase superfamily in fully sequenced plants and algae. *BMC Plant Biology*, **9**, 99.

Zhang, D., Hrmova, M., Wan, C. H., Wu, C., Balzen, J., Cai, W., Wang, J., Densmore, L. D., Fincher, G. B., Zhang, H., and Haigler, C. H. (2004) Members of a new group of chitinase-like genes are expressed preferentially in cotton cells with secondary walls. *Plant Molecular Biology*, **54**, 353–372.

Zhong, R., Burk, D. H., Morrison, W. H. 3rd,, and Ye, Z. H. (2002) A kinesin-like protein is essential for oriented deposition of cellulose microfibrils and cell wall strength. *Plant Cell*, **14**, 3101–3117.

Zuo, J., Niu, Q. W., Nishizawa, N., Wu, Y., Kost, B., and Chua, N. H. (2000) KORRIGAN, an Arabidopsis endo-1,4-beta-glucanase, localizes to the cell plate by polarized targeting and is essential for cytokinesis. *Plant Cell*, **12**, 1137–1152.

4 Cortical Microtubule Array Organization and Plant Cell Morphogenesis

Sidney L. Shaw and Laura Vineyard
Department of Biology, Indiana University, Bloomington, IN

Introduction

Microtubules are a highly conserved component of the eukaryotic cytoskeleton. These relatively long cellular polymers, comprising an ordered lattice of tubulin subunits, form a vectorally polarized fiber with chemically distinct ends. Organizing these fibers into distinct arrays provides the cell with a mechanism to spatially segregate specific cellular components. In mitosis, the microtubules form a bipolar spindle apparatus capable of correctly positioning chromosomes to the cell midplane and then guiding separated chromatids to opposite ends of the cell. During interphase, the microtubules are critical for stratifying the secretory system in animal cells and orchestrating cell wall synthesis in plant cells. The mechanisms that evolved in plants for transitioning interphase microtubules from the centralized radial array found in lower eukaryotic cells to variously patterned arrays at the cell cortex remains a subject of intense interest. In this chapter, we explore what is currently known about the cortical microtubule cytoskeleton in flowering plants and how various patterns are created that affect plant cell morphogenesis.

Microtubule Polymers

Microtubules are hollow polymers, 25 nm in diameter, composed of highly conserved α/β-tubulin dimers arranged head-to-tail in 12–14 protofilaments (Desai and Mitchison, 1997). Plants typically have extended alpha and beta tubulin gene families, expressing various tubulin combinations in different tissue types (Ludwig *et al.*, 1987; Oppenheimer *et al.*, 1988). The α/β-tubulin subunit is an obligate dimer where a complex of chaperone-like proteins are required for correct tubulin dimer assembly (Tian *et al.*, 1999) and may be

Plant Cell Wall Patterning and Cell Shape, First Edition. Edited by Hiroo Fukuda.

a site of regulation for the microtubule cytoskeleton (Steinborn *et al*., 2002; Locascio *et al*., 2013). Once assembled, tubulin dimers are contributed to a pool of ostensibly free subunits for assembly into microtubules.

Microtubules can nucleate spontaneously in solution at high tubulin concentration. However, nucleation is controlled in cells primarily by a complex

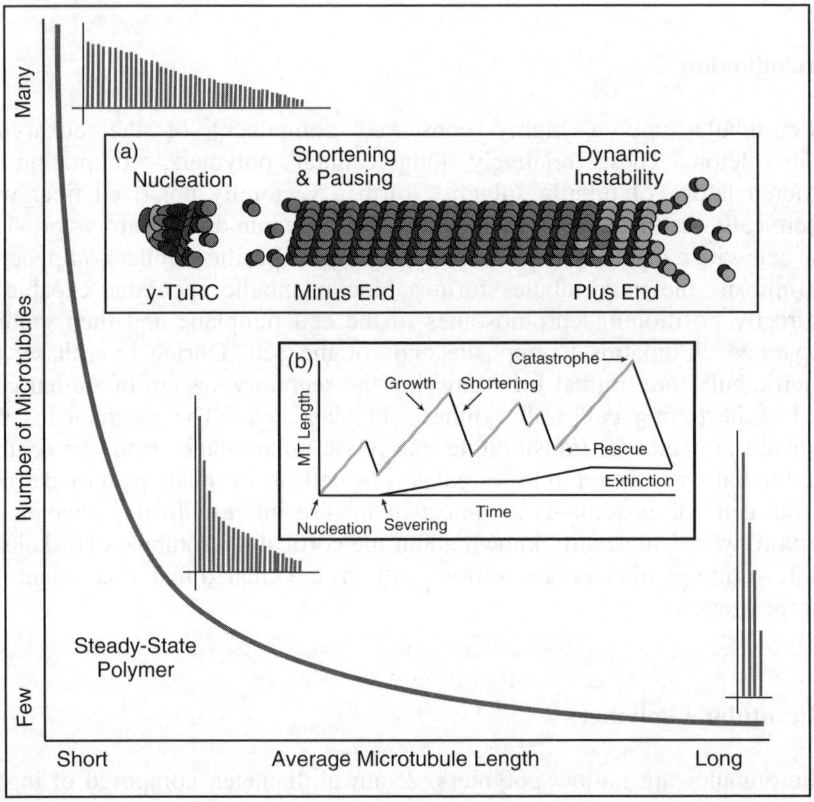

Figure 4.1 Microtubule structure, dynamics, and steady-state length distribution. (a) Microtubules in cells nucleate from templates containing a ring of γ-tubulin and associated proteins. Dimeric tubulin subunits line up head-to-tail in 13 offset rows, termed protofilaments, forming a tube. Microtubules in flowering plants are severed from the γ-TuRC and undergo polymer treadmilling at the cell cortex. (b) The microtubule minus end undergoes switching between slow depolymerization and pausing. The plus end switches between states of more rapid growth, accumulating GTP-bound tubulin dimers (light blue), and very rapid shortening when GTP hydrolysis catches up to the microtubule end. The combined dynamic instability at the plus end and depolymerization at the minus end will produce a roughly constant total polymer length in the cell. The microtubule dynamics, coupled with additional severing, will partition the polymer into a number of microtubules with an average length that form a length distribution within the cell. (See plate section for the color version of this figure.)

of proteins including γ-tubulin (Fig. 4.1) and the γ-tubulin-associated proteins (GCPs), together termed the γ-tubulin ring complex (γ-TuRC; Horio *et al.*, 1991). The proteins form a ring of γ-tubulins that template a 13 protofilament microtubule (Zheng *et al.*, 1995). The γ-TuRC effectively lowers the thermodynamic constraints for nucleating a microtubule, and hence prevents extraneous nucleations and allows cells to control where nucleation occurs.

Once nucleated, the more active 'plus end' of the microtubule stochastically switches between periods of growth (polymerization) and shortening (depolymerization) in a process termed 'dynamic instability' (Desai and Mitchison, 1997; Fig. 4.1). The phenomenon arises from the GTPase activity of the beta tubulin subunit. When free in solution, tubulin rapidly exchanges bound GDP for free GTP in the cytoplasm (Howard and Hyman, 2003). The nucleotide sits at the binding interface between tubulin subunits where the GTP-bound beta-tubulin has a higher binding affinity for alpha-tubulin than the GDP-bound (Downing and Nogales, 1998; Nogales *et al.*, 1998). Hydrolysis of GTP to GDP is very slow in the free dimer but is accelerated when the GTP-bound dimer assembles at a microtubule end. This transition sets up a rate-dependent 'GTP cap' on the microtubule (Caplow and Shanks, 1996) where small fluctuations in polymer assembly or GTPase activity lengthen or shorten the number of GTP-bound tubulins at the microtubule end (Fig. 4.1a). If by chance the GTP hydrolysis catches up to the growing microtubule end, eliminating the 'GTP cap', it is hypothesized that the microtubule undergoes a 'catastrophe' and switches to a shortening phase. If a GTP-bound subunit manages to recap the shortening microtubule, a 'rescue' event occurs switching the microtubule back to growth. In the absence of rescue, all subunits can be lost leading to the 'extinction' of the polymer.

Microtubule Behaviors in Plant Cells

Microtubules in the interphase plant cell are mostly restricted to the plasma membrane or cell cortex (Fig. 4.2). Plant cells are typically large and maintain a cytoplasmic layer just beneath the cell wall where the majority of cell volume is occupied by a vacuole. The cortical microtubules therein occupy an effectively two-dimensional (2D) space or layer in the 3D cell. Modulation of the microtubule cytoskeleton occurs throughout the life of the cell. Microtubule-associated proteins (MAPs) help to organize the polymers into functional arrays and harness the intrinsic dynamic properties of the microtubule to accomplish different tasks in the cell. Critical points of control for determining microtubule array properties include microtubule nucleation, polymer dynamics, and the interactions between microtubules and other elements of the cell.

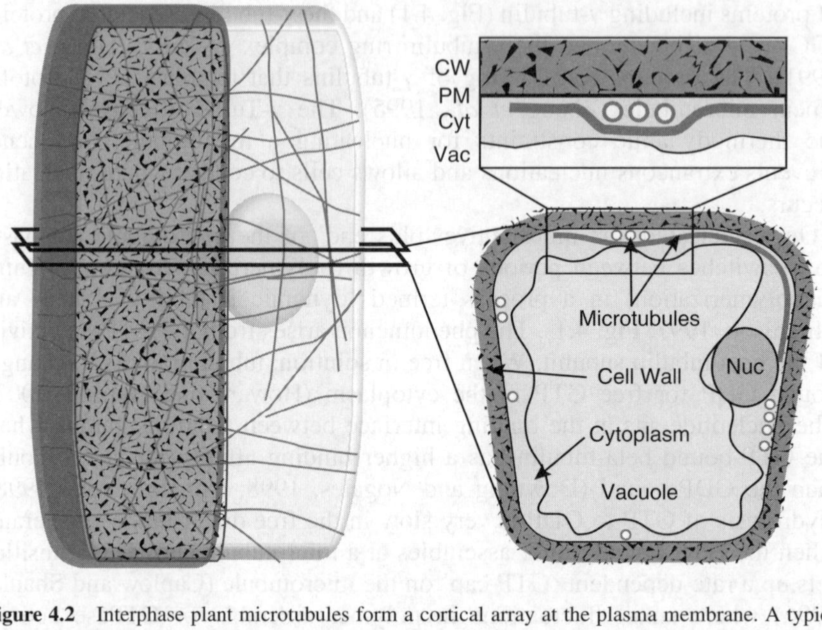

Figure 4.2 Interphase plant microtubules form a cortical array at the plasma membrane. A typical epidermal plant cell is encased within a cell wall (CW) with the majority of the internal volume occupied by a large central vacuole (Vac). The cytoplasm (Cyt) and nucleus (Nuc) exist in a relatively thin layer around the inner cell perimeter. The cortical array of microtubules laterally associates with the plasma membrane (PM) to form a 2D layer around the inner surface of the cell. Organization of the microtubules underlying the plasma membrane contributes to the patterning of the cell wall for morphogenesis.

Microtubule Nucleation

Creating new microtubules is initially accomplished through nucleation although plant cells also use polymer severing as a means of increasing microtubule number. Nucleation of new microtubules involves the activity of γ-tubulin and recruitment of the following proteins: γ-tubulin protein complex proteins 1-4 (GCP1-4); neural precursor cell-expressed developmentally down-regulated 1 (NEDD1); GCP3 Interacting protein 1 (GIP1); and AUGMIN complex proteins (Chan *et al*., 2009; Zeng, 2009; Hotta, 2012; Nakamura *et al*., 2012). Identification of microtubule nucleation complex proteins and observations of microtubule activity *in vivo* show that new interphase microtubules are primarily, but not exclusively, nucleated at the cell cortex from the sides of existing microtubules (Shaw *et al*., 2003; Murata *et al*., 2005). The majority of new microtubules appear to be created with the nucleation site associated with another microtubule with only 1–2% of nucleation events occurring unanchored from other microtubules (Chan *et al.*, 2009; Nakamura *et al.*, 2010). Prior evidence that γ-TuRC exists on the nuclear membrane and gives rise to microtubules that populate the cell (Liu *et al.*, 1994; Seltzer *et al.*, 2007) has mostly been ignored in light of finding origins at the cell cortex, but this nuclear-derived polymer may provide a key population of microtubules under certain circumstances.

There is presently very little understood about what regulates the nucleation of microtubules in plant or animal cells. The γ-TuRC provides a template for nucleation but, to date, it is not known how the number of complexes is specified in cells or how they are made competent or incompetent for nucleating activity. It is presumed that the number of nucleation complexes will be limiting. AUGMIN complex proteins are believed to interact with γ-TuRC or potentially localize the γ-TuRC to critical sites for nucleation, and evidence now shows the existence of these proteins in flowering plant spindles. NEDD1 proteins in mammalian cells (Luders *et al.*, 2006) may play a role in associating a γ-TuRC complex to the centrosome in animal cells where a NEDD1 ortholog in plants shows some promise for a role in localizing the γ-TuRC (Zeng *et al.*, 2009). No evidence has been presented to indicate how a plant cell would specify an increase or decrease in nucleation rate.

In the case of microtubule-based nucleation at the cell cortex, new microtubules can either nucleate away from the parent microtubule at an angle or be created alongside the parent to form a microtubule bundle (reviewed by Wasteneys and Ambrose, 2008; Chan *et al.*, 2009; Nakamura *et al.*, 2010). Branching microtubules polymerize away from the parent microtubule at a 40° angle with the range of potential angles falling between 20–60° (Murata *et al.*, 2005; Chan *et al.*, 2009; Nakamura *et al.*, 2010). It is not known how microtubules specify nucleation direction or angle, but Fishel and Dixit (2013) propose that modulating the players, such as TON2, and their interactions within the γ-TuRC may influence the orientation of initial polymerization. Recent work suggests that the protein phosphatase TONNEAU2 (TON2) regulates nucleation geometry (Kirik *et al.*, 2012) without influencing the nucleation rate itself.

Due to the maintenance of free minus ends in the plant cell, severing also serves as a significant mechanism for increasing polymer number. KATANIN, a microtubule-severing ATPase (Hartman *et al.*, 1998), creates a secondary plus end from an already established microtubule (Ehrhardt and Shaw, 2006). KATANIN appears to function at several points in interphase arrays, at positions of microtubule crossover, and when recruited to nucleation sites where cortical microtubules are severed within 60 seconds of nucleation (Nakamura *et al.*, 2010). KATANIN mutants have atypical cortical arrays and altered growth phenotypes with short, thick, fragile stems (Burk *et al.*, 2001).

Microtubule Association with the Plasma Membrane

A second key cortical microtubule characteristic is a strong association to the plasma membrane. Despite the force of cytoplasmic streaming, cortical microtubules remain stably attached in place during interphase and show no evidence of lateral sliding (Shaw *et al.*, 2003). Detachment of a microtubule plus end typically results in a portion of the microtubule contorting rapidly in the streaming cytoplasm and either re-adhering to the cell cortex or switching to depolymerization (Shaw *et al.*, 2003). The constant lateral attachment creates an essentially 2D surface for the microtubules leading to extensive microtubule–microtubule interactions.

It is worth noting that the association to the cell cortex changes dramatically through the cell cycle, with most microtubules disappearing from the cortex sometime after S phase. The appearance of the pre-prophase band of microtubules, presaging the eventual cell division site, suggests that the number and density of cortical associations can be managed over a significant range. Few genes have been shown to contribute pre-prophase band formation to date, with the exception of the TONNEAU2/FASS gene (Torres-Ruiz and Jurgens, 1994; Azimzadeh *et al.*, 2008) which encodes for a protein phosphatase 2A subunit (Camilleri *et al.*, 2002). Cortical microtubules return soon after mitosis and during the period when the phragmoplast is associated with the cell cortex for eventual cell division. Recent evidence suggests that the returning microtubules do not appear randomly but can show patterns related to the site of division (Lindeboom *et al.*, 2013).

The molecular nature of the microtubule association to the cell cortex is not clearly established. While a PHOSPHOLIPASE D delta (PLDd) protein co-purifies with microtubules and membranes, and alcohols known to interfere with PLDd activity disrupt microtubule associations (Gardiner *et al.*, 2001, 2003), PLDd mutants do not appear to have significant defects in cell growth and microtubule-dependent activities (Li *et al.*, 2004). Mutations in the AtCLASP1 protein appear to cause an increase in the detachment of microtubule plus ends, though no increase in microtubule sliding or lateral movement was reported (Ambrose and Wasteneys, 2008). It appears likely that the association of the microtubules to the cell cortex involves multiple mechanisms and may not be easily addressed with plants having mutations in only a single gene. Discovery of how the microtubules associate with the cell cortex should provide significant information about how the cell organizes the array.

Dynamic Properties

Unlike their animal interphase counterparts, plant cortical microtubules routinely release from their nucleation sites (Nakamura *et al.*, 2012) leaving a presumably uncapped microtubule minus end. The plus end shows dynamic instability, biased toward polymer accumulation. The minus end shows relatively slow depolymerization alternating with pausing phases. When coupled and related to the fixed lateral association of the microtubule to the membrane, these dynamic activities produce an apparent unidirectional movement of the microtubules on the cell cortex (Shaw *et al.*, 2003). The balance of dynamic properties that gives rise to this form of polymer treadmilling is likely to be controlled by a host of factors influencing the inherent association and dissociation rates of the tubulin subunits (Ehrhardt and Shaw, 2006).

Several studies have established the basic dynamic properties of microtubules in Arabidopsis cortical arrays. Microtubule plus ends polymerize *in vivo* at rates of 4–8 μm/min (Shaw and Lucas, 2011), which is relatively slow by comparison to animal systems. The velocity of shortening plus ends is commonly more than twice the rate of polymerization and anywhere within the range 10–20 μm/min. Unlike dynamic measurements made for animal tissue culture cells (which must be kept near 37°C), it is presumed that plant cells will create microtubules with a wider range of steady-state dynamics values to accommodate broader swings in ambient temperature and cellular metabolism. Distinct for plant microtubules is the *in vivo* measurement of rates for the microtubule minus end. The minus end shortens at *c.* 1.1 μm/min and rarely exhibits growth (Shaw and Lucas, 2011).

In addition to the association and dissociation rates, the dynamic properties of microtubules are usually specified with the related catastrophe and rescue frequencies. The available measured values typically come from averaged measurements of the duration that a microtubule spends in growth, shortening or paused phases. The switching frequencies so far measured, when coupled to their related growth and shortening rates, are not likely to be an accurate reflection of the true probabilities. If taken directly, the frequencies suggest that the plant microtubules are adding about twice as much tubulin as polymer than they are losing (Shaw *et al.*, 2003; Shaw and Lucas, 2011). These estimates likely arise because microtubules tend to be born on the outer cell face where the measurements are taken, and often move off of the outer array where they are no longer visible before extinction (Vineyard *et al.*, 2013). A still useful characterization for the switching rates is often given as the percent of time spent in each kinetic phase. Plant microtubule plus ends spend about 75% of their time growing, 20% shrinking, and less than 5% in a paused phase showing no growth or shortening on the outer periclinal cell face. The minus ends spend 70% of their time shrinking, 25% in pause, and less than 5% in growth (Shaw and Lucas, 2011). The growing and shortening activities of the plus end constitute a dynamic instability behavior biased toward polymer accumulation while the slow processive depolymerization and pausing at the minus end yield a net loss of polymer, indicative of classic polymer treadmilling (Shaw *et al.*, 2003; Ehrhardt and Shaw, 2006). Taken together, the rates and times spent in each phase produce a hybrid mechanism of treadmilling.

Although the polymerization rate of plant microtubules is slower than in typical animal cells, plant arrays recover from localized photobleaching 3–4 times faster. This is partly due to the plant microtubule plus ends only pausing around 5% of the time (Shaw *et al.*, 2003), while animal plus ends pause closer to 60% of the time (Rusan *et al.*, 2001). The plant cell microtubule is slow and steady, whereas the animal microtubule appears dynamic in short rapid bursts. It is believed that minus-end depolymerization fosters reduced rates of pausing by recycling free tubulin back into the global pool of free subunits (Ehrhardt and Shaw, 2006).

The dynamic properties of the microtubules should produce a steady state where the total amount of polymer and total concentration of free tubulin subunits are roughly constant over time. Since the microtubules require the energy of GTP to maintain this state, the system is not in a true chemical equilibrium. How plant cells divide up the total accumulated polymer length into a specific length distribution is not understood. The microtubules in plants are in some sense different from those in most animal cells in that the minus ends persist after release from the nucleation site or after a breakage event. In that context, the number of plant microtubules cannot be directly related to the number of nucleation sites in the cell, as is often assumed for animal and yeast systems and computational models. Since multiple microtubules can be created from a single nucleation event, the length distribution of polymers will be more than just a property of the plus-end dynamic instability parameters related to the number of nucleation sites giving rise to microtubules. The radial array created by a typical animal tissue culture cell is proposed to have a nearly exponential distribution of microtubule length, with caveats for crowding near the centrosome and induced catastrophes at the cell cortex (Verde *et al.*, 1992; Margolin *et al.*, 2006). A similar expectation for plant microtubules cannot be made because the number of microtubules cannot be easily estimated from nucleation events and the dynamic properties of both microtubule ends must be taken into account. The length distribution should have significant consequences for the lifetime of a given microtubule and for the length of time it occupies a given position in the interphase cortical array.

In principle, the dynamic properties governing global nucleation rate, growth and shortening velocities, and catastrophe and rescue frequencies should be interrelated by the resulting concentration of free subunits when finding a steady state. Increasing the catastrophe frequency, for example, would therefore be expected to release more subunits into the pool, and potentially increase the polymerization rate and support rescue. This general relationship between dynamic properties makes predicting potential outcomes on array properties from biochemical analyses of protein activity somewhat unreliable. It is evident that flowering plants regulate many of the microtubule dynamic properties with auxillary proteins; it remains to be seen how these proteins are used specifically to create a desired microtubule length distribution for different cortical array patterns.

Several proteins have been identified in plants that could regulate polymer dynamics directly. The plus-tip binding protein END BINDING 1 (EB1) binds microtubule sidewalls but has a higher affinity for plus ends, resulting in local enrichment (Mimori-Kiyosue *et al.*, 2000; Chan *et al.*, 2003; Mathur *et al.*, 2003). This protein is believed to facilitate association of free tubulin to the plus end and is considered a 'proofreader' that promotes depolymerization in microtubules with lattice defects (Vitre *et al.*, 2008). It is also thought to facilitate the binding of CLASP, another plus-tip binding protein (reviewed by Wittmann, 2008). CLASP is required for cell expansion and division and is thought to be critical to attachment of the plus end to the plasma membrane (Ambrose *et al.*, 2007). In the same way as KATANIN mutants, CLASP mutants show atypical cortical arrays and produce cells with reduced anisotropic growth, a standard cell wall organization defect (Ambrose *et al.*, 2007; Kirik, 2007).

MOR1 is an XMAP215 homologue that associates with microtubules via the highly conserved TOG domain. CLASP in fact shares this same microtubule-interacting domain (reviewed by Al-Bassam and Chang, 2011). MOR1 forms potential cross-bridges between neighboring microtubules, promotes rapid microtubule growth and shrinking, and reduces the frequency and duration of pause events (Kawamura and Wasteneys, 2008). Cortical arrays in the temperature-sensitive mutant *mor1-1* normally organize at 21°C. Once these plants are exposed to 31°C however, their arrays exhibit disorganized patterns with shortened microtubules (Whittington *et al.*, 2001; Kawamura and Wasteneys, 2008). This suggests that sustaining the wild-type plus-end dynamics is key to maintaining normal microtubule length.

Microtubule–microtubule Interactions

Microtubules are rigid tubes that polymerize in a straight line until they encounter an obstacle, such as another microtubule (Shaw *et al.*, 2003; Dixit and Cyr, 2004). The plant cortical microtubules are tightly associated with the plasma membrane; thus, any microtubule that crosses paths with another microtubule while polymerizing will physically interact in an essentially 2D environment (reviewed by Ehrhardt and Shaw, 2006). Within the environment of the cell cortex, microtubules either crossover, depolymerize, or bundle in an angle-dependent manner (Dixit and Cyr, 2004). Upon plus-end contact with the sidewall of an already established microtubule, the physical interaction appears to have an effect on microtubule dynamics. *In vitro* studies on the impact of polymerizing against a barrier on microtubule dynamics showed that microtubules flexed at the point of contact and exhibited a 20-fold increase in catastrophe rates (Janson *et al*., 2003). The resistive force exerted by the barrier was believed to be reducing the on-rate of new tubulin dimers, which in turn increased the probability of the GTP cap hydrolyzing and destabilizing the microtubule (Janson *et al*., 2003). The delay in polymerization that is the product of the resistive force of encountering another microtubule at the cortex therefore has the potential to promote catastrophe *in vivo*. Microtubule polymerization rates have been shown to slow briefly upon microtubule encounters, indicating either a direct steric hindrance or other signal to slow polymerization (Wightmann and Turner, 2007).

The frequencies of events at microtubule–microtubule interactions are dependent upon the angle of interaction. Encounters are hypothesized to exert more opposing force as the angle of those encounters approaches 90°. In tobacco BY2 culture cells, at angles greater than 40° about 60% of those microtubules underwent catastrophe while 40% of microtubules crossed over (Dixit and Cyr, 2004). In another study in Arabidopsis coaligned petiole cell arrays and net-like cotyledon epidermal cell arrays, Wightmann and Turner (2007) observed 91% and 76% crossovers in steep angle collisions, respectively. Other studies examining treadmilling of unbundled microtubules (Shaw *et al*., 2003) suggest that microtubule crossover at steep angles in Arabidopsis hypocotyl cells is common and catastrophe is relatively uncommon, leading to speculation that this event could be regulated under different circumstances. Wightmann and Turner argue that a high rate of severing at crossovers favors more coaligned arrays in petiole cells because the posterior portion of the severed microtubule typically depolymerizes. They reason that this compliments the activity of treadmilling in 'deleting' unaligned microtubules from coaligned arrays (Wightmann and Turner, 2007). Conversely, severing when coupled to effective rescue of both polymers could act as a means to increase the number of polymers.

Microtubules that do not depolymerize after contact must either transiently detach from the plasma membrane and cross over the blocking microtubule or reorient the polymerization trajectory to continue in coalignment with that microtubule (Shaw and Lucas, 2011). This reorientation of polymerization along an existing microtubule is termed 'bundling'. In different cell types, the probability of either outcome is largely dependent on the angle but not the polarity of the microtubules involved in the encounter (Dixit and Cyr, 2004; Shaw and Lucas, 2011). For example, in tobacco BY2 culture cells, if the angle is less than 40° the microtubule is likely to bundle (Dixit and Cyr, 2004) regardless of whether the two microtubules are parallel or anti-parallel (Lucas *et al*., 2011). If the angle is greater than 40°, then the microtubule will cross over and reattach to the plasma membrane on the other side or undergo the aforementioned pause or catastrophe scenarios (Dixit and Cyr, 2004).

If the microtubule redirects the trajectory to form a bundle, the lagging unbundled portion of the microtubule will eventually disappear as the minus end depolymerizes (Shaw *et al.*, 2003). It is important to note this phenomenon because if the lagging end did not depolymerize, then the array would rapidly become an interlocked meshwork (Ehrhardt and Shaw, 2006). While bundling does not alter the dynamics of individual microtubules (Shaw and Lucas, 2011), a property of a bundle containing two or more microtubules is that the duration of polymer occupation at that given space in the cell cortex is likely to be longer than if it were only a single polymer (Chan *et al.*, 2007). By that rationale, an array with a greater degree of bundling is more likely to stably maintain a particular pattern for longer than an array with less bundling.

Bundling does not appear to have a stabilizing effect on individual microtubule polymers within bundles (Shaw and Lucas, 2011). The microtubule–plasma- membrane contact is, however, important to maintaining polymerization. Bundled microtubules have not been observed to detach from the plasma membrane, but unbundled microtubules that detach exhibit rapid distortions thought to be the product of the forces of cytoplasmic streaming (Shaw *et al.*, 2003). Mechanical stress from those distortions may destabilize the microtubule lattice; in fact, in observations of detached microtubule plus ends, more than half underwent catastrophe (Shaw *et al.*, 2003). Plus ends that successfully reattached after releasing from the plasma membrane were typically oriented in a different trajectory than prior to detachment (Shaw *et al.*, 2003). One function for bundles may therefore be to protect polymers from being dislodged from the membrane.

The molecular mechanism responsible for cortical microtubule bundling is not clear. The nine genes in the MAP65 family of Arabidopsis (reviewed by Hussey *et al.*, 2002; Sasabe and Machida, 2006) are thought of as potential regulators of cortical array organization, given their biochemically defined role in mediating microtubule bundling (Chan *et al.*, 1999; Smertenko *et al.*, 2000). Although MAP65 dimers bundle microtubules *in vitro* (Chan *et al.*, 1996; Smertenko *et al.*, 2004; Gaillard *et al.*, 2008) and *in vivo* (Van Damme *et al.*, 2004; Mao *et al.*, 2005), the exact role of each family member is not well understood. Several lines of evidence suggest that these proteins are involved in holding microtubules together during mitosis when microtubules are not associated laterally to the plasma membrane (Ho *et al.*, 2011). Recent work by Lucas *et al.* (2011) demonstrated that MAP65-1 and MAP65-2, a distinct clade within the MAP65 family, clearly bind interphase cortical microtubule bundles *in vivo* concurrent with microtubule polymerization. Interestingly, while double *map65-1/2* mutants showed reduced overall growth, the patterning of the cortical microtubule arrays appeared unaffected under a number of circumstances. MAP65-1 and MAP65-2 therefore recognize anti-parallel bundles but may have more of a role in relaying the pattern of the cortical array to cell wall construction rather than defining array organization itself (Lucas *et al.*, 2011). It will be interesting to learn how the other MAP65 family members participate in microtubule bundling and cortical array regulation.

Summary of Microtubule Behaviors

The life history of a cortical array microtubule on the outer periclinal array of the cell has been reasonably well defined. The typical polymer is nucleated on the cell cortex from a γ-tubulin complex associated with the sidewall of an existing microtubule. The minus end is severed from the nucleation complex after a brief period, and the polymer exhibits dynamic instability at the plus end with intermittent depolymerization and pausing at the minus end. A constant lateral association with the plasma membrane results in the apparent movement of polymer through treadmilling. Interactions with other microtubules result in catastrophe, slight deflections, or a reorienting of the polymerization trajectory through bundling. The lifetime and length of the polymer are governed by the intrinsic dynamics and modulated by associated proteins, where the microtubule could be severed into smaller microtubules or become extinct through complete depolymerization from the plus end. All data collected so far show a continuously treadmilling system with no stabilization of polymers to form specific tasks.

Microtubule Functions in Plant Cells

As polymers with variable lengths, microtubules are well suited for involvement in a variety of transport and cell-shaping processes within cells. In animal cells, cell cargo vesicles are transported along microtubules using microtubule-motors such as kinesins and dyneins (Hirokawa, 1998). This activity further leads to the stratification of the secretory system through Golgi construction and the reticulation of the endoplasmic reticulum. Microtubules are also the machinery that cells use to align and specifically separate chromosomes in mitosis/meiosis. In conjunction with the actin cytoskeleton, microtubules regulate the formation and retraction of lamellipodia in animal fibroblasts and neurons (reviewed by Small *et al.*, 2002).

Microtubules within plant cells play similar roles during mitosis, but evidence for concurrent roles during interphase are far less compelling. The plant microtubules do not appear to play a role in reticulating or maintaining the endoplasmic reticulum and plant Golgi do not require microtubules for formation or maintenance. Further, little evidence exists for vesicle traffic moving on plant cortical microtubules even though electron microscopy shows clear indications of vesicle trafficking during phragmoplast construction at cytokinesis (Segui-Simarro *et al.*, 2004). If plant cortical microtubules act as roadways for vesicle movement, the anti-parallel organization of the microtubule bundle polarities does not suggest a clear means of specifying anterograde and retrograde transport. Evidence has been presented that microtubules may act as attachment sites for cortical vesicles prior to secretion, but an overt role in vesicle trafficking has not been documented.

As for animal cells the microtubules in plants are involved in cell morphogenesis, but it is through a very different process than typically observed for nerve cells or fibroblasts. The cell wall is far too rigid to be affected directly by the force of microtubule sliding or polymerization; rather, microtubules influence cell wall construction, which in turn impacts cell shape and the degree of growth anisotropy (reviewed by Baskin, 2005).

Long before microtubules could be visualized in living cells, Green (1962) proposed the idea that some colchicine- (a microtubule-inhibiting drug) sensitive cytoplasmic machinery was responsible for patterning wall construction. Since colchicine depolymerized the spindle filaments, it was hypothesized that filaments of a similar character might underlie the plasma membrane in cells, directing the cell wall chemistry. Shortly thereafter, Ledbetter and Porter (1963) identified cortical microtubules for the first time in higher plant cells. Experiments using known microtubule-inhibiting drugs resulted in disturbed cellulose deposition, bolstering the idea that microtubules can influence microfibril organization (Pickett-Heaps, 1967). Additional research showed that microtubules were rigid tracks along which other cellular components moved (reviewed by Newcomb, 1969; Margulis, 1973). Electron microscopy work by Giddings *et al.* (1980) showed granules with microfibrils attached, confirming their role as cellulose-producing structures. Giddings and Staehelin (1991) formally proposed that cortical microtubules restricted the direction of cellulose synthase (CES) complex movement under normal conditions. Significant work from several groups established a practical basis for suggesting that microtubules at the cell cortex could potentially direct the trajectories of the cellulose synthase complexes in the plasma membrane, and thus provide patterning information for the cell wall (Baskin, 2001). This so-called 'alignment hypothesis' was later elaborated upon and subsequently shown to have mechanistic support through direct imaging studies of microtubules and CES movement (Giddings and Staehelin, 1991; Paredez *et al.*, 2006).

It was shown that if microtubules are removed by chemical treatment, CES complexes maintain their velocity and eventually develop coaligned migration patterns (Paredez *et al.*, 2006). These data suggest that the interaction between CES complexes and microtubules may largely be a passive steric restriction rather than an active link. While this process is generally key to normal cell growth, microtubule organization does not limit the amount of total cell growth (reviewed by Lucas and Shaw, 2008), nor is microtubule-based regulation always important for morphogensis in all plant cell types (reviewed by Baskin, 2001).

Several works (e.g., Crowell *et al.*, 2009; Gutierrez *et al.*, 2009) have additionally shown that microtubules may regulate cellulose organization by targeting the new CES complexes to the plasma membrane. Associating the Golgi-derived vesicles containing CES to microtubules prior to insertion would provide a means of distributing new complexes throughout the plasma membrane and introducing them at positions that could mandate trajectory. It has also been proposed that microtubules assist in controlling the length of cellulose microfibrils (Kazama *et al.*, 2004). The theory is that short microfibrils have fewer sites to crosslink with neighboring microfibrils; selectively building a wall with a greater proportion of short microfibrils would therefore result in a wall that has less potential for anisotropic growth (Kazama *et al.*,2004).

Although progress has been made in explaining the role of the cortical microtubule array in patterning cell wall construction, many basic questions remain unanswered about the mechanism. While it is clear that microtubules influence CES complex movement, it is not known if they do so through a direct interaction or through auxiliary proteins. One idea is that the connection between the microtubules and the CES in the membrane works through the CSI1 protein (Bringmann *et al.*, 2012). Genetic evidence has been presented showing the disruption of CES tracking on microtubules when CSI1 is mutated. No biochemical data have been generated to support a direct role, although localization studies clearly show the proteins acting in the correct place at the correct time.

In summary, the animal cortical microtubules play a major role in orchestrating the deposition and orientation of cellulose microfibrils. These functions come at the expense of other microtubule roles more commonly associated with animal and fungal microtubules, such as vectoral membrane trafficking. The cortical microtubules in animal cells appear to help target new CES complexes for insertion into the membrane and subsequently act to direct CES trajectories during cellulose production.

Organizing Transverse Plant Microtubule Arrays

The coalignment hypothesis, stating that microtubules act as guides to pattern the deposition of cellulose microfibrils, requires that the cell creates specific microtubule array patterns to affect specific morphogenetic patterns. This postulate has been taken mostly from axially growing cells where the formation of microtubule arrays that are coaligned transverse to the cell axis are correlated with anisotropic cell expansion (Fig. 4.3). The most direct evidence in favor of this idea comes from seminal work showing that tubulin mutants have defined

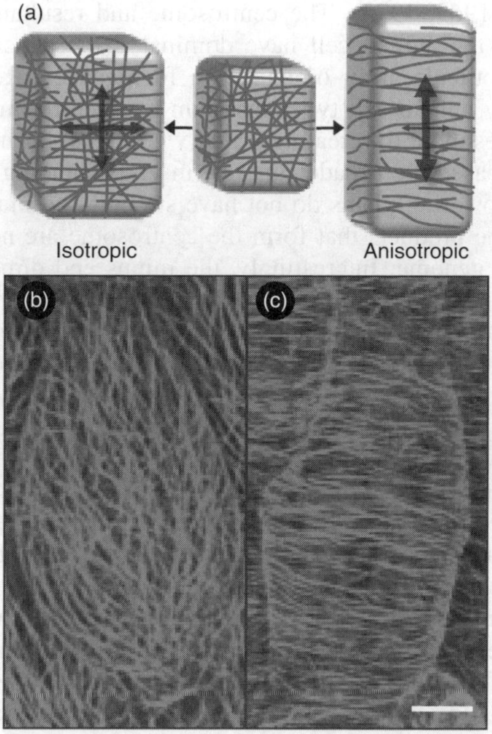

Figure 4.3 The pattern of cortical microtubules influences the directionality of cell expansion. Cellulose synthase complexes use cortical microtubules as guides for microfibril deposition. (a) Unstructured microtubule patterns tend to yield cells with more isotropic cell expansion while coaligned microtubule patterns produce anisotropic growth. Microtubules in living hypocotyl cells expressing a GFP-tubulin transgene exhibiting (b) an unstructured array pattern and (c) a transversely coaligned pattern.

effects on plant morphology. A series of Arabidopsis tubulin mutants, with inter-protein mutations and mutations at the tubulin dimer interface, showed defects in microtubule dynamics and the inability to form transverse array patterns (Ishida *et al.*, 2007a, b). Mutations that advanced dynamics led to left-handed oblique arrays and roots that rotated to the right, whereas mutations that retarded dynamics had right-handed arrays and roots that skewed left. This work demonstrates how single mutations leading to mispatterning of the microtubule array ultimately lead to global changes in plant morphology.

How plants create specific array patterns has been the subject of study since the first microtubules were discovered in plants. Of significant interest across the field of cell biology is the noted absence of centrioles or a centralized microtubule organizing center (MTOC) in flowering plants

(Pickett-Heaps, 1969, 1975). The centrosome and resultant radial array of microtubules in the animal cell have dominated the general discussion of interphase microtubule array organization for decades. Recent work indicating that many animal cell types maintain non-centrosomal microtubules, and use centrioles more for creating primary cilia than for maintaining radial arrays, has increased the broader interest in understanding non-radial array organization. Flowering plants do not have structurally intact centrioles and many of the gene products that form the centrosome are not represented in the Arabidopsis genome. Interestingly, the minus end directed microtubule motor, dynein, and some kinesin motors will drive formation of a radial array in animal cells with non-functional centrosomes (Vorobjev *et al.*, 2001; Basto *et al.*, 2006; Hornick *et al.*, 2011). Dynein is not clearly represented in the flowering plant genomes, and the lateral association of plant microtubules to the cell cortex appears to preclude the formation of radial arrays in the cytoplasm that would mimic an animal MTOC.

The connection between interphase microtubule array pattern and cell function has been most keenly explored in the epidermal cells of the axially extending root and hypocotyl (Baskin, 2001, 2005). The connection between microtubule patterning and axial extension has historically been linked to the progressive change in array pattern from transverse to longitudinal across the growth zone of primary roots (Hashimoto, 2003). Immunofluorescence images of the cortical microtubules clearly indicate the presence of transversely coaligned microtubule arrays in the smaller box-shaped cells just distal to the root tip prior to their axial expansion. The pattern remains intact until the transition zone where the axially extended cells slow to near-zero growth rate. Here, the microtubules transit to an oblique coalignment and then to a longitudinal pattern before becoming less coaligned.

A similar relation between axial growth and microtubule array order is observed in hypocotyl cells. The Arabidopsis hypocotyl is composed of vertical files of approximately 19 cells each (Gendreau *et al.*, 1997). Upon germination, the hypocotyl grows away from gravity's pull until light is sensed, at which time axial extension slows and the hypocotyl must expand radially to support the growing plant. The hypocotyl experiences little to no cell division and can axially extend each cell up to 500 μm in length while maintaining a width of 10–20 μm. The hypocotyl exhibits a gradient of extension with cells near the root–hypocotyl junction reaching their mature state earlier than cells closer to the cotyledons (Le *et al.*, 2005; Sambade *et al.*, 2012). Microtubule arrays in dark-grown hypocotyl cells tend to be strongly coaligned and transverse to the plant growth axis in the regions of the plant experiencing axial growth. Upon exposure to light, and consequent with de-etiolation, the cells stop their axial extension and the microtubules rapidly rearrange into a pattern that is coaligned with the longitudinal axis of the cell. Similar to root cells, the longitudinal arrays eventually yield to a more diverse set of patterns likely to be involved with later secondary cell wall synthesis.

Due to the direct relationship established between the transverse coaligned cortical microtubule arrays and the axial extension of the hypocotyl and root cells, this array pattern has been the focus of most research attempting to understand microtubule array organization (Lloyd and Chan, 2002, 2004, 2008; Hashimoto, 2003; Ehrhardt and Shaw, 2006; Wasteneys and Ambrose, 2009; Shaw and Lucas, 2011). Comparative work between the two systems has been difficult because of significant technical limitations. Immunofluorescence techniques work extremely well for root tissues while hypocotyl tissues have been intractable to this technique. Conversely, GFP-tubulin-expressing plant lines show relatively high incorporation of the fluorescent probe into the microtubules of hypocotyl cells, but extremely poor incorporation into root cell microtubules. Consequently, nearly all work on dynamics to date has been from hypocotyl cells while important mutant studies and screens have been performed in roots. The current hypotheses generated to explain mechanisms of transverse array organization tend to take observations from both systems into account, with little need at this early phase of discovery for differentiation between them.

Computational Models of Array Organization

Models explaining how microtubules in the plant cortex become transversely coaligned have mostly focused on using the observations made on microtubules in steady-state light-grown cells. Using quantitative data for bundling angle or nucleation type, for example, much work has been done to explain how these potentially self-organizing phenomena could bias the cortical array into specific orientations. The use of computer simulations to advance these models beyond elaborated cartoons has allowed many in this field to make specific mechanistic proposals for how the arrays organize. Models exploring organization through more specific physical principles can be found in the work of Deinum and Tindemans (Hawkins *et al.*, 2010; Tindemans *et al.*, 2010; Tindemans and Mulder, 2010; Deinum *et al.*, 2011).

Organization of the cortical microtubule array in the context of these models is defined principally by two components: the degree of microtubule coalignment and the degree of orientation of that coalignment to the long axis of the cell. On the spectrum of coalignment, transverse, oblique, and longitudinal arrays exhibit maximal coalignment, while more random arrays represent a minimum. As there is no dominant pattern of coalignment, random arrays do not have an 'orientation' relative to the plant axis. In contrast, transverse, oblique, and longitudinal arrays can all be coaligned to a similar degree, but the axis of their coalignment relative to the axis of the plant distinguishes them from each other. The modeling papers described in the following text attempt to determine which parameters of microtubule dynamics and interactions with their local environment may be responsible for patterning cortical array coalignment and orientation.

In 2004, Dixit and Cyr published a paper characterizing the frequency of bundling versus catastrophe as potential outcomes of microtubule–microtubule interactions depending on the angle of the encounter. At angles <40°, 90% of microtubules in BY2 tobacco culture cells reoriented and bundled with the other microtubule, while the rest experienced catastrophe (Dixit and Cyr, 2004). At encounter angles >40°, 60% of microtubules depolymerized. In both cases, the outcomes were not microtubule-polarity dependent. Using this information, Dixit and Cyr designed a computer simulated cortical array with just 20 microtubules. This simplified system used average rates for microtubule plus-end growth and minus-end shortening without adding any new microtubules via nucleation over the course of the experiment. Using the angle-dependent microtubule-interaction outcome frequencies as rules for microtubule–microtubule encounters in this system, they found that both catastrophe-inducing collisions at steep angles and bundling at shallow angles act synergistically and are each necessary for eventually developing a coaligned array. The direction of coalignment varied from one experiment to another, suggesting that establishing the orientation of a coaligned pattern requires more inputs than just these basic microtubule-interaction rules (Dixit and Cyr, 2004).

Baulin *et al.* (2007) investigated the role of microtubule-encounter-induced pausing in promoting cortical array coalignment. Baulin *et al.* accounted for new microtubule nucleation in this model (100 nucleations per μm^2 per second) but restricted all microtubules to pause upon encountering another microtubule, rather than allow for bundling, catastrophe, or crossover. Average growth and shrinking rates were applied to the plus and minus ends respectively; thus, any microtubule that was blocked long enough experienced extinction as the minus end depolymerized to the point of contact. As expected, domains of coaligned microtubules emerged and eventually competed with each other for dominance of the array. This process appeared to be the result of selectively eliminating microtubules that ran at angles deviating too far from the emerging local coalignment. As in Dixit and Cyr's model (2004), this simulation showed no bias for coalignment in any particular direction until additional regulating factors were added. Baulin *et al.* (2007) argue that adding a simulated, gravity-induced bias for initial nucleation direction rapidly increased the speed of coalignment and, furthermore, led to a specifically transverse orientation. They found the same results if one wall of the simulation was a catastrophe-inducing boundary or if one microtubule was fixed in position. In these scenarios, all surviving microtubules would eventually orient in the same direction as that microtubule or wall (Baulin *et al.* 2007).

Allard *et al.* (2010) developed a more complex (still two-dimensional) model that incorporated more cortical array 'rules' than that of the Dixit and Cyr (2004) and Baulin *et al.* (2007) models. Rather than assume average rates for growth and shortening of the plus and minus ends, respectively, this group used the rates and stochastic switching properties of microtubules based on the experimental results collected by Shaw *et al.* (2003). Allard *et al.* also incorporated angle-dependent microtubule bundling or catastrophe outcomes based on Wightmann and Turner (2007) and Dixit and Cyr's (2004) *in vivo* data. As previously mentioned, the rate of catastrophe varied within the range 9–60% for steep-angle collisions, but this paper does not specify what percentage was used for each simulation. They allowed for microtubule-independent and microtubule-dependent nucleation, where all of the dependent nucleations occurred at a 40° angle relative to the parent. For the baseline simulation, all minus ends are assumed to switch between pause and shortening; thus, microtubule severing from the point of nucleation is inherent to the model. They found that collision-induced catastrophe is not sufficient to coalign an array, while pause-inducing collisions are. Bundling is sufficient to induce coalignment with or without collision-induced catastrophe. This is slightly different from Dixit and Cyr's 2004 findings, in which both bundling and catastrophe-inducing collisions were necessary to develop coaligned arrays. This may be a product of using catastrophe-inducing collision rates closer to Wightmann and Turner's (2007) 9% rather than Dixit and Cyr's (2004) 60%.

Allard *et al.* (2010) then used this model to test hypotheses for the sources of cortical array disorganization in known mutants. MOR1 promotes microtubule shrinking and growing velocities (Kawamura and Wasteneys, 2008). In the temperature-sensitive *mor1-1* mutant, at high temperatures those velocities slow and the microtubules within the array become truncated and disorganized. By applying these velocities to the model, Allard *et al.* (2010) found shortened microtubules (0.5 μm on average, compared to 3 μm), lower polymer density, and reduced coalignment. Interestingly, introducing 'static' minus ends that did not depolymerize (normally a delaying factor for coalignment in a wild-type simulation) improved the *mor1-1* model's ability to coalign. It is believed this was a product of generating slightly longer microtubules that allowed for more microtubule–microtubule interactions. They also used this system to test the prediction that greater permissivity for bundling, or a wider 'entrainment angle', is what fosters the seemingly greater degree of coalignment observed in the plus-tip *clasp-1* mutant. Their results did not show an increase in coalignment, indicating that a greater range in interaction angles that allow bundling may not be the source of the *clasp-1* mutant's array defect.

In vitro studies have shown repeatedly that microtubule bending can destabilize microtubules; thus, encountering sharp corners at cell edges could induce microtubule catastrophe. Plant cells might facilitate microtubule passage through corners by targeting at that region proteins that stabilize microtubules against a resistive boundary or curvature-induced depolymerization. Complementary to Baulin *et al.*'s (2007) findings, Allard *et al.* (2010) showed that incorporating boundaries that induced microtubule catastrophe into the simulation was enough to establish an oriented and coaligned array. Furthermore, these edges were enough to induce a biased coalignment, even in the absence of microtubule–microtubule interactions. These results suggest that plant cells may establish transverse coalignment by specifically making apical and basal cell edges that induce microtubule catastrophe (Allard *et al.*, 2010).

A subsequent paper by Eren *et al.* (2010) built on these previous studies and developed a 3D modeling system where simulations ran on the surface of a 50 μm long, 30 μm circumference cylinder. The same plus- and minus-end dynamics as in the Allard *et al.* (2010) paper were applied, as were angle-dependent bundling and catastrophe upon microtubule–microtubule interaction parameters. However, they simplified these interactions by stating that all encounters at angles of <40° resulted in bundling, while all steep-angle encounters had a 30% chance of catastrophe or a 70% chance of crossover. This simulation did not consider crossover-related severing. Depending on the scenario being tested, apical and basal cell edges of the 3D system (the top and bottom of the cylinder) were considered reflective or catastrophe-inducing. In the case of a reflective boundary, the microtubule was redirected back into the array at an angle dependent on the initial encounter angle rather than being depolymerized. Simulations using experimentally observed reduced rates of polymerization and depolymerization characteristic to the *mor1-1* temperature-sensitive mutant recapitulated the array defects previously observed *in vivo* and *in silica*. Further, by simulating the absence of microtubule severing at the origin of nucleation as in the KATANIN mutant *fra2*, Eren *et al.* (2010) also observed the expected failure to coalign microtubules as seen in the mutant phenotype.

Eren *et al.* (2010) found that the ability to bundle is a major contributing factor to coalignment while catastrophe-inducing collisions and boundaries are only minor contributors to coalignment. The pitch of coalignment could be manipulated by selectively switching branch-form nucleation on or off, or by biasing frequency and/or angle of branch-nucleation on one side of the parent microtubule. Eren *et al.* (2010) also found that making only one end wall catastrophe-inducing increased the probability of developing oblique coaligned arrays.

These simulation papers, exploring potential mechanisms for array organization, provide theory for hypothesis-based experimental tests of specific self-organizing properties observed for the cortical microtubules. The outcomes suggest that microtubule–microtubule bundling interactions can have a powerful impact on the degree of coalignment within the array. They also clearly show a variety of dynamic and interacting factors that act in concert to influence the self-organizing properties of the cortical array. These include (but are not limited to): nucleation rates; growth, shortening, and pause rates for the microtubule ends; angle-dependent frequency of microtubule–microtubule interaction outcomes (bundling, pause, catastrophe, crossover, and severing); and branching biases or edge-induced catastrophes.

Observations of Live Cells

Live cell imaging of microtubule arrays organizing into a distinct pattern are becoming increasingly important for examining hypotheses about array organization. In early live-cell imaging work, cells were microinjected with rhodamine-conjugated brain tubulin to visualize the cortical array. Yuan *et al.* (1995) used this technique and found that the cortical array did not completely depolymerize and then repolymerize to create new patterns. Instead, pattern transitions emerged as small regions of new organization that overtook the rest of the array (Yuan *et al.* 1995). Time-lapse imaging of coaligned patterns gave the impression of coaligned cortical arrays 'rotating' from one coaligned orientation to the next. Cortical arrays were even observed to rotate from longitudinal to transverse through oblique patterns cyclically over time (Mayumi and Shibaoka, 1996; Takesue and Shibaoka, 1999; Hejnowicz, 2005). With Shaw *et al.* (2003) demonstrating in photobleaching experiments that cortical microtubules do not translate or slide on the plasma membrane, it became clear that the 'rotation' of the microtubule array had to be a product of treadmilling motility rather than polymer translation.

Chan *et al.* (2007) found in Arabidopsis END BINDING 1 (EB1) and tubulin co-labeling studies that regions of microtubule coalignment would migrate as the individual microtubules polymerized in parallel along self-sustaining tracks. These subdomains of coalignment could rotate their orientation either clockwise or counter-clockwise and could also change direction. They argued that this higher-order group behavior was integral to modifying the pattern of the cortical array over time (Chan *et al.*, 2007).

Examination of defects in plant mutants for the AtCLASP1 microtubule end-binding protein lead Ambrose and Wasteneys (2011) to propose a specific mechanism for eliminating non-transverse microtubules from the outer periclinal array. In this model, the AtCLASP1 protein preferentially associates with microtubules at the corners of the cell along the long axis to facilitate the polymerization across this trans-facial boundary. The absence of AtCLASP1 at the apical and basal edges of the outer periclinal array leads to an induced catastrophe of microtubules treadmilling into these regions, hence a bias for extinction. The net result over time is a strong bias for the transversely aligned microtubules in the array. This proposal constitutes the first model for an explicit molecular control of the array organization and is mostly independent of the self-organizing properties highlighted in the computational simulations.

Another model for molecular-based control of array organization comes not from hypocotyl or root cells, but from cells programmed to become vascular vessel elements. Oda and Fukuda (2012) found that the MIDD1 protein works in concert with other proteins tied to the ROP GTPase family to locally induce microtubule depolymerization in regions that will not be targeted for construction of the thick secondary wall formations indicative of vessel cells (Oda and Fukuda, 2012). The finding of a protein that works to specifically exclude microtubules from areas of the plasma membrane provides a significant clue to how plants may have evolved to locally organize microtubules in the absence of an MTOC.

Sambade *et al.* (2012) imaged hypocotyl cells in light-insensitive mutants to facilitate the observation of array transitions from longitudinal to transverse orientations. Using an EB1-reporter line, they observed a strong bipolarity in the direction of microtubule polymerization of longitudinal arrays. This means that the majority of microtubules were polymerizing apically or basally away from a central point on the outer periclinal array. As arrays transitioned from longitudinal to transverse, they observed a transient 'star' pattern in the array of new microtubules appearing specifically near the mid-zone of the cell. These microtubule bundles appeared to contain more anti-parallel microtubules than those at the apical or basal cell ends. Further, the microtubule density at the ends of the cells seemed lower than in the mid-zone as the new microtubules appeared in the star pattern. This work suggested that plant cells may have the ability to regulate array organization by tuning the density of microtubules by cell region as well as adjusting the localization and degree of anti-parallel bundling permitted for new microtubules (Sambade *et al.*, 2012). Moreover, these authors suggest that the transverse array in these hypocotyl cells is remodeled just prior to axial extension through the invasion of new microtubules from the lateral side faces of the cell into the outer periclinal array. This model infers a direct biological control of the reorganization, and therefore differs from computational models which mostly explore self-organizing properties.

Capitalizing on many years of work showing that plant-derived hormones altered the microtubule array organization in hypocotyl cells when exogenously applied (Mita and Shibaoka, 1983; Mita and Katsumi, 1986; Ishida and Katsumi, 1992; Shibaoka, 1993; Mizuno, 1994; Lloyd *et al.*, 1995; Wenzel *et al.*, 2000), Vineyard *et al.* (2013) used a combination of plant hormones to induce light-grown cells to synchronously reorganize to a transverse coalignment. The quantitative evaluation of tubulin and Eb1-based probes leads to a cell biological model for array organization with two generalizable themes (Fig. 4.4). First, prior to hormone treatment, it was observed that there are typically three microtubules polymerizing edgeward, off of the outer periclinal

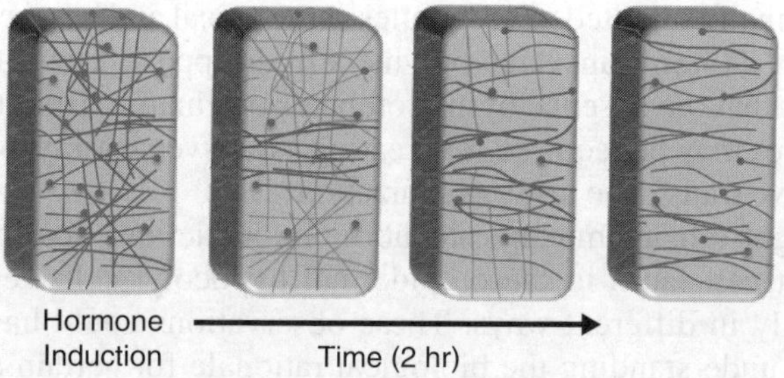

Figure 4.4 Reorganization of the cortical array microtubules to a transverse coalignment after hormone induction. Hypocotyl cells grown in the light show an apparent bias for microtubules created on the outer periclinal array (circles). Cells treated with a combination of auxin and gibberellic acid undergo a progressive reorganization to a transverse microtubule coalignment. Transverse microtubules initially appear from the lateral side faces of the cell and reach across the cell after an initial decrease in the microtubule number on the outer cell face. Microtubules appear around the mid-zone of the cell and reorganize toward the apical and basal regions.

array and into the anticlinal cell faces, for every one microtubule polymerizing into the periclinal cell face. Further, microtubules rarely originated from the apical and basal cell faces to move into the outer periclinal cell face. These observations strongly suggest that there is a bias for microtubule creation on the outer periclinal cell face in normal light-grown hypocotyl cells. This conclusion is supported by the observations that, upon hormone treatment, the total number of microtubule plus ends on the outer periclinal cell face declines by 30–40% over the first 30 min and this level is sustained during array reorganization. Consistent with observations from Sambade *et al.* (2012), these data suggest that the variety of array patterns observed on the outer periclinal array in light-grown plants may be controlled locally on that cell face.

The second general phenomenon observed with hormone addition was the appearance of transverse polymers originating at the mid-zone of the cell, with transverse organization progressing bidirectionally toward the apex and base of the cell over time (Vineyard *et al.*, 2013). These observations, consistent with the observations from light-sensing mutants (Sambade *et al.*, 2012), also showed an invasion of microtubules into the outer periclinal cell face from the lateral anticlinal cell faces. Importantly, the new microtubules appeared after the hormone-induced loss of plus ends that was observed over the entire cell face. In effect, the absence of a strong nucleation bias on the outer periclinal cell face appears to be permissive for the invasion of new transverse microtubules from the lateral cell faces. The invasion appeared spatially regulated, originating from the cell's mid-zone but not extending into the apical or basal faces of the cell. Both the apical and basal cell faces remained strongly retarded for creating new microtubules after hormone addition (Vineyard *et al.*, 2013).

The longitudinally oriented microtubules at the apical and basal cell faces were typically the last non-transverse polymers to disappear. These final observations suggest that the absence of microtubules originating from the apical and basal cell faces may be required for creating a transverse array, but they do not appear to drive transverse array organization.

Both the light-sensing mutants and hormone-induction experiments indicate that different cell faces in these epidermal hypocotyl cells regulate microtubules locally in different ways. These observations could have substantial meaning for understanding the biological rationale for certain array patterns and why they change. For example, if the outer periclinal array in light-grown hypocotyl cells must organize in concert with outer periclinal arrays in adjacent cells to affect radial hypocotyl swelling, local control might produce arrays with less reference to other arrays in the same cell. However, if the same cell is forced back into axial extension, as might be the case for tropic bending or shade avoidance, the microtubule arrays across the long cell faces could be required to coordinate in order to make cellulose fibrils of the appropriate length and orientation around the circumference of the cell.

The observations of array reorganization to transverse coalignment do not directly yield a molecular mechanism, and the molecular mechanisms thus far proposed do not apply entirely to the existing live cell observations. Regardless, plant cell biology is on the cusp of determining what genes are contributing discretely to organization and defining what mechanisms provide the key information to the cell for microtubule array patterning.

References

Al-Bassam, J. and Chang, F. (2011) Regulation of microtubule dynamics by TOG-domain proteins XMAP215/Dis1 and CLASP. *Trends in Cell Biology*, **21**, 604–614.

Allard, J.F., Wasteneys, G.O., and Cytrynbaum, E.N. (2010) Mechanisms of self-organization of cortical microtubules in plants revealed by computational simulations. *Molecular Biology of the Cell*, **21**, 278–286.

Ambrose, J.C. and Wasteneys, G.O. (2008) CLASP modulates microtubule-cortex interaction during self-organization of acentrosomal microtubules. *Molecular Biology of the Cell*, **19**, 4730–4737.

Ambrose, J.C. and Wasteneys, G.O. (2011) Cell edges accumulation gamma tubulin complex components and nucleate microtubules following cytokinesis in Arabidopsis thaliana. *PLOS One*, **6**(11), e27423.

Ambrose, J.C., Shoji, T., Kotzer, A.M., Pighin, J.A., and Wasteneys, G.O. (2007) The Arabidopsis CLASP gene encodes a microtubule-associated protein involved in cell expansion and division. *Plant Cell*, **19**, 2763–2775.

Azimzadeh, J., Nacry, P., Christodoulidou, A., Drevensek, S., Camilleri, C., Amiour, N., Parcy, F., Pastuglia, M., and Bouchez, D. (2008) Arabidopsis TONNEAU1 proteins are essential for preprophase band formation and interact with centrin. *Plant Cell*, **20**, 2146–2159.

Baskin, T.I. (2001) On the alignment of cellulose microfibrils by cortical microtubules: a review and a model. *Protoplasma*, **215**, 150–171.

Baskin, T.I. (2005) Anisotropic expansion of the plant cell wall. *Annual Reviews of Cell and Developmental Biology*, **21**, 203–222.

Basto, R., Lau, J., Vinogradova, T., Gardiol, A., Woods, C.G., Khodjakov, A., and Raff, J.W. (2006) Flies without centrioles. *Cell*, **125**, 1375–1386.

Baulin, V., Marques, C., and Thalmann, F. (2007) Collision induced spatial organization of microtubules. *Biophysical Chemistry*, **128**, 231–244.

Bringmann, M., Li, E., Sampathkumar, A., Kocabek, T., Hauser, M.T., and Persson, S. (2012) POM-POM2/cellulose synthase interacting1 is essential for the functional association of cellulose synthase and microtubules in Arabidopsis. *Plant Cell*, **24**, 163–177.

Burk, D.H., Liu, B., Zhong, R., Morrison, W.H., and Ye, Z.H. (2001) A katanin-like protein regulates normal cell wall biosynthesis and cell elongation. *Plant Cell*, **13**, 807–827.

Camilleri, C., Azimzadeh, J., Pastuglia, M., Bellini, C., Grandjean, O., and Bouchez, D. (2002) The Arabidopsis TONNEAU2 gene encodes a putative novel protein phosphatase 2A regulatory subunit essential for the control of the cortical cytoskeleton. *Plant Cell*, **14**, 833–845.

Caplow, M. and Shanks, J. (1996) Evidence that a single monolayer tubulin-GTP cap is both necessary and sufficient to stabilize microtubules. *Molecular Biology of the Cell*, **7**, 663–675.

Chan, J., Rutten, T., and Lloyd, C. (1996) Isolation of microtubule-assciated proteins from carrot cytoskeletons: a 120 kDa map decorates all four microtubule arrays and the nucleus. *Plant Journal*, **10**, 251–259.

Chan, J., Jensen, C., Jensen, L., Bush, M., and Lloyd, C. (1999) The 65-kDa carrot microtubule-associated protein forms regularly arranged filamentous cross-bridges between microtubules. *PNAS*, **96**, 14931–14936.

Chan, J., Calder, G.M., Doonan, J.H., and Lloyd, C.W. (2003) EB1 reveals mobile microtubule nucleation sites in Arabidopsis. *Nature Cell Biology*, **5**, 967–971.

Chan, J., Calder, G., Fox, S., and Lloyd, C. (2007) Cortical microtubule arrays undergo rotary movements in Arabidopsis hypocotyl epidermal cells. *Nature Cell Biology*, **9**, 171–175.

Chan, J., Sambade, A., Calder, G., and Lloyd, C. (2009) Arabidopsis cortical microtubules are initiated along, as well as branching from, existing microtubules. *Plant Cell*, **21**, 2298–2306.

Crowell, E.F., Bischoff, V., Desprez, T., Rolland, A., Stierhof, Y.D., Schumacher, K., Gonneau, M., Hofte, H., and Vernhettes, S. (2009) Pausing of Golgi bodies on microtubules regulates secretion of cellulose synthase complexes in Arabidopsis. *Plant Cell*, **21**, 1141–1154.

Deinum, E.E., Tindemans, S.H., and Mulder, B.M. (2011) Taking directions: the role of microtubule-bound nucleation in the self-organization of the plant cortical array. *Physical Biology*, **8**, 056002.

Desai, A. and Mitchison, T.J. (1997) Microtubule polymerization dynamics. *Annual Reviews in Cell and Developmental Biology*, **13**, 83–117.

Dixit, R. and Cyr, R. (2004) Encounters between dynamic cortical microtubules promote ordering of the cortical array through angle-dependent modifications of microtubule behavior. *Plant Cell*, **16**, 3274–3284.

Downing, K.H. and Nogales, E. (1998) Tubulin structure: insights into microtubule properties and functions. *Current Opinion in Structural Biology*, **8**, 785–791.

Ehrhardt, D.W. and Shaw, S. L. (2006) Microtubule dynamics and organization in the plant cortical array. *Annul Reviews in Plant Biology*, **57**, 859–875.

Eren, E.C., Dixit, R., and Gautam, N. (2010) A three-dimensional computer simulation model reveals the mechanisms for self-organization of plant cortical microtubules into oblique arrays. *Molecular Biology of the Cell*, **21**, 2674–2684.

Fishel, E. and Dixit, R. (2013) Role of nucleation in cortical microtubule array organization: variations on a theme. *The Plant Journal*, **10**, 1–8.

Gaillard, J., Neumann, E., Van Damme, D., Stoppin-Mellet, V., Ebel, C., Barbier, E., Geelen, D., and Vantard, M. (2008) Two microtubule-associated proteins of Arabidopsis MAP65a promote antiparallel microtubule bundling. *Molecular Biology of the Cell*, **19**, 4534–4544.

Gardiner, J., Harper, J.D., Weerakoon, N.D., Collings, D.A., Ritchie, S., Gilroy, S., Cyr, R.J., and Marc, J. (2001) A 90-kD phospholipase D from tobacco binds to microtubules and the plasma membrane. *Plant Cell*, **13**, 2143–2158.

Gardiner, J., Collings, D.A., Harper, J.D., and Marc, J. (2003) The effects of the phospholipase D-antagonist 1-butanol on seedling development and microtubule organisation in Arabidopsis. *Plant Cell Physiology*, **44**, 687–696.

Gendreau, E., Trass, J., Desnos, T., Grandjean, O., Caboche, M., and Hofte, H. (1997) Cellular basis of hypocotyl growth in Arabidopsis thaliana. *Plant Physiology*, **114**, 295–305.

Giddings, T. and Staehelin, L. (1991) The cytoskeletal basis of plant cell growth and form: microtubule-mediated control of microfibril deposition: a re-examination of the hypothesis, in *The Cytoskeletal Basis of Plant Growth and Form* (ed. C.W. Lloyd), Academic Press, San Diego, pp. 85–100.

Giddings, T., Brower, D., and Staehelin, L.A. (1980) Visualization of particle complexes in the plasma membrane of Micrasterias denticulata associated with the formation of cellulose fibrils in primary and secondary cell walls. *Journal of Cell Biology*, **84**, 329–339.

Green, P.B. (1962) Mechanism for plant cellular morphogenesis. *Science*, **138**, 1404–1405.

Gutierrez, R., Lindeboom, J.J., Paredez, A.R., Emons, A.M., and Ehrhardt, D.W. (2009) Arabidopsis cortical microtubules position cellulose synthase delivery to the plasma membrane and interact with cellulose synthase trafficking compartments. *Nature Cell Biology*, **11**, 797–806.

Hartman, J.J., Mahr, J., McNally, K., Okawa, K., Iwamatsu, A., Thomas, S., Cheesman, S., Heuser, J., Vale, R.D., and McNally, F.J. (1998) Katanin, a microtubule-severing protein, is a novel AAA ATPase that targets to the centrosome using a WD40-containing subunit. *Cell*, **93**, 277–287.

Hashimoto, T. (2003) Dynamics and regulation of plant interphase microtubules: a comparative view. *Current Opinion in Plant Biology*, **6**, 568–576.

Hawkins, R.J., Tindemans, S.H., and Mulder, B.M. (2010) Model for the orientational ordering of the plant microtubule cortical array. *Physical Review E: Statistical, Nonlinear, and Soft Matter Physics*, **82**, 011911.

Hejnowicz, Z. (2005) Autonomous changes in the orientation of cortical microtubules underlying the helicoidal cell wall of the sunflower hypocotyl epidermis: spatial variation translated into temporal changes. *Protoplasma*, **92**(3), 391–402.

Hirokawa, N. (1998) Kinesin and dynein superfamily proteins and the mechanism of organelle transport. *Science*, **279**, 519–526.

Ho, C.M., Hotta, T., Guo, F., Roberson, R.W., Lee, Y.R., and Liu, B. (2011) Interaction of antiparallel microtubules in the phragmoplast is mediated by the microtubule-associated protein MAP65-3 in Arabidopsis. *Plant Cell*, **23**, 2909–2923.

Horio, T., Uzawa, S., Jung, M.K., Oakley, B.R., Tanaka, K., and Yanagida, M. (1991) The fission yeast gamma-tubulin is essential for mitosis and is localized at microtubule organizing centers. *Journal of Cell Science*, **99**(4), 693–700.

Hornick, J.E., Mader, C.C., Tribble, E.K., Bagne, C.C., Vaughan, K.T., Shaw, S.L., and Hinchcliffe, E.H. (2011) Amphiastral mitotic spindle assembly in vertebrate cells lacking centrosomes. *Current Biology*, **21**, 598–605.

Hotta, T., Kong, Z., Ho, C., Zeng, C., Horio, T., Fong, S., Vuong, T., Lee, Y., and Liu, B. (2012) Characterization of the Arabidopsis Augmin complex uncoverse its critical function in the assembly of the acentrosomal spindle and phragmoplast microtubule arrays. *Plant Cell*, **24**, 1494–1509.

Howard, J. and Hyman, A.A. (2003) Dynamics and mechanics of the microtubule plus end. *Nature*, **422**, 753–758.

Hussey, P., Hawkins, T., Igarashi, H., Kaloriti, D., and Smertenko, A. (2002) The plant cytoskeleton:recent advances in the study of the plant microtubule-associated proteins MAP-65, MAP-190 and the Xenopus MAP215-like protein, MOR1. *Plant Molecular Biology*, **50**, 915–924.

Ishida, K. and Katsumi, M. (1992) Effects of gibberellin and abscisic acid on the cortical microtubule orientation in hypocotyl cells of light-grown cucumber seedlings. *International Journal of Plant Sciences*, **153**, 155–163.

Ishida, T., Thitamadee, S., and Hashimoto, T. (2007a) Twisted growth and organization of cortical microtubules. *Journal of Plant Research*, **120**, 61–70.

Ishida, T., Kaneko, Y., Iwano, M., and Hashimoto, T. (2007b) Helical microtubule arrays in a collection of twisting tubulin mutants of Arabidopsis thaliana. *Proceedings of National Academy of Science, USA*, **104**, 8544–8549.

Janson, M.E., de Dood, M. E., and Dogterom, M. (2003) Dynamic instability of microtubules is regulated by force. *Journal of Cell Biology*, **161**, 1029–1034.

Kawamura, E. and Wasteneys, G. (2008) MOR1, the Arabidopsis thaliana homologue of Xenopus MAP215, promotes rapid growth and shrinkage, and suppresses the pausing of microtubules in vivo. *Journal of Cell Science*, **121**, 4114–4123.

Kazama, H., Dan, H., Imaseki, H., and Wasteneys, G.O. (2004) Transient exposure to ethylene stimulates cell division and alters the fate and polarity of hypocotyl epidermal cells. *Plant Physiology*, **134**, 1614–1623.

Kirik, A., Ehrhardt, D.W., and Kirik, V. (2012) TONNEAU2/FASS regulates the geometry of microtubule nucleation and cortical array organization in interphase Arabidopsis cells. *Plant Cell*, **24**, 1158–1170.

Kirik, V., Herrmann, U., Parupalli, C., Sedbrook, J. C., Ehrhardt, D. W., and Hulskamp, M. (2007) CLASP localizes in two discrete patterns on cortical microtubules and is required for cell morphogenesis and cell division in Arabidopsis. *Journal of Cell Science*, **120**, 4416–4425.

Le, J., Vandenbussche, F., De Cnodder, T., Van Der Straeten, D., and Verbelen, J.P. (2005) Cell elongation and microtubule behavior in the Arabidopsis Hypocotyl: Responses to ethylene and auxin. *Journal of Plant Growth Regulation*, **24**, 166–178.

Ledbetter, M. and Porter, K. (1963) A "microtubule" in plant cell fine structure. *Journal of Cell Biology*, **19**, 239–250.

Li, W., Li, M., Zhang, W., Welti, R., and Wang, X. (2004) The plasma membrane-bound phospholipase Ddelta enhances freezing tolerance in Arabidopsis thaliana. *Nature Biotechnology*, **22**, 427–433.

Lindeboom, J.J., Lioutas, A., Deinum, E.E., Tindemans, S.H., Ehrhardt, D.W., Emons, A.M., Vos, J.W., and Mulder, B.M. (2013) Cortical microtubule arrays are initiated from a nonrandom prepattern driven by atypical microtubule initiation. *Plant Physiology*, **161**, 1189–1201.

Liu, B., Joshi, H.C., Wilson, T.J., Silflow, C.D., Palevitz, B.A., and Snustad, D.P. (1994) gamma-Tubulin in Arabidopsis: gene sequence, immunoblot, and immunofluorescence studies. *Plant Cell*, **6**, 303–314.

Lloyd, C. and Chan, J. (2002) Helical microtubule arrays and spiral growth. *Plant Cell*, **14**, 2319–2324.

Lloyd, C. and Chan, J. (2004) Microtubules and the shape of plants to come. *Nature Reviews Molecular Cell Biology*, **5**, 13–22.

Lloyd, C. and Chan, J. (2008) The parallel lives of microtubules and cellulose microfibrils. *Current Opinion in Plant Biology*, **11**, 641–646.

Lloyd, C., Shaw, P.J., Warn, R.M., and Yuan, M. (1995) Gibberellic-acid-induced reorientation of cortical microtubules in living plant cells. *Journal of Microscopy*, **181**, 140–144.

Locascio, A., Blazquez, M.A., and Alabadi, D. (2013) Dynamic regulation of cortical microtubule organization through prefoldin-DELLA interaction. *Current Biology*, **23**, 804–809.

Lucas, J. and Shaw, S.L. (2008) Cortical microtubule arrays in the Arabidopsis seedling. *Current Opinion in Plant Biology*, **11**, 94–98.

Lucas, J.R., Courtney, S., Hassfurder, M., Dhingra, S., Bryant, A., and Shaw, S.L. (2011) Microtubule-associated proteins MAP65-1 and MAP65-2 positively regulate axial cell growth in etiolated arabidopsis hypocotyls. *Plant Cell*, **23**, 1889–1903.

Luders, J., Patel, U.K., and Stearns, T. (2006) GCP-WD is a gamma-tubulin targeting factor required for centrosomal and chromatin-mediated microtubule nucleation. *Nature Cell Biology*, **8**, 137–147.

Ludwig, S.R., Oppenheimer, D.G., Silflow, C.D., and Snustad, D.P. (1987) Characterization of the alpha-tubulin gene family of Arabidopsis thaliana. *Proceedings of National Academy of Science, USA*, **84**, 5833–5837.

Mao, T., Jin, L., Li, H., Liu, B., and Yuan, M. (2005) Two microtubule-associated proteins of the Arabidopsis MAP65 family function differently on microtubules. *Plant Physiology*, **138**, 654–662.

Margolin, G., Gregoretti, I.V., Goodson, H.V., and Alber, M.S. (2006) Analysis of a mesoscopic stochastic model of microtubule dynamic instability. *Physical Review E: Statistical, Nonlinear, and Soft Matter Physics*, **74**, 041920.

Margulis, L. (1973) Colchcine-sensitive microtubules. *International Review of Cytology*, **34**, 333–361.

Mathur, J., Mathur, N., Kernebeck, B., Srinivas, B., and Hulskamp, M. (2003) A novel localization pattern for an EB1-like protein links microtubule dynamics to endomembrane organization. *Current Biology*, **13**, 1991–1997.

Mayumi, K. and Shibaoka, H. (1996) The cyclic reorientation of cortical microtubules on walls with crossed polylamellate structure: Effects of plant hormones and an inhibitor of protein kinases on the progression of the cell cycle. *Protoplasma*, **195**, 112–122.

Mimori-Kiyosue, Y., Shiina, N., and Tsukita, S. (2000) The dynamic behavior of the APC-binding protein EB1 on the distal ends of microtubules. *Current Biology*, **10**, 865–868.

Mita, T. and Shibaoka, H. (1983) Changes in microtubules in onion leaf sheath cells during bulb development. *Plant Cell Physiology*, **24**, 109–117.

Mita, T. and Katsumi, M. (1986) Gibberellin control of microtubule arrangement in the mesocotyl epidermal cells of the d5 mutant of Zea mays. *Plant Cell Physiology*, **27**, 651–659.

Mizuno, K. (1994) Inhibition of gibberellin-induced elongation, reorientation of cortical microtubules and change of isoform of tubulin in epicotyl segments of azuki bean by protein kinase inhibitors. *Plant Cell Physiology*, **35**, 1149–1157.

Murata, T., Sonobe, S., Baskin, T. I., Hyodo, S., Hasezawa, S., Nagata, T., Horio, T., and Hasebe, M. (2005) Microtubule-dependent microtubule nucleation based on recruitment of gamma-tubulin in higher plants. *Nature Cell Biology*, **7**, 961–968.

Nakamura, M., Ehrhardt, D.W., and Hashimoto, T. (2010) Microtubule and katanin-dependent dynamics of microtubule nucleation complexes in the acentrosomal Arabidopsis cortical array. *Nature Cell Biology*, **12**, 1064–1070.

Nakamura, M., Yagi, N., Kato, T., Fujita, S., Kawashima, N., Erhardt, D., and Hashimoto, T. (2012) Arabidopsis GCP3-interacting protein 1/MOTZART 1 is an integral component of the gamma-tubulin-containing microtubule nucleating complex. *Plant Journal*, **71**, 216–225.

Newcomb, E. (1969) Plant microtubules. *Annual Reviews in Plant Physiology*, **20**, 253–288.

Nogales, E., Wolf, S.G., and Downing, K.H. (1998) Structure of the alpha beta tubulin dimer by electron crystallography. *Nature*, **391**, 199–203.

Oda, Y. and Fukuda, H. (2012) Initiation of cell wall pattern by a Rho- and microtubule-driven symmetry breaking. *Science*, **337**, 1333–1336.

Oppenheimer, D.G., Haas, N., Silflow, C.D., and Snustad, D.P. (1988) The beta-tubulin gene family of Arabidopsis thaliana: preferential accumulation of the beta 1 transcript in roots. *Gene*, **63**, 87–102.

Paredez, A., Somerville, C., and Ehrhardt, D. (2006) Visualization of cellulose synthase demonstrates a functional association with microtubules. *Science*, **312**, 1491–1495.

Pickett-Heaps, J. (1967) The effects of colchicine on the ultrastructure of dividing plant cells, xylem wall differentiation and distribution of cytoplasmic microtubules. *Developmental Biology*, **15**, 206–236.

Pickett-Heaps, J.D. (1969) The evolution of the mitotic apparatus: An attempt at comparative ultrastructural cytology in dividing plant cells. *Cytobios*, **3**, 257–280.

Pickett-Heaps, J.D. (1975) Aspects of spindle evolution. *Annals of the New York Academy of Sciences*, **253**, 352–361.

Rusan, N., Fagerstrom, C., Yvon, A., and Wadsworth, P. (2001) Cell cycle-dependent changes in microtubule dynamics in living cells expressing green fluorescent protein-alpha tubulin. *Molecular Biology of the Cell*, **12**, 971–980.

Sambade, A., Pratap, A., Buschmann, H., Morris, R.J., and Lloyd, C. (2012) The influence of light on microtubule dynamics and alignment in the Arabidopsis hypocotyl. *Plant Cell*, **24**, 192–201.

Sasabe, M. and Machida, Y. (2006) MAP65: a bridge linking MAP kinase to microtubule turnover. *Current Opinion in Plant Biology*, **9**, 563–570.

Segui-Simarro, J.M., Austin, J.R., 2nd,, White, E.A., and Staehelin, L.A. (2004) Electron tomographic analysis of somatic cell plate formation in meristematic cells of Arabidopsis preserved by high-pressure freezing. *Plant Cell*, **16**, 836–856.

Seltzer, V., Janski, N., Canaday, J., Herzog, E., Erhardt, M., Evrard, J.L., and Schmit, A.C. (2007) Arabidopsis GCP2 and GCP3 are part of a soluble gamma-tubulin complex and have nuclear envelope targeting domains. *Plant Journal*, **52**, 322–331.

Shaw, S.L., and Lucas, J. (2011) Intrabundle microtubule dynamics in the Arabidopsis cortical array. *Cytoskeleton (Hoboken)*, **68**, 56–67.

Shaw, S.L., Kamyar, R., and Ehrhardt, D.W. (2003) Sustained microtubule treadmilling in Arabidopsis cortical arrays. *Science*, **300**, 1715–1718.

Shibaoka, H. (1993) Regulation of gibberellins of the orientation of cortical microtubules in plant cells. *Australian Journal of Plant Physiology*, **20**, 461–470.

Small, J., Geiger, B., Kaverina, I., and Bershadsky, A. (2002) How do microtubules guide migrating cells? *Molecular Cell Biology*, **3**, 957–964.

Smertenko, A., Saleh, N., Igarashi, H., Mori, H., Hauser-Hahn, I., Jiang, C.J., Sonobe, S., Lloyd, C.W., and Hussey, P.J. (2000) A new class of microtubule-associated proteins in plants. *Nature Cell Biology*, **2**, 750–753.

Smertenko, A.P., Chang, H.Y., Wagner, V., Kaloriti, D., Fenyk, S., Sonobe, S., Lloyd, C., Hauser, M.T., and Hussey, P.J. (2004) The Arabidopsis microtubule-associated protein AtMAP65-1: molecular analysis of its microtubule bundling activity. *Plant Cell*, **16**, 2035–2047.

Steinborn, K., Maulbetsch, C., Priester, B., Trautmann, S., Pacher, T., Geiges, B., Kuttner, F., Lepiniec, L., Stierhof, Y.D., Schwarz, H., Jurgens, G., and Mayer, U. (2002) The Arabidopsis PILZ group genes encode tubulin-folding cofactor orthologs required for cell division but not cell growth. *Genes & Development*, **16**, 959–971.

Takesue, K., and Shibaoka, H. (1999) Auxin-induced longitudinal-to-transverse reorientation of cortical microtubules in nonelongating epidermal cells of azuki bean epicotyls. *Protoplasma*, **206**, 27–30.

Tian, G., Bhamidipati, A., Cowan, N.J., and Lewis, S.A. (1999) Tubulin folding cofactors as GTPase-activating proteins. *GTP hydrolysis and the assembly of the alpha/beta-tubulin heterodimer. Journal of Biological Chemistry*, **274**, 24054–24058.

Tindemans, S.H., and Mulder, B.M. (2010) Microtubule length distributions in the presence of protein-induced severing. *Physical Reviews E: Statistical, Nonlinear, and Soft Matter Physics*, **81**, 031910.

Tindemans, S.H., Hawkins, R.J., and Mulder, B.M. (2010) Survival of the aligned: ordering of the plant cortical microtubule array. *Physical Review Letters*, **104**, 058103.

Torres-Ruiz, R.A. and Jurgens, G. (1994) Mutations in the FASS gene uncouple pattern formation and morphogenesis in Arabidopsis development. *Development*, **120**, 2967–2978.

Van Damme, D., Poucke, K., Boutant, E., Ritzenhaler, C., Inze, D., and Geelen, D. (2004) In vitro dynamics and differential microtubule-binding activities of MAP65 proteins. *Plant Physiology*, **136**, 3956–3967.

Verde, F., Dogterom, M., Stelzer, E., Karsenti, E., and Leibler, S. (1992) Control of microtubule dynamics and length by cyclin A- and cyclin B-dependent kinases in Xenopus egg extracts. *Journal of Cell Biology*, **118**, 1097–1108.

Vineyard, L., Elliott, A., Dhingra, S., Lucas, J.R., and Shaw, S.L. (2013) Progressive transverse microtubule array organization in hormone-induced Arabidopsis hypocotyl cells. *Plant Cell*, **25**, 662–676.

Vitre, B., Coquelle, F., Heichette, C., Garnier, C., Chretien, D., and Arnal, I. (2008) EB1 promotes microtubule dynamics and regulates tubulin sheet closure in vitro. *Nature Cell Biology*, **10**, 415–421.

Vorobjev, I., Malikov, V., and Rodionov, V. (2001) Self-organization of a radial microtubule array by dynein-dependent nucleation of microtubules. *Proceedings of National Academy of Science, USA*, **98**, 10160–10165.

Wasteneys, G. and Ambrose, C. (2008) Spatial organization of plant cortical microtubules: close encounters of the 2D kind. *Trends in Cell Biology*, **19**, 62–71.

Wasteneys, G.O. and Ambrose, J.C. (2009) Spatial organization of plant cortical microtubules: close encounters of the 2D kind. *Trends in Cell Biology*, **19**, 62–71.

Wenzel, C.L., Williamson, R.E., and Wasteneys, G.O. (2000) Gibberellin-induced changes in growth anisotropy precede gibberellin-dependent changes in cortical microtubule orientation in developing epidermal cells of barley leaves. Kinematic and cytological studies on a gibberellin-responsive dwarf mutant, M489. *Plant Physiology*, **124**, 813–822.

Whittington, A.T., Vugrek, O., Wei, K.J., Hasenbein, N.G., Sugimoto, K., Rashbrooke, M.C., and Wasteneys, G.O. (2001) MOR1 is essential for organizing cortical microtubules in plants. *Nature*, **411**, 610–613.

Wightmann, R. and Turner, S. (2007) Severing at sites of microtubule crossover contributes to microtubule alignment in cortical arrays. *Plant Journal*, **52**, 742–751.

Wittmann, T. (2008) EBs clip CLIPs to growing microtubule ends. *Journal of Cell Biology*, **183**, 1183–1185.

Yuan, M., Warn, R.M., Shaw, P.J., and Lloyd, C.W. (1995) Dynamic microtubules under the radial and outer tangential walls of microinjected pea epidermal cells observed by computer reconstruction. *Plant Journal*, **7**, 17–23.

Zeng, C.J., Lee, Y.R., and Liu, B. (2009) The WD40 repeat protein NEDD1 functions in microtubule organization during cell division in Arabidopsis thaliana. *Plant Cell*, **21**, 1129–1140.

Zheng, Y., Wong, M.L., Alberts, B., and Mitchison, T. (1995) Nucleation of microtubule assembly by a gamma-tubulin-containing ring complex. *Nature*, **378**, 578–583.

5 Actin Filament Dynamics and their Role in Plant Cell Expansion

Jiejie Li[1], Ruthie Arieti[1,2] and Christopher J. Staiger[1,3]

[1] *Department of Biological Sciences, Purdue University, 201 S. University Street, West Lafayette, IN*

[2] *Purdue University Life Sciences Graduate Program (PULSe), Purdue University, West Lafayette, IN*

[3] *Bindley Bioscience Center, Purdue University, West Lafayette, IN*

Introduction

Actin filaments (F-actin) are polymeric structures assembled from monomeric subunits, referred to as globular or G-actin. The best-characterized protein is the skeletal muscle α-actin isoform, which has been subjected to detailed biochemical and biophysical scrutiny for more than 40 years (Pollard, 2000; Pollard *et al.*, 2000). In plants, pollen is the best source of assembly-competent actin, a 42 kDa polypeptide, where it may comprise 5–10% of total protein (Yen *et al.*, 1995; Ren *et al.*, 1997). Actin monomers are asymmetric molecules with four subdomains organized around a deep cleft that contains nucleotide and divalent cation-binding sites (Kabsch *et al.*, 1990; Kabsch and Vanderkerckhove, 1992). In the test tube, actin assembles spontaneously in the presence of ATP, physiological salt concentration, and a divalent cation such as Mg^{2+} (Pollard *et al.*, 2000). The rate-limiting step for assembly is formation of a trimer of subunits to generate a seed or nucleus. After nucleation, elongation occurs rapidly with subunit addition occurring at both ends of the filament. Filaments are 7–9 nm in diameter and comprise two strands of protein subunits with a right-handed helical twist (Oda *et al.*, 2009). Polarity is conferred by the head-to-tail arrangement of monomers within each strand of the filament and by the nucleotide-loaded status of subunits along the filament. The plus or barbed end of a filament prefers to add ATP–actin monomers. Nucleotide hydrolysis followed by release of P_i occurs seconds to minutes after assembly of monomers, so the older portions of F-actin contain subunits with bound ADP. The minus or pointed end of a filament tends to prefer loss of these

Plant Cell Wall Patterning and Cell Shape, First Edition. Edited by Hiroo Fukuda.

ADP-loaded subunits. At equilibrium in the test tube, a dynamic steady-state regime is established whereby subunit addition is equally matched by subunit loss in a process referred to as 'treadmilling'. Popular models of actin turnover within cells, for example the dendritic array at the leading edge of crawling animal cells, describe the mechanism as a form of treadmilling (Pollard and Borisy, 2003; Pollard and Cooper, 2009; Bugyi and Carlier, 2010). However, cells are not at equilibrium and they contain hundreds of accessory proteins known as actin-binding proteins; these modulate the behavior of actin monomers and filaments (Pollard *et al.*, 2000; Dos Remedios *et al.*, 2003).

Dozens of actin-binding proteins have been described and partially characterized from plants (reviewed by Hussey *et al.*, 2002; Staiger and Hussey, 2004; Staiger and Blanchoin, 2006; Ren and Xiang, 2007; Hussey and Hashimoto, 2008; Thomas *et al.*, 2009; Blanchoin *et al.*, 2010; Staiger *et al.*, 2010; Thomas, 2012). These can be grouped into general classes based on their biochemical activities, including: (1) nucleation factors; (2) monomer-binding proteins; (3) end-capping proteins; (4) filament severing and dynamizing proteins; and (5) bundling or cross-linking proteins. Many actin-binding proteins are multidomain polypeptides and thus harbor more than one activity.

Most work has been done with proteins that are evolutionarily conserved across several kingdoms, but plant-specific actin-binding proteins have also been uncovered. For example, the NET proteins (Deeks *et al.*, 2012), CHUP1 (Oikawa *et al.*, 2003, 2008; Kong *et al.*, 2013), SCAB1 (Zhao *et al.*, 2011), VHA1 (Ma *et al.*, 2012) and THRUMIN1 (Whippo *et al.*, 2011) have been recently identified as novel players in cytoskeletal remodeling. Moreover, proteins that were previously thought to be involved with microtubules have also been shown to interact with actin filaments, e.g. MAP18 (Zhu *et al.*, 2013) or *vice versa*, such as formins (Deeks *et al.*, 2010; Li *et al.*, 2010; Rosero *et al.*, 2013) and kinesins (Preuss *et al.*, 2004; Suetsugu *et al.*, 2010; Klotz and Nick, 2011). A final category of actin-associated accessory proteins is the myosin family. In plants, these belong to myosin VIII and myosin XI classes. Myosins are molecular motors that use energy from ATP hydrolysis to translocate cargo molecules, such as vesicles and organelles, along actin filaments (reviewed by Lee and Liu, 2004; Yokota and Shimmen, 2010; Sparkes, 2012; Tominaga and Nakano, 2012).

Uncovering the exact cellular function of these actin-binding proteins requires combining knockout mutant lines with reporters of cytoskeletal organization and state-of-the-art fluorescence-imaging methods. In the past, gross perturbations to the actin cytoskeleton have been described for several knockout or knockdown mutants (Dong *et al.*, 2001; Szymanski, 2001, 2005; Frank and Smith, 2002; Ketelaar *et al.*, 2004b; Deeks and Hussey, 2005; Deeks *et al.*, 2007), but only recently have imaging techniques allowed the description of individual filament turnover with molecular precision.

The Actin Cytoskeleton and Cell Morphogenesis

In animal cells actin filaments control cell shape and power cell crawling, whereas microtubules serve as tracks for long-distance transport (Pollard and Cooper, 2009). By comparison, in plants actin filaments provide the principal tracks on which organelles and vesicles move through a cell. From delivering materials, to expanding the cell wall, to selectively enabling and/or inhibiting endocytosis, among other activities, there is substantial genetic and pharmacological evidence that the dynamic plant actin cytoskeleton plays an important role in cell morphogenesis (Smith, 2003; Smith and Oppenheimer, 2005; Szymanski and Cosgrove, 2009). The actin tracks that support intracellular trafficking comprise both fine arrays of actin filaments as well as more prominent, higher-order structures called actin bundles. It is however necessary to note that the precise mechanisms that drive actin organization and behavior, and thereby guide and regulate cell growth and cell shape determination, remain poorly understood.

Pharmacological and Genetic Evidence for Actin Function in Cell Morphogenesis

Drugs that alter actin stability and polymerization result in changes to plant cell shape, demonstrating that actin exerts some control over cell morphogenesis (Smith, 2003; Smith and Oppenheimer, 2005). However, identifying changes in a cell's phenotypic response to actin-disrupting drugs only informs us that actin has a role in the response, that is, drugs can indicate whether or not actin is a player in a process but cannot show how actin organization or dynamics give rise to the phenotype. Aberrant cell morphologies in response to drug treatments occur among diverse cell types. For example, increases in Arabidopsis root hair girth, lily pollen tube tip swelling, and eventual dose-dependent cessation of tip-growth in both cell types are reported outcomes of actin disruption by cytochalasins (Vidali *et al.*, 2001; Ketelaar *et al.*, 2003). Cytochalasin exposure results in trichomes with swollen bases and asymmetrical and twisted branching, as well as significant but incomplete growth inhibition (Szymanski *et al.*, 1999). Latrunculins, which inhibit actin polymerization (Gibbon *et al.*, 1999), induce shorter cell lengths and organs, petite plants, and swollen hypocotyls across various cell types in Arabidopsis, rye and maize (Baluska *et al.*, 2001). Latrunculin B (LatB) treatment of Arabidopsis roots inhibits not only cell elongation, but also cell division (Rahman *et al.*, 2007), and cytochalasin B leads to prominent swelling or isotropic expansion of epidermal cells (Baskin and Bivens, 1995). Following treatments with the actin-stabilizing compound jasplakinolide (JASP), slow development and diminished shape differentiation were observed in *Micrasterias denticulata*, a single-celled green alga (Holzinger and Meindl, 1997), as well as Arabidopsis root hairs (Voigt *et al.*, 2005).

Imaging and immunolocalization techniques provide indisputable evidence that actin is present in the apex of tip-growing cells (Smith and Oppenheimer, 2005; Samaj *et al.*, 2006), is necessary for tip growth (Gibbon *et al.*, 1999; Vidali *et al.*, 2001; Ketelaar *et al.*, 2003), and that its presence is accompanied by and coordinates the concentration of secretory vesicles (Gu *et al.*, 2005; Lee *et al.*, 2008). Similarly, lobed epidermal pavement cells (EPC) as well as apically extending trichomes have a fine, cortical meshwork of actin filaments that correlates with the sites of expansion (Szymanski *et al.*, 1999; Fu *et al.*, 2002, 2005). Plant cells treated with actin-destabilizing drugs exhibit altered or halted vesicle distribution (Gu *et al.*, 2005; Voigt *et al.*, 2005; Lee *et al.*, 2008). Growing cells were found to have more diffuse F-actin arrays, whereas cells at late growth stages (mature cells) have less diffuse F-actin and more bundles (Szymanski *et al.*, 1999; Ketelaar *et al.*, 2003). Endosomes (Voigt *et al.*, 2005) and Golgi (Akkerman *et al.*, 2011) were seen near patches of F-actin, and faster moving 'actin-associated spherical bodies' (Szymanski *et al.*, 1999) and Golgi (Akkerman *et al.*, 2011) were found to associate along bundles. Furthermore, Golgi stacks exhibit differential motility depending on the type of actin array on which they reside: 'wiggling' movements occur in regions of dense F-actin and very fast and directional motility is observed along actin bundles (Akkerman *et al.*, 2011), further indicating that actin organization might regulate vesicle trafficking. Significantly, actin-disrupting drugs alter cell shape development only in growing cells, supporting the argument that an intact actin cytoskeleton is necessary for normal cell morphogenesis; perhaps correct vesicle trafficking is most important during the expansion phase. If cell expansion occurred before drug treatment, the resultant disruptions in actin organization did not alter cell shape (Szymanski *et al.*, 1999). Whether this accumulation of dense actin arrays at sites of cell expansion acts as a strict inhibitor of vesicle fusion, a strict promoter of vesicle fusion, or behaves as a physical block to keep vesicles at the plasma membrane, is not entirely clear (Smith and Oppenheimer, 2005).

Somewhat more compelling evidence for the involvement of actin in cell morphogenesis comes from forward-genetic approaches. Genetic screens for altered cell shape in maize and Arabidopsis EPC as well as in Arabidopsis trichomes uncovered key regulators of actin polymerization: the SCAR/WAVE and ARP2/3 complexes. The *brick1* (*brk1*) mutant affects EPC morphogenesis in maize leaves (Frank and Smith, 2002; Frank *et al.*, 2003, 2004). The phenotype results from a single base-pair substitution in a gene for the smallest subunit of the WAVE/SCAR complex, necessary for activation of the ARP2/3 complex (Frank *et al.*, 2004). Wild-type maize leaves have lobed rectangular EPC and, while the *brk1* mutant exhibits normal elongation and organ development, EPC lobes fail to develop (Frank and Smith, 2002; Frank *et al.*, 2003).

In wild type, lobes begin to grow from regions where F-actin is enriched; *brk1* lacks these regions of enrichment (Frank and Smith, 2002), suggesting that F-actin might coordinate lobe outgrowth. Another set of mutants that exhibit abnormal EPC lobe morphology are four *arp2/3* mutants of Arabidopsis (Li *et al.*, 2003). In the *arp* mutant plants, EPC are less lobed compared to wild type and, like *brk1*, lack actin-enriched patches at lobe bases. In these cells, F-actin is distributed throughout the cell. The *arp* mutant trichomes have short branches and swollen bases – phenocopied by cytochalasin treatment of wild-type trichomes – are extensively bundled, and lack diffuse F-actin. Li *et al.* (2003) conclude that cell shape differentiation requires an ARP2/3 complex to correctly localize F-actin nucleation.

As described above, cell shape is determined during only the first few stages of growth. ROP GTPases are signaling proteins that regulate actin polymerization during this period (Fu *et al.*, 2002, 2005; Basu *et al.*, 2008). EPC in a constitutively active (CA) *rop2* mutant of Arabidopsis were supersized, circular, and the lobes did not form; F-actin was distributed throughout cells (Fu *et al.*, 2002). Dominant-negative (DN) *rop2* mutants showed smaller EPC and weak lobing, and the fluorescent signal from F-actin appeared faint. Like *brk1* and the *arp* mutants, *CA-rop2* and *DN-rop2* lacked actin-enriched patches at the base of lobes, *CA-rop2* because actin was distributed throughout and *DN-rop2* because these cells appeared to contain little actin. The lack of enrichments might have caused polarized elongation in these regions to fail (Fu *et al.*, 2002). Cell morphogenesis was shown to occur during the first 2–3 days of cell growth in EPC, and afterwards cells elongated in normal proportions no matter their genotype (Fu *et al.*, 2002). It therefore appears that ROP2 influences cell shape only during the initial period (Fu *et al.*, 2002). Fu *et al.* (2005) and a more recent paper by Nagawa *et al.* (2012) describe a potential mode of action for ROP2: ROP2 signals to RIC4, a ROP effector that stimulates regional F-actin accumulation, promoting growth in that region. In Fu *et al.* (2002), *CA-rop2* EPC appeared super-sized because the evenly distributed actin filaments promoted abnormal expansion. How exactly RIC4 generates fine arrays of actin filaments remains to be established.

A Role for Actin in Trafficking and Cell Wall Deposition

Plant cell shape is governed by cell wall flexibility (which is determined by material composition) and turgor pressure (Baskin, 2005; Smith and Oppenheimer, 2005; Szymanski and Cosgrove, 2009). The rate of plant cell expansion is a function of cell wall deposition (Kotzer and Wasteneys, 2006). Actin controls cell wall composition by anisotropic delivery of non-cellulosic polysaccharides to the wall (Szymanski and Cosgrove, 2009). In growing pollen tubes and root hairs, materials are delivered down the longitudinal axis to the tip apex; in puzzle-shaped leaf EPC, delivery occurs differentially around the perimeter of the cell.

As the major component of cell walls providing tensile strength, cellulose microfibrils constrain turgor pressure, so it is clear why cellulose placement is a key determinant of cell shape (Baskin, 2005; Smith and Oppenheimer, 2005; Fayant *et al.*, 2010; Wightman and Turner, 2010). Cellulose synthase (CesA) complexes travel in Golgi-derived vesicles to the plasma membrane where they synthesize cellulose microfibrils. Crowell *et al.* (2009) used cytochalasin D to destabilize actin filaments in Arabidopsis hypocotyl cells and Gutierrez *et al.* (2009) used Lat B to inhibit actin polymerization, also in hypocotyl epidermal cells, to provide evidence that actin filaments are the tracks on which Golgi are trafficked in cells. Both groups observed unusually distributed clusters of Golgi and increased GFP-CesA signal at the plasma membrane in the vicinity of these clusters. Because they observed CesA at the plasma membrane, albeit unevenly distributed, Gutierrez *et al.* concluded that actin guides, rather than traffics, CesA to correct locations on the plasma membrane. Sampathkumar *et al.* (2013) report that Golgi-localized CesA in JASP-treated Arabidopsis seedlings follows 'disorganized trajectories'. They also reported a significant change in cell wall assembly following actin disruption, by measuring less cellulose per unit dry weight versus controls in an *act2 act7* double mutant and in LatB-treated wild-type Arabidopsis hypocotyls. Surprisingly, this group did not observe significantly slower CesA movement at the plasma membrane in the *act2 act7* mutant or in the Lat B-treated hypocotyls, but they did report a reduction in both the delivery and recovery of CesA to/from the plasma membrane following actin disruption (Sampathkumar *et al.*, 2013).

It is a commonly held view that non-cellulosic polysaccharides travel in Golgi-derived vesicles along actin filament bundles to the plasma membrane where they deliver cell wall materials during expansion (Carpita, 2012), but the actual evidence for actin-dependent trafficking is sparse (Smith and Oppenheimer, 2005; Szymanski and Cosgrove, 2009). Leucci *et al.* (2007) found that cytochalasin D treatment results in large reductions in materials incorporated into the cell wall: a 65% reduction in α-cellulose and a 60% reduction in non-cellulosic polysaccharides. Although all these studies clearly demonstrate that the presence of actin correlates with CesA and non-cellulosic polysaccharide delivery to the cell wall, they do not identify exactly how actin drives delivery, that is, whether it guides or carries materials.

Although actin is generally thought to provide tracks for organelle and vesicle trafficking, there are instances where this model has been challenged. For example, the scarcity of actin filaments in the very tips of *Medicago truncatula* root hair cells where endosomal vesicles were observed to be 'particularly dynamic' caused some reservations about the universality of the track model (Voigt *et al.*, 2005). Although they dismiss the idea of actin as tracks, Voigt *et al.*'s alternative model maintains that vesicle trafficking is actin-dependent, but differs in that constant actin polymerization is required. This would be analogous to the comet-tail type motility that propels the bacterium *Listeria monocytogenes* and endogenous endosomes in the cytoplasm of mammalian cells. Sampathkumar *et al.* (2013) observed instances of CesA-containing vesicles in Arabidopsis aligning with actin filaments for upwards of 40 s. Vesicles continued to traffic, but movement paths were in disarray after exposure to JASP. Because CesA-containing vesicles moved despite actin stabilization, Sampathkumar's observation seems to support the evidence for actin as tracks: JASP treatment stabilized existing actin so the tracks remained and vesicles moved along actin.

Cell growth and morphogenesis require regulated endocytosis at certain areas of the cell but how actin regulates or guides endocytosis is not well understood and models describing the role of actin differ (Baisa *et al.*, 2013). Nevertheless, there is some evidence for recycling of cell wall pectins by actin-dependent endocytosis (Baluska *et al.*, 2006). There is agreement that F-actin density increases near regions of growth, and the roles of specific actin-binding proteins implicated in receptor-mediated endocytosis appear to be roughly consistent across kingdoms (Smith and Oppenheimer, 2005; Pollard and Cooper, 2009). However, there is disagreement about other aspects of the participation of actin in uptake from the plasma membrane, for example, whether the dense F-actin inhibits or facilitates endocytosis (Beck *et al.*, 2012; Nagawa *et al.*, 2012; Baisa *et al.*, 2013). Further study into the role of actin in endocytosis is important given a recent finding that plants use clathrin-mediated endocytosis to manage CesA complex activity at the plasma membrane (Bashline *et al.*, 2013).

Live-Cell Imaging and New Fluorescent Reporters for Actin Dynamics Studies

New and Innovative Imaging Modalities

Innovations in both fluorescence imaging modalities and actin cytoskeleton reporters have been instrumental for dissecting actin filament turnover in living plant cells (Ehrhardt and Frommer, 2012; Shaw and Ehrhardt, 2013). Although modern confocal laser scanning microscopes (CLSM) are faster and more sensitive than earlier generations of the instrument, and they do permit a detailed visualization of some actin rearrangements (Higaki *et al.*, 2007, 2010b; Smertenko *et al.*, 2010), two forms of microscopy facilitate high spatial and temporal imaging of the cortical cytoplasm in epidermal cells.

Variable-angle epifluorescence microscopy, or VAEM, is a variation of total internal fluorescence reflection (TIRF) microscopy that illuminates a single optical plane by directing low-angle light onto the specimen (Fujimoto *et al.*, 2007; Konopka and Bednarek, 2008a; Shaw and Ehrhardt, 2013). This low-angle light excites fluorescent molecules in a restricted plane of the cortical cytoplasm and gives a strong signal with low background; VAEM is therefore capable of producing very high-contrast images. Moreover, because there is no optical sectioning involved or rastering of a point-source of light onto the specimen, VAEM can also be extremely fast, especially when coupled to a back-thinned EM-CCD camera as a detector. This form of fluorescence microscopy has been used to capture time-lapse series of vesicle trafficking events, membrane protein dynamics, and cytoskeletal rearrangements in plant cells (Wang *et al.*, 2006; Fujimoto *et al.*, 2007; Konopka *et al.*, 2008; Konopka and Bednarek, 2008a, b; Zheng *et al.*, 2009; Vizcay-Barrena *et al.*, 2011). For individual filament dynamics (see Table 5.1), this methodology was applied to epidermal cells from the hypocotyl and petioles of dark-grown Arabidopsis seedlings (Staiger *et al.*, 2009; Henty *et al.*, 2011; Li *et al.*, 2012, 2014; Tóth *et al.*, 2012), in tobacco BY-2 cell protoplasts (Smertenko *et al.*, 2010), and moss protonema (Augustine *et al.*, 2011). Remarkably, the individual actin filaments visualized by VAEM appear quite similar to and have many properties in common with *in vitro* actin filament dynamics, especially when combined with plant actin-binding proteins (Michelot *et al.*, 2005, 2006; Khurana *et al.*, 2010; Zhang *et al.*, 2010; Henty *et al.*, 2011). One disadvantage of the VAEM imaging modality is that it does not readily allow for optical sectioning. However, on large vacuolated plant cells, the ability to illuminate depths up to 500 nm is sufficient to visualize cortical cytoplasm events.

A second fluorescence imaging modality, spinning-disk confocal (SDC) microscopy, has enjoyed more widespread use among plant cell biologists (Ehrhardt and Frommer, 2012; Shaw and Ehrhardt, 2013) and was originally employed to reveal cortical microtubule dynamics (reviewed by Ehrhardt and Shaw, 2006; Lucas and Shaw, 2008). In particular, this form of microscopy gave us the first view of dynamic instability at microtubule ends in plant cells (Shaw *et al.*, 2003). It was also used to demonstrate directly the trajectory of cellulose synthase complex (CesA) movements in the plane of the plasma membrane and the orientation of membrane-associated cortical microtubules by two-color, live-cell imaging of hypocotyl epidermal cells (Paredez *et al.*, 2006). Unlike CLSM, the SDC platform uses a set of microlenses arrayed in a pinwheel pattern on a rotating disk (a Nipkow disk) to uniformly and simultaneously illuminate a single optical plane. To generate confocality across an entire optical plane, emitted light is captured through an array of pinholes in a second disk that spins at the same speed as the Nipkow disk. Images are typically captured with an EM-CCD, which gives greatest sensitivity for faint fluorescence and high speed for dynamic events. Time-lapse SDC microscopy has been used to visualize protoplasts of rice sheath and stem cells (Shi *et al.*, 2013) as well as Arabidopsis pollen tubes (Qu *et al.*,

Table 5.1 Actin dynamic parameters.

	Arabidopsis epidermal cell					Tobacco		Rice	Arabidopsis	Moss
	Hypocotyl[a,d,g]	Petiole[a,d,h]	Cotyledon[b,e,i]	Root[b,e,j]	Guard cell[a,e,k]	Guard cell[c,e,l]	BY-2[b,d,m]	protoplast[a,f,n]	pollen[b,f,o]	protonema[b,d,p]
Elongation rate (µm/s)	1.7	1.4	1.7	0.9	0.3	0.05	0.6	–	0.3	2.3
Depolymerization rate (µm/s)	0.2	–	–	–	–	–	–	0.002	0.2	–
Shortening rate (µm/s)	–	–	1.2	0.7	0.4	0.02	0.2	–	–	2.8
Severing frequency (breaks/µm/s)	1.1×10^{-2}	1.9×10^{-2}	6.4×10^{-5}	5.1×10^{-5}	4.9×10^{-3}	2.0×10^{-3}	–	1.7×10^{-2}	3.4×10^{-2}	–
(events/µm²/s)	–	–	–	–	–	–	–	–	–	3.3×10^{-3}
Max. filament length (µm)	14.8	12.1	–	–	–	–	–	–	2.5	–
Max. filament lifetime (s)	15.3	15.6	–	–	–	–	–	114.5	20.2	–
Regrowth of severed ends (%)	2.9	–	–	–	–	–	–	–	–	–
Annealing of ends (%)	2.1	–	–	–	–	–	–	–	–	–
Dynamic ends (ends/µm²/s)	–	–	–	–	–	–	–	–	–	4.2×10^{-3}
Filament origin; de novo/ends/side (% per cell)	28/22/50	–	–	–	–	–	–	–	–	–
Convolutedness	1.6	–	1.1	1.1	–	–	–	–	–	–

Transgenic plants, expressing the actin cytoskeleton reporters GFP-fABD2[a], GFP-Lifeact[b], or GFP-mouse Talin[c], were imaged with variable-angle epifluorescence microscopy (VAEM)[d], confocal laser scanning microscopy (CLSM)[e], or spinning disk confocal (SDC)[f] microscopy. For Arabidopsis epidermal cells, 5-day-old dark-grown hypocotyls (Staiger *et al.*, 2009; Henty *et al.*, 2011; Li *et al.*, 2012, 2014; Tóth *et al.*, 2012)[g]; 5-day-old dark-grown petioles (Henty *et al.*, 2011)[h]; 10-day-old light-grown cotyledons (Smertenko *et al.*, 2010)[i]; 4-week light-grown root elongation zone (Smertenko *et al.*, 2010)[j]; or rosette leaf guard cells (Li *et al.*, 2013)[k] were imaged. In tobacco, guard cells from the abaxial surface of 3-week-old leaves (Wang *et al.*, 2011)[l] or protoplasts prepared from BY-2 suspension cells (Smertenko *et al.*, 2010)[m] were used for imaging. In rice, protoplasts prepared from 14-day-old dark-grown rice sheaths and stems (Shi *et al.*, 2013)[n] were used for experiments. As examples of tip growth, Arabidopsis pollen tubes germinated *in vitro* were imaged (Qu *et al.*, 2013; Zheng *et al.*, 2013)[o], as were protonemal cells from 7-day-old light-grown moss plants, *Physcomitrella patens* (Augustine *et al.*, 2011)[p]. Metrics for measuring single actin filament turnover are defined in the literature (e.g., Staiger *et al.*, 2009; Smertenko *et al.*, 2010; Augustine *et al.*, 2011; Henty *et al.*, 2011; Li *et al.*, 2012) and are discussed in the text. The table was modified and extended from data presented in Henty-Ridilla *et al.* (2013).

2013; Zheng *et al.*, 2013) to quantitatively analyze actin filament dynamic properties (see Table 5.1). One notable advantage of this imaging modality compared to VAEM is the ability to collect multiple z-series images, thereby increasing the depth of focus in the cortical cytoplasm or permitting deep imaging in large cells that lack a single, central vacuole. Although we expect no differences in the dynamic properties of actin filaments when imaged by these two modalities, this has never been tested head-to-head, in the same laboratory, on the same cell type expressing a single type of actin reporter.

Live-cell Reporters for the Actin Cytoskeleton

The development of fluorescent fusion proteins to decorate the dynamic actin cytoskeleton has resulted in great leaps forward. Although there is at least one report of a direct GFP-actin fusion protein used in plants (Lo *et al.*, 2011), most reporters are indirect and variations of actin-binding proteins or actin-binding modules (reviewed by Henty-Ridilla *et al.*, 2013). In yeasts, fluorescent proteins that are prepared as translational fusions with actin do not fully complement actin knockout mutants and must be used judiciously (Doyle and Botstein, 1996; Wu and Pollard, 2005). Similarly, the mungbean actin-GFP reporter in plants has not been demonstrated to rescue an *actin* mutant and the reporter decorates mainly mitochondrial-associated actin (Lo *et al.*, 2011). The earliest actin cytoskeletal reporters used in plants were often variants of vertebrate talin (Tn), an actin-binding and -stabilizing protein, which were used to show long-term (minutes and hours) rearrangements in actin arrays (Kost *et al.*, 1998; Fu *et al.*, 2001; Lemichez *et al.*, 2001). While these reporters still enjoy occasional use in some laboratories (Mathur *et al.*, 2010; Denecke *et al.*, 2012), and they have been used to measure actin dynamics in tobacco guard cells (Wang *et al.*, 2011; see Table 5.1), they have fallen from favor due to reported plant growth defects and for inducing actin stabilization and/or bundling in some cell types (Ketelaar *et al.*, 2004a; Sheahan *et al.*, 2004). Tn-based reporters were replaced by an actin-binding module from the Arabidopsis FIMBRIN1 protein, referred to as fABD2 (Ketelaar *et al.*, 2004b; Sheahan *et al.*, 2004; Wang *et al.*, 2004; Voigt *et al.*, 2005). These reporters have fewer detrimental effects on plant growth and decorate both stable and dynamic filament arrays in a wide variety of plant cells (Sheahan *et al.*, 2004; Higaki *et al.*, 2007; Wang *et al.*, 2008). Perhaps because they are derived from a plant gene, reporters based on fABD2 suffer from gene silencing and have to be reintroduced or reselected frequently in wild-type and mutant lines (our unpublished data). There are also indications that fABD2 has genetic interactions with certain actin-binding protein mutants and is sometimes not tolerated in combination with actin-bundling protein knockouts (van der Honig *et al.*,

2012; our unpublished observations). Nevertheless, GFP-fABD2 is the reporter of choice for monitoring single actin filament turnover in epidermal cells of dark-grown Arabidopsis hypocotyls (Staiger *et al.*, 2009; Henty *et al.*, 2011; Li *et al.*, 2012, 2014; Tóth *et al.*, 2012). It has also been used to quantify filament dynamics (see Table 5.1) in petiolar epidermal cells (Henty *et al.*, 2011), guard cells (Li *et al.*, 2013), and rice protoplasts (Shi *et al.*, 2013).

A new reporter, called Lifeact, is a 17-amino-acid actin-binding module from yeast Abp140p (Riedl *et al.*, 2008). It has been introduced into Arabidopsis (Era *et al.*, 2009; Mathur *et al.*, 2010; Smertenko *et al.*, 2010), tobacco (Vidali *et al.*, 2009a; Smertenko *et al.*, 2010), lily (Vidali *et al.*, 2009a), moss (Vidali *et al.*, 2009a), and the liverwort *Marchantia* (Era *et al.*, 2009, 2013), where it reports both actin filament bundle rearrangements and single actin filament dynamics. There even exists a green-to-red photoconvertible form of this reporter, LIFEACT-mEosFP (Mathur *et al.*, 2010). Lifeact does not suffer from gene-silencing, but may cause some plant growth defects when over-expressed (van der Honig *et al.*, 2011). When expression is kept to minimal levels, as should always be the case in live-cell imaging (Gilroy, 1997; Dixit *et al.*, 2006; Fricker *et al.*, 2006), Lifeact is a superb reporter of individual actin filament dynamic parameters. Quantitative data (Table 5.1) have been obtained by expressing fluorescent fusion proteins with Lifeact in EPCs of Arabidopsis cotyledons (Smertenko *et al.*, 2010) and root epidermal cells (Smertenko *et al.*, 2010); tobacco BY-2 cells (Smertenko *et al.*, 2010); Arabidopsis pollen tubes (Qu *et al.*, 2013; Zheng *et al.*, 2013); and moss protonema (Augustine *et al.*, 2011). To determine whether Lifeact or fABD2 positively or negatively influence single actin filament dynamics, a direct head-to-head comparison in the same cell type is required.

A New View of Actin Turnover in Plants

When snapshots of the cortical actin cytoskeleton taken just a few seconds apart are colorized individually and digitally overlayed, it immediately becomes obvious that a majority of the filaments undergo rearrangements over a very short timescale (Higaki *et al.*, 2007; Staiger *et al.*, 2009; Henty-Ridilla *et al.*, 2013; Hoffmann *et al.*, 2014; Fig. 5.1). The imaging modalities and fluorescent fusion proteins described above have allowed detailed descriptions of these rearrangements (Staiger *et al.*, 2009; Smertenko *et al.*, 2010; Augustine *et al.*, 2011). Moreover, establishing metrics for filament growth, translocations, and turnover, has yielded a quantitative description of actin dynamics (see Table 5.1; reviewed in Staiger *et al.*, 2010; Day *et al.*, 2011; Thomas, 2012; Henty-Ridilla *et al.*, 2013).

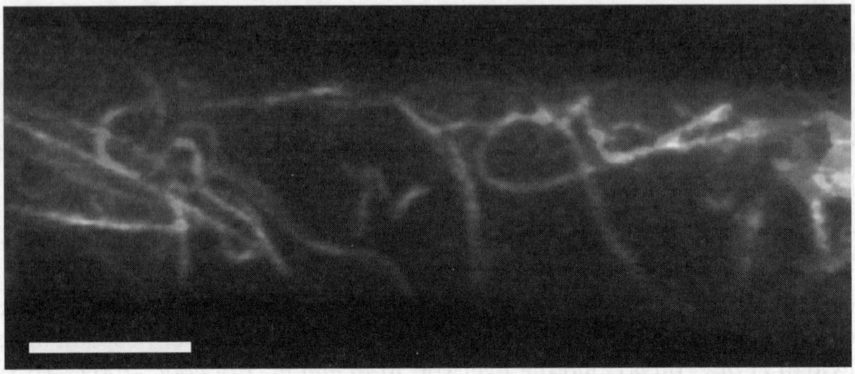

Figure 5.1 Actin organization and remodeling in epidermal cells is complex and fast. A time-series projection for an epidermal cell from the dark-grown hypocotyl of Arabidopsis expressing GFP-fABD2 is shown. Time-lapse images were collected by variable-angle epifluorescence microscopy (VAEM). Three consecutive images, separated by 3 s each, were colored red, green and blue and the three images superimposed. A white color indicates stable actin structures, such as actin filament bundles. Colored filaments are fainter and more ephemeral structures; they grow, disappear and translocate over the elapsed time of 6 s. These latter structures are presumed to be individual actin filaments based on their fluorescence intensity and dynamic properties (Staiger *et al.*, 2009). Scale bar: 10 μm. (See plate section for the color version of this figure.)

Two Populations of Actin Filaments in the Cortical Array of Epidermal Cells

Epidermal cells from the dark-grown hypocotyl of Arabidopsis seedlings expressing GFP-fABD2 were selected initially because of the excellent optical properties of these cells (Staiger *et al.*, 2009). In addition, this is an ideal system for studying links between cortical cytoskeletal dynamics and cell growth because these cells elongate by axial cell expansion, the hypocotyl undergoes little or no cell division during organ expansion, and there is an acropetal gradient of cell expansion (Gendreau *et al.*, 1997; Refrégier *et al.*, 2004; Derbyshire *et al.*, 2007; Lucas and Shaw, 2008). Three- to five-days post-germination, epidermal cells in the apical third of the hypocotyl undergo fast axial elongation. As for other epidermal cells, the cortical actin cytoskeleton of hypocotyl cells comprises two populations of filaments: actin bundles and individual actin filaments (Higaki *et al.*, 2007; Staiger *et al.*, 2009). Actin bundles are brighter, straighter, and longer-lived than individual actin filaments (Staiger *et al.*, 2009). The actin bundles are predominantly axially oriented in cells, whereas the single filaments appear randomly arrayed (Fig. 5.1). Single actin filaments are defined by dynamic behaviors (see next section) and are measured as the filamentous structures with the lowest level of fluorescence intensity (Staiger *et al.*, 2009). When these filaments extend across another single filament, their fluorescence intensity doubles at the point of crossover (Staiger *et al.*, 2009). Some researchers have therefore estimated filament number within bundles by presuming that fluorescence is additive (Smertenko *et al.*, 2010). In the cortical cytoplasm near the nucleus, actin filaments and bundles are extremely dense, making quantitative analyses of individual filament behavior difficult if not impossible. In other regions on the outer periclinal face of epidermal cells, however, the actin array is sparser, permitting the investigator to track filaments for tens of seconds and monitor the behavior of filament ends (Fig. 5.2). Bundles can be generated by the 'zippering' together of individual actin filaments (Khurana *et al.*, 2010; Hoffmann *et al.*, 2014) and disassembly occurs by the 'peeling off' of filaments and by severing activity (Smertenko *et al.*, 2010; Zheng *et al.*, 2013).

Clearly, the behavior of bundled filaments is an area that needs further study to fully characterize the underlying molecular mechanisms; additional metrics for bundle dynamics combined with live-cell imaging and genetic approaches would be beneficial toward this end (Thomas, 2012). One recently described measure of filament zippering frequency has been shown to increase three-fold in an AtWLIM1 over-expressing line (Hoffmann *et al.*, 2014). Moreover, an *adf7* mutant has reduced debundling frequency in Arabidopsis pollen tubes (Zheng *et al.*, 2013).

Dynamic Behaviors of Single Actin Filaments

Two predominant types of cytoskeletal remodeling occur in the cortical array: filament and bundle translocation, and filament growth and disassembly (Staiger *et al.*, 2009; Smertenko *et al.*, 2010). Depending on the cell type, the relative contribution of each is variable. Filament and bundle translocation results in bending and waving of these structures. Bending and waving can be quantified with a unit-less metric, termed convolutedness, which is defined as the length of the filament or bundle divided by the length of the longest axis of a bounding rectangle (Staiger *et al.*, 2009). The greater the value of convolutedness, the more 'wavy' an object; straight filaments will have a value of 1. In hypocotyl epidermal cells, filament bundles are less convoluted than individual actin filaments; bundles have average convolutedness values of 1.1, whereas individual actin filaments have values of 1.6–2.3 (Table 5.1; Staiger *et al.*, 2009). The degree of convolutedness for individual filaments changes over time, and the rate of change can be described as the difference between adjacent frames averaged over time (Table 5.1; Staiger *et al.*, 2009). In epidermal cells, this value is on average $0.08\,s^{-1}$ (Staiger *et al.*, 2009). Because the rate of change of convolutedness is significantly reduced by treatments with the putative myosin inhibitor 2,3-butanedione monoxime (BDM; McCurdy, 1999), it was speculated that filament waving and straightening is

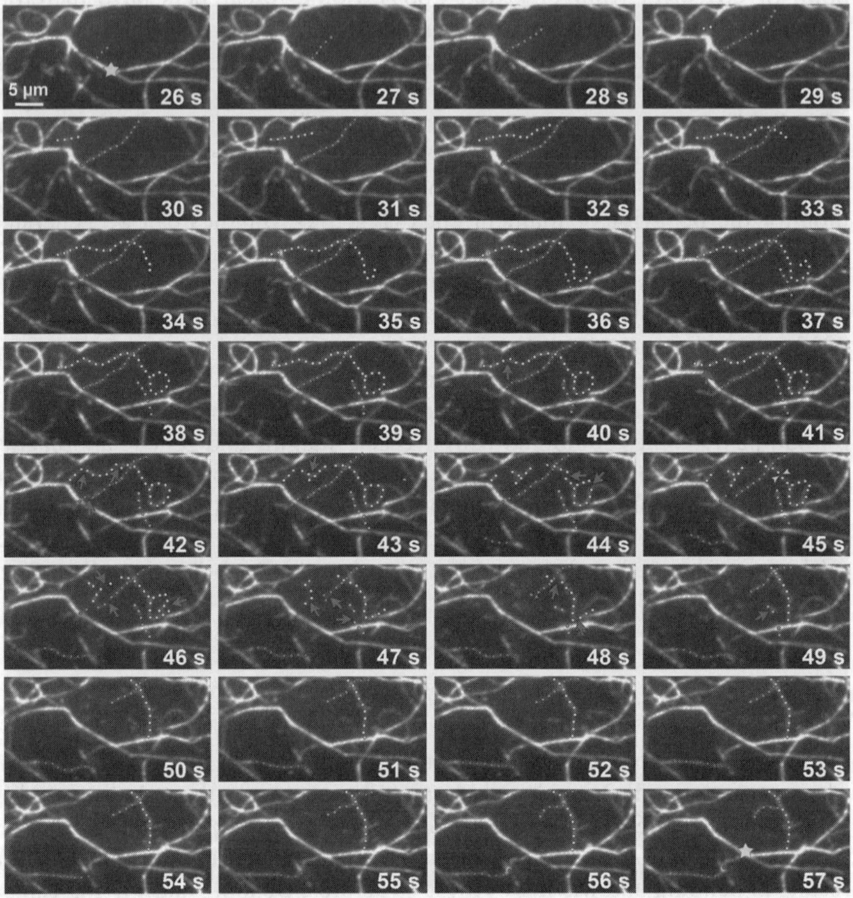

Figure 5.2 Dynamic behavior of individual actin filaments in the cortical array of Arabidopsis. Time-lapse VAEM was used to image the cortical actin cytoskeleton of dark-grown Arabidopsis hypocotyl epidermal cells expressing GFP-fABD2 (Staiger *et al.*, 2009). Two populations of actin filaments exist in the cortical array: single actin filaments and actin filament bundles. Representative single actin filaments originate from three locations: the side of a pre-existing filament or bundle (frame 26, green dots and frame 44, purple dots); *de novo* in the cytoplasm (frame 29, yellow dots) or the end of a pre-existing fragment (frame 36, blue dots). These filaments elongate rapidly and are disassembled by severing events (red arrows). Newly created ends occasionally anneal with another filament (yellow arrowheads). By contrast, a representative actin bundle (yellow star) remains relatively stationary throughout the elapsed time of approximately 30 s. Images were recorded from wild-type epidermal cells at 1 s intervals. Scale bar: 5 μm. (See plate section for the color version of this figure.)

myosin-dependent (Staiger *et al.*, 2009). The mechanism can be envisioned as filament–filament sliding, and/or filament sliding along the plasma membrane or organelle-localized myosin motors. To date, there is no genetic evidence in Arabidopsis for this hypothesis, and filament rearrangements were not suppressed in Physcomitrella protonemal cells when levels of the two myosin XI isovariants were silenced by RNAi (Vidali *et al.*, 2010).

The second major type of individual filament remodeling in the cortical array of epidermal cells involves assembly and disassembly reactions, as shown in Figure 5.2. This mechanism in plants has been termed 'stochastic dynamics' to reflect fundamental differences from the textbook model of filament treadmilling (reviewed by Staiger *et al.*, 2009, 2010; Day *et al.*, 2011; Thomas, 2012; Henty-Ridilla *et al.*, 2013). In the Staiger lab, quantitative analyses of hypocotyl and petiolar epidermal cells reveal that individual actin filaments grow at incredible rates and are turned over by prolific severing activity (see Table 5.1; Staiger *et al.*, 2009; Henty *et al.*, 2011; Li *et al.*, 2012, 2014). It should be noted that most measurements are made by selecting growing filaments in the cortical array, rather than by tracking all single filaments over time. New growing filaments originate from three locations in the cortical cytoplasm (Table 5.1; Fig. 5.2): from the ends of pre-existing filaments or fragments that were previously not growing (ends); from the side of another filament or bundle (side); and from a region of the cytoplasm apparently devoid of other structures (*de novo*). In hypocotyl epidermal cells, the three types of origin or nucleation events occur in roughly equal proportions (22/50/28%; ends/side/*de novo*), with branched growth from the side of a mother filament somewhat preferred (Table 5.1; Staiger *et al.*, 2009; Li *et al.*, 2012).

Once a new barbed end is created, elongation growth occurs at a remarkable rate. In hypocotyl cells, the values for individual filament growth range from 1.7 to 2.4 $\mu m\,s^{-1}$ (Table 5.1; Staiger *et al.*, 2009; Henty *et al.*, 2011; Li *et al.*, 2012, 2014; Tóth *et al.*, 2012; Hoffmann *et al.*, 2014). This speed allows an individual filament to grow from one sidewall to another across the short axis of an epidermal cell in approximately 10 s. From a biophysical perspective, given that 1 μm of actin filament comprises 370 actin subunits (Pollard, 1986; Pollard *et al.*, 2000), this means plant actin filaments are adding about 630 subunits each second ($1.7\ \mu m\,s^{-1} \times 370\ \mu m^{-1}$)! This is not so far-fetched. Because the rate of filament elongation *in vitro* is proportional to the G-actin concentration multiplied by k_+ for ATP-G-actin addition onto barbed ends (Pollard, 1986; Pollard *et al.*, 2000), this rate could be supported by as little as 54 μM monomer pool *in vivo* (Blanchoin *et al.*, 2010; Staiger *et al.*, 2010), certainly achievable in some plant cells (Gibbon *et al.*, 1999; Snowman *et al.*, 2002). Because ATP associated with each monomer is hydrolyzed to $ADP{\cdot}P_i$ following addition to the filament, this level of filament growth seems rather energy intensive. Assuming that the cortical cytoplasm has just 100 growing filaments at a given time, these growth rates would consume approximately 4 million ATP molecules per cell each minute! This estimate does not even include the ATP that is likely used for myosin-based organelle movements or filament–filament sliding. Collectively, actin remodeling in the cortical array is an expensive proposition, with a large proportion of that energy invested toward explosive rates of filament growth.

Evidence that a large pool of actin monomers adding onto free barbed ends underpins these filament elongation rates came initially from pharmacological approaches (Staiger *et al.*, 2009). The marine macrolide compound LatB binds to monomeric actin and inhibits its assembly onto filaments *in vitro* (Gibbon *et al.*, 1999). Short-term treatments of hypocotyls with LatB reduced the elongation rate on filament ends in a dose-dependent manner (Staiger *et al.*, 2009). Interestingly, single actin filaments were the first population of cytoskeletal element impacted by LatB treatment, suggesting more dynamicity than actin filament bundles (Gibbon *et al.*, 1999; Staiger *et al.*, 2009). Perhaps the bundles are stabilized or grow slowly only from the monomer pool. Moreover, binning the growing ends of single actin filaments into distinct populations (slow $<1\ \mu m\ s^{-1}$; moderate $1–2\ \mu m\ s^{-1}$; fast $>2\ \mu m\ s^{-1}$), reveals different types of ends that make unique contributions to the average growth rate (Li *et al.*, 2014). This is similar to fission yeast actin bundles where diverse formin isoforms and myosin-dependent translocation contribute to bundles with distinct elongation rates (Chesarone-Cataldo *et al.*, 2011; Yu *et al.*, 2011). Future work in plant cells should focus on defining the molecular nature and the regulation of these different filament ends.

Filament–filament annealing, or the joining of pre-existing filament ends, was suggested to provide a dominant and energy-inexpensive mechanism to elongate filaments in plant cells (Smertenko *et al.*, 2010). Although examples of annealing are observed in the cortical array of cotyledon and root epidermal cells (Smertenko *et al.*, 2010), this does not appear to be a prominent mechanism in hypocotyl epidermal cells (yellow arrowheads, Fig. 5.2; Li *et al.*, 2012, 2014). Only 2% of filament ends show visible filament–filament annealing events in the cortical array of wild-type hypocotyl cells (Table 5.1; Li *et al.*, 2012, 2014).

If treadmilling occurs in the cortical actin array of plant cells, one can predict that the rate of subunit loss should match the rate of subunit addition onto a single actin filament. Depolymerization of long filaments does not make a major contribution to disassembly in hypocotyl cells. Loss of subunits and shrinkage at the pointed end of filaments is observed occasionally, but the rate is nearly an order of magnitude slower than elongation at the barbed end, or $0.2–0.3\ \mu m\ s^{-1}$ (Table 5.1; Staiger *et al.*, 2009). Rather, filament disassembly is executed by prolific severing activity that turns growing filaments into short fragments (red arrows, Fig. 5.2). This can be quantified as the number of breaks, per unit length of filament, as a function of time (Staiger *et al.*, 2009).

In hypocotyl epidermal cells, the frequency of severing is approximately 0.011 breaks $\mu m^{-1}\ s^{-1}$ (Staiger *et al.*, 2009). Another way to describe this value is by writing that a 10-µm-long filament would suffer about 7 breaks each minute. Severing does not occur immediately or uniformly along growing filaments, but is delayed by many seconds and first occurs on the older region of a filament (Fig. 5.2). The fate of the fragments generated by severing is not entirely clear. Most of them disappear by moving from the field of view within seconds; others undergo slow depolymerization from their ends, and a few anneal back onto the original filament or a nearby neighbor (yellow arrowheads, Fig. 5.2). It is quite remarkable that very few of these newly created ends resume growth, with regrowth occurring at a frequency of about 3% of filament barbed ends (Table 5.1; Li *et al.*, 2012). Thus, severing is the dominant means of filament disassembly in hypocotyl epidermal cells, and perhaps slow depolymerization of short fragments contributes to the return of subunits to the monomer pool (Staiger *et al.*, 2009).

The stochastic dynamics turnover mechanism shows individual filaments in the cortical array with modest lengths and lifetimes (Table 5.1). Maximum filament length is defined as the greatest length of a growing filament before it suffers appreciable breakage events; for hypocotyl cells, the average value is approximately 15 µm (Staiger *et al.*, 2009; Henty *et al.*, 2011; Li *et al.*, 2012; Tóth *et al.*, 2012). Maximum filament lifetime is defined as the difference between the time a growing filament originates until the time all of its resulting fragments completely disappear. For hypocotyl epidermal cells, maximum filament lifetime is on the order of 15–20 s (Table 5.1).

A Comparison of Dynamics between Different Cell Types and Organisms

Since the first two publications on individual filament turnover in epidermal cells of Arabidopsis (Staiger *et al.*, 2009; Smertenko *et al.*, 2010), parameters of actin dynamics have been measured in several additional organisms and cell types (Table 5.1). For the most part, the mechanisms appear similar to those described above although several differences are obvious. Quantitative differences between cell types and among organisms may be due to true differences in the measured behavior of filaments, differences in the metrics and methods used, or both. *Physcomitrella patens*, a moss, is a model genetic system with powerful RNAi and knockout technologies (Bezanilla *et al.*, 2003) that can be used to dissect the mechanism of actin filament array organization and turnover (reviewed by Blanchoin *et al.*, 2010; Blanchoin and Staiger, 2010; Henty-Ridilla *et al.*, 2013). By combining Lifeact-eGFP with VAEM imaging, it was possible to measure several parameters in the dense actin array of tip-growing protonemal cells (Augustine *et al.*, 2011). Elongation rates are quite high (approximately 2.3 $\mu m\ s^{-1}$), but so is the apparent shrinkage rate (2.8 μm s^{-1}; Table 5.1), suggesting that treadmilling might be possible in some organisms. However, these authors (Augustine *et al.*, 2011) and others (Smertenko *et al.*, 2010) do not describe measuring depolymerization *per se*, but may be including severing-induced disassembly as part of their metric for shortening. Augustine *et al.* (2011) also introduced a welcome innovation for monitor severing frequency and dynamic filament ends: both metrics were quantified as the number of events per area (Table 5.1). Another notable difference, even between epidermal cells of Arabidopsis, is that severing frequency is two–three orders of magnitude slower in cotyledon and root cells (Smertenko *et al.*, 2010) compared with hypocotyl and petiolar cells (Staiger *et al.*, 2009; Henty *et al.*, 2011). Similarly, elongation rates are substantially slower in tip-growing pollen tubes (0.3 μm s^{-1}; Qu *et al.*, 2013; Zheng *et al.*, 2013) and guard cells of Arabidopsis (0.3 μm s^{-1}; Li *et al.*, 2013) and tobacco (0.05 μm s^{-1}; Wang *et al.*, 2011) compared to values for hypocotyl and petiolar cells. Perhaps this is because these cells have a considerably different ratio of monomer to filamentous actin, a different constellation of actin-binding proteins, or both.

Actin-binding Proteins Modulate Stochastic Dynamics

Based on the biochemical properties of plant actin-binding proteins, as well as the reconstitution of key aspects of the stochastic dynamic behavior of individual filaments *in vitro* (Michelot *et al.*, 2007; Roland *et al.*, 2008), a simple model for regulation of actin turnover has been proposed. Because this has been extensively reviewed and schematic diagrams published several times recently (Staiger *et al.*, 2010; Day *et al.*, 2011; Thomas, 2012; Henty-Ridilla *et al.*, 2013), we only briefly outline the key features and players here.

Factors that Regulate Filament Assembly and Availability of Ends

The size of the total actin pool in some plant cells is predicted to be quite large, with semi-quantitative immunoblotting of pollen grain and tube extracts giving values of 50–200 μM (Vidali and Hepler, 1997; Gibbon *et al.*, 1999; Snowman *et al.*, 2002). However, only a small percentage of this available pool appears to be present in filaments at any given time. In other words, the G- to F-actin ratio is rather high, perhaps as much as 10:1 (Gibbon *et al.*, 1999; Snowman *et al.*, 2002; Wang *et al.*, 2005; Blanchoin *et al.*, 2010; Staiger *et al.*, 2010). Monomeric actin is buffered by equally abundant profilin, a small protein that forms a 1:1 complex with G-actin (Valenta *et al.*, 1993; Vidali and Hepler, 1997; Gibbon *et al.*, 1998). Profilin suppresses spontaneous nucleation and inhibits subunit addition onto the pointed ends of filaments

(Huang *et al.*, 2003; Michelot *et al.*, 2005). This limits filament growth from the profilin-actin pool to filament barbed ends, making creation of new barbed ends through nucleation events a critical facet in the generation of actin arrays. Two types of nucleation factors are conserved in plants: the ARP2/3 complex and formins (reviewed by Deeks and Hussey, 2005; Szymanski, 2005; Blanchoin and Staiger, 2010; Cvrckova, 2013). In other organisms, the ARP2/3 complex nucleates filaments from the side of a mother filament and facilitates the construction of a dendritic network of short actin filaments (Pollard and Cooper, 2009). Whether this occurs in plants or not remains to be definitively established. On the other hand, formins nucleate unbranched filaments that typically comprise bundled actin arrays. Ectopic overexpression of Arabidopsis FORMIN1, FORMIN3, or FORMIN5 (AFH1, AFH3, and AFH5, respectively) appears to support the concept of formin-generated actin bundles or cables (Cheung and Wu, 2004; Ye *et al.*, 2009; Cheung *et al.*, 2010). However, evidence from knockout *afh1* plants appears to contradict this model, since mutant cells have more, not fewer, actin cables (Rosero *et al.*, 2013). Similarly, overexpression of AFH1 in tobacco leaf epidermal cells results in an excess of single filaments and loss of bundles (Martiniére *et al.*, 2011). Each plant formin therefore may need to be examined on a case-by-case basis. In addition to nucleating actin filament formation, some formins are able to bundle filaments directly and others exhibit severing activity (Yi *et al.*, 2005; Michelot *et al.*, 2006; Xue *et al.*, 2011; Yang *et al.*, 2011; Zhang *et al.*, 2011b). Finally, some formins aid elongating actin filament growth by remaining at the barbed end of a filament and feeding profilin-actin onto the growing filament at rates faster than actin alone. This has been shown most elegantly for a Physcomitrella Type II formin *in vitro* (Vidali *et al.*, 2009b). Growing filaments in cells could therefore have 'naked' barbed ends, have a formin dimer attached, or both situations could exist in the population. Naked barbed ends can be inhibited from subsequent growth by association of a capping factor. The best characterized capper in plants is the heterodimeric capping protein, CP, which is an ortholog of muscle CapZ (Huang *et al.*, 2003, 2006; Pleskot *et al.*, 2012). This protein complex, with an extremely high affinity for filament ends and a slow dissociation rate, serves to limit the growth of filaments to a certain length and/or funnels growth from monomers onto newly created barbed ends. CP also inhibits filament growth by suppressing filament–filament annealing events (Huang *et al.*, 2003).

Factors that Regulate Filament Disassembly

A key facet of the stochastic dynamics model is that filament breakage or severing events make a major contribution to disassembly, whereas loss of subunits from pointed ends makes a modest contribution. Two families of filament severing factors are likely responsible for this activity: ADF/cofilin and villin/gelsolin. Members of both families of protein from Arabidopsis are capable of breaking the filament backbone *in vitro* (Khurana *et al.*, 2010; Zhang *et al.*, 2010, 2011a; Henty *et al.*, 2011). Moreover, as noted above some formins are also capable of severing events (Yi *et al.*, 2005). Because ADF/cofilin family members preferentially target ADP-loaded filaments (Blanchoin and Pollard, 1999; Andrianantoandro and Pollard, 2006; Suarez *et al.*, 2011) and older regions of growing filaments in cells appear to be dismantled first, it is likely that ADF/cofilin plays a substantial role in disassembly *in vivo*. The disassembly activity of ADF is enhanced in the test-tube when combined with another protein known as AIP1 (Allwood *et al.*, 2002; Shi *et al.*, 2013). One potential mechanism to explain this synergistic activity is that AIP1 caps barbed ends and alters filament turnover in concert with ADF by altering the filament twist (Ketelaar *et al.*, 2004b; Shi *et al.*, 2013). Alternative mechanisms are also possible. Because filament ends created by severing activity rarely regrow, capping is inferred to occur simultaneously with or soon after breakage events. This could be due to CP, AIP1, or other barbed-end cappers. Alternatively, the severing protein itself may also have a capping function, as is the case for some Arabidopsis villin proteins (Khurana *et al.*, 2010; Zhang *et al.*, 2010; Bao *et al.*, 2012). The small fragments that are generated by prolific severing are also likely to be capped, preventing regrowth and suppressing annealing events. These small fragments are thought to be recycled into the subunit pool by slow depolymerization and the combined activities of three monomer binding proteins: ADF, CAP or cyclase-associated protein, and profilin (Chaudhry *et al.*, 2007; Deeks *et al.*, 2007).

Factors that Determine Higher-order Structure

Although short-lived and dynamic, single filaments can be stabilized by their incorporation into higher-order arrays. Several classes of filament-bundling or cross-linking proteins are known in plants, including: fimbrin, villin, LIM, THRUMIN, SCAB1, ADF9, EF-1α and certain formins (reviewed by Thomas *et al.*, 2009; Higaki *et al.*, 2010a; Thomas, 2012). Several of these have been shown to alter filament dynamics by inhibiting the severing caused by ADF *in vitro* (Huang *et al.*, 2005; Khurana *et al.*, 2010). What additional factors contribute to the construction or stabilization of filaments within bundles is poorly understood; a more detailed, quantitative analysis of bundle generation and turnover in combination with mutant knockout of or knockdown lines is required.

Genetic Evidence for the Stochastic Dynamics Model

In addition to the advantages for imaging and quantifying cytoskeletal dynamics as elaborated above, the ability to reduce or eliminate specific actin-binding proteins with facile genetic techniques makes plants an ideal system for dissecting the molecular mechanisms that underpin actin filament organization and turnover. Both Arabidopsis and Physcomitrella have been used to generate genetic evidence for the role of several key players in the actin stochastic dynamics model (Augustine *et al.*, 2011; Henty *et al.*, 2011; Li *et al.*, 2012, 2014; Qu *et al.*, 2013; Zheng *et al.*, 2013). For example, the function of ADF/cofilin was first demonstrated by examining the phenotypes associated with a knockout mutant of *ADF4* (Tian *et al.*, 2009), a widely expressed isoform in Arabidopsis (Ruzicka *et al.*, 2007). Loss of ADF4 results in a 2.5- to 3-fold reduction in severing frequency, but has little impact on the depolymerization rate for single filaments in hypocotyl epidermal cells (Henty *et al.*, 2011). As a consequence, maximum filament lengths and lifetimes of growing filaments are increased significantly (Henty *et al.*, 2011). Amazingly, some individual filaments were observed to grow to 40–70 μm in length before suffering a modest number of breaks (Fig. 5.3a). In Arabidopsis pollen tubes, loss of the pollen-specific isoform ADF7 also perturbs actin dynamics (Zheng *et al.*, 2013). The *adf7* knockout pollen tubes have single filaments in the shank region that suffer significantly fewer breaks, have a modestly reduced depolymerization rate, and maintain maximum lifetimes that are greater than in wild-type pollen tubes (Zheng *et al.*, 2013). A mutated version of ADF7 that fails to sever, but does retain G-actin binding activity *in vitro*, did not complement the *adf7* mutant (Zheng *et al.*, 2013). Although it is now widely accepted that ADF/cofilin functions in eukaryotic cells primarily via the severing of actin filaments, these studies on plant cells provide the first direct genetic evidence supporting this hypothesis (Henty *et al.*, 2011; Zheng *et al.*, 2013). Moreover, this work demonstrates that redundancy within multigene families is not an issue at the subcellular level. Even the loss of a weakly expressed isoform (e.g. *ADF4* in dark-grown hypocotyls; Ma *et al.*, 2005) can result in a measurable phenotype if precise and accurate metrics for cytoskeletal dynamics are employed (Henty *et al.*, 2011).

Actin-interacting protein (AIP1) is known to enhance the filament disassembly properties of ADF/cofilin *in vitro* (Allwood *et al.*, 2002; Shi *et al.*, 2013). Validation for this model was first demonstrated in living plant cells by targeting the single *AIP1* gene in *P. patens* (Augustine *et al.*, 2011). The Δ*aip1* knockout plants created by homologous recombination accumulate actin bundles and have markedly altered filament dynamics. In particular, the frequency of severing events per unit area of cortical cytoplasm, as well as the

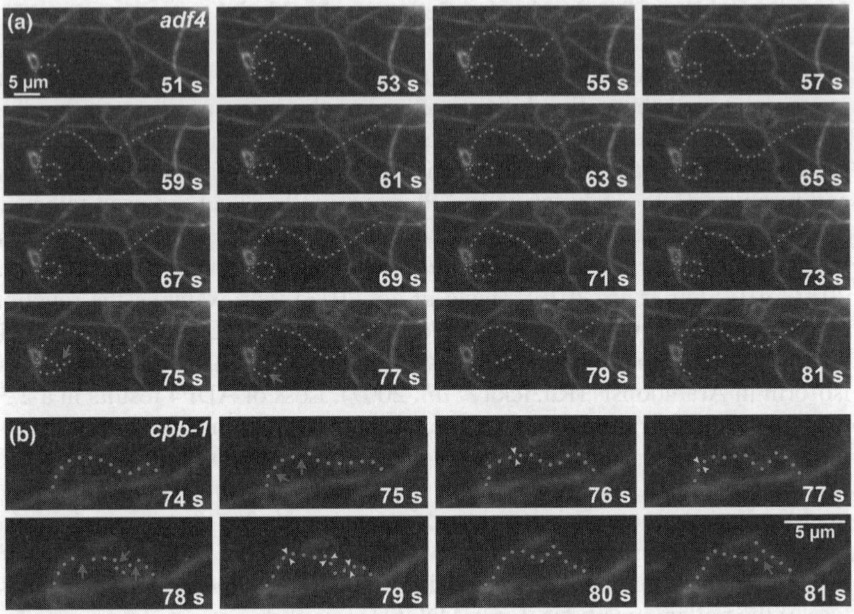

Figure 5.3 Individual actin filament dynamic properties are altered in actin-binding protein mutants. (a) Time-lapse VAEM series from an epidermal cell of *adf4* knockout mutant shows alteration of single filament dynamics. The frequency of filament severing is reduced in *adf4* epidermal cells. The representative filament (green dots) elongates rapidly and grows up to 42 μm in length (frame 73). It also persists for approximately 20 s before a few instances of severing occur (red arrows). Images were taken from a non-elongating epidermal cell of an 11-day-old dark-grown *adf4* mutant hypocotyl. Scale bar: 5 μm. (b) The frequency of filament–filament annealing is enhanced in *cpb-1* mutant cells. The highlighted filament suffers many severing events; however, several newly created ends join together within approximately 1 s. Five annealing events (yellow arrowheads) occur in the sequential time-lapse frames. Images were recorded from an epidermal cell in the non-growing region of 5-day-old *cpb-1* mutant hypocotyl by time-lapse VAEM. Scale bar: 5 μm. (See plate section for the color version of this figure.)

number of dynamic filament ends, is reduced 3.5-fold. The Δ*aip1* phenotypes are partially rescued by overexpression of ADF. By contrast, knocking down expression of the single *ADF* gene in Physcomitrella (Augustine *et al.*, 2008) almost completely obliterates actin filament dynamics (Augustine *et al.*, 2011). This study further demonstrates the critical nature of filament severing in the dynamic remodeling of actin cortical arrays.

Another class of proteins that can sever actin filaments *in vitro* is the villin/gelsolin family. VILLIN belongs to a small family of genes (five in Arabidopsis and rice) for actin filament severing and bundling proteins (Klahre *et al.*, 2000; Huang *et al.*, 2005; Khurana *et al.*, 2010; Zhang *et al.*, 2010, 2011a). Evidence that VILLINS are able to sever filaments *in vivo* comes from recent work with a *vln2 vln5* double mutant in Arabidopsis (Qu *et al.*, 2013). In addition to a greater than 10-fold reduction in severing frequency and doubling of the maximum filament lifetime, the *vln2 vln5* pollen tubes have actin filaments that are more wavy or buckled compared to the wild type (Qu *et al.*, 2013). The authors therefore infer that VILLINs confer rigidity to actin filaments on one hand, but on the other hand they are also responsible for filament turnover by severing activity (Qu *et al.*, 2013).

A new player in the game, at least with a newly described role in actin remodeling, is the microtubule-binding protein MAP18. Recently, AtMAP18 was demonstrated to sever actin filaments *in vitro*, albeit not as potently as AtVLN2 or AtVLN3, but at least as efficiently as AtADF1 (Khurana *et al.*, 2010; Zhu *et al.*, 2013). *MAP18 over-expression* (*OX*) and *map18* knockout lines display altered actin organization in pollen tubes of Arabidopsis; however, single filament parameters were not recorded (Zhu *et al.*, 2013). Nevertheless, the frequency of short actin fragments is reduced in *map18* pollen and increased in *MAP18 OX*. The authors infer that these fragments are generated via MAP18 severing activity. They further demonstrate that a mutated form of MAP18 that has lost severing function *in vitro* is unable to complement the *map18* pollen tube growth phenotype. Three players – ADF7 (Zheng *et al.*, 2013), VLN2 and VLN5 (Qu *et al.*, 2013), and MAP18 (Zhu *et al.*, 2013) – therefore appear to facilitate actin turnover in the shank and apex of pollen tubes by severing individual actin filaments.

The availability of filament barbed ends is predicted to be a key factor for funneling growth to particular actin filaments and for keeping the abundance of F-actin low in plant cells (Huang *et al.*, 2006; Staiger and Blanchoin, 2006). CP has end-capping activity *in vitro* (Huang *et al.*, 2003, 2006). Recently, knockdown mutants with insertions in Arabidopsis *CP* α- and *β*-subunit genes (*cpa* and *cpb*, respectively) were used to characterize the role of filament capping *in vivo* (Li *et al.*, 2012). As expected, reducing CP levels led to an increase in actin filament abundance in hypocotyl and root epidermal cells, consistent with an overall enhancement of polymerization from profilin-actin due to the greater availability of barbed ends (Li *et al.*, 2012; Pleskot *et al.*, 2013). Stochastic dynamics parameters also reveal a significant increase in filament ends, including more filaments that originate from pre-existing ends and up to six-fold increases in filament–filament annealing (Li *et al.*, 2012; Fig. 5.3b). Reduced CP levels also result in longer maximum filament lengths and increased lifetimes for the growing filaments. In contrast, overexpressing both subunits of CP (*CP OX* lines) results in the exact opposite phenotypes (Li *et al.*, 2014). Filament abundance in the cortical array of *CP OX* epidermal cells is reduced overall, and the filament–filament annealing frequency drops to zero (Li *et al.*, 2014). Although elongation rate and severing frequency are not affected, both maximum filament length and lifetime are significantly shorter in *CP OX* cells

(Li *et al.*, 2014). Collectively, these data show that CP functions *in vivo* to keep actin filaments short and short-lived.

Finally, the population of growing filaments which is nucleated by the ARP2/3 complex *versus* formin in plant cells remains an open question. Although knockout or knockdown mutants for formin in Arabidopsis and rice have been described (Ingouff *et al.*, 2005; Ye *et al.*, 2009; Cheung *et al.*, 2010; Yang *et al.*, 2011; Zhang *et al.*, 2011b; Rosero *et al.*, 2013), none of these has been characterized for single filament parameters. Because the gene family is so large in Arabidopsis (Deeks *et al.*, 2002), formin function and nucleation activity might be better studied in Physcomitrella (van Gisbergen *et al.*, 2012). By contrast, analysis of stochastic dynamics parameters in guard cells of *arp2* and *arp3* mutants of Arabidopsis has been reported, although the results are a bit confusing (Li *et al.*, 2013). One might expect that loss of the ARP2/3 complex would reduce the number of nucleation events in general, and likely have greatest effect on the subpopulation of growing filaments that originate from the side of a mother filament. Unfortunately, these metrics were not reported in the guard cell study (Li *et al.*, 2013). What was revealed, however, is an overall increase in actin dynamics, with elevated assembly and disassembly rates and a nearly two-fold increase in severing frequency (Li *et al.*, 2013). Further work will be required to establish whether this represents compensation by the mutant plants, or if these changes are indeed a direct consequence of loss of ARP2/3 nucleation activity.

Filament Length and Lifetime: Key Parameters Linking Actin Dynamics with Axial Cell Expansion?

In the hypocotyl system, alterations to actin dynamic properties have an impact on axial cell expansion of epidermal cells. When stochastic dynamics parameters in elongating cells at the apex of dark-grown hypocotyls were compared to non-elongating cells at the base of wild-type hypocotyls, albeit at a different developmental age, few differences were observed (Staiger *et al.*, 2009). Elongation rates and severing frequencies were virtually identical; however, maximum filament lengths were almost 50% longer in cells that are actively expanding (Staiger *et al.*, 2009). Analysis of key actin-binding proteins gives deeper insight into how single filaments may regulate axial cell expansion. Loss of either ADF4 or CP results in mutant hypocotyls and epidermal cells that are significantly longer than in wild-type seedlings (Henty *et al.*, 2011; Li *et al.*, 2012). In contrast, *CP OX* lines have significantly shorter cells in the elongation zone (Li *et al.*, 2014). None of these mutants have altered cell widths, however. The extent of actin filament bundling does not correlate with axial expansion because *adf4* cells are more bundled (Henty *et al.*, 2011), whereas *cpb-1* mutant cells are less bundled compared to wild-type epidermal cells (Li *et al.*, 2012, 2014). The overall density of cortical actin arrays also does not correspond with axial cell expansion, because *adf4* has a lower percentage of occupancy and *cpb-1* has increased filament array density compared to wild type (Henty *et al.*, 2011; Li *et al.*, 2012). Finally, the overall dynamicity of the cortical actin cytoskeleton, as measured by the change in correlation coefficient at all temporal intervals (Vidali *et al.*, 2010), is the opposite in these two actin-binding protein mutants. Cells of *adf4* show reduced dynamicity (Henty-Ridilla *et al.*, unpublished data), whereas *cpb-1* mutant cells show enhanced dynamicity (Li *et al.*, 2014).

Hypocotyl epidermal cells with enhanced rates of axial elongation have two properties in common: they have significantly longer maximum filament lengths and increased filament lifetimes (Henty *et al.*, 2011; Li *et al.*, 2012, 2014). To solidify the relevance of this observation, it is necessary to compare the actin-based phenotype of *CP OX* cells which have reduced axial elongation. Here, the cortical arrays contain single actin filaments with significantly reduced maximum lengths and lifetimes (Li *et al.*, 2014). As discussed in the following paragraph, we hypothesize that the length and stability of individual actin filaments influence their ability to function as tracks for materials transport (Li *et al.*, 2014). Specifically, long filaments and those with a greater lifetime before destruction could theoretically serve as more efficient tracks for vesicle trafficking.

As noted earlier, during active cell growth the actin cytoskeleton plays a role in the trafficking of CesA complexes, including populations that reside in the Golgi as well as in SmaCCs/MASCs (Sampathkumar *et al.*, 2013). Longer filaments could be more efficient at delivering these compartments to the cortical microtubules that are essential for proper insertion into the plasma membrane (Crowell *et al.*, 2009; Gutierrez *et al.*, 2009). Individual filaments, rather than actin bundles, might also play a role in delivering non-cellulosic polysaccharides and proteins to the cell wall via secretory vesicles (Leucci *et al.*, 2007). These individual filaments, as part of the fine actin array in the lobes of EPC, may influence the localization of PIN auxin transporters and thereby regulate local auxin levels (Xu *et al.*, 2010; Nagawa *et al.*, 2012). Long-lived and lengthier actin filaments could also influence uptake of materials from the plasma membrane and cell wall by endocytosis (Baisa *et al.*, 2013). Nagawa *et al.* (2012) showed that a dense array of fine actin filaments at the tip of lobes in EPC suppressed internalization of PIN1 in a ROP2–RIC4 dependent pathway. Generating greater lengths and lifetimes of individual filaments could be a key facet of this ROP signaling network. In this regard, we note that both *cpb-1* and *adf4* have reduced fluid-phase uptake of FM4-64 dye, consistent with inhibition of endocytosis or endosome formation (Li *et al.*, unpublished data). Moreover, *act2 act7* mutants and latrunculin-treated hypocotyls show decreased endocytic uptake of CesA, consistent with a requirement for actin filaments in the endocytic mechanism (Sampathkumar *et al.*, 2013). The role of actin in endocytosis remains poorly understood, and cortical actin filaments might have different roles in fluid-phase endocytosis, constitutive uptake of integral membrane proteins, and receptor-mediated endocytosis (Baisa *et al.*, 2013). These and other hypotheses await direct tests of exo- and endocytosis in relevant actin-binding protein mutants.

Prospects for the Future

Advances in fluorescence microscopy, live-cell imaging and actin-based reporters have illuminated the dynamic behavior of cortical actin filaments in diverse plant cells. By applying carefully designed metrics and investigating key actin-binding proteins, a model for actin filament turnover has been generated. The key facets of this model are the strict regulation of filament end availability, remarkably fast filament elongation, and disassembly via prolific severing activity. Future tests of this actin stochastic dynamics model should include studying mutants of nucleation factors and additional end cappers. Moreover, this level of analysis can be used to predict the biochemical properties of novel actin-binding proteins before putting them in the test-tube with purified actin. Finally, a hypothesis about single actin filaments as tracks for transport has been put forward and now needs to be tested by combining actin-binding protein mutants and reporters of exo- and endocytosis.

Acknowledgements

We dedicate this article to the memory of Dr Michael Sheahan (Newcastle University, AUS), an excellent scientist and remarkable individual who left us far too early. The authors wish to thank past and present members of the Staiger laboratory, as well as Dr Laurent Blanchoin (UJF/CEA/CNRS, Grenoble) and Dr Clément Thomas (CRP-Santé, Luxembourg) for valuable feedback and input on actin stochastic dynamics studies. Research in the Staiger laboratory is supported by grants from the Physical Biosciences Program of the Office of Basic Energy Sciences at the US Department of Energy (DE-FG02-09ER15526) and the US National Science Foundation-Arabidopsis 2010 Program (IOS-1021185). The VAEM imaging facility was supported in part by funds from the Bindley Bioscience Center, Discovery Park at Purdue University.

References

Akkerman, M., Overdijk, E.J.R., Schel, J.H.N., Emons, A.M.C., and Ketelaar, T. (2011) Golgi body motility in the plant cell cortex correlates with actin cytoskeleton organization. *Plant and Cell Physiology*, **52**, 1844–1855.

Allwood, E.G., Anthony, R.G., Smertenko, A.P., Reichelt, S., Drøbak, B.K., Doonan, J.H., Weeds, A.G., and Hussey, P.J. (2002) Regulation of the pollen-specific actin-depolymerizing factor LlADF1. *Plant Cell*, **14**, 2915–2927.

Andrianantoandro, E., and Pollard, T.D. (2006) Mechanism of actin filament turnover by severing and nucleation at different concentrations of ADF/cofilin. *Molecular Cell*, **24**, 13–23.

Augustine, R.C., Vidali, L., Kleinman, K.P., and Bezanilla, M. (2008) Actin depolymerizing factor is essential for viability in plants, and its phosphoregulation is important for tip growth. *Plant Journal*, **54**, 863–875.

Augustine, R.C., Pattavina, K.A., Tüzel, E., Vidali, L., and Bezanilla, M. (2011) Actin Interacting Protein1 and Actin Depolymerizing Factor drive rapid actin dynamics in *Physcomitrella patens*. *Plant Cell*, **23**, 3696–3710.

Baisa, G.A., Mayers, J.R., and Bednarek, S.Y. (2013) Budding and braking news about clathrin-mediated endocytosis. *Current Opinion in Plant Biology*, **16**, 718–725.

Baluska, F., Jasik, J., Edelmann, H.G., Salajová, T., and Volkmann, D. (2001) Latrunculin B-induced plant dwarfism: Plant cell elongation is F-actin-dependent. *Developmental Biology*, **231**, 113–124.

Baluska, F., Baroja-Fernandez, E., Pozueta-Romero, J., Hlavacka, A., Etxeberria, E., and Samaj, J. (2006) Endocytic uptake of nutrients, cell wall molecules and fluidized cell wall portions into heterotrophic plant cells, in *Plant Endocytosis* (eds J. Samaj, F. Baluska, and D. Menzel), Springer-Verlag, Berlin, pp. 19–35.

Bao, C., Wang, J., Zhang, R., Zhang, B., Zhang, H., Zhou, Y., and Huang, S. (2012) Arabidopsis *VILLIN2* and *VILLIN3* act redundantly in sclerenchyma development via bundling of actin filaments. *Plant Journal*, **71**, 962–975.

Bashline, L., Li, S., Anderson, C.T., Lei, L., and Gu, Y. (2013) The endocytosis of cellulose synthase in Arabidopsis is dependent on μ2, a clathrin-mediated endocytosis adaptin. *Plant Physiology*, **163**, 150–160.

Baskin, T.I. (2005) Anisotropic expansion of the plant cell wall. *Annual Review of Cell and Developmental Biology*, **21**, 203–222.

Baskin, T.I., and Bivens, N.J. (1995) Stimulation of radial expansion in arabidopsis roots by inhibitors of actomyosin and vesicle expansion but not by various inhibitors of metabolism. *Planta*, **197**, 514–521.

Basu, D., Le, J., Zakharova, T., Mallery, E.L., and Szymanski, D.B. (2008) A SPIKE1 signaling complex controls actin-dependent cell morphogenesis through the heteromeric WAVE and ARP2/3 complexes. *Proceedings of the National Academy of Science, USA*, **105**, 4044–4049.

Beck, M., Zhou, J.M., Faulkner, C., MacLean, D., and Robatzek, S. (2012) Spatio-temporal cellular dynamics of the *Arabidopsis* flagellin receptor reveal activation status-dependent endosomal sorting. *Plant Cell*, **24**, 4205–4219.

Bezanilla, M., Pan, A., and Quatrano, R.S. (2003) RNA interference in the moss *Physcomitrella patens*. *Plant Physiology*, **133**, 470–474.

Blanchoin, L., and Pollard, T.D. (1999) Mechanism of interaction of *Acanthamoeba* actophorin (ADF/cofilin) with actin filaments. *Journal of Biological Chemistry*, **274**, 15538–15546.

Blanchoin, L., and Staiger, C.J. (2010) Plant formins: Diverse isoforms and unique molecular mechanism. *Biochimica and Biophysica Acta: Molecular Cell Research*, **1803**, 201–206.

Blanchoin, L., Boujemaa-Paterski, R., Henty, J.L., Khurana, P., and Staiger, C.J. (2010) Actin dynamics in plant cells: a team effort from multiple proteins orchestrates this very fast-paced game. *Current Opinion in Plant Biology*, **13**, 714–723.

Bugyi, B., and Carlier, M.-F. (2010) Control of actin filament treadmilling in cell motility. *Annual Review of Biophysics*, **39**, 449–470.

Carpita, N.C. (2012) Progress in the biological synthesis of the plant cell wall: new ideas for improving biomass for bioenergy. *Current Opinion in Biotechnology*, **23**, 330–337.

Chaudhry, F., Guérin, C., von Witsch, M., Blanchoin, L., and Staiger, C.J. (2007) Identification of *Arabidopsis* Cyclase-Associate Protein 1 as the first nucleotide exchange factor for plant actin. *Molecular Biology of the Cell*, **18**, 3002–3014.

Chesarone-Cataldo, M., Guérin, C., Yu, J.H., Wedlich-Soldner, R., Blanchoin, L., and Goode, B.L. (2011) The myosin passenger protein Smy1 controls actin cable structure and dynamics by actin as a formin damper. *Developmental Cell*, **21**, 217–230.

Cheung, A.Y., and Wu, H.-M. (2004) Overexpression of an Arabidopsis formin stimulates supernumerary actin cable formation from pollen tube cell membrane. *Plant Cell*, **16**, 257–269.

Cheung, A.Y., Niroomand, S., Zou, Y., and Wu, H.-M. (2010) A transmembrane formin nucleates subapical actin assembly and controls tip-focused growth in pollen tubes. *Proceedings of the National Academy of Science, USA*, **107**, 16390–16395.

Crowell, E.F., Bischoff, V., Desprez, T., Rolland, A., Stierhof, Y.-D., Schumacher, K., Gonneau, M., Höfte, H., and Vernhettes, S. (2009) Pausing of Golgi bodies on microtubules regulates secretion of cellulose synthase complexes in *Arabidopsis*. *Plant Cell*, **21**, 1145–1154.

Cvrckova, F. (2013) Formins and membranes: anchoring cortical actin to the cell wall and beyond. *Frontiers in Plant Science*, **4**, e436, doi: 410.3389/fpls.2013.00436.

Day, B., Henty, J.L., Porter, K.J., and Staiger, C.J. (2011) The pathogen-actin connection: A platform for defense signaling in plants. *Annual Review of Phytopathology*, **49**, 489–506.

Deeks, M.J., and Hussey, P.J. (2005) Arp2/3 and scar: Plants move to the fore. *Nature Reviews Molecular Cell Biology*, **6**, 954–964.

Deeks, M.J., Hussey, P.J., and Davies, B. (2002) Formins: intermediates in signal transduction cascades that affect cytoskeletal reorganization. *Trends in Plant Science*, **7**, 492–498.

Deeks, M.J., Rodrigues, C., Dimmock, S., Ketelaar, T., Maciver, S.K., Malhó, R., and Hussey, P.J. (2007) *Arabidopsis* CAP1—a key regulator of actin organisation and development. *Journal of Cell Science*, **120**, 2609–2618.

Deeks, M.J., Fendrych, M., Smertenko, A., Bell, K.S., Oparka, K., Cvrckova, F., Žárský, V., and Hussey, P.J. (2010) The plant formin AtFH4 interacts with both actin and microtubules, and contains a newly identified microtubule-binding domain. *Journal of Cell Science*, **123**, 1209–1215.

Deeks, M.J., Calcutt, J.R., Ingle, E.K.S., Hawkins, T.J., Chapman, S., Richardson, A.C., Mentlak, D.A., Dixon, M.R., Cartwright, F., Smertenko, A.P., Oparka, K., and Hussey, P.J. (2012) A superfamily of actin-binding proteins at the actin-membrane nexus of higher plants. *Current Biology*, **22**, 1595–1600.

Denecke, J., Aniento, F., Frigerio, L., Hawes, C., Hwang, I., Mathur, J., Neuhaus, J.-M., and Robinson, D.G. (2012) Secretory pathway research: The more experimental systems the better. *Plant Cell*, **24**, 1316–1326.

Derbyshire, P., Findlay, K., McCann, M.C., and Roberts, K. (2007) Cell elongation in *Arabidopsis* hypocotyls involves dynamic changes in cell wall thickness. *Journal of Experimental Botany*, **58**, 2079–2089.

Dixit, R., Cyr, R., and Gilroy, S. (2006) Using intrinsically fluorescent proteins for plant cell imaging. *Plant Journal*, **45**, 599–615.

Dong, C.-H., Xia, G.-X., Hong, Y., Ramachandran, S., Kost, B., and Chua, N.-H. (2001) ADF proteins are involved in the control of flowering and regulate F-actin organization, cell expansion, and organ growth in Arabidopsis. *Plant Cell*, **13**, 1333–1346.

Dos Remedios, C.G., Chhabra, D., Kekic, M., Dedova, I.V., Tsubakihara, M., Berry, D.A., and Nosworthy, N.J. (2003) Actin binding proteins: Regulation of cytoskeletal microfilaments. *Physiological Reviews*, **83**, 433–473.

Doyle, T., and Botstein, D. (1996) Movement of yeast cortical actin cytoskeleton visualized in vitro. *Proceedings of the National Academy of Science, USA*, **93**, 3886–3891.

Ehrhardt, D.W., and Shaw, S.L. (2006) Microtubule dynamics and organization in the plant cortical array. *Annual Review of Plant Biology*, **57**, 859–875.

Ehrhardt, D.W., and Frommer, W.B. (2012) New technologies for 21st century plant science. *Plant Cell*, **24**, 374–394.

Era, A., Tominaga, M., Ebine, K., Awai, C., Saito, C., Ishizaki, K., Yamamoto, K.T., Kohchi, T., Nakano, A., and Ueda, T. (2009) Application of Lifeact reveals F-actin dynamics in *Arabidopsis thaliana* and the liverwort, *Marchantia polymorpha*. *Plant and Cell Physiology*, **50**, 1041–1048.

Era, A., Kutsuna, N., Higaki, T., Hasezawa, S., Nakano, A., and Ueda, T. (2013) Microtubule stability affects the unique motility of F-actin in Marchantia polymorpha. *Journal of Plant Research*, **126**, 113–119.

Fayant, P., Girlanda, O., Chebli, Y., Aubin, C.-E., Villemure, I., and Geitmann, A. (2010) Finite element model of polar growth in pollen tubes. *Plant Cell*, **22**, 2579–2593.

Frank, M.J., and Smith, L.G. (2002) A small, novel protein highly conserved in plants and animals promotes the polarized growth and division of maize leaf epidermal cells. *Current Biology*, **12**, 849–853.

Frank, M.J., Cartwright, H.N., and Smith, L.G. (2003) Three *Brick* genes have distinct functions in a common pathway promoting polarized cell division and cell morphogenesis in the maize leaf epidermis. *Development*, **130**, 753–762.

Frank, M., Egile, C., Dyachok, J., Djakovic, S., Nolasco, M., Li, R., and Smith, L.G. (2004) Activation of Arp2/3 complex-dependent actin polymerization by plant proteins distantly related to Scar/WAVE. *Proceedings of the National Academy of Science, USA*, **101**, 16379–16384.

Fricker, M., Runions, J., and Moore, I. (2006) Quantitative fluorescence microscopy: from art to science. *Annual Review of Plant Biology*, **57**, 79–107.

Fu, Y., Wu, G., and Yang, Z. (2001) Rop GTPase-dependent dynamics of tip-localized F-actin controls tip growth in pollen tubes. *Journal of Cell Biology*, **152**, 1019–1032.

Fu, Y., Li, H., and Yang, Z. (2002) The ROP2 GTPase controls the formation of cortical fine F-actin and the early phase of directional cell expansion during Arabidopsis organogenesis. *Plant Cell*, **14**, 777–794.

Fu, Y., Gu, Y., Zheng, Z., Wasteneys, G., and Yang, Z. (2005) *Arabidopsis* interdigitating cell growth requires two antagonistic pathways with opposing action on cell morphogenesis. *Cell*, **120**, 687–700.

Fujimoto, M., Arimura, S.-i., Nakazono, M., and Tustusmi, N. (2007) Imaging of plant dynamin-related proteins and clathrin around the plasma membrane by variable incidence angle fluorescence microscopy. *Plant Biotechnology*, **24**, 449–455.

Gendreau, E., Traas, J., Desnos, T., Grandjean, O., Caboche, M., and Höfte, H. (1997) Cellular basis of hypocotyl growth in *Arabidopsis thaliana*. *Plant Physiology*, **114**, 295–305.

Gibbon, B.C., Zonia, L.E., Kovar, D.R., Hussey, P.J., and Staiger, C.J. (1998) Pollen profilin function depends on interaction with proline-rich motifs. *Plant Cell*, **10**, 981–994.

Gibbon, B.C., Kovar, D.R., and Staiger, C.J. (1999) Latrunculin B has different effects on pollen germination and tube growth. *Plant Cell*, **11**, 2349–2363.

Gilroy, S. (1997) Fluorescence microscopy of living plant cells. *Annual Review of Plant Physiology and Plant Molecular Biology*, **48**, 165–190.

Gu, Y., Fu, Y., Dowd, P., Li, S., Vernoud, V., Gilroy, S., and Yang, Z. (2005) A Rho family GTPase controls actin dynamics and tip growth via two counteracting downstream pathways in pollen tubes. *Journal of Cell Biology*, **169**, 127–138.

Gutierrez, R., Lindeboom, J.J., Paredez, A.R., Emons, A.M.C., and Ehrhardt, D.W. (2009) *Arabidopsis* cortical microtubules position cellulose synthase delivery to the plasma membrane and interact with cellulose synthase trafficking compartments. *Nature Cell Biology*, **11**, 797–806.

Henty, J.L., Bledsoe, S.W., Khurana, P., Meagher, R.B., Day, B., Blanchoin, L., and Staiger, C.J. (2011) *Arabidopsis* Actin Depolymerizing Factor 4 modulates the stochastic dynamic behavior of actin filaments in the cortical array of epidermal cells. *Plant Cell*, **23**, 3711–3726.

Henty-Ridilla, J.L., Li, J., Blanchoin, L., and Staiger, C.J. (2013) Actin dynamics in the cortical array of plant cells. *Current Opinion in Plant Biology*, **16**, 678–687.

Higaki, T., Sano, T., and Hasezawa, S. (2007) Actin microfilament dynamics and actin side-binding proteins in plants. *Current Opinion in Plant Biology*, **10**, 549–552.

Higaki, T., Kojo, K.H., and Hasezawa, S. (2010a) Critical role of actin bundling in plant cell morphogenesis. *Plant Signaling and Behavior*, **5**, 484–488.

Higaki, T., Kutsuna, N., Sano, T., Kondo, N., and Hasezawa, S. (2010b) Quantification and cluster analysis of actin cytoskeletal structures in plant cells: role of actin bundling in stomatal movement during diurnal cycles in Arabidopsis guard cells. *Plant Journal*, **61**, 156–165.

Hoffmann, C., Moes, D., Dieterle, M., Neumann, K., Moreau, F., Dumas, D., Steinmetz, A., and Thomas, C. (2014) Live cell imaging approaches reveal actin cytoskeleton-induced self-association of the actin-bundling protein WLIM1. *Journal of Cell Science*, **127**, 583–598.

Holzinger, A., and Meindl, U. (1997) Jasplakinolide, a novel actin targeting peptide, inhibits cell growth and induces actin filament polymerization in the green alga *Micrasterias*. *Cell Motility and the Cytoskeleton*, **38**, 365–372.

Huang, S., Blanchoin, L., Kovar, D.R., and Staiger, C.J. (2003) *Arabidopsis* capping protein (AtCP) is a heterodimer that regulates assembly at the barbed ends of actin filaments. *Journal of Biological Chemistry*, **278**, 44832–44842.

Huang, S., Robinson, R.C., Gao, L.Y., Matsumoto, T., Brunet, A., Blanchoin, L., and Staiger, C.J. (2005) Arabidopsis VILLIN1 generates actin filament cables that are resistant to depolymerization. *Plant Cell*, **17**, 486–501.

Huang, S., Gao, L., Blanchoin, L., and Staiger, C.J. (2006) Heterodimeric capping protein from *Arabidopsis* is regulated by phosphatidic acid. *Molecular Biology of the Cell*, **17**, 1946–1958.

Hussey, P.J., and Hashimoto, T. (2008) The cytoskeleton and signal transduction: Role and regulation of plant actin- and microtubule-binding proteins. *Annual Review of Plant Biology*, **33**, 244–272.

Hussey, P.J., Allwood, E.G., and Smertenko, A.P. (2002) Actin-binding proteins in the *Arabidopsis* genome database: properties of functionally distinct plant actin-depolymerizing factors/cofilins. *Philosophical Transactions of the Royal Society, London, B*, **357**, 791–798.

Ingouff, M., Fitz Gerald, J.N., Guérin, C., Robert, H., Sørensen, M.B., Van Damme, D., Geelen, D., Blanchoin, L., and Berger, F. (2005) Plant formin AtFH5 is an evolutionarily conserved actin nucleator involved in cytokinesis. *Nature Cell Biology*, **7**, 374–380.

Kabsch, W., and Vanderkerckhove, J. (1992) Structure and function of actin. *Annual Review of Biophysics and Biomolecular Structure*, **21**, 49–76.

Kabsch, W., Mannherz, H.G., Suck, D., Pai, E.F., and Holmes, K.C. (1990) Atomic structure of the actin:DNase I complex. *Nature*, **347**, 37–44.

Ketelaar, T., de Ruijter, N.C.A., and Emons, A.M.C. (2003) Unstable F-actin specifies the area and microtubule direction of cell expansion in Arabidopsis root hairs. *Plant Cell*, **15**, 285–292.

Ketelaar, T., Anthony, R.G., and Hussey, P.J. (2004a) Green fluorescent protein-mTalin causes defects in actin organization and cell expansion in Arabidopsis and inhibits actin depolymerization factor's actin depolymerizing activity in vitro. *Plant Physiology*, **136**, 3990–3998.

Ketelaar, T., Allwood, E.G., Anthony, R., Voigt, B., Menzel, D., and Hussey, P.J. (2004b) The actin-interacting protein AIP1 is essential for actin organization and plant development. *Current Biology*, **14**, 145–149.

Khurana, P., Henty, J.L., Huang, S., Staiger, A.M., Blanchoin, L., and Staiger, C.J. (2010) Arabidopsis VILLIN1 and VILLIN3 have overlapping and distinct activities in actin bundle formation and turnover. *Plant Cell*, **22**, 2727–2748.

Klahre, U., Friederich, E., Kost, B., Louvard, D., and Chua, N.-H. (2000) Villin-like actin-binding proteins are expressed ubiquitously in Arabidopsis. *Plant Physiology*, **122**, 35–47.

Klotz, J., and Nick, P. (2011) A novel actin–microtubule cross-linking kinesin, NtKCH, functions in cell expansion and division. *New Phytologist*, **193**, 576–589.

Kong, S.-G., Arai, Y., Suetsugu, N., Yanagida, T., and Wada, M. (2013) Rapid severing and motility of chloroplast-actin filaments are required for the chloroplast avoidance response in *Arabidopsis*. *Plant Cell*, **25**, 572–590.

Konopka, C.A., and Bednarek, S.Y. (2008a) Variable-angle epifluorescence microscopy: a new way to look at protein dynamics in the plant cell cortex. *Plant Journal*, **53**, 186–196.

Konopka, C.A., and Bednarek, S.Y. (2008b) Comparison of the dynamics and functional redundancy of the Arabidopsis dynamin-related isoforms, DRP1A and DRP1C, during plant development. *Plant Physiology*, **147**, 1590–1602.

Konopka, C.A., Backues, S.K., and Bednarek, S.Y. (2008) Dynamics of Arabidopsis dynamin-related protein 1C and a clathrin light chain at the plasma membrane. *Plant Cell*, **20**, 1363–1380.

Kost, B., Spielhofer, P., and Chua, N.-H. (1998) A GFP-mouse talin fusion protein labels plant actin filaments *in vivo* and visualizes the actin cytoskeleton in growing pollen tubes. *Plant Journal*, **16**, 393–401.

Kotzer, A.M., and Wasteneys, G.O. (2006) Mechanisms behind the puzzle: microtubule–microfilament cross-talk in pavement cell formation. *Canadian Journal of Botany*, **84**, 594–603.

Lee, Y.-R.J., and Liu, B. (2004) Cytoskeletal motors in Arabidopsis. Sixty-one kinesins and seventeen myosins. *Plant Physiology*, **136**, 3877–3883.

Lee, Y.J., Szumlanski, A., Nielsen, E., and Yang, Z. (2008) Rho-GTPase–dependent filamentous actin dynamics coordinate vesicle trafficking and exocytosis during tip growth. *Journal of Cell Biology*, **181**, 1155–1168.

Lemichez, E., Wu, Y., Sanchez, J.-P., Mettouchi, A., Mathur, J., and Chua, N.-H. (2001) Inactivation of AtRac1 by abscisic acid is essential for stomatal closure. *Genes and Development*, **15**, 1808–1816.

Leucci, M.R., Di Sansebastiano, G.-P., Giganti, M., Dalessandro, G., and Piro, G. (2007) Secretion marker proteins and cell-wall polysaccharides move through different secretory pathways. *Planta*, **225**, 1001–1017.

Li, L.-J., Ren, F., Gao, X.-Q., Wei, P.-C., and Wang, X.-C. (2013) The reorganization of actin filaments is required for vacuolar fusion of guard cells during stomatal opening in *Arabidopsis*. *Plant, Cell and Environment*, **36**, 484–497.

Li, J., Henty-Ridilla, J.L., Huang, S., Wang, X., Blanchoin, L., and Staiger, C.J. (2012) Capping protein modulates the dynamic behavior of actin filaments in response to phosphatidic acid in *Arabidopsis*. *Plant Cell*, **24**, 3742–3754.

Li, J., Staiger, B.H., Henty-Ridilla, J.L., Abu-Abied, M., Sadot, E., Blanchoin, L., and Staiger, C.J. (2014) The availability of filament ends modulates actin stochastic dynamics in live plant cells. *Molecular Biology of the Cell*, **25**, 1263–1275.

Li, S., Blanchoin, L., Yang, Z., and Lord, E.M. (2003) The putative Arabidopsis Arp2/3 complex controls leaf cell morphogenesis. *Plant Physiology*, **132**, 2034–2044.

Li, Y., Shen, Y., Cai, C., Zhong, C., Zhu, L., Yuan, M., and Ren, H. (2010) The Type II *Arabidopsis* Formin14 interacts with microtubules and microfilaments to regulate cell division. *Plant Cell*, **22**, 2710–2726.

Lo, Y.-S., Cheng, N., Hsiao, L.-J., Annamalai, A., Jauh, G.-Y., Wen, T.-N., and Dai, H. (2011) Actin in mung bean mitochondria and implications for its function. *Plant Cell*, **23**, 3727–3744.

Lucas, J., and Shaw, S.L. (2008) Cortical microtubule arrays in the Arabidopsis seedling. *Current Opinion in Plant Biology*, **11**, 94–98.

Ma, B., Qian, D., Nan, Q., Tan, C., An, L., and Xiang, Y. (2012) *Arabidopsis* vacuolar H^+-ATPase (V-ATPase) B subunits are involved in actin cytoskeleton remodeling via binding to, bundling, and stabilizing F-actin. *Journal of Biological Chemistry*, **287**, 19008–19017.

Ma, L., Sun, N., Liu, X., Jiao, Y., Zhao, H., and Deng, X.W. (2005) Organ-specific expression of Arabidopsis genome during development. *Plant Physiology*, **138**, 80–91.

Martiniére, A., Gayral, P., Hawes, C., and Runions, J. (2011) Building bridges: FORMIN1 of Arabidopsis forms a connection between the cell wall and the actin cytoskeleton. *Plant Journal*, **66**, 354–365.

Mathur, J., Radhamony, R., Sinclair, A.M., Donoso, A., Dunn, N., Roach, F., Radford, D., Mohaghegh, S.M., Logan, D.C., Kokolic, K., and Mathur, N. (2010) mEosFP-based green-to-red photoconvertible subcellular probes for plants. *Plant Physiology*, **154**, 1573–1587.

McCurdy, D.W. (1999) Is 2,3-butanedione monoxime an effective inhibitor of myosin-based activities in plant cells? *Protoplasma*, **209**, 120–125.

Michelot, A., Guérin, C., Huang, S., Ingouff, M., Richard, S., Rodiuc, N., Staiger, C.J., and Blanchoin, L. (2005) The formin homology 1 domain modulates the actin nucleation and bundling activity of Arabidopsis FORMIN1. *Plant Cell*, **17**, 2296–2313.

Michelot, A., Derivery, E., Paterski-Boujemaa, R., Guérin, C., Huang, S., Parcy, F., Staiger, C.J., and Blanchoin, L. (2006) A novel mechanism for the formation of actin-filament bundles by a non-processive formin. *Current Biology*, **16**, 1924–1930.

Michelot, A., Berro, J., Guérin, C., Boujemaa-Paterski, R., Staiger, C.J., Martiel, J.-L., and Blanchoin, L. (2007) Actin-filament stochastic dynamics mediated by ADF/cofilin. *Current Biology*, **17**, 825–833.

Nagawa, S., Xu, T., Lin, D., Dhonukshe, P., Zhang, X., Friml, J., Scheres, B., Fu, Y., and Yang, Z. (2012) ROP GTPase-dependent actin microfilaments promote PIN1 polarization by localized inhibition of clathrin-dependent endocytosis. *PLoS Biology*, **10**, e1001299.

Oda, T., Iwasa, M., Aihara, T., Maéda, Y., and Narita, A. (2009) The nature of the globular- to fibrous-actin transition. *Nature*, **457**, 441-445.

Oikawa, K., Kasahara, M., Kiyosue, T., Kagawa, T., Suetsugu, N., Takahashi, F., Kanegae, T., Niwa, Y., Kadota, A., and Wada, M. (2003) CHLOROPLAST UNUSUAL POSITIONING1 is essential for proper chloroplast positioning. *Plant Cell*, **15**, 2805–2815.

Oikawa, K., Yamasato, A., Kong, S.-G., Kasahara, M., Nakai, M., Takahashi, F., Ogura, Y., Kagawa, T., and Wada, M. (2008) Chloroplast outer envelope protein CHUP1 is essential for chloroplast anchorage to the plasma membrane and chloroplast movement. *Plant Physiology*, **148**, 829–842.

Paredez, A.R., Somerville, C.R., and Ehrhardt, D.W. (2006) Visualization of cellulose synthase demonstrates functional association with microtubules. *Science*, **312**, 1491–1495.

Pleskot, R., Pejchar, P., Žárský, V., Staiger, C.J., and Potocký, M. (2012) Structural insights into the inhibition of actin-capping protein by interactions with phosphatidic acid and phosphatidylinositol (4,5)-bisphosphate. *PLoS Computational Biology*, **8**, e1002765.

Pleskot, R., Li, J., Žárský, V., Potocký, M., and Staiger, C.J. (2013) Regulation of cytoskeletal dynamics by phospholipase D and phosphatidic acid. *Trends in Plant Science*, **18**, 496–504.

Pollard, T.D. (1986) Rate constants for the reactions of ATP- and ADP-actin with the ends of actin filaments. *Journal of Cell Biology*, **103**, 2747–2754.

Pollard, T.D. (2000) Reflections on a quarter century of research on contractile systems. *Trends in Biochemical Science*, **25**, 607–611.

Pollard, T.D., and Borisy, G.G. (2003) Cellular motility driven by assembly and disassembly of actin filaments. *Cell*, **112**, 453–465.

Pollard, T.D., and Cooper, J.A. (2009) Actin, a central player in cell shape and movement. *Science*, **27**, 1208–1212.

Pollard, T.D., Blanchoin, L., and Mullins, R.D. (2000) Molecular mechanisms controlling actin filament dynamics in nonmuscle cells. *Annual Review of Biophysics and Biomolecular Structure*, **29**, 545–576.

Preuss, M.L., Kovar, D.R., Lee, Y.-R.J., Staiger, C.J., Delmer, D.P., and Liu, B. (2004) A plant-specific kinesin binds to actin microfilaments and interacts with cortical microtubules in cotton fibers. *Plant Physiology*, **136**, 3945–3955.

Qu, X., Zhang, H., Xie, Y., Wang, J., Chen, N., and Huang, S. (2013) *Arabidopsis* villins promote actin turnover at pollen tube tips and facilitate the construction of actin collars. *Plant Cell*, **25**, 1803–1817.

Rahman, A., Bannigan, A., Sulaman, W., Pechter, P., Blancaflor, E.B., and Baskin, T.I. (2007) Auxin, actin and growth of the *Arabidopsis thaliana* primary root. *Plant Journal*, **50**, 514–528.

Refrégier, G., Pelletier, S., Jaillard, D., and Höfte, H. (2004) Interaction between wall deposition and cell elongation in dark-grown hypocotyl cells in Arabidopsis. *Plant Physiology*, **135**, 959–968.

Ren, H., Gibbon, B.C., Ashworth, S.L., Sherman, D.M., Yuan, M., and Staiger, C.J. (1997) Actin purified from maize pollen functions in living plant cells. *Plant Cell*, **9**, 1445–1457.

Ren, H.-Y., and Xiang, Y. (2007) The function of actin-binding proteins in pollen tube growth. *Protoplasma*, **230**, 171–182.

Riedl, J., Crevenna, A.H., Kessenbrock, K., Yu, J.H., Neukirchen, D., Bista, M., Bradke, F., Jenne, D., Holak, T.A., Werb, Z., Sixt, M., and Wedlich-Soldner, R. (2008) Lifeact: a versatile marker to visualize F-actin. *Nature Methods*, **5**, 605–607.

Roland, J., Berro, J., Michelot, A., Blanchoin, L., and Martiel, J.-L. (2008) Stochastic severing of actin filaments by actin depolymerizing factor/cofilin controls the emergence of a steady dynamical regime. *Biophysical Journal*, **94**, 2082–2094.

Rosero, A., Zarsky, V., and Cvrcková, F. (2013) AtFH1 formin mutation affects actin filament and microtubule dynamics in *Arabidopsis thaliana*. *Journal of Experimental Botany*, **64**, 585–597.

Ruzicka, D.R., Kandasamy, M.K., McKinney, E.C., Burgos-Rivera, B., and Meagher, R.B. (2007) The ancient subclasses of Arabidopsis *ACTIN DEPOLYMERIZING FACTOR* genes exhibit novel and differential expression. *Plant Journal*, **52**, 460–472.

Samaj, J., Müller, J., Beck, M., Böhm, N., and Menzel, D. (2006) Vesicular trafficking, cytoskeleton and signalling in root hairs and pollen tubes. *Trends in Plant Science*, **11**, 1360–1385.

Sampathkumar, A., Gutierrez, R., McFarlane, H.E., Bringmann, M., Lindeboom, J., Emons, A.-M., Samuels, L., Ketelaar, T., Ehrhardt, D.W., and Persson, S. (2013) Patterning and lifetime of plasma membrane-localized cellulose synthase is dependent on actin organization in Arabidopsis interphase cells. *Plant Physiology*, **162**, 675–688.

Shaw, S.L., and Ehrhardt, D.W. (2013) Smaller, faster, brighter: Advances in optical imaging of living plant cells. *Annual Review of Plant Biology*, **64**, 351–375.

Shaw, S.L., Kamyar, R., and Ehrhardt, D.W. (2003) Sustained microtubule treadmilling in *Arabidopsis* cortical arrays. *Science*, **300**, 1715–1718.

Sheahan, M.B., Staiger, C.J., Rose, R.J., and McCurdy, D.W. (2004) A green fluorescent protein fusion to actin-binding domain 2 of Arabidopsis fimbrin highlights new features of a dynamic actin cytoskeleton in live plant cells. *Plant Physiology*, **136**, 3968–3978.

Shi, M., Xie, Y., Zheng, Y., Wang, J., Su, Y., Yang, Q., and Huang, S. (2013) *Oryza sativa* actin-interacting protein 1 is required for rice growth by promoting actin turnover. *Plant Journal*, **73**, 747–760.

Smertenko, A.P., Deeks, M.J., and Hussey, P.J. (2010) Strategies of actin reorganisation in plant cells. *Journal of Cell Science*, **123**, 3019–3028.

Smith, L.G. (2003) Cytoskeletal control of plant cell shape: getting the fine points. *Current Opinion in Plant Biology*, **6**, 63–73.

Smith, L.G., and Oppenheimer, D.G. (2005) Spatial control of cell expansion by the plant cytoskeleton. *Annual Reviews of Cell and Developmental Biology*, **21**, 271–295.

Snowman, B.N., Kovar, D.R., Shevchenko, G., Franklin-Tong, V.E., and Staiger, C.J. (2002) Signal-mediated depolymerization of actin in pollen during the self-incompatibility response. *Plant Cell*, **14**, 2613–2626.

Sparkes, I. (2012) Recent advances in understanding plant myosin function: Life in the fast lane. *Molecular Plant*, **4**, 805–812.

Staiger, C.J., and Hussey, P.J. (2004) Actin and actin-modulating proteins, in *The Plant Cytoskeleton in Cell Differentiation and Development* (ed. P.J. Hussey), Blackwell Publishers, UK, pp. 32–80.

Staiger, C.J., and Blanchoin, L. (2006) Actin dynamics: old friends with new stories. *Current Opinion in Plant Biology*, **9**, 554–562.

Staiger, C.J., Sheahan, M.B., Khurana, P., Wang, X., McCurdy, D.W., and Blanchoin, L. (2009) Actin filament dynamics are dominated by rapid growth and severing activity in the *Arabidopsis* cortical array. *Journal of Cell Biology*, **184**, 269–280.

Staiger, C.J., Poulter, N.S., Henty, J.L., Franklin-Tong, V.E., and Blanchoin, L. (2010) Regulation of actin dynamics by actin-binding proteins in pollen. *Journal of Experimental Botany*, **61**, 1969–1986.

Suarez, C., Roland, J., Boujemaa-Paterski, R., Kang, H., McCullough, B.R., Reymann, A.-C., Guérin, C., Martiel, J.-L., De La Cruz, E.M., and Blanchoin, L. (2011) Cofilin tunes the nucleotide state of actin filaments and severs at bare and decorated segment boundaries. *Current Biology*, **21**, 862–868.

Suetsugu, N., Yamada, N., Kagawa, T., Yonekura, H., Uyeda, T.Q.P., Kadota, A., and Wada, M. (2010) Two kinesin-like proteins mediate actin-based chloroplast movement in *Arabidopsis thaliana*. *Proceedings of the National Academy of Science, USA*, **107**, 8860–8865.

Szymanski, D.B. (2001) *Arabidopsis* trichome morphogenesis: A genetic approach to studying cytoskeletal function. *Journal of Plant Growth Regulation*, **20**, 131–140.

Szymanski, D.B. (2005) Breaking the WAVE complex: the point of *Arabidopsis* trichomes. *Current Opinion in Plant Biology*, **8**, 103–112.

Szymanski, D.B., and Cosgrove, D.J. (2009) Dynamic coordination of cytoskeletal and cell wall systems during plant cell morphogenesis. *Current Biology*, **19**, R800–R811.

Szymanski, D.B., Marks, M.D., and Wick, S.M. (1999) Organized F-actin is essential for normal trichome morphogenesis in Arabidopsis. *Plant Cell*, **11**, 2331–2347.

Thomas, C. (2012) Bundling actin filaments from membranes: some novel players. *Frontiers in Plant Science*, **3**, e00118.

Thomas, C., Tholl, S., Moes, D., Dieterle, M., Papuga, J., Moreau, F., and Steinmetz, A. (2009) Actin bundling in plants. *Cell Motility and the Cytoskeleton*, **66**, 940–957.

Tian, M., Chaudhry, F., Ruzicka, D.R., Meagher, R.B., Staiger, C.J., and Day, B. (2009) *Arabidopsis* actin-depolymerizing factor AtADF4 mediates defense signal transduction triggered by the *Pseudomonas syringae* effector AvrPphB. *Plant Physiology*, **150**, 815–824.

Tominaga, M., and Nakano, A. (2012) Plant-specific myosin XI, a molecular perspective. *Frontiers in Plant Science*, **3**, e211.

Tóth, R., Gerding-Reimers, C., Deeks, M.J., Menninger, S., Gallegos, R.M., Tonaco, I.A.N., Hübel, K., Hussey, P.J., Waldmann, H., and Coupland, G. (2012) Prieurianin/endosidin 1 is an actin stabilizing small molecule identified from chemical genetic screen for circadian clock effectors in *Arabidopsis thaliana*. *Plant Journal*, **71**, 338–352.

Valenta, R., Ferreira, F., Grote, M., Swoboda, I., Vrtala, S., Duchêne, M., Deviller, P., Meagher, R.B., McKinney, E., Heberle-Bors, E., Kraft, D., and Scheiner, O. (1993) Identification of profilin as an actin-binding protein in higher plants. *Journal of Biological Chemistry*, **268**, 22777–22781.

van der Honig, H.S., Bezouwen, L.S., Emons, A.-M., and Ketelaar, T. (2011) High expression of Lifeact in *Arabidopsis thaliana* reduces dynamic reorganization of actin filaments but does not affect plant development. *Cytoskeleton*, **68**, 578–587.

van der Honig, H.S., Kieft, H., Emons, A.M.C., and Ketelaar, T. (2012) Arabidopsis VILLIN2 and VILLIN3 are required for the generation of thick actin filament bundles and for directional organ growth. *Plant Physiology*, **158**, 1426–1438.

van Gisbergen, P.A.C., Li, M., Wu, S.-Z., and Bezanilla, M. (2012) Class II formin targeting to the cell cortex by binding $PI(3,5)P_2$ is essential for polarized growth. *Journal of Cell Biology*, **198**, 235–250.

Vidali, L., and Hepler, P.K. (1997) Characterization and localization of profilin in pollen grains and tubes of *Lilium longiflorum*. *Cell Motility and the Cytoskeleton*, **36**, 323–338.

Vidali, L., McKenna, S.T., and Hepler, P.K. (2001) Actin polymerization is essential for pollen tube growth. *Molecular Biology of the Cell*, **12**, 2534–2545.

Vidali, L., Rounds, C.M., Hepler, P.K., and Bezanilla, M. (2009a) Lifeact-mEGFP reveals a dynamic apical F-actin network in tip growing plant cells. *PLoS One*, **4**, e5744.

Vidali, L., van Gisenbergen, P.A.C., Guérin, C., Franco, P., Li, M., Burkart, G.M., Augustine, R.C., Blanchoin, L., and Bezanilla, M. (2009b) Rapid formin-mediated actin-filament elongation is essential for polarized plant cell growth. *Proceedings of the National Academy of Science, USA*, **106**, 13341–13346.

Vidali, L., Burkart, G.M., Augustine, R.C., Kerdavid, E., Tüzel, E., and Bezanilla, M. (2010) Myosin XI is essential for tip growth in *Physcomitrella patens*. *Plant Cell*, **22**, 1868–1882.

Vizcay-Barrena, G., Webb, S.E.D., Martin-Fernandez, M.L., and Wilson, Z.A. (2011) Subcellular and single-molecule imaging of plant fluorescent proteins using total internal reflection fluorescence microscopy (TIRFM). *Journal of Experimental Botany*, **62**, 5419–5428.

Voigt, B., Timmers, A.C.J., Samaj, J., Müller, J., Baluska, F., and Menzel, D. (2005) GFP-FABD2 fusion construct allows in vivo visualization of the dynamic actin cytoskeleton in all cells of *Arabidopsis* seedlings. *European Journal of Cell Biology*, **84**, 595–608.

Wang, H.-Y., Yu, Y., Chen, Z.-L., and Xia, G.-X. (2005) Functional characterization of *Gossypium hirsutum* profilin 1 gene (GhPFN1) in tobacco suspension cells. *Planta*, **222**, 594–603.

Wang, X., Teng, Y., Wang, Q., Li, X., Sheng, S., Zheng, M., Samaj, J., Baluska, F., and Lin, J. (2006) Imaging of dynamic secretory vesicles in living pollen tubes of Picea meyeri using evanescent wave microscopy. *Plant Physiology*, **141**, 1591–1603.

Wang, X.-L., Gao, X.-Q., and Wang, X.-C. (2011) Stochastic dynamics of actin filaments in guard cells regulating chloroplast localization during stomatal movement. *Plant, Cell and Environment*, **34**, 1248–1257.

Wang, Y.-S., Motes, C.M., Mohamalawari, D.R., and Blancaflor, E.B. (2004) Green fluorescent protein fusions to *Arabidopsis* fimbrin 1 for spatio-temporal imaging of F-actin dynamics in roots. *Cell Motility and the Cytoskeleton*, **59**, 79–93.

Wang, Y.-S., Yoo, C.-M., and Blancaflor, E.B. (2008) Improved imaging of actin filaments in transgenic *Arabidopsis* plants expressing a green fluorescent protein fusion to the C- and N-termini of the fimbrin actin-binding domain 2. *New Phytologist*, **177**, 525–536.

Whippo, C.W., Khurana, P., Davis, P.A., DeBlasio, S.L., Sloover, D., Staiger, C.J., and Hangarter, R.P. (2011) THRUMIN1 is a light-regulated actin-bundling protein involved in chloroplast motility. *Current Biology*, **21**, 59–64.

Wightman, R., and Turner, S.R. (2010) Trafficking of the plant cellulose synthase complex. *Plant Physiology*, **153**, 427–432.

Wu, J.-Q., and Pollard, T.D. (2005) Counting cytokinesis proteins globally and locally in fission yeast. *Science*, **310**, 310–314.

Xu, T., Wen, M., Nagawa, S., Fu, Y., Chen, J.-G., Wu, M.-J., Perrot-Rechenmann, C., Friml, J., Jones, A.M., and Yang, Z. (2010) Cell surface- and Rho GTPase-based auxin signaling controls cellular interdigitation in *Arabidopsis*. *Cell*, **143**, 99–110.

Xue, X.-H., Guo, C.-Q., Du, F., Lu, Q.-L., Zhang, C.-M., and Ren, H.-Y. (2011) AtFH8 is involved in root development under effect of low-dose latrunculin B in dividing cells. *Molecular Plant*, **4**, 264–278.

Yang, W., Ren, S., Zhang, X., Gao, M., Ye, S., Qi, Y., Zheng, Y., Wang, J., Zeng, L., Li, Q., Huang, S., and He, Z. (2011) *BENT UPPERMOST INTERNODE1* encodes the class II formin FH5 crucial for actin organization and rice development. *Plant Cell*, **23**, 661–680.

Ye, J., Zheng, Y., Yan, A., Chen, N., Wang, Z., Huang, S., and Yang, Z. (2009) *Arabidopsis* Formin3 directs the formation of actin cables and polarized growth in pollen tubes. *Plant Cell*, **19**, 3868–3884.

Yen, L., -F., Liu, X., and Cai, S. (1995) Polymerization of actin from maize pollen. *Plant Physiology*, **107**, 73–76.

Yi, K., Guo, C., Chen, D., Zhao, B., Yang, B., and Ren, H. (2005) Cloning and functional characterization of a formin-like protein (AtFH8) from Arabidopsis. *Plant Physiology*, **138**, 1071–1082.

Yokota, E., and Shimmen, T. (2010) Plant myosins, in *The Plant Cyoskeleton* (ed. B. Liu), Springer, New York, pp. 33–56.

Yu, J.H., Crevenna, A.H., Bettenbühl, M., Freisinger, T., and Wedlich-Söldner, R. (2011) Cortical actin dynamics driven by formins and myosin V. *Journal of Cell Science*, **124**, 1533–1541.

Zhang, H., Qu, X., Bao, C., Khurana, P., Wang, Q., Xie, Y., Zheng, Y., Chen, N., Blanchoin, L., Staiger, C.J., and Huang, S. (2010) *Arabidopsis* VILLIN5, an actin filament bundling and severing protein, is necessary for normal pollen tube growth. *Plant Cell*, **22**, 2749–2767.

Zhang, Y., Xiao, Y., Du, F., Cao, L., Dong, H., and Ren, H. (2011a) *Arabidopsis* VILLIN4 is involved in root hair growth through regulating actin organization in a Ca^{2+}-dependent manner. *New Phytologist*, **190**, 667–682.

Zhang, Z., Zhang, Y., Tan, H., Wang, Y., Li, G., Liang, W., Yuan, Z., Hu, J., Ren, H., and Zhang, D. (2011b) *RICE MORPHOLOGY DETERMINANT* encodes the type II formin FH5 and regulates rice morphogenesis. *Plant Cell*, **23**, 681–700.

Zhao, Y., Zhao, S., Mao, T., Qu, X., Cao, W., Zhang, L., Zhang, W., He, L., Li, S., Ren, S., Zhao, J., Zhu, G., Huang, S., Ye, K., Yuan, M., and Guo, Y. (2011) The plant-specific actin binding protein SCAB1 stabilizes actin filaments and regulates stomatal movement in *Arabidopsis*. *Plant Cell*, **23**, 2314–2330.

Zheng, M., Beck, M., Müller, J., Chen, T., Wang, X., Wang, F., Wang, Q., Wang, Y., Baluska, F., Logan, D.C., Samaj, J., and Lin, J. (2009). Actin turnover is required for myosin-dependent mitochondrial movements in *Arabidopsis* root hairs. *PLoS One*, **4**, e5961.

Zheng, Y., Xie, Y., Jiang, Y., Qu, X., and Huang, S. (2013). *Arabidopsis* ACTIN-DEPOLYMERIZING FACTOR7 severs actin filaments and regulates actin cable turnover to promote pollen tube growth. *Plant Cell*, **25**, 3405–3423.

Zhu, L., Zhang, Y., Kang, E., Xu, Q., Wang, M., Rui, Y., Liu, B., Yuan, M., and Fu, Y. (2013). MAP18 regulates the direction of pollen tube growth in *Arabidopsis* by modulating F-actin organization. *Plant Cell*, **25**, 851–867.

Section 2
Cellular Mechanisms Underlying Various Cell Shapes

6 The Regulation of Cell Shape Formation by ROP-dependent Auxin Signaling

Shingo Nagawa[1] and Zhenbiao Yang[1,2]

[1]*Center for Plant Stress Biology, Shanghai Institutes for Biological Sciences, Chinese Academy of Sciences, Shanghai, China*

[2]*Center for Plant Cell Biology, Department of Botany and Plant Sciences, University of California, CA*

Introduction

In multicellular plants there are various cell types with characteristic functions and, in many cases, their specialized morphologies facilitate their roles. For example, tip-growing pollen tubes form characteristic elongating cylindrical structures that enable them to go through the transmitting channel of the female tissues to deliver sperm cells to ovules. Tracheary elements in the xylem conduct water and nutrients in vascular tissues; they are derived from cells with characteristic secondary wall thickenings that enhance rigidity and strength. Differential morphogenesis in the epidermal cell layer is especially important, as it directly faces the extracellular environment and thus needs to protect the inner layers and function as a 'sensing antenna' for the outside environment. Moreover, in certain tissues, epidermal cells need to absorb extracellular materials such as water or nutrients. Morphologies and functions of epidermal cells are therefore highly diversified even within a single organism (Guimil and Dunand, 2006).

Polarity within a cell enables the cell to be anisotropic, and epidermal cells tend to have cellular polarity that underlies their characteristic morphogenesis and functions. In plants as well as in other organisms the Rho-family of small GTPases plays a major role in the establishment and maintenance of cellular polarity (Jaffe and Hall, 2005). Rho proteins are GTPases and as such switch between their GDP-bound inactive state and a GTP-bound active state. Active Rho proteins can specifically bind to downstream effector proteins and effectively serve as a molecular switch to regulate cellular functions such as cytoskeletal dynamics and vesicle trafficking. The activation of Rho proteins by upstream factors typically occurs on or in proximity of the plasma membrane, linking extracellular developmental cues to diverse downstream

Plant Cell Wall Patterning and Cell Shape, First Edition. Edited by Hiroo Fukuda.

cellular signaling events. Investigating mechanisms underlying Rho signaling, including the activation process of Rho proteins as well as the downstream pathways, is therefore critical for both cellular and developmental biology. Not surprisingly, Rho proteins are involved in a number of developmental processes of specialized epidermal cells, such as planar cell polarity in animal systems (Klein and Mlodzi, 2005) as well as epidermal cell morphogenesis in plants (Yang, 2008; Murphy and Peer, 2012).

Plants possess a single family of small GTPase proteins belonging to the Rho family, and they are called Rho of Plant (ROP) proteins. ROPs are involved in a variety of developmental processes in plants such as pollen tube tip growth (Lin *et al.* 1996; Kost *et al.* 1999; Fu *et al.*, 2001), disease resistance (Kawasaki *et al.* 1999; Yamaguchi *et al.*, 2012), and cellular/tissue/organ development such as root hair development, pollen tube tip growth, vascular development, and leaf development (Molendijk *et al.*, 2001; Fu *et al.*, 2002, 2005; Jones *et al.*, 2002; Brembu *et al.*, 2005). ROP-mediated signaling includes factors that are well conserved across kingdoms as well as plant-specific factors. Plant-specific proteins include upstream regulatory proteins such as GDP–GTP exchanging factor proteins (plant-specific ROPGEFs) and downstream effector proteins (RICs, ICR1, etc.; Wu *et al.*, 2001; Berken *et al.*, 2005; Gu *et al.*, 2006; Lavy *et al.*, 2007; Li *et al.*, 2008). The recent discovery of ROP signaling in the downstream signaling pathway of quintessential plant hormones such as auxin and abscisic acid (ABA) further emphasizes the importance of ROP signaling in diverse aspects of plant development. In this chapter, we focus on how auxin-dependent ROP signaling regulates the morphogenesis of leaf epidermal pavement cells (PCs).

Leaf PCs as a Model System to Study Cellular Morphogenesis

As in other organs or tissues of multicellular plants, the epidermal layer in plant leaves typically consists of several cell types that have different characteristics for specialized roles. Stomatal guard cells are paired forming a pore between them that works as a conductive apparatus for air and moisture. Many plant species have hair-like structures called ‘trichomes’ that are typically branched into several tips with nuclei undergoing several rounds of endoreduplication. Trichomes are thought to have protective functions against small feeding animals. The most abundant cell types on the leaf surface are called ‘pavement cells’. Leaf epidermal pavement cells have a characteristic intercalating jigsaw-piece-like shape which is seen in various plants, and serves as an attractive model system to study cellular morphogenesis (Qiu *et al.*, 2002; Fu *et al.*, 2005; Xu *et al.*, 2010).

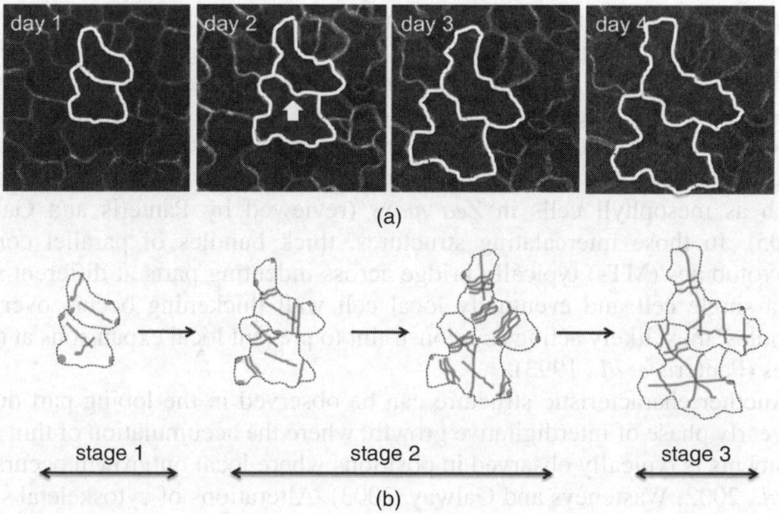

Figure 6.1 Development of Arabidopsis PCs with interdigitating structure. (a) Consecutive images of developing PC in a cotyledon from day 1 to day 4. The arrow indicates the formation of new interdigitation, which cannot be observed in day 1 cells. Note that interdigitation occurs without forming any gaps between cells, indicating the existence of coordination between neighboring cells. (b) A schematic representation of cortical f-actin accumulation (gray) and MTs (thick lines) based on Fu *et al.* (2002, 2005). Cells at stage 1 initiate multiple outgrowths with accumulation of f-actin close to the tip of lobes. At stage 2 cells, MT bundling constrains outgrowth at the neck region, while initiation of outgrowth continues to occur to increase number of interdigitations. In stage 3 cells in which production of new lobes is completed, both accumulation of cortical f-actin and MT bundling are less pronounced.

The protruding part of the intercalation is referred to as the 'lobe' and lobes develop from initial isotropic cells. The counteracting parts that face lobes of a neighboring cell are referred to as 'indentations'. As those epidermal PCs are required to fill in the surface of leaves without huge gaps, production of lobes in a cell needs to be coordinated with development of indentations in the neighboring cell (Fig. 6.1). The morphogenesis of the PC intercalating structure therefore involves the formation of polarity (determination of lobe position) and maintenance of such polarity (subsequent polar growth at the tip of lobes) within a cell as well as in a coordinated manner with neighboring cells. Although the initiation of lobe formation requires anisotropic growth in the specific protruding sites, cell expansion of later stages is thought to be isotropic for whole cell boundaries (Zhang *et al.*, 2011).

The benefits of having such a complicated intercalating structure covering the surface of PCs are still unclear. Some genetic mutants defective in PC lobing have reduced tissue integrity and have sometimes formed gaps between cells (Fu *et al.*, 2002; Basu *et al.*, 2005, 2008), suggesting mechanically important roles of the structure. However, the significance of an intercalating structure in tissue integrity remains to be investigated in more direct ways to show physical changes caused by the defects in interdigitation. Nevertheless, these characteristic morphologies have attracted the attention of many researchers for a long time. Earlier studies revealed that a correlation between cytoskeletal patterning and intercalating structure exists not only in PCs from multiple species but also in other plant cell types with similar morphologies such as mesophyll cells in *Zea mays* (reviewed by Panteris and Galatis, 2005). In those intercalating structures, thick bundles of parallel cortical microtubules (MTs) typically bridge across indenting parts at different sides in a single cell and eventually local cell wall thickening occurs over MT bundles, most likely acting as a constraint to prevent local expansions at those sites (Panteris *et al.*, 1993).

Another characteristic structure can be observed in the lobing part during the early phase of interdigitative growth, where the accumulation of thin actin filaments is typically observed in positions where local outgrowth occurs (Fu *et al.*, 2002; Wasteneys and Galway, 2003). Alterations of cytoskeletal structures by pharmacological disturbance result in loss or reduced amount of those wavy intercalating structures, strongly suggesting that cytoskeletal structures are essential determinants for the development of such structures (Panteris *et al.* 1993). The detailed molecular mechanisms underlying the development of these structures were not revealed until recently; however, considerable progress has been made in this line of research using molecular and cellular approaches and model plants such as Arabidopsis. Recent studies in Arabidopsis PCs have revealed that the phytohormone auxin activates ROPs and subsequent signaling through multiple downstream pathways; these include the regulation of endomembrane trafficking of auxin transport proteins and regulation of cytoskeletal dynamics.

Cytoskeletal Regulation of PC Morphogenesis: Counteraction between ROP-mediated Cortical Actin Microfilament and Microtubules

Signaling mediated by Rho GTPases across kingdoms typically regulates cytoskeletal dynamics through multiple downstream pathways. Likewise, multiple ROP downstream pathways regulating cytoskeletal dynamics have been shown to contribute to the interdigitative growth of Arabidopsis leaf PCs. In many cases, Rho signaling is precisely controlled by the counteraction of distinct members that belong to a different subfamily, or positive/negative feedback regulation from downstream factors (Sander *et al.*, 1999; Burridge and Wennerberg, 2004). To achieve differential growth at multiple lobing sites and indentations within a single cell in leaf epidermal PC morphogenesis, distinct members of ROPs locally activate counteracting pathways that affect the organization of two major cytoskeletal structures existing within a cell. The ROP2–RIC4 pathway regulates F-actin dynamics, while the ROP6–RIC1 pathway promotes organized microtubule (MT) bundling. Since the subcellular distribution of these cortical cytoskeletal structures is different, to gain an understanding of the growth of PCs it is of key importance to elucidate how the two ROP signaling pathways counteract each other to promote distinct targets and the significance of those cytoskeletal structures in the generation of PC shape.

Initial insight into the contribution of ROP signaling in the morphogenesis of PCs was obtained by examining the effects of dominant negative (DN) or constitutive active (CA) ROP2 in Arabidopsis plants (Fu *et al.*, 2005). *rop4 rop2* RNAi (*r4R2i*) PCs were shown to exhibit similar abnormal phenotypes as those observed in DN-ROP2 PCs, proving a contribution by ROP2 and the homologous ROP4 in the earlier phase of PC development (Fu *et al.*, 2005). ROP signaling typically requires downstream effector proteins that interact specifically with activated ROPs. ROP effector proteins belonging to the RIC family possess the conserved CDC42/Rac interactive binding (CRIB) motif to bind Rho proteins (Wu *et al.*, 2001). A yeast 2-hybrid screening identified RICs as proteins binding to CA-ROPs (Wu *et al.*, 2001). RIC4 contributes to assemble dense cortical f-actin patches at the tip of pollen tubes (Gu *et al.*, 2005). Fu and collaborators discovered that the *ric4-1* mutant has defects in PCs that are comparable to those observed in DN-ROP2 or *r4R2i* PCs (Fu *et al.*, 2005). Furthermore, they showed that RIC4 is required for the assembly of f-actin, and that active ROP2 binds to RIC4 at the tip of protruding sites in young PCs. ROP6 is a phylogenetically closely related member of ROP2/ROP4 (Yang, 2002). However, ROP6 is preferentially localized to the indenting region, while ROP2 is enriched in the tip of the lobes, suggesting that two closely related members of ROP have differential roles in PC morphogenesis. Indeed, the ROP6–RIC1 pathway has an antagonistic role in the ROP2–RIC4 pathway. Fu and colleagues found that loss of function of *ric1* induces an increase in cell width (neck), which resembles CA-ROP2 expressing plants (Fu *et al.*, 2005). Overexpression of RIC1 induced the generation of highly organized transverse MT arrays throughout PC development from the early stage (stage 1) to the late stage (stage 3), whereas MT organization is normally characteristic of PCs in the stage 2. Later studies demonstrated that ROP6 acts as an activator for RIC1 (Fu *et al.*, 2009).

Interestingly, RIC1 interacts with both activated ROP2 and activated ROP6 (Fu *et al.*, 2005, 2009). The sequence similarity and the shared downstream effector beg the question of how ROP2 and ROP6 exert differential local growth response. Although RIC1 interaction with ROP2 or ROP6 occurs on or close to PM, they exert different effects on the subsequent dynamic changes of the subcellular localization of RIC1 itself are observed. In WT cells, RIC1 localizes to the PM and cortical MTs. The recruitment of RIC1 to MTs requires ROP6 (Fu *et al.*, 2009), and this recruitment is enhanced in mutants that have reduced ROP2 activity (Fu *et al.*, 2005; Xu *et al.*, 2010). The activation of ROP6 by auxin was observed to induce the accumulation of RIC1 in vesicle-like structures on MT (Xu *et al.*, 2010). Conversely, the recruitment of RIC1 to MT is inhibited in cells having higher activity of ROP2 (Fu *et al.*, 2005). ROP2 and ROP6 therefore seem to be competitive in inhibiting or promoting the recruitment of RIC1 to MTs, which in turn contributes to the organization of MT bundling, suggesting a difference in local growth rate within a single cell.

Neither RIC4 nor RIC1 have any known functional domains except the CRIB motif, which is necessary for binding with ROPs. However, a recent study shed light on the potential mechanism behind organized MT bundling by ROP6–RIC1 signaling. From a genetic screening for altered PC shape in ROP6-OX background, Lin *et al.* (2013) identified KTN1, a p60 subunit of the MT-severing katanin complex as a direct downstream factor for the ROP6-RIC1 pathway. The katanin complex is important for generating and organizing MTs pattern, as it is proposed to sever existing strands to generate new growth sites in animal systems (McNally *et al.*, 2006; Roll-Mecak and Vale, 2006) as well as in plants (Nakamura *et al.*, 2010). Indeed, rates of branching from existing MTs were increased in RIC1OX cells, further supporting the idea that ROP6–RIC1 signaling induces ordered MT bundling through the promotion of katanin activity. This led to the question of how RIC1 affects the MT severing activity of KTN1. As RIC1 itself is recruited to MTs upon ROP6 activation (Fu *et al.*, 2009), RIC1 may act as a scaffold for recruiting the katanin complex to MTs. Alternatively, RIC1 may also affect the MT severing activity of KTN1, as RIC1 indeed increases the MT severing activity of katanin *in vitro* (Lin *et al.*, 2013).

Detailed analysis of the structural aspects of these proteins in a complex is necessary to understand the underlying mechanisms. Since ROP6–RIC1 signaling is activated by auxin in PCs (Xu *et al.*, 2010), the promotion of KTN1 activity by auxin is also expected. Future analysis will focus on whether auxin activates KTN1 activity in PCs or other tissues. The mechanisms underlying how RIC4 affects actin dynamics are currently unknown. Since it does not have any known functional domain apart from ROP-interacting CRIB motif, RIC4 may also interact with other protein(s) to exert its function. Identifying such additional factor(s) and assessing the precise mechanisms that are involved in stabilizing f-actin structures is an important aspect for future research.

Given that PCs have different ROP signaling modules for regulating distinct cytoskeletal structures, it is natural to ask whether these cortical structures play a role in generating a particular shape. As observed in other plant cells with directional growth, organized microtubule bundling in the neck region or indenting region is expected to restrict local growth. This inhibition of growth is at least partly accounted for by the local deposition of cellulose microfibrils that are deposited along with microtubules arrays (Paredez *et al.*,

2006; Bringmann *et al.*, 2012; Li *et al.*, 2012). However, cortical microtubules may play additional roles in the morphogenesis of PCs by affecting earlier phases of ROP signaling. For example, in xylem cells undergoing secondary growth, microtubules are shown to restrict active ROP11 on plasma membrane resulting in mutual inhibition wherein ROP11 effector MIDD1 induces cortical microtubule disassembly while microtubules eliminate active ROP11 from plasma membrane (Oda and Fukuda, 2012).

Similar mechanisms to restrict active ROPs, presumably ROP2 and ROP4, may exists in PCs, and this partially explains the spatially distinct accumulation of ROP2 and ROP6 in lobes and indentations, respectively. Likewise, cortical f-actin accumulation in developing lobes is expected to have multiple roles in outgrowth promotion. Cortical f-actin structures are shown to be essential for exocytic vesicle transport in tip-growing cells such as pollen tubes and root hairs (Lee *et al.*, 2008; Kroeger *et al.*, 2009). One important role of exocytosis in those cells is to deliver the materials that are necessary to sustain outgrowth, including cell wall components and enzymes regulating cell wall rigidity such as pectin methyl esterases and/or their inhibitors (Wang *et al.*, 2013). Although local varieties among cell wall stiffness within individual developing PCs are yet to be understood, expanding lobes may have similar properties to that of tip-growing cells. The existence of localized exocytosis toward outgrowing lobes and local cell wall stiffness in different regions within individual cells need to be investigated. The ROP2–RIC4 dependent f-actin accumulation inhibits the endocytosis of PIN putative auxin efflux proteins; this is likely to form a positive feedback loop of ROP2 activation by auxin, and activated ROP2 leads to the accumulation of auxin through PIN-mediated auxin polar transport at the tip of lobes (Nagawa *et al.*, 2012).

Microtubules may also affect ROP activity at the upstream signaling level, as emerging evidence suggests that microtubule-associated proteins affect endomembrane trafficking of PIN proteins at multiple steps (Ambrose *et al.*, 2013; Kakar *et al.*, 2013), and are therefore expected to cross-talk with regulation by f-actin-mediated processes. As there are multiple steps in which those cortical cytoskeletal structures seem to be playing counteractive roles, the unraveling of the precise function of specific factors associated with them at different steps is important to untangle each process regulated by cortical cytoskeletons during interdigitative growth of PCs.

Self-organizing Mechanism Controls the Localized Auxin Transport Critical for PC Morphogenesis

As Rho proteins typically bind to their downstream effector proteins only when they are in GTP-bound active state, activation of ROPs by upstream factor(s) is important in the regulation of ROP signaling. Recent findings showed that the plant hormone auxin is the key factor that activates ROPs during PC morphogenesis as well as for other developmental processes involving ROP signaling.

Auxin is a quintessential plant hormone involved in a diverse array of developmental events. Auxin is involved in the acquisition of cell identity (Dhonukshe *et al.*, 2008b; Schlereth *et al.*, 2010), tissue specification and differentiation (Hardtke and Berleth 1998; Friml *et al.*, 2002; Scarpella *et al.*, 2006), organ development (Fukaki *et al.*, 2002; Heisler *et al.*, 2005), and response to environment such as gravitropic (Muller *et al.* 1998; Abas *et al.*, 2006) and wounding responses (Kramer *et al.*, 2008). Auxin is transported across cells and/or tissues in a directional manner and is believed to generate a gradient of concentration that enables it to function as a morphogen. Members of the PIN putative auxin efflux carrier proteins typically show polar distribution to one cell end, and PIN polarity is thought to reflect the direction of auxin transport (Petrasek *et al.*, 2006; Robert and Friml, 2009). Auxin itself has a positive role in generating PIN polarity (Paciorek *et al.*, 2005), but the detailed molecular mechanisms of PIN polarity formation and maintenance of the auxin distribution pattern remained unidentified until recently. A key step in PIN polarity formation was thought to be the endosomal trafficking of PIN proteins, as genetic or pharmacologic disturbance of the endomembrane trafficking pathways led to developmental defects related to auxin transport with altered responses to auxin as well as altered PIN distribution (Geldner *et al.*, 2003; Jaillais *et al.*, 2007; Dhonukshe *et al.*, 2008b; Kleine-Vehn *et al.*, 2008; Kitakura *et al.*, 2011).

A number of studies have also shown a strong link between cytoskeletal dynamics and the regulation of auxin transport. These correlations might be due to the fact that the regulation of the PIN endomembrane trafficking is regulated by auxin through the control of cytoskeletal dynamics. A recent study using the classical coleoptile-bending system showed that actin bundling affects not only auxin transport activity of cells but also the cellular sensitivity to auxin in promoting the auxin efflux activity (Nick *et al.*, 2009). Furthermore, the subcellular distribution patterns of PIN proteins are altered by drugs targeting actin dynamics (Geldner *et al.*, 2001; Robert *et al.*, 2008; Tóth *et al.*, 2012), and classical auxin transport inhibitor chemicals such as TIBA turned out to be targeting actin dynamics even in non-plant cells (Dhonukshe *et al.*, 2008a). Microtubules are also implicated in positioning PIN proteins to the correct polar side of a cell (Boutte *et al.*, 2006). Recent reports showed that CLASP, a microtubule-binding protein with an important role in the regulation of MTs orientation across kingdoms (Inoue *et al.*, 2000; Maiato *et al.*, 2003; Ambrose *et al.*, 2007), has a role in multiple steps of the PIN endomembrane trafficking (Ambrose *et al.*, 2013; Kakar *et al.*, 2013). However, studies of the signaling mechanisms that link auxin signaling to the cytoskeleton regulation remained sparse until recently, when ROPs were demonstrated to be key players bridging the developmental auxin signaling pathway to the subcellular cytoskeletal dynamics (Xu *et al.*, 2010; Chen *et al.*, 2012; Nagawa *et al.*, 2012; Lin *et al.*, 2012, 2013).

Early studies showed that ROPs are activated by auxin in above-ground tissues (Tao *et al.*, 2002), and plants expressing either constitutive active (CA) or dominant negative (DN) form of ROPs show developmental abnormalities in roots that are reminiscent of the defects observed in auxin-related mutants such as abnormalities in root hair elongation, lateral root development, or gravitropic responses (Fu *et al.*, 2002; Li *et al.*, 2005). More recent reports showed that ROPs mediate a positive feedback loop of auxin and intracellular PIN trafficking in both leaves and roots. In an attempt to identify factors upstream of ROPs in PC morphogenesis, Xu *et al.* (2010) found that auxin is the key upstream regulator for the ROP-mediated interdigitating growth. They showed that auxin is essential for the interdigitative growth of PCs. Moreover, they discovered that auxin increases the amount of ROP2 or ROP6 that is in the active GTP-bound state. Importantly, auxin activates both ROP2 and ROP6 within one minute, suggesting that the activation of ROPs is independent of the well-characterized TIR1 auxin receptor system (Dharmasiri *et al.*, 2005; Kepinski and Leyser, 2005), which relies on transcriptional regulation mediated by AUX/IAA and ARF transcription factors. Indeed, they showed that the Auxin Binding Protein 1 (ABP1), which had been thought to be involved in auxin signaling (Badescu and Napier, 2006; Perrot-Rechenmann, 2010), is in fact responsible for activation of ROPs by auxin. This finding is key to understanding the bases for multiple roles of auxin in the plant developmental regulation, and is also in line with the classical model that auxin activates multiple signaling events within a cell (Fig. 6.2). A recent report has shown that a receptor-like kinase (TMK1) forms a complex with ABP1 and mediates activation of ROP signaling by auxin (Xu *et al.*, 2014), strengthening the idea that ABP1 acts as a cell surface auxin receptor for ROP activation.

In addition to ABP1, another key factor that is involved in the auxin-mediated PC morphogenesis is PIN1. PIN1 localizes to the tip of lobes, where ROP2 mediated signaling is activated; however, the polar PIN1 localization was lost and PIN1 protein was internalized to small aggregations in plant cells defective in ROP2 signaling (Xu *et al.*, 2010). A recent study on ROP2 signaling further revealed that the ROP2–RIC4 pathway inhibits endocytosis of PIN1 through the stabilization of dynamic f-actin structures (Nagawa *et al.*, 2012). These data suggest that the ROP2–RIC4 pathway negatively regulates PIN1 endocytosis to achieve polar localization. This raises the hypothesis of the existence of a positive-feedback loop. The ROP2 activation by auxin through the ABP1-mediated signaling accumulates PIN1 at the tip of lobes. PIN1 exports auxin and increases the local auxin concentration, which induces further activation of ROP2 (Xu *et al.*, 2010, 2011).

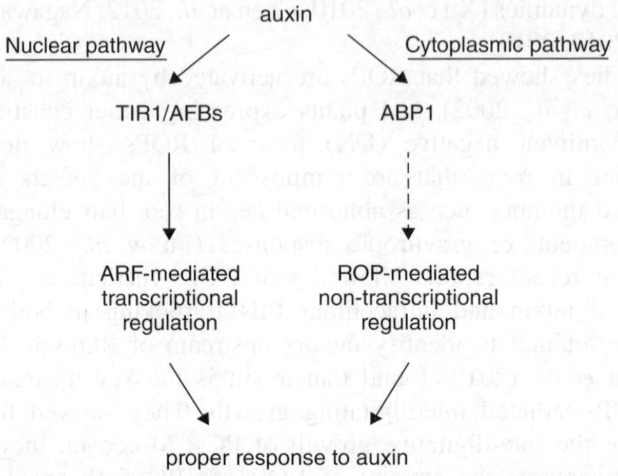

Figure 6.2 Auxin signaling pathways in plants. At least two independent auxin signaling pathways exist in plants. Auxin signaling mediated by TIR1/AFBs leads to the transcriptional regulation and is mediated by the ARF transcription factor proteins (nuclear pathways). In contrast, rapid auxin signaling is mediated by ABP1 and occurs without transcriptional regulation; cytoplasmic ROP signaling appears to be act downstream of ABP1 (cytoplasmic pathway). Both pathways are necessary for proper responses to auxin.

Robert *et al.* (2010) showed that ABP1 regulates endocytic trafficking of PIN proteins in roots, but the details of the underlying mechanisms seem to be different between leaf PCs and root cells; ABP1 appears to be the negative regulator of endocytosis of PIN1 in leaves but has been shown to promote PIN1 endocytosis in the absence of auxin in roots (Robert *et al.*, 2010; Xu *et al.*, 2010). Furthermore, the ROP6 pathway suppresses endocyotsis of PINs in roots instead of the ROP2 pathway (Chen *et al.*, 2012; Lin *et al.*, 2012). The loss of function mutants of ROP6 as well as the downstream effector RIC1 protein showed developmental abnormalities similar to *pin2* loss of function mutants such as an increased number of lateral roots and a disturbed distribution pattern of auxin upon gravity stimulus (Chen *et al.*, 2012; Lin *et al.*, 2012). Indeed, Lin *et al.* showed that auxin activates ROP6 through SPIKE1, a single DHR2-type Rho GEF protein, and activated ROP6 subsequently inhibits the endocytosis of PIN2.

The molecular relationship between auxin and ROPs is further strengthened by the findings of Chen *et al.* (2012). The ROP6–RIC1 pathway was shown to act downstream of ABP1 in mediating the suppression of clathrin-mediated endocytosis by auxin. These findings established a molecular framework for the suppression of PIN endocytosis by cytoplasmic auxin signaling that is mediated by ROPs, as in the case for leaf epidermal pavement cells. Furthermore, pharmacological analysis by Lin *et al.* (2012) suggested that the ROP6–RIC1 pathway regulates actin dynamics in roots, which is also reminiscent of the regulation for the suppression of PIN1 endocytosis in leaf PCs by the ROP2–RIC4 pathway (Xu *et al.*, 2010; Nagawa *et al.*, 2012).

However, how the PC MTs bundling regulator RIC1 affects actin dynamics in root cells is currently unclear. Moreover, unlike other mutants defective in root PIN endomembrane trafficking (Jaillais *et al.*, 2007; Dhonukshe *et al.*, 2008b), PIN polarity itself was not altered in ROP6 pathway-related mutants, suggesting that the ROP6–RIC1 pathway is unlikely to regulate the polar localization of PINs in root cells. Whether an additional ROP-mediated signaling exists to regulate PIN polarity is an open question for future studies. In both PCs and root cells, the suppression of clathrin-mediated endocytosis through the stabilization of fine f-actin partially accounts for the suppressive effect of auxin on PIN endocytosis. However, Rho GTPase mediated signaling has been shown to be essential in both promoting (Kaksonen *et al.*, 2005, 2006; Slaughter *et al.*, 2009) and inhibiting (Lamaze *et al.* 1996; Izumi *et al.*, 2004) endocytosis in different systems. Detailed analyses of the mechanical aspects will shed light on how ROP effectors affect f-actin dynamics and how f-actin dynamics influence clathrin-mediated endocytosis. Finally, as the ROP6–RIC1 pathway contributes to PC morphogenesis by restricting outgrowth through MT stabilization, it is likely that the ROP6–RIC1 pathway may have unexplored roles in PIN endocytosis through MT assembly during PC development. Further work is required to test whether the ROP6–RIC1 pathway affects PIN1 endocytosis in leaf PC development.

Exocytic trafficking of PINs to PM is also a critical step in the generation and maintenance of PIN polarity (Jaillais *et al.*, 2007; Feraru *et al.*, 2012), and ROP signaling is implicated in exocytosis well as endocytosis of PINs. The interactor of constitutive active ROPs 1/ROP interactive partner 1 (ICR1/RIP1) is a ROP effector protein identified by two independent studies (Lavy *et al.*, 2007; Li *et al.*, 2008). The loss of function *icr1-1* mutant showed diverse abnormalities related to auxin-mediated developmental processes such as abnormal PC morphogenesis, swelling of the root hair cells, and defects in the root growth (Lavy *et al.*, 2007). Lavy et al. also showed that ICR1 form complex with SEC3 which is a well-conserved component of the exocyst complex (Hala *et al.*, 2008). A recent study from the same group as Lavy et al. showed that the polar PIN2 localization toward the basal root-tip side of cortex cells is altered in the *icr1-1* mutant (Hazak *et al.*, 2010). Moreover, they showed that the endocytic recycling of PIN2 depends on ICR1, which further strengthens the idea that ROP signaling is involved in the exocytic recycling of PIN proteins besides regulating the PIN endocytosis from the PM.

Future analysis will focus on why the ICR1-mediated formation of PIN2 polarity is required only for basal root-tip side of cortex cells but not for the apical shoot-ward polar localization in the epidermal cells. It will be interesting to test whether other ROP signaling mechanisms are involved in the acquisition or maintenance of apical polarity. Since *icr1-1* mutants show strong abnormal phenotypes in PC morphogenesis (Lavy *et al.*, 2007), ICR1-mediated regulation of exocytosis – possibly PIN exocytosis – is likely to contribute to the interdigitation of PCs. The regulatory mechanism seems to be shared with tip-growing cells such as pollen tubes and root hairs. A fascinating scenario is that polar exocytosis occurs at the tip of lobes where PIN1 accumulates (Xu *et al.*, 2010; Nagawa *et al.*, 2012), and inhibition of endocytosis and delivery of PIN or other proteins by polar exocytosis coordinate to sustain outgrowth at the lobing sites. However, experimental evidence for the polar exocytosis of PIN proteins in PCs must undergo further analysis.

By either locally inhibiting endocytosis or promoting the exocytosis of PINs, the activation of ROP signaling seems to be required for the accumulation of the PIN proteins at the sites close to ROP activation, which eventually result in a local positive feedback loop of auxin accumulation and ROP activation in a self-organizing manner (Xu *et al.*, 2011). As ROP signaling is implicated in the vesicle trafficking of the PIN proteins at multiple steps including endocytosis (Chen *et al.*, 2012; Lin *et al.*, 2012; Nagawa *et al.*, 2012), recycling (Jaillais *et al.*, 2007), and exocytosis (Hazak *et al.*, 2010), understanding the precise mechanism regulating the contribution of multiple ROP signaling of each of these trafficking processes will help us decipher the complex process of intracellular PIN trafficking.

Coordination of Interdigitative Growth by Auxin

Auxin function in PC development may also extend to coordinating interdigitation between neighboring cells as well as the coordination of PC development at the whole organ level. Auxin can diffuse, and is suggested to act as a non-cell-autonomous signaling molecule. Studies in other tissues revealed roles for auxin in local cell–cell communication during developmental processes such as cell specification in early embryogenesis (Friml *et al.*, 2003; Schlereth *et al.*, 2010); auxin has also been shown to govern developmental process at whole organ level, for example by regulating multiple environmental responses (Luschnig *et al.*, 1998; Noh *et al.*, 2003). Auxin is therefore an ideal candidate molecule for mediating the coordinated growth of PCs at multiple levels, from the coordination of interdigitative growth of individual lobes/indentations between neighboring cells to the coordination of growth required by an entire field of cotyledons and leaves.

Although direct proof for the role of auxin in local cell–cell communication during PC development has yet to be obtained, there is an increasing amount of circumstantial evidence. Importantly, although both the ROP2–RIC4 and the ROP6–RIC1 pathways are activated by auxin, their activation kinetics are different. ROP2 activation saturates at a concentration of around 100 nM, while

ROP6 activation has a higher saturation concentration of around 1 μM. This differential activation kinetics, along with a self-organizing positive feedback loop of local auxin transport at the lobes and the mutual inhibitory relationship between the ROP2–RIC4 pathway and the ROP6–RIC1 pathway within single cells, may provide the basis for a hypothetical model that explains the complementary development of lobes and indentations in neighboring cells (Xu *et al.*, 2010, 2011; Fig. 6.3). In this model, ROP2 activation is expected to occur stochastically in the initial phase of PC development. This local induction of ROP2 activation will be stabilized through the positive feedback loop of ROP2 activation and the polar PIN1 accumulation. At these ROP2 activation sites, ROP6 activation is inhibited by ROP2. However, the extracellular level of auxin close to the ROP2 activation sites is increased by PIN1-mediated auxin export, quickly leading to levels that are higher than the optimal concentration for ROP2 activation, and a consequent activation of ROP6 at the complementary side of the neighboring cell. Since activated ROP6 inhibits the activation of ROP2, this complementary pattern of ROP2 activation and ROP6 activation becomes stable and then finally forms interdigitative structures between neighboring cells. External application of auxin increases interdigitations in both wild-type plants and an auxin synthesis mutant (Xu *et al.*, 2010).

This result supports the view in which the stochastic activation of local positive feedback can produce polarity from an initial uniform activation signal and resembles the generation of the protruding leading edge and the trailing edge in neutrophils under uniform attractant, which is also a self-organized process dependent on counteracting Rho signaling (Xu *et al.*, 2003; Van Keymeulen *et al.*, 2006; Wong *et al.*, 2006). Similar molecular mechanisms may therefore underlie systems with distinct features, one from plant multicellular epidermal cells and the other from isolated single cells from animals. Further investigation of the different components involved in respective signaling pathways may reveal a common strategy for self-organizing polarity acquisition.

Auxin distribution forms gradients at the tissue/organ level (Ulmasov *et al.* 1997; Petersson *et al.*, 2009; Vernoux *et al.*, 2011; Brunoud *et al.*, 2012), and development of organs relies on the formation of these gradients by polar auxin transport (Benkova *et al.*, 2003; Reinhardt *et al.*, 2003). Auxin gradient is essential for cellular morphogenesis as for example in the acquisition of planar cell polarity in root hair cells (Ikeda *et al.*, 2009), a mechanism that is also dependent on PIN-mediated auxin transport (Fischer *et al.*, 2006). Although this idea has not been tested yet, planar morphogenesis of PCs may also be regulated by auxin gradients over the surface of cotyledons/leaves. In developing young leaves auxin seems to accumulate at the tip region, and later it accumulates in the vascular precursor procambium cells and margin cells (Scarpella *et al.*, 2006; Koizumi *et al.*, 2005). An interesting future experiment is to test

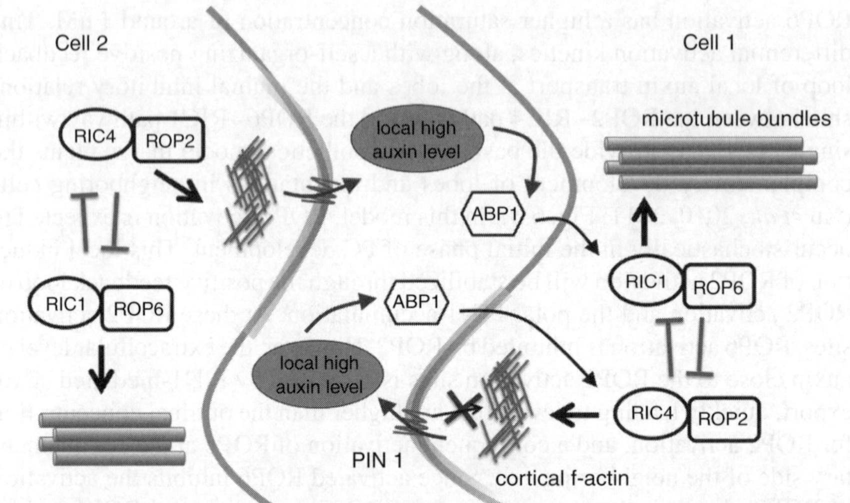

Figure 6.3 Auxin activates counteracting ROP signaling pathways by regulating distinct cytoskeletal structures. At the outgrowing lobing sites, auxin activates the ROP2–RIC4 pathway, which stabilizes cortical f-actin. Endocytosis of auxin-exporting protein PIN1 from PM is prohibited by stabilized f-actin and PIN1 accumulates at the tip of lobes, which is proposed to produce a local accumulation of auxin outside the cell and form a positive feedback loop of ROP activation and auxin accumulation. At the counteracting indenting sites, auxin activates the ROP6–RIC1 pathway, which in turn promotes bundling of microtubules to restrict outgrowth. The ROP2–RIC4 pathway and the ROP6–RIC1 pathway counteract each other and form interdigitative growth patterns. Activation of both pathways by auxin is mediated by ABP1.

whether interdigitative growing PCs show a change in pattern that reflects local auxin accumulation.

Connecting Signals to Rop GTPases: GEFs as Activators of ROP Signaling

Although auxin is the obvious candidate for a trigger of the activation of ROP signaling in PC morphogenesis, the activation of ROP signaling by auxin is most likely to be indirect; the activation of Rho signaling is typically mediated by proteins that have guanine nucleotide exchange activity (Rho GEFs) to convert Rho proteins from GDP-bound inactive state into GTP-bound active state. Several distinct families of proteins possessing Rho GEF activity have been identified. In plants, three families of proteins are shown to have GEF activity; these include DOCK Homology Region 2 (DHR2) type GEF proteins and

DH-PH type GEF proteins conserved across kingdoms, and the plant-specific PRONE GEF proteins (Berken *et al.*, 2005; Gu *et al.*, 2006; Basu *et al.*, 2008; Yamaguchi *et al.*, 2012). Some of these factors are implicated in intercalating growth of PCs or auxin-mediated growth in other tissues/cell types.

Unlike vertebrates that typically possess a large gene family encoding DHR2-domain-containing proteins (Rossman *et al.*, 2005), Arabidopsis possesses a single DHR2 protein called SPIKE1. SPIKE1 was identified through a genetic screen for mutants with defective trichome morphology. In addition to the trichome phenotypes, the *spike1* mutant displays pleiotropic abnormalities throughout the whole plant. Interestingly, *spike1* displayed severe alteration in PC shape, where intercalating structures are almost completely lost (Qiu *et al.*, 2002). Moreover, both MTs and actin filaments show various degrees of alterations in *spike1* mutant plants depending on the developmental stage of PCs, suggesting an important role of SPIKE1 signaling in cytoskeletal dynamics. A subsequent study showed that SPIKE1 has nucleotide exchange activity toward multiple members of ROPs *in vitro* and SPIKE1 preferentially binds to GDP-bound forms of ROPs, further strengthening the idea that SPIKE1 act as GEF for ROPs (Basu *et al.*, 2008).

Interestingly, the same study also showed that SPIKE1 forms a complex with a component of the WAVE/SCAR protein complex called NAP1. The WAVE/SCAR complex is required for the ARP2/3 mediated nucleation of actin filaments (Eden *et al.*, 2002; Basu *et al.*, 2005). A component of this complex (SRA1) has been shown to be a common effector protein linking Rho GTPase signaling to the WAVE/SCAR complex across kingdoms (Kobayashi *et al.* 1998; Basu *et al.*, 2004), and SRA1 physically interacts with NAP1 in plants (El-Assal *et al.*, 2004). As loss of either SRA1 or NAP1 affects the molecular mass of the SPIKE1 complex and actin organization is disturbed in the corresponding loss of function mutants, it is proposed that SPIKE1 constitutes a ROP-activating protein complex which also contains the ROP effector WAVE/SCARs to directly regulate the downstream actin organization. Although SPIKE1 does not show preferential binding towards different members of ROP *in vitro*, SRA1 shows a differential binding affinity towards certain members of ROP (Basu *et al.*, 2008), a mechanism that may provide signaling specificity to downstream events. It is tempting to speculate that the preferential binding of ROP2 or ROP4 to SRA1 may partially explain their roles in promoting actin dynamics. As there is currently no knowledge on the relationship between the RIC4 mediated f-actin stabilization and its regulation through WAVE/SCARs, it will be important to examine whether these two pathways share common or independent targets in the regulation of actin dynamics.

A more recent study showed that SPIKE1 interacts with ROPs in specific subdomains of the ER, called ER exit sites (ERES; Zhang *et al.*, 2010). As *spike1* and DN-ROP2 but not WAVE/SCARs mutants share some abnormal phenotypes in ER stress responses, Zhang *et al.* proposed that SPIKE1-ROP signaling has an additional function in ER stress responses. Since there is no clear way to distinguish the cellular morphological disturbance induced by the alteration of these two processes, it remains to be determined whether the SPIKE1-ROP signaling involved in ER stress responses contributes to the PC morphogenesis. SPIKE1 mediates the activation of ROP6 by auxin in roots (Lin *et al.*, 2012), and it is also expected that SPIKE1 may at least play a role in the auxin-ROP signaling that is involved in PC morphogenesis. Analysis of the detailed mechanisms that regulate the GEF activity of SPIKE1 will reveal if such an assumption is true.

Plants possess a unique family of proteins having Rho GEF activity, referred to as 'ROPGEFs'. ROPGEFs were identified by several independent studies. Yeast 2-hybrid screens to identify interactors of DN ROPs led two independent groups to the identification of ROPGEFs as a family of proteins characterized by the presence of a domain of unknown function (PRONE domain; Berken *et al.*, 2005; Gu *et al.*, 2006). Both studies showed that the PRONE domain mediates their GEF activity. In a separate study, a homologous protein was identified in tomato as a protein interacting with a receptor-like kinase (RLK) protein (LePRK2; Kaothien *et al.*, 2005). In plants, RLKs belong to a large family and serve as receptors for a number of signaling molecules (Shiu and Bleecker, 2001). When partial or full-length ROPGEF proteins are overexpressed in pollen tubes, abnormal phenotypes resembling the phenotypes induced by increased activity of ROPs is observed (Gu *et al.*, 2006). Furthermore, a study linking ROPGEF and the Arabidopsis RLK proteins related to LePRK2 (AtPRK2; Zhang and McCormick, 2007) showed that some ROPGEFs have regulatory domains in their C-terminus and/or N-terminus of PRONE domain. This led to the hypothesis that ROP activation is mediated by RLKs through the modulation of the inhibitory domain of ROPGEFs.

ROP activation by RLK-ROPGEF signaling does not seem to be limited to pollen growth. Duan *et al.* reported that FERONIA (FER), another RLK previously found to be involved in fertilization event (Escobar-Restrepo *et al.*, 2007), can also interact with ROPGEFs (Duan *et al.*, 2010). FER acts downstream of auxin and upstream of ROP signaling to regulate the NADPH-dependent ROS production essential for root hair development, thus linking an upstream signaling molecule to downstream events involving ROS-mediated signaling that had previously been shown to be essential for root hair growth (Takeda *et al.*, 2008).

Another recent study has reported that FER- and ROPGEF-mediated ROP signaling negatively regulate ABA response (Yu *et al.*, 2012); the signaling pathway may therefore have an important role in the cross-talk of these two essential hormones. Whether auxin directly binds to FER alone or together with a co-receptor, or indirectly affects the activity of FER to regulate the activation of ROPs, is currently unknown. The relationship between

FER-mediated activation of ROPs through the PRONE-type ROPGEFs and activation of ROPs by the DHR2-type ROPGEF SPIKE1 is also currently obscure. As auxin seems to be upstream of both pathways, it will be important to determine whether there is an overlap in the recognition step as well as cross-talk of downstream components between the two pathways. The contribution of signaling mediated by LRR-RLK and ROPGEFs in PC morphogenesis is also currently unreported. However, as plants possess a large family of RLKs and evolved abundant member of plant-specific ROPGEFs, it is expected that some signaling modules of RLK-ROPGEFs may have important roles in the developmental process of leaf PCs.

Future Perspectives

ROPs are essential for numerous important steps in the plant leaf epidermal PC development including the promotion of outgrowth at the lobes and constraints of expansion at indentations. A potential feedback loop of auxin transport and ROP activation at the tip of lobe also suggests that ROP signaling contributes to the amplification or maintenance of the initial signal for interdigitative growth. Although diffusive auxin seems to be a good candidate for mediating cell–cell communication and acts as a long-range morphogenic signal over whole cotyledons/leaves to coordinate interdigitative growth, direct evidence to support this idea is currently missing. An important future task is to directly test the non-cell-autonomous nature of the auxin effect by utilizing tools such as laser ablation, clonal analysis, and local applications of auxin.

Aside from similarities in terms of their upstream regulation by auxin and downstream effect on cytoskeletal dynamics, distinct components of ROPs, effector proteins, and upstream regulatory proteins are utilized depending on the developmental phase. It is important to determine the exact combination of ROP signaling modules in each context to reveal generality and specificity among them. Since phylogenetically close members of ROPs such as ROP2/4 and ROP6 seem to regulate different target pathways, it is important to determine the domain or residues within each ROP that confer their functional diversifications. As plants possess unique molecules for both upstream regulators and downstream effectors of ROPs, analyzing the precise molecular mechanisms behind each step – such as the mechanism regulating the activation of ROPGEFs by RLKs and how binding to ROP affects the activity of RICs or ICR1 in the regulation of downstream events – will also reveal plant-specific mechanisms of Rho-mediated signaling. Genetically encoded tools, such as live-imaging tools to detect the activation state of Rho proteins (Mochizuki *et al.*, 2001; Itoh *et al.*, 2002) or photo-activatable forms of Rho proteins (Wu *et al.*, 2009), have been developed for yeast or animal studies and are beginning to be applied to plant research (Wong *et al.*, 2007;

Kawano *et al.*, 2010; Akamatsu *et al.*, 2013). Such novel tools will be useful to expand our knowledge of ROP signaling. Finally, comparing contribution of Rho-mediated signaling in epidermal morphogenesis in the different kingdom will reveal generality of strategies underlying those highly specialized tissues with divergent shapes.

Acknowledgements

We thank Irene Lavagi and Kaori Miyawaki for critically reading the manuscript. The work in the Yang laboratory is supported by grants from National Institute of General Medical Sciences (R01GM081451 and R01GM100130).

References

Abas, L., Benjamins, R., Malenica, N., Paciorek, T., Wirniewska, J., Moulinier-Anzola, J.C., Sieberer, T., Friml, J., and Luschnig, C. (2006) Intracellular trafficking and proteolysis of the Arabidopsis auxin-efflux facilitator PIN2 are involved in root gravitropism. *Nature Cell Biology*, **8**, 249–256.

Akamatsu, A., Wong, H.L., Fujiwara, M., Okuda, J., Nishide, K., Uno, K., Imai, K., Umemura, K., Kawasaki, T., Kawano, Y., and Shimamoto, K. (2013) An OsCEBiP/OsCERK1-OsRacGEF1-OsRac1 module is an essential early component of chitin-induced rice immunity. *Cell Host and Microbe*, **13**, 465–476.

Ambrose, C., Ruan, Y., Gardiner, J., Tamblyn, L.M., Catching, A., Kirik, V., Marc, J., Overall, R., and Wasteneys, G.O. (2013) CLASP interacts with sorting nexin 1 to link microtubules and auxin transport via PIN2 recycling in Arabidopsis thaliana. *Developmental Cell*, **24**, 649–659.

Ambrose, J.C., Shoji, T., Kotzer, A.M., Pighin, J.A., and Wasteneys, G.O. (2007) The Arabidopsis CLASP gene encodes a microtubule-associated protein involved in cell expansion and division. *Plant Cell*, **19**, 2763–2775.

Badescu, G.O. and Napier, R.M. (2006) Receptors for auxin: will it all end in TIRs? *Trends in Plant Science*, **11**, 217–223.

Basu, D., El-Assal, S.E.D., Le, J., Mallery, E.L. and Szymanski, D.B. (2004) Interchangeable functions of Arabidopsis PIROGI and the human WAVE complex subunit SRA1 during leaf epidermal development. *Development*, **131**, 4345–4355.

Basu, D., Le, J., El-Essal, S.E.D., Huang, S., Zhang, C.H., Mallery, E.L., Koliantz, G., Staiger, C.J., and Szymanski, D.B. (2005) DISTORTED3/SCAR2 is a putative arabidopsis WAVE complex subunit that activates the Arp2/3 complex and is required for epidermal morphogenesis. *Plant Cell*, **17**, 502–524.

Basu, D., Le, J., Zakharova, T., Mallery, E.L., and Szymanski, D.B. (2008) A SPIKE1 signaling complex controls actin-dependent cell morphogenesis through the heteromeric WAVE and ARP2/3 complexes. *Proceedings of National Academy of Science, USA*, **105**, 4044–4049.

Benkova, E., Michniewicz, M., Sauer, M., Teichmann, T., Seifertova, D., Jurgens, G., and Friml, J. (2003) Local, efflux-dependent auxin gradients as a common module for plant organ formation. *Cell*, **115**, 591–602.

Berken, A., Thomas, C., and Wittinghofer, A. (2005) A new family of RhoGEFs activates the Rop molecular switch in plants. *Nature*, **436**, 1176–1180.

Boutte, Y., Crosnier, M.T., Carraro, N., Traas, J., and Satiat-Jeunemaitre, B. (2006) The plasma membrane recycling pathway and cell polarity in plants: studies on PIN proteins. *Journal of Cell Science*, **119**, 1255–1265.

Brembu, T., Winge, P., and Bones, A.M. (2005) The small GTPase AtRAC2/ROP7 is specifically expressed during late stages of xylem differentiation in Arabidopsis. *Journal of Experimental Botany*, **56**, 2465–2476.

Bringmann, M., Li, E.Y., Sampathkumar, A., Kocabek, T., Hauser, M.T., and Persson, S. (2012) Pom-Pom2/Cellulose Synthase Interacting1 Is Essential for the Functional Association of Cellulose Synthase and Microtubules in Arabidopsis. *Plant Cell*, **24**, 163–177.

Brunoud, G., Wells, D.M., Oliva, M., Larrieu, A., Mirabet, V., Burrow, A.H., Beeckman, T., Kepinski, S., Traas, J., Bennett, M.J., and Vernoux, T. (2012) A novel sensor to map auxin response and distribution at high spatio-temporal resolution. *Nature*, **482**, 103–132.

Burridge, K. and Wennerberg, K. (2004) Rho and Rac take center stage. *Cell*, **116**, 167–179.

Chen, X., Naramoto, S., Robert, S., Tejos, R., Lofke, C., Lin, D.S., Yang, Z.B., and Friml, J. (2012) ABP1 and ROP6 GTPase Signaling Regulate Clathrin-Mediated Endocytosis in Arabidopsis Roots. *Current Biology*, **22**, 1326–1332.

Dharmasiri, N., Dharmasiri, S., and Estelle, M. (2005) The F-box protein TIR1 is an auxin receptor. *Nature*, **435**, 441–445.

Dhonukshe, P., Grigoriev, I., Fischer, R., Tominaga, M., Robinson, D.G., Hasek, J., Paciorek, T., Petrasek, J., Seifertova, D., Tejos, R., Meisel, L.A., Zazimalova, E., Gadella, T.W., Jr.,, Stierhof, Y.D., Ueda, T., Oiwa, K., Akhmanova, A., Brock, R., Spang, A., and Friml, J. (2008a) Auxin transport inhibitors impair vesicle motility and actin cytoskeleton dynamics in diverse eukaryotes. *Proceedings of National Academy of Science, USA*, **105**, 4489–4494.

Dhonukshe, P., Tanaka, H., Goh, T., Ebine, K., Mahonen, A.P., Prasad, K., Blilou, I., Geldner, N., Xu, J., Uemura, T., Chory, J., Ueda, T., Nakano, A., Scheres, B., and Friml, J. (2008b) Generation of cell polarity in plants links endocytosis, auxin distribution and cell fate decisions. *Nature*, **456**, 962–966.

Duan, Q., Kita, D., Li, C., Cheung, A.Y., and Wu, H.M. (2010) FERONIA receptor-like kinase regulates RHO GTPase signaling of root hair development. *Proceedings of National Academy of Science, USA*, **107**, 17821–17826.

Eden, S., Rohatgi, R., Podtelejnikov, A.V., Mann, M. and Kirschner, M.W. (2002) Mechanism of regulation of WAVE1-induced actin nucleation by Rac1 and Nck. *Nature*, **418**, 790–793.

El-Assal, S.E.D., Le, J., Basu, D., Mallery, E.L., and Szymanski, D.B. (2004) Arabidopsis GNARLED encodes a NAP125 homolog that positively regulates ARP2/3. *Current Biology*, **14**, 1405–1409.

Escobar-Restrepo, J.M., Huck, N., Kessler, S., Gagliardini, V., Gheyselinck, J., Yang, W.C., and Grossniklaus, U. (2007) The FERONIA receptor-like kinase mediates male–female interactions during pollen tube reception. *Science*, **317**, 656–660.

Feraru, E., Feraru, M.I., Asaoka, R., Paciorek, T., De Rycke, R., Tanaka, H., Nakano, A., and Friml, J. (2012) BEX5/RabA1b Regulates trans-Golgi Network-to-Plasma Membrane Protein Trafficking in Arabidopsis. *Plant Cell*, **24**, 3074–3086.

Fischer, U., Ikeda, Y., Ljung, K., Serralbo, O., Singh, M., Heidstra, R., Palme, K., Scheres, B., and Grebe, M. (2006) Vectorial information for Arabidopsis planar polarity is mediated by combined AUX1, EIN2, and GNOM activity. *Current Biology*, **16**, 2143–2149.

Friml, J., Benkova, E., Blilou, I., Wisniewska, J., Hamann, T., Ljung, K., Woody, S., Sandberg, G., Scheres, B., Jurgens, G., and Palme, K. (2002) AtPIN4 mediates sink-driven auxin gradients and root patterning in Arabidopsis. *Cell*, **108**, 661–673.

Friml, J., Vieten, A., Sauer, M., Weijers, D., Schwarz, H., Hamann, T., Offringa, R., and Jurgens, G. (2003) Efflux-dependent auxin gradients establish the apical-basal axis of Arabidopsis. *Nature*, **426**, 147–153.

Fu, Y., Wu, G., and Yang, Z. (2001) Rop GTPase-dependent dynamics of tip-localized F-actin controls tip growth in pollen tubes. *Journal of Cell Biology*, **152**, 1019–1032.

Fu, Y., Li, H., and Yang, Z. (2002) The ROP2 GTPase controls the formation of cortical fine F-actin and the early phase of directional cell expansion during Arabidopsis organogenesis. *Plant Cell*, **14**, 777–794.

Fu, Y., Gu, Y., Zheng, Z., Wasteneys, G., and Yang, Z. (2005) Arabidopsis interdigitating cell growth requires two antagonistic pathways with opposing action on cell morphogenesis. *Cell*, **120**, 687–700.

Fu, Y., Xu, T., Zhu, L., Wen, M., and Yang, Z. (2009) A ROP GTPase signaling pathway controls cortical microtubule ordering and cell expansion in Arabidopsis. *Current Biology*, **19**, 1827–1832.

Fukaki, H., Tameda, S., Masuda, H., and Tasaka, M. (2002) Lateral root formation is blocked by a gain-of-function mutation in the SOLITARY-ROOT/IAA14 gene of Arabidopsis. *Plant Journal*, **29**, 153–168.

Geldner, N., Friml, J., Stierhof, Y.D., Jurgens, G., and Palme, K. (2001) Auxin transport inhibitors block PIN1 cycling and vesicle trafficking. *Nature*, **413**, 425–428.

Geldner, N., Anders, N., Wolters, H., Keicher, J., Kornberger, W., Muller, P., Delbarre, A., Ueda, T., Nakano, A., and Jurgens, G. (2003) The Arabidopsis GNOM ARF-GEF mediates endosomal recycling, auxin transport, and auxin-dependent plant growth. *Cell*, **112**, 219–230.

Gu, Y., Fu, Y., Dowd, P., Li, S., Vernoud, V., Gilroy, S., and Yang, Z. (2005) A Rho family GTPase controls actin dynamics and tip growth via two counteracting downstream pathways in pollen tubes. *Journal of Cell Biology*, **169**, 127–138.

Gu, Y., Li, S., Lord, E.M., and Yang, Z. (2006) Members of a novel class of Arabidopsis Rho guanine nucleotide exchange factors control Rho GTPase-dependent polar growth. *Plant Cell*, **18**, 366–381.

Guimil, S. and Dunand, C. (2006) Patterning of Arabidopsis epidermal cells: epigenetic factors regulate the complex epidermal cell fate pathway. *Trends in Plant Science*, **11**, 601–609.

Hala, M., Cole, R., Synek, L., Drdova, E., Pecenkova, T., Nordheim, A., Lamkemeyer, T., Madlung, J., Hochholdinger, F., Fowler, J.E., and Zarsky, V. (2008) An exocyst complex functions in plant cell growth in Arabidopsis and tobacco. *Plant Cell*, **20**, 1330–1345.

Hardtke, C.S. and Berleth, T. (1998) The Arabidopsis gene MONOPTEROS encodes a transcription factor mediating embryo axis formation and vascular development. *Embo Journal*, **17**, 1405–1411.

Hazak, O., Bloch, D., Poraty, L., Sternberg, H., Zhang, J., Friml, J., and Yalovsky, S. (2010) A rho scaffold integrates the secretory system with feedback mechanisms in regulation of auxin distribution. *Plos Biology*, **8**, e1000282.

Heisler, M.G., Ohno, C., Das, P., Sieber, P., Reddy, G.V., Long, J.A., and Meyerowitz, E.M. (2005) Patterns of auxin transport and gene expression during primordium development revealed by live imaging of the Arabidopsis inflorescence meristem. *Current Biology*, **15**, 1899–1911.

Ikeda, Y., Men, S., Fischer, U., Stepanova, A.N., Alonso, J.M., Ljung, K., and Grebe, M. (2009) Local auxin biosynthesis modulates gradient-directed planar polarity in Arabidopsis. *Nature Cell Biology*, **11**, 731–738.

Inoue, Y.H., do Carmo Avides, M., Shiraki, M., Deak, P., Yamaguchi, M., Nishimoto, Y., Matsukage, A., and Glover, D.M. (2000) Orbit, a novel microtubule-associated protein essential for mitosis in Drosophila melanogaster. *Journal of Cell Biology*, **149**, 153–166.

Itoh, R.E., Kurokawa, K., Ohba, Y., Yoshizaki, H., Mochizuki, N., and Matsuda, M. (2002) Activation of rac and cdc42 video imaged by fluorescent resonance energy transfer-based single-molecule probes in the membrane of living cells. *Molecular and Cellular Biology*, **22**, 6582–6591.

Izumi, G., Sakisaka, T., Baba, T., Tanaka, S., Morimoto, K., and Takai, Y. (2004) Endocytosis of E-cadherin regulated by Rac and Cdc42 small G proteins through IQGAP1 and actin filaments. *Journal of Cell Biology*, **166**, 237–248.

Jaffe, A.B. and Hall, A. (2005) Rho GTPases: biochemistry and biology. *Annual Review of Cell and Developmental Biology*, **21**, 247–269.

Jaillais, Y., Santambrogio, M., Rozier, F., Fobis-Loisy, I., Miege, C., and Gaude, T. (2007) The retromer protein VPS29 links cell polarity and organ initiation in plants. *Cell*, **130**, 1057–1070.

Jones, M.A., Shen, J.J., Fu, Y., Li, H., Yang, Z., and Grierson, C.S. (2002) The Arabidopsis Rop2 GTPase is a positive regulator of both root hair initiation and tip growth. *Plant Cell*, **14**, 763–776.

Kakar, K., Zhang, H.T., Scheres, B., and Dhonukshe, P. (2013) CLASP-mediated cortical microtubule organization guides PIN polarization axis. *Nature*, **495**, 529–533.

Kaksonen, M., Toret, C.P., and Drubin, D.G. (2005) A modular design for the clathrin- and actin-mediated endocytosis machinery. *Cell*, **123**, 305–320.

Kaksonen, M., Toret, C.P., and Drubin, D.G. (2006) Harnessing actin dynamics for clathrin-mediated endocytosis. *Nature Reviews Molecular Cell Biology*, **7**, 404–414.

Kaothien, P., Ok, S.H., Shuai, B., Wengier, D., Cotter, R., Kelley, D., Kiriakopolos, S., Muschietti, J., and McCormick, S. (2005) Kinase partner protein interacts with the LePRK1 and LePRK2 receptor kinases and plays a role in polarized pollen tube growth. *Plant Journal*, **42**, 492–503.

Kawano, Y., Akamatsu, A., Hayashi, K., Housen, Y., Okuda, J., Yao, A., Nakashima, A., Takahashi, H., Yoshida, H., Wong, H.L., Kawasaki, T., and Shimamoto, K. (2010) Activation of a Rac GTPase by the NLR family disease resistance protein Pit plays a critical role in rice innate immunity. *Cell Host and Microbe*, **7**, 362–375.

Kawasaki, T., Henmi, K., Ono, E., Hatakeyama, S., Iwano, M., Satoh, H., and Shimamoto, K. (1999) The small GTP-binding protein rac is a regulator of cell death in plants. *Proceedings of National Academy of Science, USA*, **96**, 10922–10926.

Kepinski, S. and Leyser, O. (2005) The Arabidopsis F-box protein TIR1 is an auxin receptor. *Nature*, **435**, 446–451.

Kitakura, S., Vanneste, S., Robert, S., Lofke, C., Teichmann, T., Tanaka, H., and Friml, J. (2011) Clathrin Mediates Endocytosis and Polar Distribution of PIN Auxin Transporters in Arabidopsis. *Plant Cell*, **23**, 1920–1931.

Klein, T.J. and Mlodzik, M. (2005) Planar cell polarization: an emerging model points in the right direction. *Annual Review of Cell and Developmental Biology*, **21**, 155–176.

Kleine-Vehn, J., Dhonukshe, P., Sauer, M., Brewer, P.B., Wisniewska, J., Paciorek, T., Benkova, E., and Friml, J. (2008) ARF GEF-dependent transcytosis and polar delivery of PIN auxin carriers in Arabidopsis. *Current Biology*, **18**, 526–531.

Kobayashi, K., Kuroda, S., Fukata, M., Nakamura, T., Nagase, T., Nomura, N., Matsuura, Y., Yoshida-Kubomura, N., Iwamatsu, A., and Kaibuchi, K. (1998) p140Sra-1 (specifically Rac1-associated protein) is a novel specific target for Rac1 small GTPase. *Journal of Biological Chemistry*, **273**, 291–295.

Koizumi, K., Naramoto, S., Sawa, S., Yahara, N., Ueda, T., Nakano, A., Sugiyama, M., and Fukuda, H. (2005) VAN3 ARF-GAP-mediated vesicle transport is involved in leaf vascular network formation. *Development*, **132**, 1699–1711.

Kost, B., Lemichez, E., Spielhofer, P., Hong, Y., Tolias, K., Carpenter, C., and Chua, N.H. (1999) Rac homologues and compartmentalized phosphatidylinositol 4,5-bisphosphate act in a common pathway to regulate polar pollen tube growth. *Journal of Cell Biology*, **145**, 317–330.

Kramer, E.M., Lewandowski, M., Beri, S., Bernard, J., Borkowski, M., Borkowski, M.H., Burchfield, L.A., Mathisen, B., and Normanly, J. (2008) Auxin gradients are associated with polarity changes in trees. *Science*, **320**, 1610.

Kroeger, J.H., Daher, F.B., Grant, M., and Gieitmann, A. (2009) Microfilament Orientation Constrains Vesicle Flow and Spatial Distribution in Growing Pollen Tubes. *Biophysics Journal*, **97**, 1822–1831.

Lamaze, C., Chuang, T.H., Terlecky, L.J., Bokoch, G.M., and Schmid, S.L. (1996) Regulation of receptor-mediated endocytosis by Rho and Rac. *Nature*, **382**, 177–179.

Lavy, M., Bloch, D., Hazak, O., Gutman, I., Poraty, L., Sorek, N., Sternberg, H., and Yalovsky, S. (2007) A Novel ROP/RAC effector links cell polarity, root-meristem maintenance, and vesicle trafficking. *Current Biology*, **17**, 947–952.

Lee, Y.J., Szumlanski, A., Nielsen, E., and Yang, Z. (2008) Rho-GTPase-dependent filamentous actin dynamics coordinate vesicle targeting and exocytosis during tip growth. *Journal of Cell Biology*, **181**, 1155–1168.

Li, L., Xu, J., Xu, Z.H., and Xue, H.W. (2005) Brassinosteroids stimulate plant tropisms through modulation of polar auxin transport in Brassica and Arabidopsis. *Plant Cell*, **17**, 2738–2753.

Li, S., Gu, Y., Yan, A., Lord, E., and Yang, Z.B. (2008) RIP1 (ROP Interactive Partner 1)/ICR1 marks pollen germination sites and may act in the ROP1 pathway in the control of polarized pollen growth. *Molecular Plant*, **1**, 1021–1035.

Li, S.D., Lei, L., Somerville, C.R., and Gu, Y. (2012) Cellulose synthase interactive protein 1 (CSI1) links microtubules and cellulose synthase complexes. *Proceedings of the National Academy of Science, USA*, **109**, 185–190.

Lin, D.S., Nagawa, S., Chen, J.S., Cao, L.Y., Chen, X., Xu, T.D., Li, H.J., Dhonukshe, P., Yamamuro, C., Friml, J., Scheres, B., Fu, Y., and Yang, Z.B. (2012) A ROP GTPase-Dependent Auxin Signaling Pathway Regulates the Subcellular Distribution of PIN2 in Arabidopsis Roots. *Current Biology*, **22**, 1319–1325.

Lin, D.S., Cao, L.Y., Zhou, Z.Z., Zhu, L., Ehrhardt, D., Yang, Z.B. and Fu, Y. (2013) Rho GTPase Signaling Activates Microtubule Severing to Promote Microtubule Ordering in Arabidopsis. *Current Biology*, **23**, 290–297.

Lin, Y., Wang, Y., Zhu, J.K., and Yang, Z. (1996) Localization of a Rho GTPase Implies a Role in Tip Growth and Movement of the Generative Cell in Pollen Tubes. *Plant Cell*, **8**, 293–303.

Luschnig, C., Gaxiola, R.A., Grisafi, P., and Fink, G.R. (1998) EIR1, a root-specific protein involved in auxin transport, is required for gravitropism in Arabidopsis thaliana. *Genes and Development*, **12**, 2175–2187.

Maiato, H., Fairley, E.A.L., Rieder, C.L., Swedlow, J.R., Sunkel, C.E., and Earnshaw, W.C. (2003) Human CLASP1 is an outer kinetochore component that regulates spindle microtubule dynamics. *Cell*, **113**, 891–904.

McNally, K., Audhya, A., Oegema, K., and McNally, F.J. (2006) Katanin controls mitotic and meiotic spindle length. *Journal of Cell Biology*, **175**, 881–891.

Mochizuki, N., Yamashita, S., Kurokawa, K., Ohba, Y., Nagai, T., Miyawaki, A., and Matsuda, M. (2001) Spatio-temporal images of growth-factor-induced activation of Ras and Rap1. *Nature*, **411**, 1065–1068.

Molendijk, A.J., Bischoff, F., Rajendrakumar, C.S., Friml, J., Braun, M., Gilroy, S., and Palme, K. (2001) Arabidopsis thaliana Rop GTPases are localized to tips of root hairs and control polar growth. *EMBO Journal*, **20**, 2779–2788.

Muller, A., Guan, C., Galweiler, L., Tanzler, P., Huijser, P., Marchant, A., Parry, G., Bennett, M., Wisman, E., and Palme, K. (1998) AtPIN2 defines a locus of Arabidopsis for root gravitropism control. *EMBO Journal*, **17**, 6903–6911.

Murphy, A.S. and Peer, W.A. (2012) Vesicle trafficking: ROP-RIC roundabout. *Current Biology*, **22**, R576–578.

Nagawa, S., Xu, T.D., Lin, D.S., Dhonukshe, P., Zhang, X.X., Friml, J., Scheres, B., Fu, Y., and Yang, Z.B. (2012) ROP GTPase-Dependent Actin Microfilaments Promote PIN1 Polarization by Localized Inhibition of Clathrin-Dependent Endocytosis. *Plos Biology*, **10**, e1001299.

Nakamura, M., Ehrhardt, D.W., and Hashimoto, T. (2010) Microtubule and katanin-dependent dynamics of microtubule nucleation complexes in the acentrosomal Arabidopsis cortical array. *Nature Cell Biology*, **12**, 1064–1049.

Nick, P., Han, M.J., and An, G. (2009) Auxin stimulates its own transport by shaping actin filaments. *Plant Physiology*, **151**, 155–167.

Noh, B., Bandyopadhyay, A., Peer, W.A., Spalding, E.P., and Murphy, A.S. (2003) Enhanced gravi- and phototropism in plant mdr mutants mislocalizing the auxin efflux protein PIN1. *Nature*, **423**, 999–1002.

Oda, Y. and Fukuda, H. (2012) Initiation of Cell Wall Pattern by a Rho- and Microtubule-Driven Symmetry Breaking. *Science*, **337**, 1333–1336.

Paciorek, T., Zazimalova, E., Ruthardt, N., Petrasek, J., Stierhof, Y.D., Kleine-Vehn, J., Morris, D.A., Emans, N., Jurgens, G., Geldner, N., and Friml, J. (2005) Auxin inhibits endocytosis and promotes its own efflux from cells. *Nature*, **435**, 1251–1256.

Panteris, E. and Galatis, B. (2005) The morphogenesis of lobed plant cells in the mesophyll and epidermis: organization and distinct roles of cortical microtubules and actin filaments. *New Phytologist*, **167**, 721–732.

Panteris, E., Apostolakis, P., and Galatis, B. (1993) Microtubules and morphogenesis in ordinary epidermal cells of Vigna sinensis leaves. *Protoplasma*, **174**, 91–100.

Paredez, A.R., Somerville, C.R., and Ehrhardt, D.W. (2006) Visualization of cellulose synthase demonstrates functional association with microtubules. *Science*, **312**, 1491–1495.

Perrot-Rechenmann, C. (2010) Cellular Responses to Auxin: Division versus Expansion. *CSH Perspectives in Biology*, **2**, a001466.

Petersson, S.V., Johansson, A.I., Kowalczyk, M., Makoveychuk, A., Wang, J.Y., Moritz, T., Grebe, M., Benfey, P.N., Sandberg, G., and Ljung, K. (2009) An auxin gradient and maximum in the Arabidopsis root apex shown by high-resolution cell-specific analysis of IAA distribution and synthesis. *Plant Cell*, **21**, 1659–1668.

Petrasek, J., Mravec, J., Bouchard, R., Blakeslee, J.J., Abas, M., Seifertova, D., Wisniewska, J., Tadele, Z., Kubes, M., Covanova, M., Dhonukshe, P., Skupa, P., Benkova, E., Perry, L., Krecek, P., Lee, O.R., Fink, G.R., Geisler, M., Murphy, A.S., Luschnig, C., Zazimalova, E., and Friml, J. (2006) PIN proteins perform a rate-limiting function in cellular auxin efflux. *Science*, **312**, 914–918.

Qiu, J.L., Jilk, R., Marks, M.D., and Szymanski, D.B. (2002) The Arabidopsis SPIKE1 gene is required for normal cell shape control and tissue development. *Plant Cell*, **14**, 101–118.

Reinhardt, D., Pesce, E.R., Stieger, P., Mandel, T., Baltensperger, K., Bennett, M., Traas, J., Friml, J., and Kuhlemeier, C. (2003) Regulation of phyllotaxis by polar auxin transport. *Nature*, **426**, 255–260.

Robert, H.S. and Friml, J. (2009) Auxin and other signals on the move in plants. *Nature Chemical Biology*, **5**, 325–332.

Robert, S., Chary, S.N., Drakakaki, G., Li, S., Yang, Z., Raikhel, N.V., and Hicks, G.R. (2008) Endosidin1 defines a compartment involved in endocytosis of the brassinosteroid receptor BRI1 and the auxin transporters PIN2 and AUX1. *Proceedings of National Academy of Science, USA*, **105**, 8464–8469.

Robert, S., Kleine-Vehn, J., Barbez, E., Sauer, M., Paciorek, T., Baster, P., Vanneste, S., Zhang, J., Simon, S., Covanova, M., Hayashi, K., Dhonukshe, P., Yang, Z., Bednarek, S.Y., Jones, A.M., Luschnig, C., Aniento, F., Zazimalova, E., and Friml, J. (2010) ABP1 mediates auxin inhibition of clathrin-dependent endocytosis in Arabidopsis. *Cell*, **143**, 111–121.

Roll-Mecak, A. and Vale, R.D. (2006) Making more microtubules by severing: a common theme of noncentrosomal microtubule arrays? *Journal of Cell Biology*, **175**, 849–851.

Rossman, K.L., Der, C.J., and Sondek, J. (2005) GEF means go: turning on RHO GTPases with guanine nucleotide-exchange factors. *Nat Rev Molecular and Cellular Biology*, **6**, 167–180.

Sander, E.E., ten Klooster, J.P., van Delft, S., van der Kammen, R.A., and Collard, J.G. (1999) Rac downregulates Rho activity: Reciprocal balance between both GTPases determines cellular morphology and migratory behavior. *Journal of Cell Biology*, **147**, 1009–1021.

Scarpella, E., Marcos, D., Friml, J., and Berleth, T. (2006) Control of leaf vascular patterning by polar auxin transport. *Genes and Development*, **20**, 1015–1027.

Schlereth, A., Moller, B., Liu, W.L., Kientz, M., Flipse, J., Rademacher, E.H., Schmid, M., Jurgens, G., and Weijers, D. (2010) MONOPTEROS controls embryonic root initiation by regulating a mobile transcription factor. *Nature*, **464**, 913–128.

Shiu, S.H. and Bleecker, A.B. (2001) Plant receptor-like kinase gene family: diversity, function, and signaling. *Science STKE*, **2001**, RE22.

Slaughter, B.D., Das, A., Schwartz, J.W., Rubinstein, B., and Li, R. (2009) Dual Modes of Cdc42 Recycling Fine-Tune Polarized Morphogenesis. *Developmental Cell*, **17**, 823–835.

Takeda, S., Gapper, C., Kaya, H., Bell, E., Kuchitsu, K., and Dolan, L. (2008) Local positive feedback regulation determines cell shape in root hair cells. *Science*, **319**, 1241–1244.

Tao, L.Z., Cheung, A.Y., and Wu, H.M. (2002) Plant Rac-like GTPases are activated by auxin and mediate auxin-responsive gene expression. *Plant Cell*, **14**, 2745–2760.

Tóth, R., Gerding-Reimers, C., Deeks, M.J., Menninger, S., Gallegos, R.M., Tonaco, I.A.N., Hübel, K., Hussey, P.J., Waldmann, H., and Coupland, G. (2012) Prieurianin/endosidin 1 is an actin-stabilizing small molecule identified from a chemical genetic screen for circadian clock effectors in Arabidopsis thaliana. *Plant Journal*, **71**, 338–352.

Ulmasov, T., Murfett, J., Hagen, G., and Guilfoyle, T.J. (1997) Aux/IAA proteins repress expression of reporter genes containing natural and highly active synthetic auxin response elements. *Plant Cell*, **9**, 1963–1971.

Van Keymeulen, A., Wong, K., Knight, Z.A., Govaerts, C., Hahn, K.M., Shokat, K.M., and Bourne, H.R. (2006) To stabilize neutrophil polarity, PIP3 and Cdc42 augment RhoA activity at the back as well as signals at the front. *Journal of Cell Biology*, **174**, 437–445.

Vernoux, T., Brunoud, G., Farcot, E., Morin, V., Van den Daele, H., Legrand, J., Oliva, M., Das, P., Larrieu, A., Wells, D., Guedon, Y., Armitage, L., Picard, F., Guyomarc'h, S., Cellier, C., Parry, G., Koumproglou, R., Doonan, J.H., Estelle, M., Godin, C., Kepinski, S., Bennett, M., De Veylder, L., and Traas, J. (2011) The auxin signalling network translates dynamic input into robust patterning at the shoot apex. *Molecular Systems Biology*, **7**, 508.

Wang, H., Zhuang, X., Cai, Y., Cheung, A.Y., and Jiang, L. (2013) Apical F-actin-regulated exocytic targeting of NtPPME1 is essential for construction and rigidity of the pollen tube cell wall. *Plant Journal*, **76**, 367–379.

Wasteneys, G.O. and Galway, M.E. (2003) Remodeling the cytoskeleton for growth and form: An overview with some new views. *Annual Review of Plant Biology*, **54**, 691–722.

Wong, H.L., Pinontoan, R., Hayashi, K., Tabata, R., Yaeno, T., Hasegawa, K., Kojima, C., Yoshioka, H., Iba, K., Kawasaki, T., and Shimamoto, K. (2007) Regulation of rice NADPH oxidase by binding of Rac GTPase to its N-terminal extension. *Plant Cell*, **19**, 4022–4034.

Wong, K., Pertz, O., Hahn, K., and Bourne, H. (2006) Neutrophil polarization: Spatiotemporal dynamics of RhoA activity support a self-organizing mechanism. *Proceedings of the National Academy of Science, USA*, **103**, 3639–3644.

Wu, G., Gu, Y., Li, S., and Yang, Z. (2001) A genome-wide analysis of Arabidopsis Rop-interactive CRIB motif-containing proteins that act as Rop GTPase targets. *Plant Cell*, **13**, 2841–2856.

Wu, Y.I., Frey, D., Lungu, O.I., Jaehrig, A., Schlichting, I., Kuhlman, B., and Hahn, K.M. (2009) A genetically encoded photoactivatable Rac controls the motility of living cells. *Nature*, **461**, 104–108.

Xu, J., Wang, F., Van Keymeulen, A., Herzmark, P., Straight, A., Kelly, K., Takuwa, Y., Sugimoto, N., Mitchison, T., and Bourne, H.R. (2003) Divergent signals and cytoskeletal assemblies regulate self-organizing polarity in neutrophils. *Cell*, **114**, 201–214.

Xu, T., Wen, M., Nagawa, S., Fu, Y., Chen, J.G., Wu, M.J., Perrot-Rechenmann, C., Friml, J., Jones, A.M., and Yang, Z. (2010) Cell Surface- and Rho GTPase-Based Auxin Signaling Controls Cellular Interdigitation in Arabidopsis. *Cell*, **143**, 99–110.

Xu, T., Nagawa, S., and Yang, Z. (2011) Uniform auxin triggers the Rho GTPase-dependent formation of interdigitation patterns in pavement cells. *Small GTPases*, **2**, 227–232.

Xu, T., Dai, N., Chen, J., Nagawa, S., Cao, M., Li, H., Zhou, Z., Chen, X., De Rycke, R., Rakusova, H., Wang, W., Jones, A.M., Friml, J., Patterson, S.E., Bleecker, A.B., and Yang, Z. (2014) Cell surface ABP1-TMK auxin-sensing complex activates ROP GTPase signaling. *Science*, **343**, 1025–1028.

Yamaguchi, K., Imai, K., Akamatsu, A., Mihashi, M., Hayashi, N., Shimamoto, K., and Kawasaki, T. (2012) SWAP70 functions as a Rac/Rop guanine nucleotide-exchange factor in rice. *Plant Journal*, **70**, 389–397.

Yang, Z. (2002) Small GTPases: versatile signaling switches in plants. *Plant Cell*, **14** Suppl, 375–388.

Yang, Z. (2008) Cell polarity signaling in Arabidopsis. *Annual Review of Cell and Developmental Biology*, **24**, 551–575.

Yu, F., Qian, L., Nibau, C., Duan, Q., Kita, D., Levasseur, K., Li, X., Lu, C., Li, H., Hou, C., Li, L., Buchanan, B.B., Chen, L., Cheung, A.Y., Li, D., and Luan, S. (2012) FERONIA receptor kinase pathway suppresses abscisic acid signaling in Arabidopsis by activating ABI2 phosphatase. *Proceedings of National Academy of Science, USA*, **109**, 14693–14698.

Zhang, C., Kotchoni, S.O., Samuels, A.L., and Szymanski, D.B. (2010) SPIKE1 signals originate from and assemble specialized domains of the endoplasmic reticulum. *Current Biology*, **20**, 2144–2149.

Zhang, C., Halsey, L.E., and Szymanski, D.B. (2011) The development and geometry of shape change in Arabidopsis thaliana cotyledon pavement cells. *BMC Plant Biology*, **11**, 27.

Zhang, Y. and McCormick, S. (2007) A distinct mechanism regulating a pollen-specific guanine nucleotide exchange factor for the small GTPase Rop in Arabidopsis thaliana. *Proceedings of National Academy of Science, USA*, **104**, 18830–18835.

7 Xylem Cell Wall Pattern Formation Regulated by Microtubule-associated Proteins and ROP GTPases

Yoshihisa Oda[1,2,3] and Hiroo Fukuda[4]

[1] *Center for Frontier Research, National Institute of Genetics, Shizuoka, Japan*
[2] *Department of Genetics, Graduate University for Advanced Studies (SOKENDAI), Shizuoka, Japan*
[3] *PRESTO, Japan Science and Technology Agency (JST), Saitama, Japan*
[4] *Department of Biological Sciences, Graduate School of Science, University of Tokyo, Tokyo, Japan*

Introduction

Land plants require a highly organized vascular system to provide both mechanical support and efficient transport of water and nutrients throughout the plant. Vascular plants have developed xylem tissues that are composed of xylem vessels, xylem fibers, and xylem parenchyma. Xylem vessels and fibers deposit highly lignified secondary cell walls between the plasma membrane and primary cell walls that are thick and impermeable (Fig. 7.1a). The secondary cell walls form in distinct patterns such as annular, spiral, reticulate, and pitted, and also have a perforation at their end wall (Fig. 7.1b). Xylem vessel cells eventually undergo programmed cell death; cell contents are lysed and hollow water-conductive tubes are formed. In addition, secondary wall pits in xylem vessels allow lateral transport of xylem sap. Xylem fibers also deposit secondary walls at their cell surface but, although fibers have pits on their secondary cell walls, these are small and become stretched during cell elongation so do not contribute to lateral water transport. The xylem vessel is an assembly of tracheary elements that are joined end-to-end, forming a long conductive network. Protoxylem vessels occur in young growing tissue and deposit secondary walls in annular and spiral patterns. These secondary walls become stretched during elongation of the surrounding tissue and finally collapse. By contrast, metaxylem vessels form in elongated tissue and develop reticulate, scalariform, or pitted secondary walls that are more robust than those of the

Plant Cell Wall Patterning and Cell Shape, First Edition. Edited by Hiroo Fukuda.

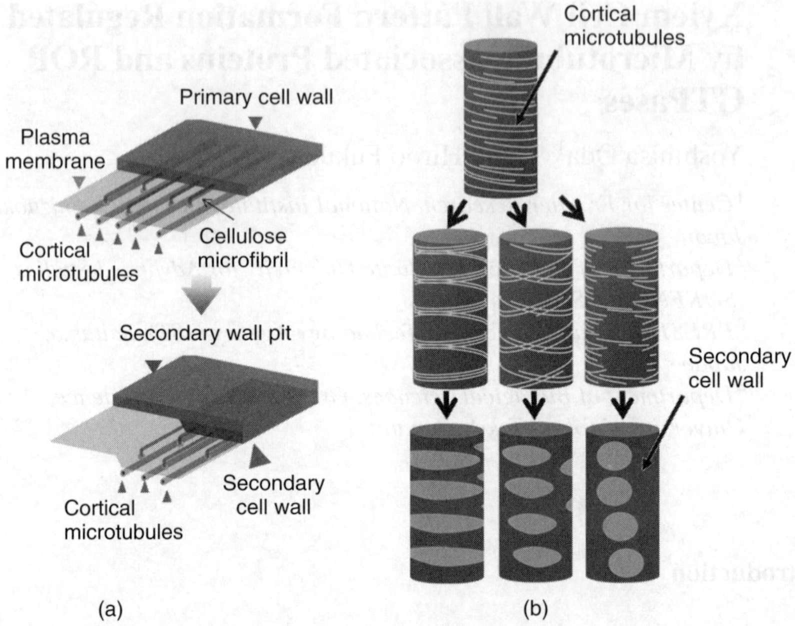

Figure 7.1 Cortical microtubules regulate secondary cell wall patterning. (a) The cortical microtubule array is locally assembled to induce local deposition of secondary cell walls between the plasma membrane and primary cell walls. (b) Cortical microtubules are arranged into distinct patterns that guide the secondary wall deposition.

protoxylem vessels. Reticulate, scalariform, and pitted secondary wall patterns also develop in the secondary xylem, and proper patterning of secondary wall deposition is essential for the function of the xylem cells (Esau, 1977).

Xylem cell fate is determined by NAM/ATAF/CUC (NAC) group transcription factors. VASCULAR-RELATED NAC DOMAIN (VND)6 and VND7 were identified as master transcription factors for the differentiation of metaxylem and protoxylem vessels, respectively (Kubo *et al.*, 2005); however, subsequent studies revealed that VND7 is involved in differentiation of all types of xylem cells (Yamaguchi *et al.*, 2008). SECONDARY WALL-ASSOCIATED NAC DOMAIN PROTEIN (SND)1/NAC SECONDARY WALL THICKENING PROMOTING FACTOR (NST)3 and NST1 were identified as master regulators for xylem fiber differentiation (Mitsuda *et al.*, 2007; Zhong *et al.*, 2007). Overexpression of VND6, VND7, and SND1 induced ectopic secondary wall deposition in pitted, spiral/reticulate, and flat patterns, respectively (Kubo *et al.*, 2005; Oda *et al.*, 2010; Ohashi-Ito *et al.*, 2010; Yamaguchi *et al.*, 2010, 2011). These transcription factors directly induce expression of a broad spectrum of genes that are related to secondary wall synthesis and cytoskeleton organization (Ohashi-Ito *et al.*, 2010; Zhong *et al.*, 2010; Yamaguchi *et al.*, 2011). It therefore appears that secondary wall patterns are genetically programmed by the transcription factors.

Secondary wall deposition patterns are dependent upon cortical microtubules, although the precise control mechanism remains unclear. Experiments are hampered by the difficulty of access to living xylem cells, which reside deep within the tissue. However, several novel genes that are essential for secondary wall development have been identified through Arabidopsis genetic screens of mutants with defects in xylem development and through gene profiling combined with reverse genetics. Furthermore, live imaging of differentiating xylem cells from xylogenic *in vitro* Arabidopsis cell cultures revealed that distinct microtubule-associated proteins (MAPs) and Rho of Plants (ROP) GTPases play critical roles in secondary cell wall patterning. In this chapter, we first introduce cortical microtubule associations with secondary wall development, and discuss the recently established experimental systems contributing to cellular analysis in living xylem cells. Secondly, we summarize recent insights into the molecular mechanisms underlying the regulation of secondary cell wall patterns in xylem fibers and protoxylem and metaxylem vessels, and focus on the emerging interaction between microtubules and ROP GTPases.

Microtubules and Secondary Wall Development in Xylem Cells

In non-xylem cells, cortical microtubules determine the deposition pattern of cellulose microfibrils by regulating the trajectory of the cellulose synthase (CESA) complex (Paredez *et al.*, 2006) and its targeting to the plasma membrane (Crowell *et al.*, 2009; Gutierrez *et al.*, 2009). CELLULOSE SYNTHASE INTERACTING PROTEIN (CSI) was found to partly bridge cortical microtubules and CESAs by binding to both microtubules and CESAs (Gu *et al.*, 2010; Bringmann *et al.*, 2012; Li *et al.*, 2012). In xylem cells, early studies using electron microscopy in xylem tissue sections (Hepler and Newcomb, 1964; Wooding and Northcote, 1964; Pickett-Heaps, 1966, 1967; Pickett-Heaps and Northcote, 1966; Hepler and Fosket, 1971; Hardham and Gunning, 1978, 1979; Hepler, 1981; Gunning and Hardham, 1982; Haigler and Brown, 1986) and indirect fluorescent immunostaining in cultured zinnia mesophyll cells (Falconer and Seagull, 1985, 1986, 1988; Kobayashi *et al.*, 1987, 1988) to examine the effects of various pharmaceuticals demonstrated that cortical microtubules are arranged into distinct patterns that guide secondary wall deposition. In secondary xylems, the cortical microtubules are accumulated along the edges of secondary walls in the bordered pits (Abe *et al.*, 1995a, b; Chaffey *et al.*, 1997, 1999; Funada *et al.*, 1997, 2001; Chaffey and Barlow, 2002), regions in which secondary walls are deposited most extensively. Cortical microtubules along the edges of secondary walls were also observed in tracheary elements derived from Arabidopsis cultured cells, which suggests that the cortical microtubules may direct cell wall deposition at the sides of secondary wall thickening to promote lateral development of the secondary wall thickening (Oda *et al.*, 2005; Oda and Hasezawa, 2006). These studies are discussed in detail in a separate review (Oda and Hasezawa, 2006).

In Arabidopsis, the genes encoding three CESA members – IRREGULAR XYLEM (IRX)1, IRX3, and IRX5 – are responsible for cellulose synthesis in secondary walls (Turner and Somerville, 1997; Taylor *et al.*, 2000, 2003). Cellulose synthase complexes marked with YFP-IRX3 co-localized with cortical microtubule bundles in protoxylem cells in Arabidopsis roots, and pharmacological experiments indicated that this localization was dependent on the cortical microtubules (Gardiner *et al.*, 2003; Wightman and Turner, 2008). Movement of CESA particles was not visualized as observed in primary walls. This was however due to the high density of CESAs beneath secondary cell walls, and fluorescent recovery after photo-bleaching (FRAP) analysis revealed motility of CESA along cortical microtubules (Wightman *et al.*, 2009). This confirms that cortical microtubules guide the moving CESA complex during secondary wall development.

Cortical microtubules exhibit highly dynamic behavior due to their growth and shrinkage (Shaw *et al.*, 2003), bundling, collision (Dixit and Cyr, 2004), and nucleation from the sides of cortical microtubules (Murata *et al.*, 2005; Chan *et al.*, 2009; Nakamura *et al.*, 2010). The dynamics and interactions of the cortical microtubules are regulated by various MAPs and lead to the self-organization of parallel cortical microtubules (Wasteneys and Ambrose, 2009). In the epidermis of the Arabidopsis hypocotyl, the parallel cortical microtubule array exhibits rotary movement (Chan *et al.*, 2007). The global orientation of cortical microtubules appears to be regulated by various mechanisms, including a cell-edge-dependent mechanism (Ambrose *et al.*, 2011), a cell edge-independent mechanism (Lindeboom *et al.*, 2013a), and a katanin-dependent mechanism (Lindeboom *et al.*, 2013b). The mechanism governing the overall organization of the cortical microtubule array is still not fully understood, however.

During tracheary element differentiation in which protoxylem-like secondary walls are developed, cortical microtubules are gradually assembled to form bundles; in Arabidopsis cultured cells, this subsequently leads to secondary wall thickening (Oda *et al.*, 2005; Oda and Hasezawa, 2006). In metaxylem cells, cortical microtubules are locally disassembled to leave secondary wall pits. This process is carried out by the ROP-MICROTUBULE DEPLETION DOMAIN 1 (MIDD1)-Kinesin-13A pathway (Oda *et al.*, 2010; Oda and Fukuda, 2012, 2013). The mechanisms underlying microtubule bundling in protoxylem vessels are not known, but MAP65 and MAP70 proteins are thought to be involved. In xylem fibers, katanin KTN1 is required for proper arrangement of cortical microtubules, while FRAGILE FIBER1 (FRA1) kinesin is required for cellulose deposition along cortical microtubules.

Several studies suggest that actin microfilaments are also involved in secondary cell wall patterning. In isolated zinnia mesophyll cells, thick actin cables are observed prior to secondary cell walls becoming apparent. Actin microfilaments are subsequently observed exclusively in the secondary cell wall pits and are finally found co-localized with cortical microtubules along the edge of secondary walls. Disruption of actin microfilaments inhibits reorientation of cortical microtubules from longitudinal to transverse, and affects the direction of cortical microtubule bundles. Actin microfilaments therefore might promote redirection of cortical microtubules at the early stages of differentiation (Kobayashi *et al.*, 1987, 1988).

In living xylem vessels of Arabidopsis roots, actin microfilaments organize longitudinal cables and transverse fine filaments. Disruption of actin microfilaments results in loss of cellulose synthase complex at the plasma membrane, suggesting that these actin cables contribute to the transport of vesicles containing cell wall enzymes and materials to the plasma membrane marked with cortical microtubules (Wightman and Turner, 2008, 2010a, b; Wightman *et al.*, 2009).

Three genes have been identified that are involved both in actin organization and secondary cell wall deposition, *FRA3*, *AtSAC1*, and *RHD3*, which encode a Type II Inositol Polyphosphate 5-Phosphatase, phosphoinositide phosphatase, and a dynamin-like GTPase, respectively. Loss-of-function mutants of these genes exhibit dramatically thin secondary cell wall thickenings in fibers and xylem vessels, and abnormally thick actin bundles (Hu *et al.*, 2003; Zhong *et al.*, 2004, 2005). Although the relationship between actin organization and secondary wall defects is unclear, these studies suggest that actin organization is tightly coupled to secondary wall deposition probably via vesicle transport.

In conifer, actin microfilaments are co-localized with cortical microtubules along the edges of secondary walls in bordered pits (Chaffey and Barlow, 2002); the function of these actin filaments is however unclear.

Experimental Systems using Arabidopsis for Study of Xylem Cell Differentiation

In vitro xylem cell cultures have been developed and used to investigate various cellular events in differentiating xylem cells. One culture system, using isolated zinnia mesophyll cells (Fukuda and Komamine, 1980), has long been used for cellular and molecular analyses including those on cytoskeleton dynamics (Falconer and Seagull, 1985, 1986, 1988; Kobayashi *et al.*, 1987,

1988; Fukuda and Kobayashi, 1989; Roberts *et al.*, 2004; Mao *et al.*, 2006), programmed cell death (Obara *et al.*, 2001; Ito and Fukuda, 2002), gene expression profiling (Milioni *et al.*, 2001, 2002), microarray analysis (Demura *et al.*, 2002), and intercellular signaling (Motose *et al.*, 2004; Ito *et al.*, 2006). More recently, several *in vitro* experimental systems for xylem differentiation have been developed using Arabidopsis tissues and cultured cells (Kubo *et al.*, 2005; Oda *et al.*, 2005, 2010; Sawa *et al.*, 2005; Pesquet *et al.*, 2010; Yamaguchi *et al.*, 2010). Arabidopsis is experimentally tractable, allowing techniques such as transcriptome analysis, Agrobacterium-based transformation, and high-resolution imaging. Arabidopsis-based *in vitro* systems can therefore be exploited to explore the molecular mechanisms underlying xylem cell differentiation. Indeed, recent studies using Arabidopsis cultured cells have contributed to our understanding of secondary wall patterning in xylem cells.

Some experimental systems use hormone treatment to induce xylem cell differentiation. For example, Sawa *et al.* (2005) induced xylem vessel differentiation in the hypocotyl of Arabidopsis seedlings by supplying cytokinin and brassinosteroid. Oda *et al.* (2005) reported *in vitro* xylem vessel cell differentiation in Arabidopsis cultured cells by adding brassinosteroid and removing auxin from the cell culture. In this latter system, about 30% of the cells differentiated into xylem cells within 72 hours. In addition, transgenic GFP-tubulin was used as a microtubule marker, and allowed visualization of the rearrangement of cortical microtubules in living xylem cells (Oda *et al.*, 2005; Oda and Hasezawa, 2006). Another group reported that *in vitro* differentiation of xylem cells occurred after the addition of brassinosteroid and boric acid to Arabidopsis suspension cells (Kubo *et al.*, 2005). In this culture system, 50% of the cells differentiated into xylem cells within 7 days. Microarray analysis of the differentiated cells allowed identification of the NAC transcription factors VND6 and VND7 as master transcription factors for metaxylem and protoxylem vessel differentiation, respectively. Kwon *et al.* (2010) used similar methods to induce xylem differentiation *in vitro*, and demonstrated that RabG3b promotes xylem cell differentiation and autophagy (Kwon *et al.*, 2010). Pesquet *et al.* (2010) used Arabidopsis cultured cells that do not require auxin for their proliferation, and then induced cell culture differentiation into xylem cells by the addition of auxin, cytokinin, and brassinosteroid. This system produced the discovery that MAP70-5 is upregulated in xylem cells and regulates secondary wall patterning (Pesquet *et al.*, 2010).

Several other *in vitro* systems have been developed using transcription factors to induce differentiation of xylem cells. These systems display substantially improved xylem differentiation rates, synchronicity, and productivity when compared to hormone-induced differentiation. Oda *et al.* (2010) developed transgenic Arabidopsis cultured cells harboring estradiol-inducible VND6 (Oda *et al.*, 2010). In this system, 80% of the cells differentiate into metaxylem cells within 48 hours (Fig. 7.2a). Facile Agrobacterium-based transformation and protein tagging allowed high-throughput live imaging to analyze protein localization and dynamics and to detect *in vivo* protein–protein interactions by bimolecular complementation (BiFC) assay and fluorescent resonance energy transfer (FRET). Using this system, the research group revealed that the ROP11-MIDD1-Kinesin-13A pathway determines secondary wall patterns in metaxylem vessels (Oda *et al.*, 2010; Oda and Fukuda, 2012, 2013). A similar system for xylem fiber was developed by using SND1, a NAC transcription factor that can induce fiber differentiation (Ohashi-Ito *et al.*, 2010). Yamaguchi *et al.* (2008) developed a similar transgenic BY-2 cell line, harboring estradiol-inducible VND7, for the stimulation of protoxylem differentiation (Yamaguchi *et al.*, 2008). This research group developed transgenic Arabidopsis plants expressing VND6 or VND7 fused with VP16, a transcription activator, and with a glucocorticoid receptor. In these transgenic plants, dexamethasone triggers translocation of VND6-VP16 or VND7-VP16 from the cytoplasm into the nucleus, where it strongly induces metaxylem or protoxylem differentiation, respectively (Yamaguchi *et al.*, 2010). A wide range of Arabidopsis mutants are available, and these systems will continue to be key for investigating gene functions in xylem differentiation.

In summary, development of a number of Arabidopsis xylem cell differentiation systems contributed to the discovery of master xylem differentiation transcription factors. Discovery of these transcription factors allowed further development of inducible *in vitro* systems, which are accelerating studies of secondary wall development.

Secondary Cell Wall Patterning in Xylem Cells

Next, we summarize the current understanding of the molecular mechanisms underlying secondary wall patterning in xylem fibers and protoxylem and metaxylem vessels.

Patterning of Cellulose Microfibrils in Xylem Fibers

In xylem fibers, a kinesin motor protein FRA1 and a microtubule severing protein KTN1 are required for proper arrangement of the cellulose microfibrils. The microfibrils are crucial for the mechanical strength of xylem fibers and FRA1 and KTN1 are therefore critical; however, the two proteins are widely expressed and their functions are likely not confined to xylem fibers.

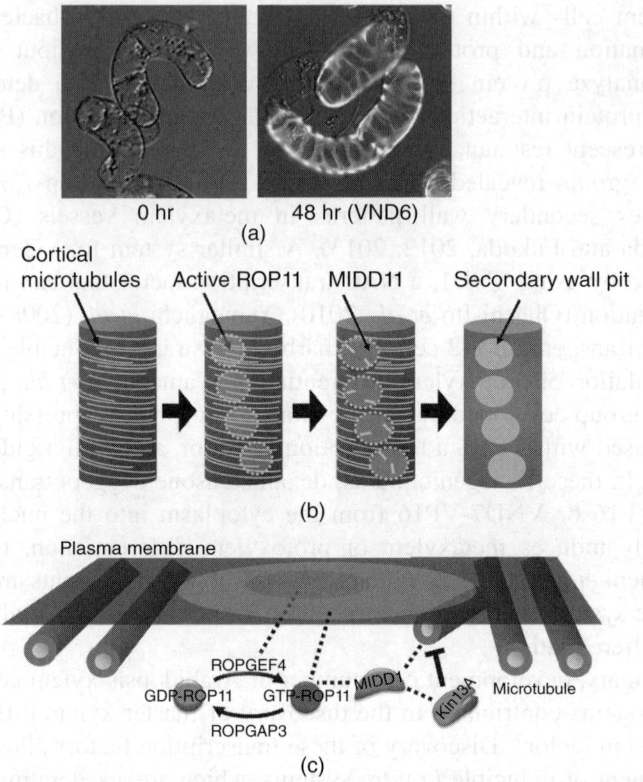

Figure 7.2 ROP1-MIDD1-Kinesin-13A pathway regulates the formation of secondary wall pits. (a) Cultured Arabidopsis cells differentiate into metaxylem cells upon induction with VND6. Secondary walls are formed after 48 hours (0 hours, bright field; 48 hours, fluorescent signal from secondary walls stained with Alexa594-conjugated wheat germ agglutinin). (b) ROP11 is locally activated on the plasma membrane to recruit MIDD1. MIDD1 promotes disassembly of cortical microtubules through Kinesin-13A, which results in the formation of secondary wall pits. (c) A ROP-MIDD1-Kinesin-13A network in secondary wall pits. ROPGEF4 and ROPGAP3 mediate local activation of ROP11. Activated ROP11 recruits MIDD1 to the plasma membrane domain. MIDD1 directly binds and depolymerizes cortical microtubules growing into the domain by recruiting Kinesin-13A. Broken lines indicate direct interaction.

FRA1 was identified in a forward genetic screen for reduced mechanical strength of the Arabidopsis stem (Zhong *et al.*, 2002). Secondary cell walls in *fra1* mutants have normal thickness and composition, and a normal cortical microtubule array. In *fra1* mutants however, the cellulose microfibrils are randomly organized which is believed to cause the reduction in their mechanical strength. Overexpression of FRA1 caused a reduction in total thickness of secondary cell walls but an increase in the number of layers within the secondary walls. Because the number of secondary wall layers is dependent upon rotation of the orientation of cellulose microfibrils, this suggests that FRA1 regulates the deposition pattern of cellulose microfibrils (Zhou *et al.*, 2007). FRA1 is a member of the kinesin-4 group within the kinesin superfamily. Recently, an *in vitro* assay using a single molecule imaging technique revealed that FRA1 forms a dimer to processively move along microtubules at a rate of 400 nm s^{-1}. This is faster than the 330 nm s^{-1} movement rate of the cellulose synthase complex along the cortical microtubules (Paredez *et al.*, 2006). The authors suggest that FRA1 might be involved in transporting cellulose synthase complexes or moving Golgi-derived vesicles carrying cellulose synthase complexes, cell wall polysaccharides, or protein components (Zhu and Dixit, 2011). A mutation in a rice *FRA1* homologue, *Brittle Culm12* (*BC12*), causes the orientation of the cellulose microfibrils in secondary cell walls of sclerenchyma fibers to be altered. The mutant fibers exhibit reduced mechanical strength, suggesting that BC12 contributes to proper cellulose patterning of secondary cell walls. However, in contrast to *fra1* mutations, *bc12* mutants exhibit dwarfism due to reduced cell numbers and reduced gibberellin synthesis levels. These data suggest that FRA1 has a common function in cellulose patterning in dicots and monocots, alongside additional cell cycle- and gibberellin-related functions in rice (Zhang *et al.*, 2010; Li *et al.*, 2011).

Arabidopsis KTN1 encodes a microtubule severing protein, katanin p60 subunit. A *ktn1* mutant was also isolated in the mutant screen for reduced mechanical strength of the inflorescence stem. The *ktn1* mutation causes reduced cellulose and hemicellulose content, with a shorter fiber length and reduced mechanical strength. In *ktn1* mutants, cortical microtubule orientation is randomized, causing a defective cellulose deposition pattern (Bichet *et al.*, 2001; Burk *et al.*, 2001; Burk and Ye, 2002; Bouquin *et al.*, 2003). Conditional overexpression of KTN1 induces degradation of cortical microtubules, suggesting that KTN1 severs cortical microtubules (Stoppin-Mellet *et al.*, 2006). KTN1 activity in epidermal cells of hypocotyls and leaves has been well characterized in recent years. The protein promotes parallel alignment of cortical microtubules by severing cortical microtubules at the nucleation site (to release the newly nucleated microtubule from the mother microtubule) (Nakamura *et al.*, 2010) and at the cross-point of cortical microtubules (Wightman and Turner, 2007; Zhang *et al.*, 2013). RIC1, which is a microtubule-associated effector of ROP6, is preferentially localized at the cross-point of cortical microtubules and interacts with KTN1 to increase its severing activity (Lin *et al.*, 2013). By contrast another MAP, SPIRAL2 (SPR2), localizes to the cross-point of cortical microtubules and inhibits severing by KTN1 (Wightman *et al.*, 2013). It therefore appears that MAPs and ROP GTPase signaling finely regulate the severing activity of KTN1. Although it is not yet clear whether KTN1 functions similarly in xylem cells, the critical ordering of cortical microtubules may be mediated by the activity of KTN1 in severing cortical microtubules at their nucleation sites and crossing-points.

Secondary Wall Patterning in Protoxylem Vessels

Annular and spiral secondary walls in protoxylem vessels are likely established by highly bundled cortical microtubules that direct secondary wall thickening. To date, two MAPs have been implicated in these secondary wall pattern developments.

MAP65-family proteins are approximately 65 kDa and are involved in cross-linking microtubules (Chan *et al.*, 1999; Mao *et al.*, 2005; Smertenko *et al.*, 2006; Gaillard *et al.*, 2008; Fache *et al.*, 2010). The Arabidopsis genome has nine *MAP65* genes, and a few family members have been functionally characterized. For example, AtMAP65-3 cross-links interdigitated microtubules in the phragmoplast (Muller *et al.*, 2004; Ho *et al.*, 2011, 2012) and AtMAP65-4 promotes the bundling and elongation of spindle microtubules (Fache *et al.*, 2010). In zinnia, *ZeMAP65-1* is expressed in stem xylem cells and upregulated during xylem differentiation in zinnia mesophyll cell culture. In Arabidopsis suspension cells, GFP-ZeMAP65-1 overexpression induced bundling of cortical microtubules (Mao *et al.*, 2006). Similarly, AtMAP65-8 is upregulated during xylem differentiation in Arabidopsis cultured cells (Kubo *et al.*, 2005). These diverse studies clearly suggest that MAP65-family proteins are involved in xylem differentiation. Recently, AtMAP65-1 and AtMAP65-4 were demonstrated to coordinately promote microtubule bundling and elongation by affecting the catastrophe and rescue rates of microtubules in the bundle (Fache *et al.*, 2010; Stoppin-Mellet *et al.*, 2013). MAP65-mediated bundling may therefore influence the formation of annular and helical secondary cell walls in protoxylem cells because long bundles of cortical microtubules are required for this cell wall patterning.

MAP70 is a plant-specific family of 70 kDa proteins with five known Arabidopsis members. MAP70-5 is thought to stabilize microtubules *in vitro*, and constitutive expression of MAP70-5 in non-xylem cultured cells causes abnormal cell shape (Korolev *et al.*, 2005, 2007). In addition, MAP70-5 was found to be upregulated during xylem differentiation in Arabidopsis cultured cells. In a yeast assay, MAP70-5 interacts with MAP70-1 suggesting that MAP70-1 is also involved in xylem differentiation (Pesquet *et al.*, 2010). Supporting this, both MAP70-1 and MAP70-5 localize along the edge of the thickened secondary wall, and overexpression of MAP70-1 or MAP70-5 increases the population of cells with secondary walls of protoxylem vessel type. By contrast, knockdown of MAP70-1 or MAP70-5 resulted in an increased population of cells with pitted secondary walls and an increased tendency for secondary walls to detach from the cell surface. Based on these results, MAP70 is believed to regulate the border of secondary walls and primary walls to confine the region of secondary wall deposition (Pesquet *et al.*, 2010). It remains unclear how MAP70s localize along the border of secondary walls or how they regulate secondary wall patterning.

The roles of MAP65 and MAP70 in protoxylem secondary cell wall patterning of whole plants remain to be determined. Further analysis of these genes is needed, as well as identification of novel protoxylem-specific factors, to elucidate the molecular mechanisms underlying formation of annular and spiral secondary walls.

Secondary Wall Patterning in Metaxylem Vessels

MIDD1-Kinesin-13A Locally Depolymerizes Microtubules: The molecular basis for secondary cell wall patterning in metaxylem vessels was revealed by recent studies using VND6-inducible Arabidopsis cultured cells (Fig. 7.2a). During metaxylem differentiation, the cortical microtubules gradually disappear at the specific areas of future secondary wall pits (Oda *et al.*, 2010). This microtubule degradation is induced by MIDD1, which preferentially localizes to areas where cortical microtubules are unstable (Oda *et al.*, 2010; Fig. 7.2b). Knockdown of *MIDD1* inhibits the local disassembly of cortical microtubules and hence disrupts the formation of secondary wall pits. Conversely, overexpression of MIDD1 in non-xylem cells decreases the amount of cortical microtubules, suggesting that MIDD1 probably induces the formation of secondary wall pits by promoting microtubule disassembly. In secondary wall pits, MIDD1 accumulates at the plus end of microtubules during shrinkage, and therefore likely regulates microtubule dynamics at their plus end (Oda *et al.*, 2010).

An *in vitro* assay showed that MIDD1 binds directly to microtubules but does not have any microtubule depolymerization activity, suggesting that MIDD1 indirectly regulates microtubule disassembly (Oda *et al.*, 2010). Recently, a kinesin motor protein Kinesin-13A was found to cooperate with MIDD1 to depolymerize cortical microtubules (Mucha *et al.*, 2010; Oda and Fukuda, 2013). Kinesin-13A belongs to the kinesin-13 family (Lu *et al.*, 2005), whose animal members are known to depolymerize microtubules (Desai *et al.*, 1999; Moores and Milligan, 2006). *In vitro*, Kinesin-13A binds directly to microtubules and depolymerizes them in an ATP-dependent manner. *In vivo*, Kinesin-13 requires MIDD1 in order to localize at cortical microtubules and to induce depolymerization (Fig. 7.2c). Knockdown of *Kinesin-13A* caused smaller secondary wall pits in metaxylem vessels. By contrast, overexpression of Kinesin-13A caused larger secondary wall pits in metaxylem vessels, suggesting that Kinesin-13A enlarges the secondary wall pits through promotion of microtubule depolymerization. Interestingly, overexpression of Kinesin-13A does not affect the spiral secondary wall pattern but does slightly enlarge the secondary wall thickenings (Oda and Fukuda, 2013), suggesting that Kinesin-13A has a different activity in protoxylem cells. Animal kinesin-13 family members are regulated by phosphorylation (Ems-McClung and Walczak, 2010), and it is therefore possible that Kinesin-13A activity could be controlled by phosphorylation or by other interacting partners in different ways in protoxylem and metaxylem cells.

ROP GTPases Determine the Position of Secondary Wall Pits: ROP GTPases are a plant-specific family of Rac/Rho GTPases and function in polarized cell growth, polarity establishment, and various phenomena including response to pathogenesis and drought stress (Yang, 2008; Nagawa *et al.*, 2010; Yang and Lavagi, 2012). In root hairs and pollen tubes, ROP GTPase accumulates at the cell tips and maintains growth by controlling vesicle trafficking and cytoskeleton development (Yang, 2008). ROP GTPases are present in two forms: a GTP-bound active form and a GDP-bound inactive form. ROP GTPases are activated by plant-specific ROP guanine nucleotide exchange factors (ROPGEFs; Berken *et al.*, 2005; Gu *et al.*, 2006) and are inactivated by ROP GTPase activating proteins (ROPGAPs; Wu *et al.*, 2000). Usually, the GTP-bound active form of GTPase interacts with effector proteins to activate specific signal transduction cascades (Yang, 2008).

MIDD1 belongs to the ROP INTERACTIVE PARTNER (RIP)/INTERACTOR OF CONSTITUTIVE ACTIVE ROPS (ICR) family, whose members possess a ROP-interacting motif at their C-terminus (Lavy *et al.*, 2007; Li *et al.*, 2008). In differentiating metaxylem vessel cells, active ROP11 is exclusively localized at the plasma membrane within the secondary wall pits and recruits MIDD1-Kinesin-13A complexes (Fig. 7.2b and c). By contrast, inactive ROP11 broadly localizes at the plasma membrane. Induction of a constitutively active form of ROP11 inhibits targeting of the MIDD1-Kinesin-13A complex to the plasma membrane domains and prevents formation of secondary wall pits. ROPGEF4 and ROPGAP3 also localize at the plasma membrane in the secondary wall pits (Oda and Fukuda, 2012, 2013). Specifically, ROPGEF4 localizes as a patch in the central region of the secondary wall pit (Fig. 7.2c), while ROPGAP3 localizes more broadly in the pits. Such patches of ROPGEF4 appeared prior to onset of secondary wall deposition. Knockout or knockdown of *ROPGEF4* causes a reduction in the density of secondary wall pits. These data suggest that ROPGEF4 may determine the formation of secondary wall pits through local activation of ROP11 (Oda and Fukuda, 2012).

Importantly, co-expression of ROPGEF4, ROPGAP3, and ROP11 in non-xylem cells results in evenly spaced patches of ROPGEF4 at the plasma membrane that are similar to the ROPGEF4 patches observed in differentiating metaxylem cells (Oda and Fukuda, 2012). The ROPGEF4 patches locally activate ROP11, which in turn recruits MIDD1 to induce local disassembly of cortical microtubules. This suggests that the experimental reconstruction mimics the local activation of ROP11 and subsequent events observed in differentiating metaxylem cells. Lack of either ROPGEF4 or ROPGAP3 completely abolishes this reconstruction. Replacing ROP11 with a constitutively activated form or a constitutively inactivated form also abolishes the formation of ROPGEF4 patches. These data suggest that local ROP11 activation is spontaneously induced by GTP-/GDP-cycling of ROP11 by ROPGEF4 and ROPGAP3 (Oda and Fukuda, 2012).

One possible mechanism to explain the spontaneous local activation of ROP11 is positive feedback between active ROP11 and ROPGEF4. In budding yeast a Rho GTPase, Cdc42p, is locally activated by positive feedback between Rho GEF and active Rho GTPase via a scaffold protein (Kozubowski *et al.*, 2008). In plants, some ROPGEFs interact with receptor-like kinases in pollen tubes, in root hairs, and in immune responses (Zhang and McCormick, 2007; Duan *et al.*, 2010; Chang *et al.*, 2012; Akamatsu *et al.*, 2013; Zhao *et al.*, 2013). AtPRK2, a receptor-like kinase that functions in polarized growth of pollen tubes interacts with both ROPGEF1 and ROP GTPases and phosphorylates ROPGEF1 to activate ROP GTPase (Chang *et al.*, 2012). Similarly, some unknown receptor-like kinases might interact with ROPGEF4 and ROP11 to initiate positive feedback and consequently establish local activation of ROP11. Another feedback mechanism between ROPGAP3 and ROPGEF4 via ROP11 may also contribute to the self organization of activated ROP11 patches. Turing (1952) demonstrated that slow-diffusible activators and fast-diffusible inactivators can break symmetry and develop various localization patterns through positive feedback between the activator and inactivator (Turing, 1952). ROPGEF4 and ROPGAP3 might play slow-diffusible activator and fast-diffusible inactivator roles, respectively. In this model, the density and size of the active ROP area may be determined by the amount, activity and balance of ROP11, ROPGEF4, and ROPGAP3 (Fig. 7.3). To summarize, the spontaneous local activation of ROP11 determines the area in which cortical microtubules are depolymerized, which establishes the secondary wall patterns in metaxylem vessels.

Microtubule-ROP Interaction May Allow Diverse Wall Patterns in Metaxylem Vessels: The ROP-MIDD1-Kinesin-13A pathway simply induces local depletion of cortical microtubules, which is insufficient to explain the diverse patterning effects such as formation of scalariform, reticulate, and pitted secondary walls. Oda and Fukuda (2012) reported that a microtubule stabilizing agent, taxol, causes elongation of secondary wall pits, suggesting that cortical microtubules can themselves affect the area of active ROP11. The study also showed that cortical microtubules limit the area of active ROP-MIDD1 complex at the plasma membrane in non-xylem cells (Oda and Fukuda, 2012). The authors further examined the spatial interaction between cortical microtubules, active ROP11, and MIDD1 using non-xylem cells, and

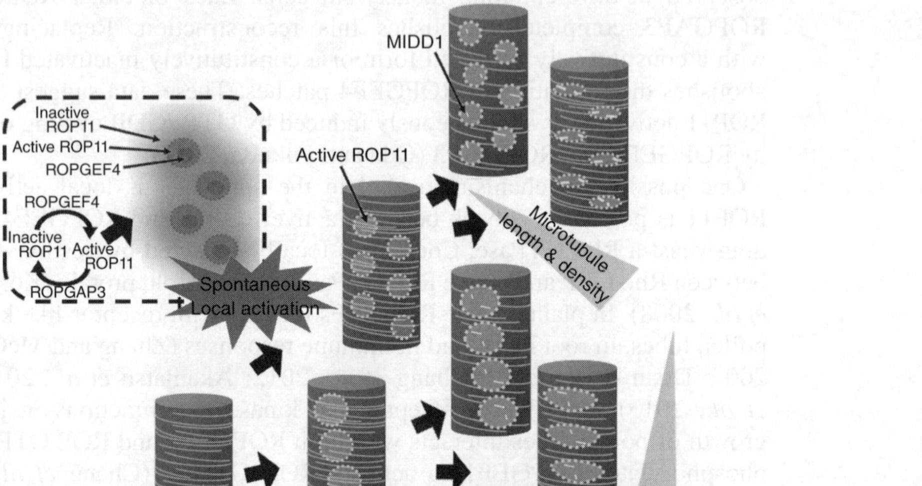

Figure 7.3 Formation of diverse wall patterns in metaxylem is ROP and microtubule dependent. Left: the GTP/GDP-cycling of ROP11 by ROPGEF4 and ROPGAP3 spontaneously generates active ROP domains. The balance between ROPGEF, ROPGAP, and ROP11 may generate active ROP domains of various density and size. Right: mutual inhibitory interactions between cortical microtubules and ROP-MIDD1 complexes generate various shapes of secondary wall pits. Long and dense cortical microtubules restrict ROP-MIDD1 complex movement more strongly, resulting in narrower secondary wall pits.

found that cortical microtubules do not affect the localization of ROP11 in the absence of MIDD1. However, when a truncated MIDD1 that lacks its microtubule-binding domain is introduced, ROP11-MIDD1 complexes are eliminated from the plasma membrane beneath the cortical microtubules. Cortical microtubules can also eliminate truncated MIDD1 that is artificially anchored to the plasma membrane by fusion with a membrane anchor domain. Based on these results, the authors concluded that cortical microtubules eliminate active ROP11 from the plasma membrane in a MIDD1-dependent manner. Cortical microtubules are closely anchored to the plasma membrane, and it is therefore possible that the active ROP11 protein complexes with MIDD1 and Kinesin-13A are unable to pass through the space between the cortical microtubules and the plasma membrane. Correspondingly, the Arabidopsis formin AtFH1 which has a large cytoplasmic domain is eliminated from the plasma membrane by cortical microtubules (Martiniere *et al.*, 2011) in a similar manner. Cortical microtubules may maintain the plasma membrane domains by restricting the localization of plasma membrane-anchored proteins.

In summary, MIDD1 mediates elimination of active ROP11 from the plasma membrane by cortical microtubules, and also mediates disassembly of cortical microtubules at their tip through Kinesin-13A. The MIDD1-mediated inhibitory interaction between cortical microtubules and plasma membrane domains may explain how metaxylem cells develop the various secondary wall pit shapes. For example, long and dense cortical microtubules surrounding the active ROP domains may strongly suppress the diffusion of ROP11-MIDD1 complexes, producing a long oval pattern of active ROP domains and resulting in reticulate or scalariform patterns of secondary walls. By contrast, short and sparse cortical microtubules around the active ROP domains may only weakly suppress the diffusion of ROP11-MIDD1 complexes, resulting in rounder secondary wall pits (Fig. 7.3). The mutual inhibitory interaction between cortical microtubules and the ROP11-MIDD1 complex may therefore explain the diverse patterns of secondary cell walls in metaxylem vessels.

Membrane Traffic and Cellulose Synthesis Activity are Essential for Secondary Wall Patterning

In addition to various MAPs and ROP GTPases, proper delivery and synthesis of secondary cell wall components appears to be required for secondary wall patterns. TED6 and TED7 were identified as novel transmembrane proteins that promote secondary wall development in cultured zinnia mesophyll cells (Endo *et al.*, 2009). TED6 and TED7 are localized at the plasma membrane beneath the secondary wall thickenings. In Arabidopsis roots, knockdown of *TED7* produces thin, spiral-shaped secondary walls in metaxylem vessels, whereas knockdown of both *TED6* and *TED7* causes loss of secondary walls. TED6 interacts with IRX3, suggesting that these proteins promote activity of cellulose synthase complexes (Endo *et al.*, 2009). Activity of cellulose synthesis may influence patterning of secondary walls.

Poplar MAP20 is a plant-specific TPX2-containing protein that is preferentially expressed in xylem cells (Rajangam *et al.*, 2008). MAP20 binds directly to microtubules *in vitro* and is associated with cortical microtubules *in vivo*. Interestingly, a cellulose synthesis inhibitor, 2,6-dichlorobenzonitrile (DCB), was found to bind to MAP20 (Rajangam *et al.*, 2008). DCB inhibits mobility of cellulose synthase complexes at the plasma membrane (DeBolt *et al.*, 2007) but does not inhibit microtubule binding of MAP20; MAP20 may therefore promote cellulose synthesis activity (Rajangam *et al.*, 2008), which has the potential to affect the secondary wall patterns.

An exocyst subunit EXO70A1 is specifically expressed in xylem cells, and knockout of *EXO70A1* causes disorganization of secondary wall thickenings and accumulation of large membrane compartments within the differentiating xylem cells. This suggests that EXO70A1 regulates secondary cell-wall-related vesicular traffic (Li *et al.*, 2013). Loss of *EXO70A1* does not inhibit secondary wall deposition, and so EXO70A1 likely regulates the targeting and/or fusion of vesicles to the areas of the plasma membrane marked with cortical microtubules.

Taken together, the data indicate that cellulose synthesis and vesicular traffic are essential processes influencing the patterned deposition of secondary cell walls onto the cortical microtubules.

Future Perspectives

Recent studies have revealed important factors and candidate genes that are involved in the development of secondary cell wall patterns. However, the spatial organization of secondary cell walls is not yet fully understood. Critical evidence confirming the involvement of MAP65 and MAP70 in secondary wall patterning in protoxylem vessels is absent. The regulatory mechanisms in metaxylem vessels that determine the position and density of secondary wall pits via local ROP11 activation remain enigmatic. In particular, secondary wall pits in one xylem cell are coupled with secondary wall pits developed in the neighboring xylem cell, forming a pit-pair. This requires local intercellular communication, the means of which are not yet understood. It is also not yet known whether the mechanism leading to secondary wall pit development in xylem fibers is the same as the mechanism in metaxylem vessels. The molecules that mediate the inhibitory interaction between cortical microtubules and active ROP domains are also currently unknown. The most intriguing question is whether a common mechanism governs all patterns of secondary walls in various xylem cells or whether distinct mechanisms independently regulate each secondary wall pattern. Expression data do not appear to be sufficient to answer this question conclusively. MIDD1 and MAP65-8 are upregulated both by VND6 and VND7 (Ohashi-Ito *et al.*, 2010; Yamaguchi *et al.*, 2011). ROPGEF4 is expressed both in protoxylem vessels and metaxylem vessels (Oda and Fukuda, 2012). However, overexpression of Kinesin-13A affects secondary wall patterns differently in protoxylem and metaxylem (Oda and Fukuda, 2013). Induction of constitutively active ROP11 affected only the secondary wall pits in metaxylem cells and not those of protoxylem cells (Oda and Fukuda, 2012). It is therefore important to examine post-translational regulation in addition to precise comparison of gene expression between different xylem cells. Proteomics approaches should be useful for future studies in this area. Another area of interest is secondary xylems in woods. Given the commercial and ecological importance of woody biomass, it is desirable to further our understanding of secondary wall development of the secondary xylem in wood. This is hampered by the extended time required for tree studies; it is therefore highly desirable to develop experimental systems for xylem differentiation using faster-growing models trees such as poplar or by using cell cultures, such as those using tree callus culture (Moller *et al.*, 2003; Yamagishi *et al.*, 2012, 2013). Use of tree NAC transcription factors (Ohtani *et al.*, 2011) may prove an effective approach to further improve these systems.

Acknowledgements

This work was supported in part by Grants-in-Aid from the Ministry of Education, Science, Sports and Culture of Japan to YO (25114507); from the Japan Society for the Promotion of Science to HF (23227001) and YO (25440128); from the NC-CARP project to HF; and from the JST, Precursory Research for Embryonic Science and Technology (PRESTO) to YO (20103).

References

Abe, H., Funada, R., Ohtani, J., and Fukazawa, K. (1995a) Changes in the arrangement of microtubules and microfibrils in differentiating conifer tracheids during the expansion of cells. *Annals of Botany (London)*, **75**, 305–310.

Abe, H., Funada, R., Imaizumi, H., Ohtani, J., and Fukazawa, K. (1995b) Dynamic changes in the arrangement of cortical microtubules in conifer tracheids during differentiation. *Planta*, **197**, 418–421.

Akamatsu, A., Wong, H. L., Fujiwara, M., Okuda, J., Nishide, K., Uno, K., Imai, K., Umemura, K., Kawasaki, T., Kawano, Y. *et al.* (2013) An OsCEBiP/OsCERK1-OsRacGEF1-OsRac1 module is an essential early component of chitin-induced rice immunity. *Cell Host and Microbe*, **13**, 465–476.

Ambrose, C., Allard, J. F., Cytrynbaum, E. N., and Wasteneys, G. O. (2011) A CLASP-modulated cell edge barrier mechanism drives cell-wide cortical microtubule organization in Arabidopsis. *Nature Communications*, **2**, 430.

Berken, A., Thomas, C., and Wittinghofer, A. (2005) A new family of RhoGEFs activates the Rop molecular switch in plants. *Nature*, **436**, 1176–1180.

Bichet, A., Desnos, T., Turner, S., Grandjean, O., and Hofte, H. (2001) BOTERO1 is required for normal orientation of cortical microtubules and anisotropic cell expansion in Arabidopsis. *Plant Journal*, **25**, 137–148.

Bouquin, T., Mattsson, O., Naested, H., Foster, R., and Mundy, J. (2003) The Arabidopsis lue1 mutant defines a katanin p60 ortholog involved in hormonal control of microtubule orientation during cell growth. *Journal of Cell Science*, **116**, 791–801.

Bringmann, M., Li, E., Sampathkumar, A., Kocabek, T., Hauser, M. T., and Persson, S. (2012) POM-POM2/cellulose synthase interacting1 is essential for the functional association of cellulose synthase and microtubules in Arabidopsis. *Plant Cell*, **24**, 163–177.

Burk, D. H. and Ye, Z. H. (2002) Alteration of oriented deposition of cellulose microfibrils by mutation of a katanin-like microtubule-severing protein. *Plant Cell*, **14**, 2145–2160.

Burk, D. H., Liu, B., Zhong, R., Morrison, W. H., and Ye, Z. H. (2001) A katanin-like protein regulates normal cell wall biosynthesis and cell elongation. *Plant Cell*, **13**, 807–827.

Chaffey, N. and Barlow, P. (2002) Myosin, microtubules, and microfilaments: co-operation between cytoskeletal components during cambial cell division and secondary vascular differentiation in trees. *Planta*, **214**, 526–536.

Chaffey, N., Barnett, J. R., and Barlow, P. W. (1997) Cortical microtubule involvement in bordered pit formation in secondary xylem vessel elements of Aesculus hippocastanum L (Hippocastanaceae): A correlative study using electron microscopy and indirect immunofluorescence microscopy. *Protoplasma*, **197**, 64–75.

Chaffey, N., Barnett, J., and Barlow, P. (1999) A cytoskeletal basis for wood formation in angiosperm trees: the involvement of cortical microtubules. *Planta*, **208**, 19–30.

Chan, J., Jensen, C. G., Jensen, L. C., Bush, M., and Lloyd, C. W. (1999) The 65-kDa carrot microtubule-associated protein forms regularly arranged filamentous cross-bridges between microtubules. *Proceedings of the National Academy of Science, USA*, **96**, 14931–14936.

Chan, J., Calder, G., Fox, S., and Lloyd, C. (2007) Cortical microtubule arrays undergo rotary movements in Arabidopsis hypocotyl epidermal cells. *Nature Cell Biology*, **9**, 171–175.

Chan, J., Sambade, A., Calder, G., and Lloyd, C. (2009) Arabidopsis cortical microtubules are initiated along, as well as branching from, existing microtubules. *Plant Cell*, **21**, 2298–2306.

Chang, F., Gu, Y., Ma, H., and Yang, Z. (2012) AtPRK2 Promotes ROP1 Activation via RopGEFs in the Control of Polarized Pollen Tube Growth. *Journal of Cell Biology*, **77**, 14–34.

Crowell, E. F., Bischoff, V., Desprez, T., Rolland, A., Stierhof, Y. D., Schumacher, K., Gonneau, M., Hofte, H., and Vernhettes, S. (2009) Pausing of Golgi bodies on microtubules regulates secretion of cellulose synthase complexes in Arabidopsis. *Plant Cell*, **21**, 1141–1154.

DeBolt, S., Gutierrez, R., Ehrhardt, D. W., and Somerville, C. (2007) Nonmotile cellulose synthase subunits repeatedly accumulate within localized regions at the plasma membrane in Arabidopsis hypocotyl cells following 2,6-dichlorobenzonitrile treatment. *Plant Physiology*, **145**, 334–338.

Demura, T., Tashiro, G., Horiguchi, G., Kishimoto, N., Kubo, M., Matsuoka, N., Minami, A., Nagata-Hiwatashi, M., Nakamura, K., Okamura, Y. *et al.* (2002) Visualization by comprehensive microarray analysis of gene expression programs during transdifferentiation of mesophyll cells into xylem cells. *Proceedings of the National Academy of Science, USA*, **99**, 15794–15799.

Desai, A., Verma, S., Mitchison, T. J., and Walczak, C. E. (1999) Kin I kinesins are microtubule-destabilizing enzymes. *Cell*, **96**, 69–78.

Dixit, R. and Cyr, R. (2004) Encounters between dynamic cortical microtubules promote ordering of the cortical array through angle-dependent modifications of microtubule behavior. *Plant Cell*, **16**, 3274–3284.

Duan, Q., Kita, D., Li, C., Cheung, A. Y., and Wu, H. M. (2010) FERONIA receptor-like kinase regulates RHO GTPase signaling of root hair development. *Proceedings of the National Academy of Science, USA*, **107**, 17821–17826.

Ems-McClung, S. C. and Walczak, C. E. (2010) Kinesin-13s in mitosis: Key players in the spatial and temporal organization of spindle microtubules. *Seminars in Cell and Developmental Biology*, **21**, 276–282.

Endo, S., Pesquet, E., Yamaguchi, M., Tashiro, G., Sato, M., Toyooka, K., Nishikubo, N., Udagawa-Motose, M., Kubo, M., Fukuda, H. *et al.* (2009) Identifying new components participating in the secondary cell wall formation of vessel elements in zinnia and Arabidopsis. *Plant Cell*, **21**, 1155–1165.

Esau, K. (1977) *Anatomy of Seed Plants*, 2nd edition. Wiley, New York.

Fache, V., Gaillard, J., Van Damme, D., Geelen, D., Neumann, E., Stoppin-Mellet, V., and Vantard, M. (2010) Arabidopsis kinetochore fiber-associated MAP65-4 cross-links microtubules and promotes microtubule bundle elongation. *Plant Cell*, **22**, 3804–3815.

Falconer, M. M. and Seagull, R. W. (1985) Xylogenesis in tissue-culture-taxol effects on microtubule reorientation and lateral association in differentiating cells. *Protoplasma*, **128**, 157–166.

Falconer, M. M. and Seagull, R. W. (1986) Xylogenesis in tissue-culture. 2. Microtubules, cell-shape and secondary wall patterns. *Protoplasma*, **133**, 140–148.

Falconer, M. M. and Seagull, R. W. (1988) Xylogenesis in tissue-culture. 3. Continuing wall deposition during tracheary element development. *Protoplasma*, **144**, 10–16.

Fukuda, H. and Komamine, A. (1980) Establishment of an experimental system for the study of tracheary element differentiation from single cells isolated from the mesophyll of Zinnia elegans. *Plant Physiology*, **65**, 57–60.

Fukuda, H. and Kobayashi, H. (1989) Dynamic organization of the cytoskeleton during tracheary-element differentiation. *Development, Growth and Differentiation*, **31**, 9–16.

Funada, R., Abe, H., Furusawa, O., Imaizumi, H., Fukazawa, K., and Ohtani, J. (1997) The orientation and localization of cortical microtubules in differentiating conifer tracheids during cell expansion. *Plant Cell Physiology*, **38**, 210–212.

Funada, R., Miura, H., Shibagaki, M., Furusawa, O., Miura, T., Fukatsu, E., and Kitin, P. (2001) Involvement of localized cortical microtubules in the formation of a modified structure of wood. *Journal of Plant Research*, **114**, 7.

Gaillard, J., Neumann, E., Van Damme, D., Stoppin-Mellet, V., Ebel, C., Barbier, E., Geelen, D., and Vantard, M. (2008) Two microtubule-associated proteins of Arabidopsis MAP65s promote antiparallel microtubule bundling. *Molecular Biology of the Cell*, **19**, 4534–4544.

Gardiner, J. C., Taylor, N. G., and Turner, S. R. (2003) Control of cellulose synthase complex localization in developing xylem. *Plant Cell*, **15**, 1740–1748.

Gu, Y., Li, S., Lord, E. M., and Yang, Z. (2006) Members of a novel class of Arabidopsis Rho guanine nucleotide exchange factors control Rho GTPase-dependent polar growth. *Plant Cell*, **18**, 366–381.

Gu, Y., Kaplinsky, N., Bringmann, M., Cobb, A., Carroll, A., Sampathkumar, A., Baskin, T. I., Persson, S., and Somerville, C. R. (2010) Identification of a cellulose synthase-associated protein required for cellulose biosynthesis. *Proceedings of the National Academy of Science, USA*, **107**, 12866–12871.

Gunning, B. E. S. and Hardham, A. R. (1982) Microtubules. *Annual Review of Plant Biology*, **33**, 651–698.

Gutierrez, R., Lindeboom, J. J., Paredez, A. R., Emons, A. M., and Ehrhardt, D. W. (2009) Arabidopsis cortical microtubules position cellulose synthase delivery to the plasma membrane and interact with cellulose synthase trafficking compartments. *Nature Cell Biology*, **11**, 797–806.

Haigler, C.H. and Brown, R.M. (1986) Transport of rosettes from the Golgi apparatus to the plasma membrane in isolated mesophyll cells of Zinnia elegans during differentiation to tracheary elements in suspension culture. *Protoplasma*, **134**, 10.

Hardham, A. R. and Gunning, B. E. S. (1978) Structure of cortical microtubule arrays in plant cells. *Molecular Plant*, **6**, 1187–1201.

Hardham, A.R. and Gunning, B.E.S. (1979) Interpolation of microtubules into cortical arrays during cell elongation and differentiation in roots of Azolla pinnata. *Journal of Cell Science*, **37**, 31.

Hepler, P.K. (1981) Morphogenesis of tracheary elements and guard cells. *Cell Biology Monographs*, **8**, 21.

Hepler, P.K. and Newcomb, E.H. (1964) Microtubules and fibrils in the cytoplasm of Coleus cells undergoing secondary wall deposition. *Journal of Cell Biology*, **20**, 5.

Hepler, P.K. and Fosket, D.E. (1971) The role of microtubules in vessel member differentiation in Coleus. *Protoplasma*, **72**, 213–236.

Ho, C. M., Hotta, T., Guo, F., Roberson, R. W., Lee, Y. R., and Liu, B. (2011) Interaction of antiparallel microtubules in the phragmoplast is mediated by the microtubule-associated protein MAP65-3 in Arabidopsis. *Plant Cell*, **23**, 2909–2923.

Ho, C. M., Lee, Y. R., Kiyama, L. D., Dinesh-Kumar, S. P., and Liu, B. (2012) Arabidopsis microtubule-associated protein MAP65-3 cross-links antiparallel microtubules toward their plus ends in the phragmoplast via its distinct C-terminal microtubule binding domain. *Plant Cell*, **24**, 2071–2085.

Hu, Y., Zhong, R., Morrison, W. H., 3rd,, and Ye, Z. H. (2003) The Arabidopsis RHD3 gene is required for cell wall biosynthesis and actin organization. *Planta*, **217**, 912–921.

Ito, J. and Fukuda, H. (2002) ZEN1 is a key enzyme in the degradation of nuclear DNA during programmed cell death of tracheary elements. *Plant Cell*, **14**, 3201–3211.

Ito, Y., Nakanomyo, I., Motose, H., Iwamoto, K., Sawa, S., Dohmae, N., and Fukuda, H. (2006) Dodeca-CLE peptides as suppressors of plant stem cell differentiation. *Science*, **313**, 842–845.

Kobayashi, H., Fukuda, H., and Shibaoka, H. (1987) Reorganization of actin-filaments associated with the differentiation of tracheary elements in zinnia mesophyll-cells. *Protoplasma*, **138**, 69–71.

Kobayashi, H., Fukuda, H., and Shibaoka, H. (1988) Interrelation between the spatial disposition of actin-filaments and microtubules during the differentiation of tracheary elements in cultured zinnia cells. *Protoplasma*, **143**, 29–38.

Korolev, A. V., Chan, J., Naldrett, M. J., Doonan, J. H., and Lloyd, C. W. (2005) Identification of a novel family of 70 kDa microtubule-associated proteins in Arabidopsis cells. *Plant Journal*, **42**, 547–555.

Korolev, A. V., Buschmann, H., Doonan, J. H., and Lloyd, C. W. (2007) AtMAP70-5, a divergent member of the MAP70 family of microtubule-associated proteins, is required for anisotropic cell growth in Arabidopsis. *Journal of Cell Science*, **120**, 2241–2247.

Kozubowski, L., Saito, K., Johnson, J. M., Howell, A. S., Zyla, T. R., and Lew, D. J. (2008) Symmetry-breaking polarization driven by a Cdc42p GEF-PAK complex. *Current Biology*, **18**, 1719–1726.

Kubo, M., Udagawa, M., Nishikubo, N., Horiguchi, G., Yamaguchi, M., Ito, J., Mimura, T., Fukuda, H., and Demura, T. (2005) Transcription switches for protoxylem and metaxylem vessel formation. *Genes and Development*, **19**, 1855–1860.

Kwon, S. I., Cho, H. J., Jung, J. H., Yoshimoto, K., Shirasu, K., and Park, O. K. (2010) The Rab GTPase RabG3b functions in autophagy and contributes to tracheary element differentiation in Arabidopsis. *Plant Journal*, **64**, 151–164.

Lavy, M., Bloch, D., Hazak, O., Gutman, I., Poraty, L., Sorek, N., Sternberg, H., and Yalovsky, S. (2007) A Novel ROP/RAC effector links cell polarity, root-meristem maintenance, and vesicle trafficking. *Current Biology*, **17**, 947–952.

Li, J., Jiang, J., Qian, Q., Xu, Y., Zhang, C., Xiao, J., Du, C., Luo, W., Zou, G., Chen, M. *et al.* (2011) Mutation of rice BC12/GDD1, which encodes a kinesin-like protein that binds to a GA biosynthesis gene promoter, leads to dwarfism with impaired cell elongation. *Plant Cell*, **23**, 628–640.

Li, S., Gu, Y., Yan, A., Lord, E., and Yang, Z. B. (2008) RIP1 (ROP Interactive Partner 1)/ICR1 marks pollen germination sites and may act in the ROP1 pathway in the control of polarized pollen growth. *Molecular Plant*, **1**, 1021–1035.

Li, S., Lei, L., Somerville, C. R., and Gu, Y. (2012) Cellulose synthase interactive protein 1 (CSI1) links microtubules and cellulose synthase complexes. *Proceedings of the National Academy of Science, USA*, **109**, 185–190.

Li, S., Chen, M., Yu, D., Ren, S., Sun, S., Liu, L., Ketelaar, T., Emons, A. M., and Liu, C. M. (2013) EXO70A1-mediated vesicle trafficking is critical for tracheary element development in Arabidopsis. *Plant Cell*, **25**, 1774–1786.

Lin, D., Cao, L., Zhou, Z., Zhu, L., Ehrhardt, D., Yang, Z., and Fu, Y. (2013) Rho GTPase signaling activates microtubule severing to promote microtubule ordering in Arabidopsis. *Current Biology*, **23**, 290–297.

Lindeboom, J. J., Lioutas, A., Deinum, E. E., Tindemans, S. H., Ehrhardt, D. W., Emons, A. M., Vos, J. W., and Mulder, B. M. (2013a) Cortical microtubule arrays are initiated from a nonrandom prepattern driven by atypical microtubule initiation. *Plant Physiology*, **161**, 1189–1201.

Lindeboom, J. J., Nakamura, M., Hibbel, A., Shundyak, K., Gutierrez, R., Ketelaar, T., Emons, A. M., Mulder, B. M., Kirik, V., and Ehrhardt, D. W. (2013b) A mechanism for reorientation of cortical microtubule arrays driven by microtubule severing. *Science*, published online 6 December 2013b, doi: 10.1126/science.1245533.

Lu, L., Lee, Y. R., Pan, R., Maloof, J. N., and Liu, B. (2005) An internal motor kinesin is associated with the Golgi apparatus and plays a role in trichome morphogenesis in Arabidopsis. *Molecular Biology of the Cell*, **16**, 811–823.

Mao, G., Buschmann, H., Doonan, J. H., and Lloyd, C. W. (2006) The role of MAP65-1 in microtubule bundling during Zinnia tracheary element formation. *Journal of Cell Science*, **119**, 753–758.

Mao, T., Jin, L., Li, H., Liu, B., and Yuan, M. (2005) Two microtubule-associated proteins of the Arabidopsis MAP65 family function differently on microtubules. *Plant Physiology*, **138**, 654–662.

Martiniere, A., Gayral, P., Hawes, C., and Runions, J. (2011) Building bridges: formin1 of Arabidopsis forms a connection between the cell wall and the actin cytoskeleton. *Plant Journal*, **66**, 354–365.

Milioni, D., Sado, P. E., Stacey, N. J., Domingo, C., Roberts, K., and McCann, M. C. (2001) Differential expression of cell-wall-related genes during the formation of tracheary elements in the Zinnia mesophyll cell system. *Plant Molecular Biology*, **47**, 221–238.

Milioni, D., Sado, P. E., Stacey, N. J., Roberts, K., and McCann, M. C. (2002) Early gene expression associated with the commitment and differentiation of a plant tracheary element is revealed by cDNA-amplified fragment length polymorphism analysis. *Plant Cell*, **14**, 2813–2824.

Mitsuda, N., Iwase, A., Yamamoto, H., Yoshida, M., Seki, M., Shinozaki, K., and Ohme-Takagi, M. (2007) NAC transcription factors, NST1 and NST3, are key regulators of the formation of secondary walls in woody tissues of Arabidopsis. *Plant Cell*, **19**, 270–280.

Moller, R., McDonald, A. G., Walter, C., and Harris, P. J. (2003) Cell differentiation, secondary cell-wall formation and transformation of callus tissue of Pinus radiata D. Don. *Planta*, **217**, 736–747.

Moores, C. A. and Milligan, R. A. (2006) Lucky 13-microtubule depolymerisation by kinesin-13 motors. *Journal of Cell Science*, **119**, 3905–3913.

Motose, H., Sugiyama, M., and Fukuda, H. (2004) A proteoglycan mediates inductive interaction during plant vascular development. *Nature*, **429**, 873–878.

Mucha, E., Hoefle, C., Huckelhoven, R., and Berken, A. (2010) RIP3 and AtKinesin-13A – a novel interaction linking Rho proteins of plants to microtubules. *European Journal of Cell Biology*, **89**, 906–916.

Muller, S., Smertenko, A., Wagner, V., Heinrich, M., Hussey, P. J., and Hauser, M. T. (2004) The plant microtubule–associated protein AtMAP65-3/PLE is essential for cytokinetic phragmoplast function. *Current Biology*, **14**, 412–417.

Murata, T., Sonobe, S., Baskin, T. I., Hyodo, S., Hasezawa, S., Nagata, T., Horio, T., and Hasebe, M. (2005) Microtubule-dependent microtubule nucleation based on recruitment of gamma-tubulin in higher plants. *Nature Cell Biology*, **7**, 961–968.

Nagawa, S., Xu, T., and Yang, Z. (2010) RHO GTPase in plants: Conservation and invention of regulators and effectors. *Small GTPases*, **1**, 78–88.

Nakamura, M., Ehrhardt, D. W., and Hashimoto, T. (2010) Microtubule and katanin-dependent dynamics of microtubule nucleation complexes in the acentrosomal Arabidopsis cortical array. *Nature Cell Biology*, **12**, 1064–1070.

Obara, K., Kuriyama, H., and Fukuda, H. (2001) Direct evidence of active and rapid nuclear degradation triggered by vacuole rupture during programmed cell death in Zinnia. *Plant Physiology*, **125**, 615–626.

Oda, Y. and Hasezawa, S. (2006) Cytoskeletal organization during xylem cell differentiation. *Journal of Plant Research*, **119**, 167–177.

Oda, Y. and Fukuda, H. (2012) Initiation of cell wall pattern by a Rho- and microtubule-driven symmetry breaking. *Science*, **337**, 1333–1336.

Oda, Y. and Fukuda, H. (2013) ROP GTPase signaling regulates the behavior of Kinesin-13A to establish secondary cell wall patterns. *Plant Cell*, **25**, 4439–4450.

Oda, Y., Mimura, T., and Hasezawa, S. (2005) Regulation of secondary cell wall development by cortical microtubules during tracheary element differentiation in Arabidopsis cell suspensions. *Plant Physiology*, **137**, 1027–1036.

Oda, Y., Iida, Y., Kondo, Y., and Fukuda, H. (2010) Wood cell-wall structure requires local 2D-microtubule disassembly by a novel plasma membrane-anchored protein. *Current Biology*, **20**, 1197–1202.

Ohashi-Ito, K., Oda, Y., and Fukuda, H. (2010) Arabidopsis VASCULAR-RELATED NAC-DOMAIN6 directly regulates the genes that govern programmed cell death and secondary wall formation during xylem differentiation. *Plant Cell*, **22**, 3461–3473.

Ohtani, M., Nishikubo, N., Xu, B., Yamaguchi, M., Mitsuda, N., Goue, N., Shi, F., Ohme-Takagi, M., and Demura, T. (2011) A NAC domain protein family contributing to the regulation of wood formation in poplar. *Plant Journal*, **67**, 499–512.

Paredez, A. R., Somerville, C. R., and Ehrhardt, D. W. (2006) Visualization of cellulose synthase demonstrates functional association with microtubules. *Science*, **312**, 1491–1495.

Pesquet, E., Korolev, A. V., Calder, G., and Lloyd, C. W. (2010) The microtubule-associated protein AtMAP70-5 regulates secondary wall patterning in Arabidopsis wood cells. *Current Biology*, **20**, 744–749.

Pickett-Heaps, J.D. (1966) Incorporation of radioactivity into wheat xylem walls. *Planta Medica*, **71**, 14.

Pickett-Heaps, J.D. (1967) The effects of colchicine on the ultrastructure of dividing plant cell, xylem wall differentiation and distribution of cytoplasmic microtubules. *Developmental Biology*, **15**, 31.

Pickett-Heaps, J.D. and Northcote, D.H. (1966) Relationship of cellular organelles to the formation and development of the plant cell wall. *Journal of Experimental Botany*, **17**, 7.

Rajangam, A. S., Kumar, M., Aspeborg, H., Guerriero, G., Arvestad, L., Pansri, P., Brown, C. J., Hober, S., Blomqvist, K., Divne, C. *et al.* (2008) MAP20, a microtubule-associated protein in the secondary cell walls of hybrid aspen, is a target of the cellulose synthesis inhibitor 2,6-dichlorobenzonitrile. *Plant Physiology*, **148**, 1283–1294.

Roberts, A. W., Frost, A. O., Roberts, E. M., and Haigler, C. H. (2004) Roles of microtubules and cellulose microfibril assembly in the localization of secondary-cell-wall deposition in developing tracheary elements. *Protoplasma*, **224**, 217–229.

Sawa, S., Demura, T., Horiguchi, G., Kubo, M., and Fukuda, H. (2005) The ATE genes are responsible for repression of transdifferentiation into xylem cells in Arabidopsis. *Plant Physiology*, **137**, 141–148.

Shaw, S. L., Kamyar, R., and Ehrhardt, D. W. (2003) Sustained microtubule treadmilling in Arabidopsis cortical arrays. *Science*, **300**, 1715–1718.

Smertenko, A. P., Chang, H. Y., Sonobe, S., Fenyk, S. I., Weingartner, M., Bogre, L., and Hussey, P. J. (2006) Control of the AtMAP65-1 interaction with microtubules through the cell cycle. *Journal of Cell Science*, **119**, 3227–3237.
Stoppin-Mellet, V., Gaillard, J., and Vantard, M. (2006) Katanin's severing activity favors bundling of cortical microtubules in plants. *Plant Journal*, **46**, 1009–1017.
Stoppin-Mellet, V., Fache, V., Portran, D., Martiel, J. L., and Vantard, M. (2013) MAP65 coordinate microtubule growth during bundle formation. *PloS One*, **8**.
Taylor, N. G., Laurie, S., and Turner, S. R. (2000) Multiple cellulose synthase catalytic subunits are required for cellulose synthesis in Arabidopsis. *Plant Cell*, **12**, 2529–2540.
Taylor, N. G., Howells, R. M., Huttly, A. K., Vickers, K., and Turner, S. R. (2003) Interactions among three distinct CesA proteins essential for cellulose synthesis. *Proceedings of the National Academy of Science, USA*, **100**, 1450–1455.
Turing, A. M. (1952) The chemical basis of morphogenesis. *Proceedings of the Royal Society of London, B: Biological Sciences*, **237**, 37–72.
Turner, S. R. and Somerville, C. R. (1997) Collapsed xylem phenotype of Arabidopsis identifies mutants deficient in cellulose deposition in the secondary cell wall. *Plant Cell*, **9**, 689–701.
Wasteneys, G. O. and Ambrose, J. C. (2009) Spatial organization of plant cortical microtubules: close encounters of the 2D kind. *Trends in Cell Biology*, **19**, 62–71.
Wightman, R. and Turner, S. R. (2007) Severing at sites of microtubule crossover contributes to microtubule alignment in cortical arrays. *Plant Journal*, **52**, 742–751.
Wightman, R. and Turner, S. R. (2008) The roles of the cytoskeleton during cellulose deposition at the secondary cell wall. *Plant Journal*, **54**, 794–805.
Wightman, R. and Turner, S. (2010a) Trafficking of the plant cellulose synthase complex. *Plant Physiology*, **153**, 427–432.
Wightman, R. and Turner, S. (2010b) Trafficking of the cellulose synthase complex in developing xylem vessels. *Biochemical Society Transactions*, **38**, 755–760.
Wightman, R., Marshall, R., and Turner, S. R. (2009) A cellulose synthase-containing compartment moves rapidly beneath sites of secondary wall synthesis. *Plant Cell Physiology*, **50**, 584–594.
Wightman, R., Chomicki, G., Kumar, M., Carr, P., and Turner, S. R. (2013) SPIRAL2 determines plant microtubule organization by modulating microtubule severing. *Current Biology*, **23**, 1902–1907.
Wooding, F. B. and Northcote, D. H. (1964) The development of the secondary wall of the xylem in Acer Pseudoplatanus. *Journal of Cell Biology*, **23**, 327–337.
Wu, G., Li, H., and Yang, Z. (2000) Arabidopsis RopGAPs are a novel family of rho GTPase-activating proteins that require the Cdc42/Rac-interactive binding motif for rop-specific GTPase stimulation. *Plant Physiology*, **124**, 1625–1636.
Yamagishi, Y., Sato, T., Uchiyama, H., Yoshimoto, J., Nakagawa, R., Nakaba, S., Kubo, T., and Funada, R. (2012) Tracheary elements that resemble secondary xylem in calli derived from the conifers, Torreya nucifera and Cryptomeria japonica. *Journal of Wood Science*, **58**, 557–562.
Yamagishi, Y., Yoshimoto, J., Uchiyama, H., Nabeshima, E., Nakaba, S., Watanabe, U., and Funada, R. (2013) In vitro induction of secondary xylem-like tracheary elements in calli of hybrid poplar (Populus sieboldii x P. grandidentata). *Planta*, **237**, 1179–1185.
Yamaguchi, M., Kubo, M., Fukuda, H., and Demura, T. (2008) Vascular-related NAC-DOMAIN7 is involved in the differentiation of all types of xylem vessels in Arabidopsis roots and shoots. *Plant Journal*, **55**, 652–664.
Yamaguchi, M., Goue, N., Igarashi, H., Ohtani, M., Nakano, Y., Mortimer, J. C., Nishikubo, N., Kubo, M., Katayama, Y., Kakegawa, K. *et al.* (2010) VASCULAR-RELATED NAC-DOMAIN6 and VASCULAR-RELATED NAC-DOMAIN7 effectively induce transdifferentiation into xylem vessel elements under control of an induction system. *Plant Physiology*, **153**, 906–914.
Yamaguchi, M., Mitsuda, N., Ohtani, M., Ohme-Takagi, M., Kato, K., and Demura, T. (2011) VASCULAR-RELATED NAC-DOMAIN7 directly regulates the expression of a broad range of genes for xylem vessel formation. *Plant Journal*, **66**, 579–590.

Yang, Z. (2008) Cell polarity signaling in Arabidopsis. *Annual Reviews of Cell and Developmental Biology*, **24**, 551–575.

Yang, Z. and Lavagi, I. (2012) Spatial control of plasma membrane domains: ROP GTPase-based symmetry breaking. *Current Opinion in Plant Biology*, **15**, 601–607.

Zhang, M., Zhang, B., Qian, Q., Yu, Y., Li, R., Zhang, J., Liu, X., Zeng, D., Li, J., and Zhou, Y. (2010) Brittle Culm 12, a dual-targeting kinesin-4 protein, controls cell-cycle progression and wall properties in rice. *Plant Journal*, **63**, 312–328.

Zhang, Q., Fishel, E., Bertroche, T., and Dixit, R. (2013) Microtubule severing at crossover sites by katanin generates ordered cortical microtubule arrays in Arabidopsis. *Current Biology*, **23**, 2191–2195.

Zhang, Y., and McCormick, S. (2007) A distinct mechanism regulating a pollen-specific guanine nucleotide exchange factor for the small GTPase Rop in Arabidopsis thaliana. *Proceedings of the National Academy of Science, USA*, **104**, 18830–18835.

Zhao, X. Y., Wang, Q., Li, S., Ge, F. R., Zhou, L. Z., McCormick, S., and Zhang, Y. (2013) The juxtamembrane and carboxy-terminal domains of Arabidopsis PRK2 are critical for ROP-induced growth in pollen tubes. *Journal of Experimental Botany*.

Zhong, R., Burk, D. H., Morrison, W. H., and Ye, Z. H. (2002) A kinesin-like protein is essential for oriented deposition of cellulose microfibrils and cell wall strength. *Plant Cell*, **14**, 3101–3117.

Zhong, R., Burk, D. H., Morrison, W. H., 3rd,, and Ye, Z. H. (2004) FRAGILE FIBER3, an Arabidopsis gene encoding a type II inositol polyphosphate 5-phosphatase, is required for secondary wall synthesis and actin organization in fiber cells. *Plant Cell*, **16**, 3242–3259.

Zhong, R., Burk, D. H., Nairn, C. J., Wood-Jones, A., Morrison, W. H., 3rd,, and Ye, Z. H. (2005) Mutation of SAC1, an Arabidopsis SAC domain phosphoinositide phosphatase, causes alterations in cell morphogenesis, cell wall synthesis, and actin organization. *Plant Cell*, **17**, 1449–1466.

Zhong, R., Richardson, E. A., and Ye, Z. H. (2007) Two NAC domain transcription factors, SND1 and NST1, function redundantly in regulation of secondary wall synthesis in fibers of Arabidopsis. *Planta*, **225**, 1603–1611.

Zhong, R., Lee, C., and Ye, Z. H. (2010) Global analysis of direct targets of secondary wall NAC master switches in Arabidopsis. *Molecular Plant*, **3**, 1087–1103.

Zhou, J. L., Qiu, J., and Ye, Z. H. (2007) Alteration in secondary wall deposition by overexpression of the fragile Fiber1 kinesin-like protein in Arabidopsis. *Journal of Integrative Plant Biology*, **49**, 1235–1243.

Zhu, C. and Dixit, R. (2011) Single molecule analysis of the Arabidopsis FRA1 kinesin shows that it is a functional motor protein with unusually high processivity. *Molecular Plant*, **4**, 879–885.

8 ROP Signaling and the Cytoskeleton in Pollen Tube Growth

Lei Zhu and Ying Fu

State Key Laboratory of Plant Physiology and Biochemistry, College of Biological Sciences, China Agricultural University, Beijing, China

Introduction

Cell morphogenesis is of great importance for plant cells to fulfill their biological functions. Immobile plant cells achieve their final shapes through polarized cell growth, which is under precise developmental and environmental regulation. The pollen tube system is one of the most-studied models for cell morphogenesis in plants. Pollen grains hydrate and germinate after landing on the stigmatic surface of the pistil, and produce a cylinder-like cell shape through an extreme form of polar–tip growth. Tip-growing pollen tubes are directed through the stigma surface, grow within and emerge from the transmitting track, and finally target ovules in the distance to deliver sperm cells for double fertilization (Johnson and Preuss, 2002). Understanding the mechanisms by which pollen tubes extend and are directed toward the ovule is a fundamentally important question in plant sexual reproduction (Franklin-Tong, 1999a, 2002). Pollen tube tip growth guided by cues from female tissues has recently been reviewed (Lord and Russell, 2002; Lord, 2003; Dresselhaus, 2006; Dresselhaus and Franklin-Tong, 2013), and will not be covered here. Because pollen grains from many species can germinate and elongate rapidly *in vitro*, advanced cell biological techniques, such as live cell imaging, can be applied to observe pollen tube growth patterns, subcellular events, and responses to external cues. In addition, pollen tube growth is genetically controlled by the haploid genome, which allows lethal mutations or pollen-specific genes to be maintained in heterozygous plants. These advantages make pollen tubes a very useful model system in which to study the regulatory mechanisms of polar cell growth and cell morphogenesis (Yang, 2008).

A wealth of previous evidences has demonstrated important roles of the actin cytoskeleton in regulating plant cell growth and cell morphogenesis. Dynamic actin filaments (F-actin) are involved in cytoplasmic streaming and vesicle

Plant Cell Wall Patterning and Cell Shape, First Edition. Edited by Hiroo Fukuda.

trafficking in pollen tubes to support tip growth (Vidali *et al.*, 2001; Samaj *et al.*, 2006; Chen *et al.*, 2007). The organization and dynamics of this fundamental cellular structure depend on multiple actin-binding proteins (ABPs). Therefore, transduction of signal cues to ABPs and F-actin is a general principle involved in spatial and temporal regulation of tip growth.

Rho-family small guanine nucleotide-binding proteins (GTPases) are well known for their conserved function in regulating the actin cytoskeleton in all eukaryotic cells (Hall, 1998; Wennerberg *et al.*, 2005; Brennwald and Rossi, 2007; Fu *et al.*, 2008). A plant-specific subfamily of Rho GTPases called ROPs (Rho-like GTPases from plants) has been identified, including 11 members in Arabidopsis, 7 in rice, and 9 in maize (Winge *et al.*, 1997; Li *et al.*, 1998; Christensen *et al.*, 2003). Several ROP GTPases from various plants have been found specifically or highly expressed in pollen tubes, and target F-actin. In regulating the cytoskeleton, ROP signaling pathways involve specific regulators and effectors unique to plants in addition to conserved components that are shared with their counterparts in other organisms. In turn, F-actin provides feedback regulation of ROP GTPases. Together with the oscillating tip-focused calcium gradient, ROPs and F-actin constitute a complex signaling network, referred to as LENS (localization enhancing network, self-sustaining) by Cole and Fowler (2006).

This chapter highlights recent advances evaluating F-actin-based molecular mechanisms that underlie pollen tube growth, and emphasizes the ROP-dependent regulation of actin organization and dynamics as well as feedback regulation of ROP activity by the actin cytoskeleton.

Structural Basis for Pollen Tube Tip Growth

Evidence from *in vivo* and *in vitro* studies has demonstrated that pollen tubes elongate rapidly. For instance, tobacco pollen tubes may reach 4.5 cm within the pistil in approximately 30 hours (Cheung *et al.*, 1995). It takes about 10–15 hours for Arabidopsis pollen tubes to penetrate a pistil approximately 2.5 mm long and arrive at the most distal ovules (Schiott *et al.*, 2004). *In vitro* cultured pollen tubes are able to maintain their polarity and grow synchronously. The growth rate of maize pollen tubes reaches 1 cm hr^{-1} when they grow through style (Barnabas and Fridvalszky, 1984), and lily pollen tubes can grow at 200–300 nm s^{-1} in culture media (Hepler *et al.*, 2001). The growth rates of *in vitro*-germinated tobacco pollen tubes have been reported to range from c. 25–30 to c. 80–100 nm s^{-1} (Kost *et al.*, 1998; Chen *et al.*, 2002; Hwang *et al.*, 2005).

Previous observations reveal a highly polarized, dynamic intracellular organization pattern in pollen tubes required for tip growth (Hepler *et al.*, 2001; Cheung and Wu, 2007). In general, the most proximal region (within 50 μm

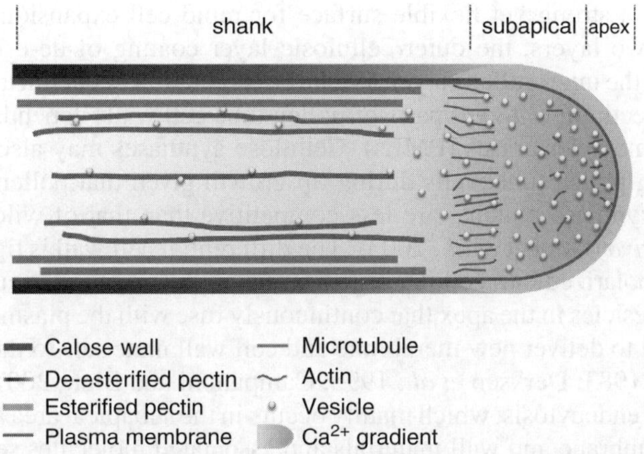

Figure 8.1 Polarized structural elements in pollen tubes. The orange shading represents the tip-focused (Ca2+) gradient. Exocytic vesicles are accumulated to a inverted cone-shaped domain at the apex of pollen tubes. Highly dynamic short F-actin is presented in the apical zone, and organized into a ring/collar structure in the subapical region. Longitudinal thick F-actin cables (red lines) and microtubule cables (green lines) exist in the shank region. The cell wall structure also varies in distinct zones. The apex contains a single soft esterified pectin apical wall, whereas the shank wall contain an outer cellulosic layer coating of de-esterified hard pectin and an inner callose layer. (See plate section for the color version of this figure.)

of the tip) of a growing pollen tube displays three distinct zones: the apex, the subapical zone, and the shank region (the latter was subdivided into a vacuolated zone and a nuclear zone by Qin and Yang, 2011; see also Hepler *et al.*, 2001; Cheung and Wu, 2007). It is quite noticeable that an abundance of exocytic vesicles are accumulated to an inverted cone-shaped domain at the apex for tip-targeted exocytosis. There is also a tip-focused cytoplasmic Ca^{2+} gradient in the apical zone. The subapical zone is enriched with secretory and biosynthetic organelles derived from an extensive endoplasmic reticulum (ER) in pollen tube. Long F-actin cables and microtubules are both prominent and aligned axially in the shank region, but are excluded from the apical and subapical zones. However, another form of F-actin structures has been reported to exist in the subapical region, which has been described as a ring, collar, mesh, or a fringe (Cheung and Wu, 2008). Additionally, a population of highly dynamic short F-actin bundles has also been revealed by fluorescent protein tagged actin markers, such as GFP-talin (Fu *et al.*, 2001; Hwang *et al.*, 2005), and GFP-NtADF1 (Chen *et al.*, 2002). Schematic diagrams of internal zonation and structural elements are included in Figure 8.1 and previous reviews (Cheung and Wu, 2008; Qin and Yang, 2011).

Interestingly, the cell wall structure of pollen tubes varies in distinct zones. The apex contains a single soft esterified pectin apical wall, which is assumed to provide a strong yet flexible surface for rapid cell expansion. The shank wall has two layers, the outer cellulosic layer coating of de-esterified hard pectin and the inner callose layer (Fig. 8.1). Regulation of chemical differences between pectin in different parts of pollen tube cell walls depends on a group of pectin methylesterases (PMEs). Cellulose synthases may also play a role in building pollen tube walls during tip growth given that pollen tubes from cellulose synthase mutants are less competitive than that of wild type when grown *in vivo* (Goubet *et al.*, 2003). The differential cell wall is tightly associated with polarized intracellular organization of pollen tubes. Accumulation of exocytic vesicles in the apex that continuously fuse with the plasma membrane is required to deliver new membrane and cell wall materials to the tip (Picton and Steer, 1983; Derksen *et al.*, 1995; Campanoni and Blatt, 2007). This is in contrast to endocytosis, which mainly occurs in the subapical area and retrieves excess membrane and wall materials and associated molecules secreted from female tissues for recycling, storage, or degradation (Picton and Steer, 1983; Goldraij *et al.*, 2006; Kim *et al.*, 2006; Dhonukshe *et al.*, 2007; Moscatelli *et al.*, 2007). Practically, coordination of polarized exocytosis and endocytosis is a common mechanism in eukaryotes for establishing and maintaining cell polarity and is functionally related to proper organization and dynamics of the actin cytoskeleton (Eitzen, 2003; Yang, 2008).

The Cytoskeleton

Functions of Cytoskeletal Elements in Pollen Tubes

Pollen tubes contain both microtubules and F-actin. Microtubule cables exist in the shank of pollen tubes, but are excluded from the apex. In pollen tubes, the role of microtubules in movement of the male germ unit has been demonstrated (Åström *et al.*, 1995; Raudaskoski *et al.*, 2001). Microtubules are believed to serve as tracks for mitochondria and Golgi vesicles. Assisted by microtubule motors proteins (kinesins and dyneins), organelles move slowly and continuously along the microtubules (Romagnoli *et al.*, 2007). It has been proposed that microtubules may appropriately place organelles in the correct location (Cai and Cresti, 2006). Microtubules are a target of self-incompatibility (SI) signals and mediate SI-induced programmed cell death (Poulter *et al.*, 2008). Microtubules were recently suggested to be involved in deposition of callose synthase in distal regions of pollen tubes and synthesis of callose plugs (Cai, 2011; Cai *et al.*, 2011), because disruption of microtubules was shown to strongly inhibit formation of callose plugs (Laitiainen *et al.*, 2002). However, pharmacological treatment demonstrated that microtubules are not directly essential for pollen tube tip growth, but affect directionality (Joos *et al.*, 1995; Cai, 2011; Foissner *et al.*, 2002; Gossot and Geitmann, 2007). Because a decline in pollen tube length was observed after disruption of microtubules (Gossot and Geitmann, 2007), they may be indirectly involved in pollen tube tip growth. There is evidence that the integrity of microtubules depends on the configuration of F-actin (Gossot and Geitmann, 2007). A recent study showed that microtubule integrity is required for exo-endocytosis events in the pollen tube tip, although F-actin plays a key role in plasma membrane internalization. Depolymerization of microtubules in the apex and shank led to misallocation of the internalized plasma membrane and affected the rate of exocytosis in the central domain of the apical plasma membrane (Idilli *et al.*, 2013). The roles of microtubules in pollen tube growth need to be carefully characterized in the future.

F-actin is known to play a major role in pollen tube tip growth. In the shank of pollen tubes, axially aligned long F-actin cables (Fig. 8.1) provide the main tracks for transporting organelles and vesicles and contribute to cytoplasmic streaming (Staiger *et al.*, 1994; Cai and Cresti, 2009), which may rapidly bring exocytic vesicles to the tips. In the subapical zone, a group of shorter, thinner actin cables assemble into collar, ring, fringe, or funnel-like structures (Fig. 8.1). A lack of this highly dynamic F-actin structure leads to abnormal twisting tubes. A subapical actin fringe was recently correlated with tip growth velocity in pollen tubes (Dong *et al.*, 2012). It was proposed that the subapical F-actin structure could capture vesicles released from long cables in the shank region and transport them to the apex. In the apex, a population of fine F-actin has been demonstrated to promote accumulation of exocytic vesicles in an inverted cone-shaped pattern. Correct tip growth of pollen tubes requires proper organization of multiple forms of F-actin structures (Fig. 8.1).

Organizing F-actin in Pollen Tubes

Plant actin gene families often contain multiple genes that encode various actin isovariants. For instance, there are 8 actin genes in Arabidopsis that can be grouped into two major classes: vegetative and reproductive (McDowell *et al.*, 1996; Meagher *et al.*, 1999). Mutations in the vegetative gene *ACT7* caused defects in germination and root growth (Gilliland *et al.*, 2003), while mutations in *ACT2* induced shorter bulged root hairs (Ringli *et al.*, 2002). The reproductive genes *ACT1*, *ACT3*, and *ACT11* are strongly expressed in mature pollen and growth pollen tubes, yet no consequences of mutations in these genes have been reported to date. However, pollen tube tip growth can be inhibited by applying low concentrations of actin depolymerizing drugs (Gibbon *et al.*, 1999; Fu *et al.*, 2001), suggesting that actin polymerization is essential for forward extension of pollen tubes (Vidali *et al.*, 2001). Additionally, treatment with the actin-stabilizing drug jasplakinolide implied that actin polymerization promotes reverse streaming in lily pollen tubes (Cardenas *et al.*, 2005).

As in other eukaryote cells, F-actin organization and dynamics in plant cells are regulated by a large group of actin-binding proteins (ABPs), which physically bind to actin monomers or filaments and organize the actin cytoskeleton (Staiger, 2000; Ren and Xiang, 2007; Chen *et al.*, 2009; see also Chapter 5). Some ABPs bind to actin monomers to regulate the size and activity of the actin subunit pool, and some act as nucleation factors to promote actin assembly. Some ABPs can sever filaments to generate new ends for assembly or disassembly and other proteins cap the ends of F-actin to prevent subunit loss and addition. There are also ABPs that bind to the side of actin filaments, stabilizing and/or bundling and cross-linking adjacent F-actin to form higher-order structures (Hussey *et al.*, 2006; Staiger *et al.*, 2010). To date, the functions of several major ABP families – including profilin, CAP (adenylate cyclase-associated protein), ADF, CP (capping protein), fimbrins, villins, formins, and LIM domain proteins – in pollen tubes have been well documented in several excellent reviews (Ren and Xiang, 2007; Chen *et al.*, 2009; Staiger *et al.*, 2010). Herein we highlight the ABPs that exert a distinct function on organizing various F-actin structures in different pollen tube zones (Fig. 8.2).

Formins are F-actin nucleation factors that contain a conserved proline-rich formin-homology 1 (FH1) domain and a FH2 domain. Using an RNAi approach to knock down the Arabidopsis pollen-specific formin AtFH3, Ye *et al.* (2009) demonstrated that AtFH3 is required for generation of

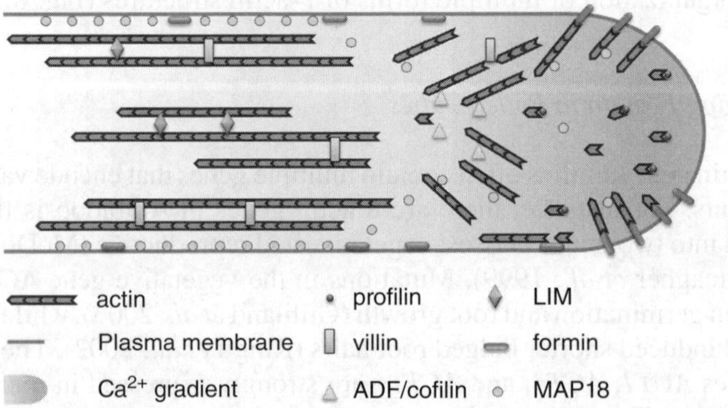

Figure 8.2 Major actin-binding proteins (ABPs) in pollen tubes. Formins are F-actin nucleation factors; membrane-anchored formins are able to generate F-actin from the plasma membrane of pollen tubes. LIMs belong to a class of actin-bundling proteins and have the potential role to regulate F-actin organization and dynamics through binding and bundling F-actin. ADFs/cofilins depolymerize actin filaments from their slow-growing end and promote actin turnover. In the pollen tube apex, the tip-focused Ca2+ gradient increases the activities of various ABPs, including villin/gelsolin/fragmin superfamily proteins, profilin, and MAP18, to regulate actin organization and dynamics.

long F-actin cables in the shank that support cytoplasmic streaming and tip growth. Interestingly, overexpression of another Arabidopsis formin, AtFH1, in pollen tubes also induced formation of actin cables (Cheung and Wu, 2004). The authors also found that the N-terminal domains target AtFH1 to the cell membrane, and stimulate actin assembly along the pollen tube plasma membrane. However, AtFH1-induced supernumerary actin cables are more evident in apical and subapical regions (Cheung and Wu, 2004). AtFH5 is another transmembrane formin. GFP-tagged AtFH5 displayed an apical localization pattern in pollen tubes; FH5-GFP-labeled vesicles accumulated in the clear zone and were also targeted to the apical and subapical membrane of pollen tubes. AtFH5 has been proposed to stimulate actin assembly from the subapical membrane and also plays an important role in establishing subapical actin structures and apical vesicular organization in pollen tube tip growth (Cheung *et al.*, 2010).

Actin-depolymerizing factors (ADFs) are actin-binding proteins that can bind to both actin monomer and F-actin. ADFs perform complex actions on actin organization. There is evidence suggesting that ADFs are recruited to older actin filaments and increase the dissociation rate of monomers from the minus end (Carlier *et al.*, 1997). Recently, Andrianantoandro and Pollard (2006) demonstrated the severing activity of ADF to disassemble F-actin. They also showed the capacity of ADF to nucleate actin when present at high concentrations (Andrianantoandro and Pollard, 2006). In lily and tobacco pollen tubes, ADF was found to decorate F-actin and accumulate in the cortex of the subapical region. It was proposed to play a central role in regulating subapical actin fringe turnover in a pH- dependent manner by enhancing actin polymerization at an alkaline pH and destabilizing F-actin at neutral or acidic pH conditions (Chen *et al.*, 2003; Lovy-Wheeler *et al.*, 2006; Wilsen et al, 2006).

Plant villins are a group of ABPs first identified from lily pollen. Lily 135-ABP, 110-ABP, and 115-ABP bundle F-actin in a calcium- and calmodulin- sensitive manner (Yokota and Shimmen Ki, 1998; Yokota *et al.*, 2000; Yokota *et al.*, 2003). Immunolocalization analysis revealed that villins localize along actin cables in lily pollen tubes (Vidali *et al.*, 1999; Yokota *et al.*, 2003) and may be involved in bundling F-actin in the shank of pollen tubes. In Arabidopsis, there are 5 full-length *VILLIN* genes. AtVLN1 simply bundles F-actin, and its activity is not dependent on Ca^{2+} or calmodulin. AtVLN3 displays Ca^{2+} activated F-actin severing activity *in vitro* however, contributing to actin bundle formation and turnover (Khurana *et al.*, 2010). AtVLN5 is highly and preferentially expressed in pollen and exhibits a full range of villin activities, including barbed-end capping, bundling, and Ca^{2+}-dependent severing *in vitro*. Although loss-of-function of AtVLN5 did not cause obvious differences in F-actin intensity and organization between wild-type and mutant pollen tubes, the growth of *vln5* pollen tubes and the actin cytoskeleton were more sensitive to the actin disruption drug LatB in mutant pollen tubes (Zhang *et al.*, 2010). Recently, villin-related polypeptides such as ABP29 and ABP41 were identified from lily. They are likely splice variants of villins or proteolytically processed villin isoforms. Interestingly, ABP41 appears to accumulate in the tip rather than localize to actin cables. Overexpression of villin-related polypeptides led to disruption of actin cables, organization of the apical zone in pollen tubes, and tip growth (Fan *et al.*, 2004; Xiang *et al.*, 2007; Wang *et al.*, 2008b).

Another class of actin-bundling proteins, LIMs, has the potential to regulate actin organization and dynamics in pollen. GFP-fused lily LIM associates with various actin structures, such as shank cables, subapical actin fringe, and apical fine F-actin. Similarly, GFP-fused tobacco LIM, NtPLIM2b, decorates both axial actin cables and subapical F-actin structures. LIMs likely regulate various actin arrays in pollen tubes through binding and bundling F-actin, which have been demonstrated *in vitro* (Cheung *et al.*, 2008; Wang *et al.*, 2008a).

In addition to well-known ABPs that are conserved in eukaryotic cells, novel ABPs have recently been identified and several are specific to plants. Hussey *et al.* (2006) reported a superfamily of ABPs at actin-membrane interaction sites in Arabidopsis, called NETs (networked). Thirteen NET proteins from the family share a novel NAB (NET actin-binding) domain. NET2A is specifically expressed in pollen. Interestingly, native promoter-driven NET2A-GFP associates with the plasma membrane as punctate dots, but is absent from the apical plasma membrane in pollen tubes (Deeks *et al.*, 2012). How this protein regulates the behavior of the actin network in pollen tubes remains a mystery.

Several proteins originally characterized as microtubule-associated proteins (MAPs) were recently reported to bind and regulate F-actin organization and dynamics in pollen. *Solanum Berthaultii* SB401 binds and bundles F-actin *in vitro* (Huang *et al.*, 2007). Although it still primarily decorates cortical microtubules in pollen tubes, it may facilitate coordination of microtubules and F-actin during pollen tube tip growth. In comparison, MAP18 has a clear and important role in regulating apical F-actin and growth direction of pollen tubes. MAP18 exhibits Ca^{2+}-dependent F-actin-severing activity *in vitro*. *In vivo*, it localizes to the flank plasma membrane of growing pollen tubes and appears at the apical plasma membrane when pollen tubes stop growing. Abnormal expression of MAP18 in knockdown mutant *map18* and the overexpression line *MAP18 OX* led to disorganization of apical F-actin and subsequent aberrant pollen tube growth patterns and inaccurate micropyle targeting. Site mutations that abolished F-actin-severing activity demonstrated that MAP18 regulates the pollen tube growth direction by modulating apical F-actin (Zhu *et al.*, 2013). Given the role of MAP18 in regulating cortical microtubule organization and polar cell growth as a microtubule destabilizing factor in vegetative tissues (Wang *et al.*, 2007), MAP18 is likely a bifunctional protein and its action on F-actin or microtubules is precisely and specifically regulated in different cell types.

An increasing amount of evidence supports direct physical interactions between F-actin and microtubules involving proteins, such as motor protein kinesins, formins, phospholipase D delta, and more. Integrating the functions of F-actin and microtubules is important for intracellular transport and formation of the mitotic and cytokinetic apparatus, which are essential for polar cell growth and cell division (see reviews by Collings, 2008; Petrasek and Schwarzerova, 2009). Although similar studies in pollen tubes are lacking, identifying new proteins involved in F-actin and microtubule interactions, and illustrating the roles of these proteins in regulating cytoskeletal elements and underlying regulatory mechanisms, will provide new insights into our current understanding of plant cell growth and development, including pollen tube tip growth.

ROP GTPases in Pollen Tubes

Previous studies have demonstrated that pollen tubes grow at uneven speeds, exhibiting a pronounced oscillatory growth pattern with alternate rapid and slow growth phases (Feijo *et al.*, 2001; Holdaway-Clarke and Hepler, 2003; Iwano *et al.*, 2004). This is attributable to precisely regulated spatial and temporal coordination of many cellular events, including ion influxes, organization and dynamics of F-actin, vesicular trafficking, and wall synthesis, assembly, and remodeling (Franklin-Tong, 1999b; Yang, 2002). Regulation of the actin cytoskeleton and its interface with the ROP (Rho-like GTPase from plants) signaling network are fundamental mechanisms underlying pollen tube tip growth.

Similar to other small GTPases in the Rho family, ROP GTPases are important signaling molecules. ROPs cycle between GTP-bound and GDP-bound states to turn 'ON' or 'OFF' and downstream signaling pathways regulate and coordinate a variety of cellular processes, such as cytoskeleton organization and dynamics and membrane trafficking (Fig. 8.3; Etienne-Manneville and Hall, 2002; Vernoud *et al.*, 2003; Mucha *et al.*, 2011). To date, ROP subfamily members have been widely identified in lower plant moss, gymnosperms, and angiosperms (Winge *et al.*, 2000; Brembu *et al.*, 2006), including 11 ROPs in Arabidopsis, 7 in rice, and 9 in maize (Li *et al.*, 1998; Winge *et al.*, 2000; Christensen *et al.*, 2003). Some ROPs are specifically or widely expressed in pollens and pollen tubes, and are essential for pollen development and pollen tube growth (Table 8.1). The first indication that ROPs may be involved in pollen tube tip growth was apical plasma membrane localization of pea ROP shown by immunostaining (Lin *et al.*, 1996). Microinjection of anti-ROP1Ps led to rapid arrest of pea pollen tube growth (Lin and Yang, 1997). Arabidopsis ROP1, ROP3, and ROP5 are closely related ROPs expressed in pollen (Li *et al.*, 1998). Overexpression of wild-type ROP1 or

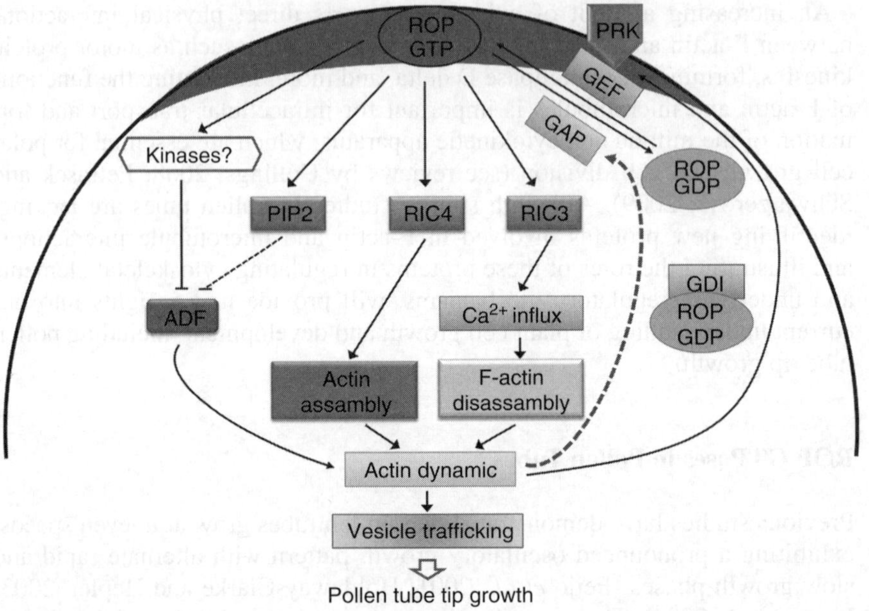

Figure 8.3 ROP GTPase signaling to F-actin during pollen tube growth. Rho-related GTPase from plants (ROP) is locally activated in the PM to determine the site of exocytosis and growth direction. The activity of ROP is regulated by guanine nucleotide exchange factors (GEFs), GTPase activating proteins (GAPs), and guanine nucleotide dissociation inhibitors (GDIs). Pollen-specific receptor kinase (PRK) locates at the plasma membrane which activates RhoGEFs. Active ROP activates two counteracting pathways mediated by RIC4 and RIC3. RIC4 promotes the assembly of tip F-actin and RIC3 leads to Ca2+ influx and promotes the disassembly of the tip F-actin. The dynamic of F-actin in the apex is critical for pollen tube tip growth through proper regulated vesicle trafficking. It has been proposed that ROP might regulate ADF's function on F-actin through phosphorylation. In addition, a link between ROP and ADF may also be mediated by PIP2. In turn, F-actin dynamics is involved in the feedback regulation of ROP activities.

ROP5, or their constitutively active mutants induced depolarized pollen tube growth (Kost *et al.*, 1999; Li *et al.*, 1999; Fu *et al.*, 2001). Hwang *et al.* demonstrated that apical ROP1 activity oscillates with growth of the pollen tube and that disruption of the dynamics of ROP1 activity causes growth depolarization (Hwang *et al.*, 2005, 2008). ROP8, ROP9, ROP10, and ROP11 are also expressed in pollen (Li *et al.*, 1998; Gu *et al.*, 2004). Overexpression of ROP9/AtRac7 in pollen tubes causes diverse growth defects and reduced seed set (Cheung *et al.*, 2003). Interestingly, GFP-AtRac7 exhibited strong apical membrane localization patterns only in growth-arrested pollen tubes, but was excluded from the tip membrane of growing pollen tubes (Cheung *et al.*, 2003). Recently, Li *et al.* reported that ROP9 was targeted to the invaginated pollen tube plasma membrane by its C-terminal hypervariable domain

Table 8.1 ROPs functioning in pollen and pollen tube.

Protein name	Accession no.	Species	Function	Reference(s)
ROP1/AtRac11	AT3G51300	*Arabidopsis thaliana*	Localizes to the apical plasma membrane and regulates F-actin dynamics and Ca^{2+} gradient at the tip; expression highest in pollen tubes and mature pollen	Kost *et al.*, 1999; Li *et al.*, 1999; Fu *et al.*, 2001
ROP3/ARac1/ AtGP2	AT2G17800	*A. thaliana*	Functionally redundant to ROP1; expression highest in pollen.	Li *et al.*, 1998
ROP5/AtRAC6/ AtRac2	AT4G35950	*A. thaliana*	Functionally redundant to ROP1; expression highest in pollen.	Li *et al.*, 1998
ROP9/AtRac7	AT4G28950	*A. thaliana*	Involves in intracellular organization of pollen tube and seed set; expression in pollen and vegetative organs	Cheung *et al.*, 2003
NtRac1	AY029330	*Nicotiana tabacum*	Modulates NtADF1 activity through phosphorylation at Ser-6 to regulate actin dynamics.	Chen *et al.*, 2003
NtRac5	AJ250174	*N. tabacum*	Maintains pollen tube polarity	Klahre and Kost, 2006
ZmROP2	AAD34356	*Zea mays*	Involved in male gametophyte development	Arthur *et al.*, 2003
ZmROP9	AAO41289	*Z. mays*	Highly conserved duplicate gene of ZmROP2	Arthur *et al.*, 2003
LLP-ROP1	EF140700	*Lilium longiflorum Thunb.*	Overexpression of LLP-ROP1 inhibits pollen tube elongation and forms a ballooned tip.	Hsu *et al.*, 2010
ROP1Ps	L19093	*Pisum sativum Linn*	Controls the actin-dependent pollen tubes tip growth; involves in the movement of the generative cell; expressed predominantly in the pollen and pollen tubes	Yang and Watson, 1993; Lin *et al.*, 1996; Lin and Yang, 1997

(Li *et al.*, 2013). These distinct localization patterns suggest a different function of ROP9 in pollen tubes compared to ROP1, ROP3, and ROP5. It has been proposed that ROP8 through ROP11 may participate in sensing signals from the pistil to guide pollen tubes toward the ovule or releasing sperm from the pollen tube (Gu *et al.*, 2004). NtRac5 is a tobacco (*Nicotiana tabacum*) ROP GTPase specifically expressed in pollen tubes and is closely related to ROP1 and ROP5 in Arabidopsis, and ROP1Ps in peas. It is believed that spatial restriction of NtRac5 to pollen tube apex is important for sustaining tip growth (Kieffer *et al.*, 2000; Klahre and Kost, 2006). Overexpression of another tobacco pollen-specific ROP, NtRac1, also induced depolarized pollen tube (Chen *et al.*, 2003), suggested multiple tobacco ROPs are involved in regulating pollen tube tip growth. In maize, 3 out of 9 ZmROPs – ZmROP2, ZmROP8, and ZmROP9 – are highly expressed in mature pollen, and ZmROP2 and ZmROP9 are in the same phylogenetic group as Arabidopsis ROP1 and ROP5 (Christensen *et al.*, 2003). Arthur *et al.* (2003) investigated the *in vivo* function of ZmROP2 by analyzing the transmission frequency of *Zmrop2::Mu* mutant alleles, and found that ZmROP2 provides a competitive advantage to male gametophytes. Recently, a lily ROP, LLP-ROP1, was reported to respond to desiccation-associated ABA signaling during pollen maturation. The authors also found that overexpression of LLP-ROP1 in pollen tubes causes depolarized growth (Hsu *et al.*, 2010).

All of these observations consistently reveal the importance of ROPs in controlling pollen tube growth in various plant species.

ROPs regulate F-actin Organziation and Dynamics

Rho GTPases in eukaryotic kingdoms share a conserved function in regulating the actin cytoskeleton, which serves as a unifying mechanism in regulating cell polarity and cell morphogenesis. In plants, ROP signaling to F-actin is critical to polar cell growth in different cell types, including tip growth in pollen tubes (Cheung and Wu, 2007; Cole and Fowler, 2006). Specifically, ROPs regulate the dynamics of a group of fine F-actin associated with the polar site of cell growth (Fu *et al.*, 2001; Jones *et al.*, 2002; Gu *et al.*, 2003, 2005; Lee *et al.*, 2008).

Plants adopt a unique mechanism of ROP regulation of F-actin in pollen tubes. Members from a plant-specific family of ROP effectors, ROP-interactive CRIB motif-containing proteins (RICs), mediate ROP1 signaling to F-actin (Fu *et al.*, 2001; Wu *et al.*, 2001; Gu *et al.*, 2005; Hwang *et al.*, 2005). During tip growth, RIC4 is activated by ROP1 to promote assembly of apical F-actin. Active ROP1 also activates RIC3, which promotes depolymerization of F-actin by regulating the Ca^{2+} gradient in the tip region (Fig. 8.3). These two counteracting ROP1 signaling pathways coordinate F-actin dynamics and tip growth (Gu *et al.*, 2005). Research by Lee *et al.* provides insight into the mechanism by which ROP1 regulates F-actin-dependent tip growth. The authors demonstrated that ROP1-regulated F-actin dynamics in the apex are essential for polarized exocytosis. ROP1-RIC4 promotes assembly of apical F-actin, which facilitates accumulation of exocytic vesicles at the apex. In contrast, ROP1-RIC3-Ca^{2+} promotes disassembly of apical F-actin, which is required for docking and/or fusion of exocytic vesicles for delivery of new plasma membrane and cell wall components (Lee *et al.*, 2008; Lee and Yang, 2008).

A phosphatidylinositol monophosphate kinase (PIPK) was found to physically interact with ROPs. Its product, phosphatidylinositol 4,5-bisphosphate (PI4,5-P2 or PIP2), co-localizes with ROPs at the apical plasma membrane (Kost *et al.*, 1999). Because PIP2 has been reported to inhibit a lily LlADF1 binding to F-actin (Allwood *et al.*, 2002), it could act in ROP signaling pathways to regulate ADF. Tobacco NtADF1 has also been shown to play an important role in maintaining normal F-actin organization and polar growth in pollen tubes and overexpression of NtADF1 was able to suppress the NtRac1-induced phenotype, which indicated a functional interaction between NtADF1 and NtRac1. It was proposed that NtRac1 might regulate the activity of NtADF1 through phosphorylation of NtADF1, since overexpression of constitutively active NtRac1 dramatically increased the ratio of phosphorylated to non-phosphorylated NtADF1 in pollen grains (Chen *et al.*, 2003).

Many ABPs have shown calcium-dependent activities in regulating actin assembly and dynamics. Given that ROP1 regulates oscillation of the apical Ca^{2+} gradient, ROPs probably also have an impact on these ABPs to regulate the actin cytoskeleton and pollen tube growth.

Regulation of ROP Activity in Pollen Tubes

Upstream Regulators of ROP GTPases

As a binary molecular switch that transmits extracellular signals to intracellular biological events, rapid conversion of ROPs between GTP-bound form and GDP-bound form makes them ideal regulators of cellular processes that require quick responses. To interpret different signals, there are many positive and negative regulatory proteins for small GTPases in eukaryotic genomes. Three major classes of regulatory proteins are GEFs (guanine nucleotide exchange factors), GAPs (GTPase-activating proteins), and GDIs (guanine nucleotide dissociation inhibitors) (DerMardirossian and Bokoch, 2005; Siderovski and Willard, 2005; Fu *et al.*, 2008; Nagawa *et al.*, 2010). They all play important roles in regulating ROP activities and functions in pollen tubes (Fig. 8.3).

Several ROPs exhibit polar localization to the apical plasma membrane. Recruitment and activation of ROP1 to the apical plasma membrane are critical in establishing the tip growth domain. Distribution of active ROP1 is correlated with pollen tube growth oscillation (Hwang *et al.*, 2005). Growth depolarization induced by overexpression of *CA-ROP1* or wild-type *ROP1* is associated with dramatically increased active ROP1 in the membrane. During pollen tube tip growth, active ROP1 in the apical tip oscillates and is dynamically self-organized to maintain a proper size, which involves several GEFs, GAPs, and GDIs (Hwang *et al.*, 2010; Qin and Yang, 2011).

GEFs convert GDP-bound inactive form of ROPs into GTP-bound active form. KPP is a tomato homolog to RopGEFs, which interacts with the receptor-like kinase LePRK1 expressed in pollen (Kaothien *et al.*, 2005). Overexpression of KPP induced the same depolarization growth in pollen tube as ROP1 overexpression (Li *et al.*, 1999; Kaothien *et al.*, 2005). Arabidopsis AtRopGEF1 belongs to the plant-specific RopGEF family that localizes to the apical plasma membrane corresponding to the site of active ROP1. Overexpression of AtRopGEF1 caused severe depolarization of pollen tubes. However, this induction can be suppressed by coexpression of RopGAP1 or DN-rop1, suggesting that AtRopGEF1 acts as a positive regulator of ROP1 in controlling pollen tube tip growth. An autoinhibitory mechanism regulates AtRopGEF1 activity (Hwang *et al.*, 2005; Gu *et al.*, 2006). This indicates that an upstream activator, most likely a receptor-like kinase, is needed to release autoinhibition of AtRopGEF1 activity. AtRopGEF12 is a pollen-specific RopGEF that interacts with AtPRK2a through its C terminus in pollen tubes. Co-expression of AtROPGEF12 and AtPRK2 induced pollen tube isotropic growth, which was significantly more pronounced than expression of AtRopGEF12 alone. AtPRK2a has been proposed to release autoinhibition of the catalytic PRONE (plant-specific Rop nucleotide exchanger) domain by the C termini of AtRopGEF12 (Gu *et al.*, 2006; Zhang and McCormick, 2007). RopGEFs and RLKs may therefore form a plant-specific signaling module to respond to specific extracellular signals (Yang and Fu, 2007).

RopGAPs and RopGDIs are negative regulators of ROPs (Wu *et al.*, 2000; Brembu *et al.*, 2006). Both are needed to restrict ROP activity to the apical plasma membrane. RopGAPs inactivate ROPs by promoting GTP hydrolysis. In addition to the conserved catalytic GAP domain, RopGAPs contain a CDC42/Rac–interactive binding (CRIB) motif, which is absent from animal and fungal counterparts. Tobacco NtRopGAP1 localizes to the flanks of the pollen tube tip, suppresses the NtRac5 overexpression phenotype, and may restrict NtRac5 activity to the apex (Klahre and Kost, 2006). Arabidopsis RopGAP1 has been shown to regulate ROP1 activity in pollen tubes (Wu *et al.*, 2000; Fu *et al.*, 2001; Hwang *et al.*, 2005, 2010;). Hwang *et al.* demonstrated that global inhibition of ROP1 activity by RopGAP1 and RhoGDI1 could balance lateral propagation of apical ROP1 activation induced by overexpressing

ROP1 (Hwang *et al.*, 2010). Transcription analysis and transient expression experiments in tobacco pollen tubes suggested that several other RopGAPs and RhoGDIs, including RhoGDI2a, RhoGDI2b, and RopGAP3, may regulate ROPs in pollen tubes (Hwang *et al.*, 2010). Interestingly, knocking out pollen-expressed RopGAP only slightly affected pollen tube growth polarity in Arabidopsis (Klahre and Kost, 2006; Hwang *et al.*, 2010), suggesting that other regulators may play major roles in limiting ROP signaling to the apex. A novel pleckstrin-homology (PH) domain-containing RhoGAP of Arabidopsis, REN1 (ROP enhancer1), is localized to the apical plasma membrane and exocytic vesicles; the association of REN1 with exocytic vesicles is critical for its function in restricting ROP1 activity to the pollen tube tip (Hwang *et al.*, 2008). Global inhibition of REN1 facilitates self-organization of ROP signaling in pollen tubes.

RhoGDIs are known to sequester ROPs in the cytosol to prevent them from being activated by RhoGEFs on the membrane (Klahre and Kost, 2006; Fu *et al.*, 2008; Lee and Yang, 2008). Interestingly, the tobacco GDI NtRhoGDI2 is required for proper localization of NtRac5. When binding to NtRhoGDI2 was inhibited, NtRac5 was mis-localized to the flanks of pollen tubes. This suggests that NtRhoGDI2 has a role in recycling NtRac5 to the apex to maintain the polarity of ROP signaling and pollen tube growth (Klahre and Kost, 2006).

F-actin Mediated Feedback Regulation of ROP Signaling

F-actin is a downstream target of ROP signaling to control pollen tube tip growth. Lee *et al.* provided evidence indicating that tip-targeted exocytosis is regulated by F-actin dynamics, which coordinates vesicle accumulation and docking/fusion. CA-rop1-induced growth depolarization is associated with depolarization of exocytosis. DN-rop1 expression can inhibit exocytosis (Lee *et al.*, 2008). Interestingly, ROP-dependent spatial and temporal regulation of exocytosis provides feedback regulation of ROP activity (Lee *et al.*, 2008). The upstream regulator of ROP1, REN1 RhoGAP, is localized to exocytic vesicles and the apical plasma membrane, and its accumulation in the pollen-tube apex appears to be dependent on F-actin and vesicle trafficking. The proper localization of REN1 is required for the control of negative feedback global inhibition of ROP1 and spatial restriction of ROP signaling to the growth site and its temporal oscillation during tip growth (Hwang *et al.*, 2008). Low concentrations of LatB (which only affects apical F-actin) not only disrupted accumulation of GFP-REN1 at the apex, but also induced severe growth defects in *ren1-1 Lat52:GFP-REN1* pollen tubes (Hwang *et al.*, 2008), implying that F-actin contributes to exocytosis-based negative feedback regulation of ROP1 activity.

Interestingly, apical F-actin oscillates in the same phase with ROP1 activity, and stabilizing these fine F-actin led to enhanced ROP1 activity at the pollen tube apex (Gu *et al.*, 2005; Hwang *et al.*, 2005; Lee *et al.*, 2008). F-actin is therefore also likely to be involved in the positive feedback regulation of ROP1 activity. It is proposed that F-actin-dependent exocytosis is also responsible for targeting upstream positive regulators of ROP1, such as those receptor-like kinases which can activate RopGEFs to the pollen tube tip (Kaothien *et al.*, 2005; Gu *et al.*, 2006; Zhang and McCormick, 2007; Qin and Yang, 2011).

Future Prospects

Using pollen tubes as a model system, researchers have made significant progress in studying polar cell growth and cell morphogenesis. One of the most important breakthroughs is elucidation of the ROP signaling network for controlling pollen tube tip growth, in which players involved in the cytoskeleton and vesicular trafficking form positive and negative feedback regulatory loops. In fact, it has been suggested that a similar signaling paradigm is adopted by various cell types to control the establishment and maintenance of cell polarity. In addition to ROP signaling pathways, other signaling pathways, such as Ca^{2+}, phosphoinositides, and reactive oxygen species signaling pathways, have been implicated in pollen tube growth (Potocky *et al.*, 2007; Iwano *et al.*, 2009; Yan *et al.*, 2009; Zhao *et al.*, 2010; Qin and Yang, 2011). Recent evidence suggests that different signaling pathways are integrated into a complex network to control F-actin dynamics and subsequently tip growth.

Despite new insights into the mechanisms underlying pollen tube tip growth, many questions remain to be answered. For example, what are the key 'nodes' for integration of various signaling pathways to regulate F-actin dynamics? How do these important integrators decode or interpret different internal or external cues? Due to the complexity of pollen tube tip growth regulation, we believe that more new components will be discovered. For instance, ROP1 promotes F-actin assembly through RIC4; however, there is no evidence that RIC4 directly interacts with F-actin. An ABP might therefore be a target of the ROP1-RIC4 pathway to regulate F-actin polymerization. Interestingly, several known MAPs regulate F-actin organization in pollen tubes. Are they merely multifunctional proteins that act on either microtubules or F-actin in a tissue/cell-specific manner or do they link two cytoskeleton networks and coordinate their function in pollen tubes? Is ROP1 signaling involved in the coordination of microtubules and F-actin networks?

These are interesting questions to further understanding of the role of microtubules and their interactions with F-actin in pollen and pollen tubes. To date, the advances in understanding signaling mechanisms underlying tip growth mostly originate from research performed on cultured pollen tubes *in vitro*. However, whether these same mechanisms are employed *in vivo* in the pollen tube response to signals from female tissues/cells to direct their growth is also of great importance. Despite difficulties with the technique, there is an increasing effort to elucidate the process *in vivo* when pollen tubes grow toward ovules. Several interesting questions have been raised. Are there receptors for pollen tube guidance signals? Does ROP1 regulate the F-actin-dependent changing direction of pollen tube growth in the pistil? How is ROP1 activated? Answering these challenging questions will provide new insights into understanding regulation of pollen tube tip growth and sexual reproduction in plants.

Acknowledgements

This work was supported by grants from the Natural Science Foundation of China (90817105 to YF and 31371352 to LZ).

References

Allwood, E.G., Anthony, R.G., Smertenko, A.P., Reichelt, S., Drobak, B.K., Doonan, J.H., Weeds, A.G., and Hussey, P.J. (2002) Regulation of the pollen-specific actin-depolymerizing factor LIADF1. *Plant Cell*, **14**, 2915–2927.

Andrianantoandro, E. and Pollard, T.D. (2006) Mechanism of actin filament turnover by severing and nucleation at different concentrations of ADF/cofilin. *Molecular Cell*, **24**, 13–23.

Arthur, K.M., Vejlupkova, Z., Meeley, R.B., and Fowler, J.E. (2003) Maize ROP2 GTPase provides a competitive advantage to the male gametophyte. *Genetics*, **165**, 2137–2151.

Åström, H., Sorri, O., and Raudaskoski, M. (1995) Role of microtubules in the movement of the vegetative nucleus and generative cell in tobacco pollen tubes. *Sexual Plant Reproduction*, **8**, 61–69.

Barnabas, B. and Fridvalszky, L. (1984) Adhesion and germination of differently treated maize pollen grains on the stigma. *Acta Botanica Hungarica*, **30**, 329–332.

Brembu, T., Winge, P., Bones, A.M., and Yang, Z. (2006) A RHOse by any other name: a comparative analysis of animal and plant Rho GTPases. *Cell Research*, **16**, 435–445.

Brennwald, P. and Rossi, G. (2007) Spatial regulation of exocytosis and cell polarity: yeast as a model for animal cells. *FEBS Letters*, **581**, 2119–2124.

Cai, G. (2011) How do microtubules affect deposition of cell wall polysaccharides in the pollen tube? *Plant Signal Behavior*, **6**, 732–735.

Cai, G. and Cresti, M. (2006) The microtubular cytoskeleton in pollen tubes: structure and role in organelle trafficking, in *The Pollen Tube* (ed. R. Malhó), Springer, pp. 157–175.

Cai, G. and Cresti, M. (2009) Organelle motility in the pollen tube: a tale of 20 years. *Journal of Experimental Botany*, **60**, 495508.

Cai, G., Faleri, C., Del Casino, C., Emons, A.M., and Cresti, M. (2011) Distribution of callose synthase, cellulose synthase, and sucrose synthase in tobacco pollen tube is controlled in dissimilar ways by actin filaments and microtubules. *Plant Physiology*, **155**, 1169–1190.

Campanoni, P. and Blatt, M.R. (2007) Membrane trafficking and polar growth in root hairs and pollen tubes. *Journal of Experimental Botany*, **58**, 65–74.

Cardenas, L., Lovy-Wheeler, A., Wilsen, K.L., and Hepler, P.K. (2005) Actin polymerization promotes the reversal of streaming in the apex of pollen tubes. *Cell Motility and the Cytoskeleton*, **61**, 112–127.

Carlier, M.F., Laurent, V., Santolini, J., Melki, R., Didry, D., Xia, G.X., Hong, Y., Chua, N.H., and Pantaloni, D. (1997) Actin depolymerizing factor (ADF/cofilin) enhances the rate of filament turnover: implication in actin-based motility. *Journal of Cell Biology*, **136**, 1307–1322.

Chen, C.Y., Wong, E.I., Vidali, L., Estavillo, A., Hepler, P.K., Wu, H.M., and Cheung, A.Y. (2002) The regulation of actin organization by actin-depolymerizing factor in elongating pollen tubes. *Plant Cell*, **14**, 2175–2190.

Chen, C.Y., Cheung, A.Y., and Wu, H.M. (2003) Actin-depolymerizing factor mediates Rac/Rop GTPase-regulated pollen tube growth. *Plant Cell*, **15**, 237–249.

Chen, T., Teng, N., Wu, X., Wang, Y., Tang, W., Šamaj, J., Baluška, F., and Lin, J. (2007) Disruption of actin filaments by latrunculin B affects cell wall construction in Picea meyeri pollen tube by disturbing vesicle trafficking. *Plant and Cell Physiology*, **48**, 19–30.

Chen, N., Qu, X., Wu, Y., and Huang, S. (2009) Regulation of actin dynamics in pollen tubes: control of actin polymer level. *Journal of Integrative Plant Biology*, **51**, 740–750.

Cheung, A.Y. and Wu, H.M. (2004) Overexpression of an Arabidopsis formin stimulates supernumerary actin cable formation from pollen tube cell membrane. *Plant Cell*, **16**, 257–269.

Cheung, A.Y. and Wu, H.M. (2007) Structural and functional compartmentalization in pollen tubes. *Journal of Experimental Botany*, **58**, 75–82.

Cheung, A.Y. and Wu, H.M. (2008) Structural and signaling networks for the polar cell growth machinery in pollen tubes. *Annual Review of Plant Biology*, **59**, 547–572.

Cheung, A.Y., Wang, H., and Wu, H.M. (1995) A floral transmitting tissue-specific glycoprotein attracts pollen tubes and stimulates their growth. *Cell*, **82**, 383–393.

Cheung, A.Y., Chen, C.Y., Tao, L.Z., Andreyeva, T., Twell, D., and Wu, H.M. (2003) Regulation of pollen tube growth by Rac-like GTPases. *Journal of Experimental Botany*, **54**, 73–81.

Cheung, A.Y., Duan, Q.H., Costa, S.S., de Graaf, B.H., Di Stilio, V.S., Feijo, J., and Wu, H.M. (2008) The dynamic pollen tube cytoskeleton: live cell studies using actin-binding and microtubule-binding reporter proteins. *Molecular Plant*, **1**, 686–702.

Cheung, A.Y., Niroomand, S., Zou, Y., and Wu, H.M. (2010) A transmembrane formin nucleates subapical actin assembly and controls tip-focused growth in pollen tubes. *Proceedings of National Academy of Science, USA*, **107**, 16390–16395.

Christensen, T.M., Vejlupkova, Z., Sharma, Y.K., Arthur, K.M., Spatafora, J.W., Albright, C.A., Meeley, R.B., Duvick, J.P., Quatrano, R.S., and Fowler, J.E. (2003) Conserved subgroups and developmental regulation in the monocot rop gene family. *Plant Physiology*, **133**, 1791–1808.

Cole, R.A. and Fowler, J.E. (2006) Polarized growth: maintaining focus on the tip. *Current Opinion in Plant Biology*, **9**, 579–588.

Collings, D.A. (2008) Crossed-wires: interactions and cross-talk between the microtubule and microfilament networks in plants, in *Plant Microtubules* (ed. P. Nick), Springer, 47–79.

Deeks, M.J., Calcutt, J.R., Ingle, E.K., Hawkins, T.J., Chapman, S., Richardson, A.C., Mentlak, D.A., Dixon, M.R., Cartwright, F., Smertenko, A.P., *et al.* (2012) A superfamily of actin-binding proteins at the actin-membrane nexus of higher plants. *Current Biology*, **22**, 1595–1600.

Derksen, J., Rutten, T., Lichtscheidl, I., De Win, A., Pierson, E., and Rongen, G. (1995) Quantitative analysis of the distribution of organelles in tobacco pollen tubes: implications for exocytosis and endocytosis. *Protoplasma*, **188**, 267–276.

DerMardirossian, C. and Bokoch, G.M. (2005) GDIs: central regulatory molecules in Rho GTPase activation. *Trends in Cell Biology*, **15**, 356–363.

Dhonukshe, P., Aniento, F., Hwang, I., Robinson, D.G., Mravec, J., Stierhof, Y.D., and Friml, J. (2007) Clathrin-mediated constitutive endocytosis of PIN auxin efflux carriers in Arabidopsis. *Current Biology*, **17**, 520–527.

Dong, H., Pei, W., and Haiyun, R. (2012) Actin fringe is correlated with tip growth velocity of pollen tubes. *Molecular Plant*, **5**, 1160–1162.

Dresselhaus, T. (2006) Cell-cell communication during double fertilization. *Current Opinion in Plant Biology*, **9**, 41–47.

Dresselhaus, T. and Franklin-Tong, N. (2013) Male-Female crosstalk during pollen germination, tube growth and guidance, and double fertilization. *Molecular Plant*, **6**, 1018–1036.

Eitzen, G. (2003) Actin remodeling to facilitate membrane fusion. *Biochimica et Biophysica Acta*, **1641**, 175–181.

Etienne-Manneville, S. and Hall, A. (2002) Rho GTPases in cell biology. *Nature*, **420**, 629–635.

Fan, X., Hou, J., Chen, X., Chaudhry, F., Staiger, C.J., and Ren, H. (2004) Identification and characterization of a Ca2+-dependent actin filament-severing protein from lily pollen. *Plant Physiology*, **136**, 3979–3989.

Feijo, J.A., Sainhas, J., Holdaway-Clarke, T., Cordeiro, M.S., Kunkel, J.G., and Hepler, P.K. (2001) Cellular oscillations and the regulation of growth: the pollen tube paradigm. *Bioessays*, **23**, 86–94.

Foissner, I., Grolig, F., and Obermeyer, G. (2002) Reversible protein phosphorylation regulates the dynamic organization of the pollen tube cytoskeleton: effects of calyculin A and okadaic acid. *Protoplasma*, **220**, 1–15.

Franklin-Tong, V.E. (1999a) Signaling in pollination. *Current Opinion in Plant Biology*, **2**, 490–495.

Franklin-Tong, V.E. (1999b) Signaling and the modulation of pollen tube growth. *Plant Cell*, **11**, 727–738.

Franklin-Tong, V.E. (2002) The difficult question of sex: the mating game. *Current Opinion in Plant Biology*, **5**, 14–18.

Fu, Y., Kawasaki, T., Shimamoto, K., and Yang, Z. (2008) ROP/RAC GTPases. *Annual Plant Reviews*, **33** (Intracellular Signaling in Plants), 64–99.

Fu, Y., Wu, G., and Yang, Z. (2001) Rop GTPase-dependent dynamics of tip-localized F-actin controls tip growth in pollen tubes. *Journal of Cell Biology*, **152**, 1019–1032.

Gibbon, B.C., Kovar, D.R., and Staiger, C.J. (1999) Latrunculin B has different effects on pollen germination and tube growth. *Plant Cell*, **11**, 2349–2363.

Gilliland, L.U., Pawloski, L.C., Kandasamy, M.K., and Meagher, R.B. (2003) Arabidopsis actin gene ACT7 plays an essential role in germination and root growth. *Plant Journal*, **33**, 319–328.

Goldraij, A., Kondo, K., Lee, C.B., Hancock, C.N., Sivaguru, M., Vazquez-Santana, S., Kim, S., Phillips, T.E., Cruz-Garcia, F., and McClure, B. (2006) Compartmentalization of S-RNase and HT-B degradation in self-incompatible Nicotiana. *Nature*, **439**, 805–810.

Gossot, O., and Geitmann, A. (2007) Pollen tube growth: coping with mechanical obstacles involves the cytoskeleton. *Planta*, **226**, 405–416.

Goubet, F., Misrahi, A., Park, S.K., Zhang, Z., Twell, D., and Dupree, P. (2003) AtCSLA7, a cellulose synthase-like putative glycosyltransferase, is important for pollen tube growth and embryogenesis in Arabidopsis. *Plant Physiology*, **131**, 547–557.

Gu, Y., Vernoud, V., Fu, Y., and Yang, Z. (2003) ROP GTPase regulation of pollen tube growth through the dynamics of tip-localized F-actin. *Journal of Experimental Botany*, **54**, 93–101.

Gu, Y., Wang, Z., and Yang, Z. (2004) ROP/RAC GTPase: an old new master regulator for plant signaling. *Current Opinion in Plant Biology*, **7**, 527–536.

Gu, Y., Fu, Y., Dowd, P., Li, S., Vernoud, V., Gilroy, S., and Yang, Z. (2005) A Rho family GTPase controls actin dynamics and tip growth via two counteracting downstream pathways in pollen tubes. *Journal of Cell Biology*, **169**, 127–138.

Gu, Y., Li, S., Lord, E.M., and Yang, Z. (2006) Members of a novel class of Arabidopsis Rho guanine nucleotide exchange factors control Rho GTPase-dependent polar growth. *Plant Cell*, **18**, 366–381.

Hall, A. (1998) Rho GTPases and the actin cytoskeleton. *Science*, **279**, 509–514.

Hepler, P.K., Vidali, L., and Cheung, A.Y. (2001) Polarized cell growth in higher plants. *Annual Review of Cell Development and Biology*, **17**, 159–187.

Holdaway-Clarke, T.L. and Hepler, P.K. (2003) Control of pollen tube growth: role of ion gradients and fluxes. *New Phytologist*, **159**, 539–563.

Hsu, S.W., Cheng, C.L., Tzen, T.C., and Wang, C.S. (2010) Rop GTPase and its target Cdc42/Rac-interactive-binding motif-containing protein genes respond to desiccation during pollen maturation. *Plant Cell Physiology*, **51**, 1197–1209.

Huang, S., Jin, L., Du, J., Li, H., Zhao, Q., Ou, G., Ao, G., and Yuan, M. (2007) SB401, a pollen-specific protein from Solanum berthaultii, binds to and bundles microtubules and F-actin. *The Plant Journal*, **51**, 406–418.

Hussey, P.J., Ketelaar, T., and Deeks, M.J. (2006) Control of the actin cytoskeleton in plant cell growth. *Annual Review of Plant Biology*, **57**, 109–125.

Hwang, J.U., Gu, Y., Lee, Y.J., and Yang, Z. (2005) Oscillatory ROP GTPase activation leads the oscillatory polarized growth of pollen tubes. *Molecular Biology of the Cell*, **16**, 5385–5399.

Hwang, J.U., Vernoud, V., Szumlanski, A., Nielsen, E., and Yang, Z. (2008) A tip-localized RhoGAP controls cell polarity by globally inhibiting Rho GTPase at the cell apex. *Current Biology*, **18**, 1907–1916.

Hwang, J.U., Wu, G., Yan, A., Lee, Y.J., Grierson, C.S., and Yang, Z. (2010) Pollen-tube tip growth requires a balance of lateral propagation and global inhibition of Rho-family GTPase activity. *Journal of Cell Science*, **123**, 340–350.

Idilli, A.I., Morandini, P., Onelli, E., Rodighiero, S., Caccianiga, M., and Moscatelli, A. (2013) Microtubule depolymerization affects endocytosis and exocytosis in the tip and influences endosome movement in tobacco pollen tubes. *Molecular Plant*, **6**, 1109–1130.

Iwano, M., Shiba, H., Miwa, T., Che, F.S., Takayama, S., Nagai, T., Miyawaki, A., and Isogai, A. (2004) Ca2+ dynamics in a pollen grain and papilla cell during pollination of Arabidopsis. *Plant Physiology*, **136**, 3562–3571.

Iwano, M., Entani, T., Shiba, H., Kakita, M., Nagai, T., Mizuno, H., Miyawaki, A., Shoji, T., Kubo, K., Isogai, A., *et al.* (2009) Fine-tuning of the cytoplasmic Ca2+ concentration is essential for pollen tube growth. *Plant Physiology*, **150**, 1322–1334.

Johnson, M.A. and Preuss, D. (2002) Plotting a course: multiple signals guide pollen tubes to their targets. *Developmental Cell*, **2**, 273–281.

Jones, M.A., Shen, J.J., Fu, Y., Li, H., Yang, Z., and Grierson, C.S. (2002) The Arabidopsis Rop2 GTPase is a positive regulator of both root hair initiation and tip growth. *Plant Cell*, **14**, 763–776.

Joos, U., Van Aken, J., and Kristen, U. (1995) The anti-microtubule drug carbetamide stopsNicotiana sylvestris pollen tube growth in the style. *Protoplasma*, **187**, 182–191.

Kaothien, P., Ok, S.H., Shuai, B., Wengier, D., Cotter, R., Kelley, D., Kiriakopolos, S., Muschietti, J., and McCormick, S. (2005) Kinase partner protein interacts with the LePRK1 and LePRK2 receptor kinases and plays a role in polarized pollen tube growth. *Plant Journal*, **42**, 492–503.

Khurana, P., Henty, J.L., Huang, S., Staiger, A.M., Blanchoin, L., and Staiger, C.J. (2010) Arabidopsis VILLIN1 and VILLIN3 have overlapping and distinct activities in actin bundle formation and turnover. *Plant Cell*, **22**, 2727–2748.

Kieffer, F., Elmayan, T., Rubier, S., Simon-Plas, F., Dagher, M.C., and Blein, J.P. (2000) Cloning of Rac and Rho-GDI from tobacco using an heterologous two-hybrid screen. *Biochimie*, **82**, 1099–1105.

Kim, C.S., Gibbon, B.C., Gillikin, J.W., Larkins, B.A., Boston, R.S., and Jung, R. (2006) The maize Mucronate mutation is a deletion in the 16-kDa gamma-zein gene that induces the unfolded protein response. *Plant Journal*, **48**, 440–451.

Klahre, U. and Kost, B. (2006) Tobacco RhoGTPase ACTIVATING PROTEIN1 spatially restricts signaling of RAC/Rop to the apex of pollen tubes. *Plant Cell*, **18**, 3033–3046.

Kost, B., Spielhofer, P., and Chua, N.H. (1998) A GFP-mouse talin fusion protein labels plant actin filaments in vivo and visualizes the actin cytoskeleton in growing pollen tubes. *Plant Journal*, **16**, 393–401.

Kost, B., Lemichez, E., Spielhofer, P., Hong, Y., Tolias, K., Carpenter, C., and Chua, N.H. (1999) Rac homologues and compartmentalized phosphatidylinositol 4, 5-bisphosphate act in a common pathway to regulate polar pollen tube growth. *Journal of Cell Biology*, **145**, 317–330.

Laitiainen, E., Nieminen, K.M., Vihinen, H., and Raudaskoski, M. (2002) Movement of generative cell and vegetative nucleus in tobacco pollen tubes is dependent on microtubule cytoskeleton but independent of the synthesis of callose plugs. *Sexual Plant Reproduction*, **15**, 195–204.

Lee, Y.J. and Yang, Z. (2008) Tip growth: signaling in the apical dome. *Current Opinion in Plant Biology*, **11**, 662–671.

Lee, Y.J., Szumlanski, A., Nielsen, E., and Yang, Z. (2008) Rho-GTPase-dependent filamentous actin dynamics coordinate vesicle targeting and exocytosis during tip growth. *Journal of Cell Biology*, **181**, 1155–1168.

Li, H., Wu, G., Ware, D., Davis, K.R., and Yang, Z. (1998) Arabidopsis Rho-related GTPases: differential gene expression in pollen and polar localization in fission yeast. *Plant Physiology*, **118**, 407–417.

Li, H., Lin, Y., Heath, R.M., Zhu, M.X., and Yang, Z. (1999) Control of pollen tube tip growth by a Rop GTPase–dependent pathway that leads to tip-localized calcium influx. *Plant Cell*, **11**, 1731–1742.

Li, S., Zhou, L.Z., Feng, Q.N., McCormick, S., and Zhang, Y. (2013) The C-terminal hypervariable domain targets Arabidopsis ROP9 to the invaginated pollen tube plasma membrane. *Molecular Plant*, **6**, 1362–1364.

Lin, Y., Wang, Y., Zhu, J.K., and Yang, Z. (1996) Localization of a Rho GTPase implies a role in tip growth and movement of the generative cell in pollen tubes. *Plant Cell*, **8**, 293–303.

Lin, Y. and Yang, Z. (1997) Inhibition of pollen tube elongation by microinjected anti-Rop1Ps antibodies suggests a crucial role for Rho-type GTPases in the control of tip growth. *Plant Cell*, **9**, 1647–1659.

Lord, E.M. (2003) Adhesion and guidance in compatible pollination. *Journal of Experimental Botany*, **54**, 47–54.

Lord, E.M. and Russell, S.D. (2002) The mechanisms of pollination and fertilization in plants. *Annual Review of Cell Development and Biology*, **18**, 81–105.

Lovy-Wheeler, A., Kunkel, J.G., Allwood, E.G., Hussey, P.J., and Hepler, P.K. (2006) Oscillatory increases in alkalinity anticipate growth and may regulate actin dynamics in pollen tubes of lily. *Plant Cell*, **18**, 2182–2193.

McDowell, J.M., Huang, S., McKinney, E.C., An, Y.-Q., and Meagher, R.B. (1996) Structure and evolution of the actin gene family in Arabidopsis thaliana. *Genetics*, **142**, 587–602.

Meagher, R.B., McKinney, E.C., and Vitale, A.V. (1999) The evolution of new structures: clues from plant cytoskeletal genes. *Trends in Genetics*, **15**, 278–284.

Moscatelli, A., Ciampolini, F., Rodighiero, S., Onelli, E., Cresti, M., Santo, N., and Idilli, A. (2007) Distinct endocytic pathways identified in tobacco pollen tubes using charged nanogold. *Journal of Cell Science*, **120**, 3804–3819.

Mucha, E., Fricke, I., Schaefer, A., Wittinghofer, A., and Berken, A. (2011) Rho proteins of plants–functional cycle and regulation of cytoskeletal dynamics. *European Journal of Cell Biology*, **90**, 934–943.

Nagawa, S., Xu, T., and Yang, Z. (2010) RHO GTPase in plants: conservation and invention of regulators and effectors. *Small GTPases*, **1**, 78–88.

Petrasek, J. and Schwarzerova, K. (2009) Actin and microtubule cytoskeleton interactions. *Current Opinion in Plant Biology*, **12**, 728–734.

Picton, J.M. and Steer, M.W. (1983) Membrane recycling and the control of secretory activity in pollen tubes. *Journal of Cell Science*, **63**, 303–310.

Potocky, M., Jones, M.A., Bezvoda, R., Smirnoff, N., and Zarsky, V. (2007) Reactive oxygen species produced by NADPH oxidase are involved in pollen tube growth. *New Phytologist*, **174**, 742–751.

Poulter, N.S., Vatovec, S., and Franklin-Tong, V.E. (2008) Microtubules are a target for self-incompatibility signaling in Papaver pollen. *Plant Physiology*, **146**, 1358–1367.

Qin, Y. and Yang, Z. (2011) Rapid tip growth: insights from pollen tubes. *Seminars in Cell and Developmental Biology*, **22**, 816–824.

Raudaskoski, M., Åström, H., and Laitiainen, E. (2001) Pollen tube cytoskeleton: structure and function. *Journal of Plant Growth Regulation*, **20**, 113–130.

Ren, H., and Xiang, Y. (2007) The function of actin-binding proteins in pollen tube growth. *Protoplasma*, **230**, 171–182.

Ringli, C., Baumberger, N., Diet, A., Frey, B., and Keller, B. (2002) ACTIN2 is essential for bulge site selection and tip growth during root hair development of Arabidopsis. *Plant Physiology*, **129**, 1464–1472.

Romagnoli, S., Cai, G., Faleri, C., Yokota, E., Shimmen, T., and Cresti, M. (2007) Microtubule- and actin filament-dependent motors are distributed on pollen tube mitochondria and contribute differently to their movement. *Plant Cell Physiology*, **48**, 345–361.

Samaj, J., Muller, J., Beck, M., Bohm, N., and Menzel, D. (2006) Vesicular trafficking, cytoskeleton and signalling in root hairs and pollen tubes. *Trends in Plant Science*, **11**, 594–600.

Schiott, M., Romanowsky, S.M., Baekgaard, L., Jakobsen, M.K., Palmgren, M.G., and Harper, J.F. (2004) A plant plasma membrane Ca2+ pump is required for normal pollen tube growth and fertilization. *Proceedings of National Academy of Science, USA*, **101**, 9502–9507.

Siderovski, D.P. and Willard, F.S. (2005) The GAPs, GEFs, and GDIs of heterotrimeric G-protein alpha subunits. *International Journal of Biological Sciences*, **1**, 51.

Staiger, C.J. (2000) Signaling to the actin cytoskeleton in plants. *Annual Reviews of Plant Physiology and Plant Molecular Biology*, **51**, 257–288.

Staiger, C.J., Yuan, M., Valenta, R., Shaw, P.J., Warn, R.M., and Lloyd, C.W. (1994) Microinjected profilin affects cytoplasmic streaming in plant cells by rapidly depolymerizing actin microfilaments. *Current Biology*, **4**, 215–219.

Staiger, C.J., Poulter, N.S., Henty, J.L., Franklin-Tong, V.E., and Blanchoin, L. (2010) Regulation of actin dynamics by actin-binding proteins in pollen. *Journal of Experimental Botany*, **61**, 1969–1986.

Vernoud, V., Horton, A.C., Yang, Z., and Nielsen, E. (2003) Analysis of the small GTPase gene superfamily of Arabidopsis. *Plant Physiology*, **131**, 1191–1208.

Vidali, L., Yokota, E., Cheung, A., Shimmen, T., and Hepler, P. (1999) The 135 kDa actin-bundling protein fromLilium longiflorum pollen is the plant homologue of villin. *Protoplasma*, **209**, 283–291.

Vidali, L., McKenna, S.T., and Hepler, P.K. (2001) Actin polymerization is essential for pollen tube growth. *Molecular Biology of the Cell*, **12**, 2534–2545.

Wang, H.J., Wan, A.R., and Jauh, G.Y. (2008a) An actin-binding protein, LlLIM1, mediates calcium and hydrogen regulation of actin dynamics in pollen tubes. *Plant Physiology*, **147**, 1619–1636.

Wang, T., Xiang, Y., Hou, J., and Ren, H.Y. (2008b) ABP41 is involved in the pollen tube development via fragmenting actin filaments. *Molecular Plant*, **1**, 1048–1055.

Wang, X., Zhu, L., Liu, B., Wang, C., Jin, L., Zhao, Q., and Yuan, M. (2007) Arabidopsis MICROTUBULE-ASSOCIATED PROTEIN18 functions in directional cell growth by destabilizing cortical microtubules. *Plant Cell*, **19**, 877–889.

Wennerberg, K., Rossman, K.L., and Der, C.J. (2005) The Ras superfamily at a glance. *Journal of Cell Science*, **118**, 843–846.

Wilsen, K.L., Lovy-Wheeler, A., Voigt, B., Menzel, D., Kunkel, J.G., and Hepler, P.K. (2006) Imaging the actin cytoskeleton in growing pollen tubes. *Sexual Plant Reproduction*, **19**, 51–62.

Winge, P., Brembu, T., and Bones, A.M. (1997) Cloning and characterization of rac-like cDNAs from Arabidopsis thaliana. *Plant Molecular Biology*, **35**, 483–495.

Winge, P., Brembu, T., Kristensen, R., and Bones, A.M. (2000) Genetic structure and evolution of RAC-GTPases in Arabidopsis thaliana. *Genetics*, **156**, 1959–1971.

Wu, G., Li, H., and Yang, Z. (2000) Arabidopsis RopGAPs are a novel family of rho GTPase-activating proteins that require the Cdc42/Rac-interactive binding motif for rop-specific GTPase stimulation. *Science Signaling*, **124**, 1625.

Wu, G., Gu, Y., Li, S., and Yang, Z. (2001) A genome-wide analysis of Arabidopsis Rop-interactive CRIB motif-containing proteins that act as Rop GTPase targets. *Science Signaling*, **13**, 2841.

Xiang, Y., Huang, X., Wang, T., Zhang, Y., Liu, Q., Hussey, P.J., and Ren, H. (2007) ACTIN BINDING PROTEIN 29 from Lilium pollen plays an important role in dynamic actin remodeling. *Plant Cell*, **19**, 1930–1946.

Yan, A., Xu, G., and Yang, Z.B. (2009) Calcium participates in feedback regulation of the oscillating ROP1 Rho GTPase in pollen tubes. *Proceedings of National Academy of Science, USA*, **106**, 22002–22007.

Yang, Z. (2002) Small GTPases versatile signaling switches in plants. *Plant Cell*, **14**, S375–S388.

Yang, Z. (2008) Cell polarity signaling in Arabidopsis. *Annual Review of Cell Development and Biology*, **24**, 551–575.

Yang, Z. and Watson, J.C. (1993) Molecular cloning and characterization of rho, a ras-related small GTP-binding protein from the garden pea. *Proceedings of the National Academy of Sciences*, **90**, 8732–8736.

Yang, Z. and Fu, Y. (2007) ROP/Rac GTPase signaling. *Current Opinion in Plant Biology*, **10**, 490–494.

Ye, J., Zheng, Y., Yan, A., Chen, N., Wang, Z., Huang, S., and Yang, Z. (2009) Arabidopsis formin3 directs the formation of actin cables and polarized growth in pollen tubes. *Plant Cell*, **21**, 3868–3884.

Yokota, E. and Shimmen Ki, T.T. (1998) Actin-bundling protein isolated from pollen tubes of lily. Biochemical and immunocytochemical characterization. *Plant Physiology*, **116**, 1421–1429.

Yokota, E., Muto, S., and Shimmen, T. (2000) Calcium-calmodulin suppresses the filamentous actin-binding activity of a 135-kilodalton actin-bundling protein isolated from lily pollen tubes. *Plant Physiology*, **123**, 645–654.

Yokota, E., Vidali, L., Tominaga, M., Tahara, H., Orii, H., Morizane, Y., Hepler, P.K., and Shimmen, T. (2003) Plant 115-kDa actin-filament bundling protein, P-115-ABP, is a homologue of plant villin and is widely distributed in cells. *Plant Cell Physiology*, **44**, 1088–1099.

Zhang, Y. and McCormick, S. (2007) A distinct mechanism regulating a pollen-specific guanine nucleotide exchange factor for the small GTPase Rop in Arabidopsis thaliana. *Proceedings of the National Academy of Sciences*, **104**, 18830–18835.

Zhang, H., Qu, X., Bao, C., Khurana, P., Wang, Q., Xie, Y., Zheng, Y., Chen, N., Blanchoin, L., Staiger, C.J., *et al.* (2010) Arabidopsis VILLIN5, an actin filament bundling and severing protein, is necessary for normal pollen tube growth. *Plant Cell*, **22**, 2749–2767.

Zhao, Y., Yan, A., Feijo, J.A., Furutani, M., Takenawa, T., Hwang, I., Fu, Y., and Yang, Z. (2010) Phosphoinositides regulate clathrin-dependent endocytosis at the tip of pollen tubes in Arabidopsis and tobacco. *Plant Cell*, **22**, 4031–4044.

Zhu, L., Zhang, Y., Kang, E., Xu, Q., Wang, M., Rui, Y., Liu, B., Yuan, M., and Fu, Y. (2013) MAP18 regulates the direction of pollen tube growth in Arabidopsis by modulating F-Actin organization. *Plant Cell*, **25**, 851–867.

9 Phosphoinositide Signaling in Root Hair Tip Growth

Hiroaki Kusano[1], Rumi Tominaga[2], Takuji Wada[2], Mariko Kato[3] and Takashi Aoyama[3]

[1]*Department of Biological Science and Technology, Tokyo University of Science, Tokyo, Japan*
[2]*Graduate School of Biosphere Science, Hiroshima University, Hiroshima, Japan*
[3]*Institute for Chemical Research, Kyoto University, Uji, Kyoto, Japan*

Introduction to Root Hair Study

Root hairs, cellular protuberances resulting from the polar outgrowth of root epidermal cells, have been intensively studied as a model for plant cell morphogenesis (Datta *et al.*, 2011; Tominaga-Wada *et al.*, 2011; Grebe, 2012; Jones and Dolan, 2012). Root hairs have distinct advantages in genetic and cell biological studies (Peterson and Farquhar, 1996; Schiefelbein, 2000; Dolan, 2001; Bibikova and Gilroy, 2002). First, the morphogenesis of root hairs involves many cell biological events, including the establishment and sustentation of cell polarity, which are important for plant cell morphogenesis. Second, although root hairs are requisite in the natural field, they are dispensable under experimental conditions, facilitating genetic and reverse gentic approaches. Moreover, because root hairs are transparent and protruding from the root surface, microscopic observations for their intracellular structures are easily performed.

Root hairs develop from specialized root epidermal cells called trichoblasts. In *Arabidopsis thaliana*, epidermal cells that overlie a cleft between two underlying cortical cell files become trichoblasts (Fig. 9.1; Dolan *et al.* 1994; Galway *et al.* 1994; Balkunde *et al.*, 2010; Grebe, 2012). During root hair development, a bulge is initially formed in a distal area on the trichoblast outer surface (Fig. 9.1). This bulge positioning requires a mechanism for establishing the planar polarity on the cell surface (Fischer *et al.*, 2006). Subsequently, the bulge further protrudes perpendicularly to the root surface via tip growth, highly polarized cell expansion in which cell wall deposition

Plant Cell Wall Patterning and Cell Shape, First Edition. Edited by Hiroo Fukuda.

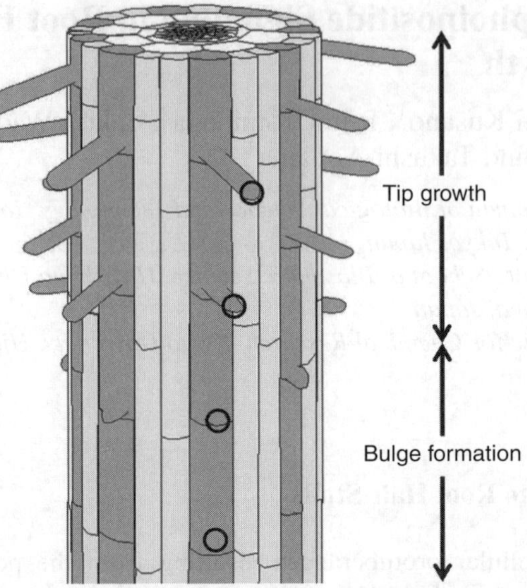

Figure 9.1 Root hair development from bulge formation to root hair tip growth on the Arabidopsis root epidermis. Root hair cell files and non-root hair cell files are indicated by dark and pale gray, respectively. Circles mark loci where bulge formation and root hair tip growth occur.

and plasma membrane expansion are limited to the tip (Fig. 9.1; Gilroy and Jones, 2000; Ryan *et al.*, 2001; Carol and Dolan, 2002; Samaj *et al.*, 2006). Because root hairs have a straight cylindrical shape, the tip-growth polarity should be sustained by some robust mechanism throughout root hair growth (Cole and Fowler, 2006).

In root hair tip growth, vesicular transport is correctly localized to the root hair apical region to deposit materials for the expanding plasma membrane and cell wall. In Arabidopsis, the regulatory mechanism of root hair tip growth has been revealed to involve signaling factors including calcium ion (Ca^{2+}), reactive oxygen species (ROS), Rho-family GTPase, and signaling phospholipids (Smith and Oppenheimer, 2005; Cole and Fowler, 2006; Lee and Yang, 2008; Cardenas, 2009; Gu and Nielsen, 2013). ROP (Rho-family GTPase of plants) localizes to the plasma membrane in the future bulge-forming site on the trichoblast outer surface and in the apical region of growing root hairs (Molendijk *et al.*, 2001; Jones *et al.*, 2002), where ROP promotes the ROS accumulation through the NADPH oxidase encoded by *ROOT HAIR DEFECTIVE 2* (*RHD2*) (Foreman *et al.*, 2003; Jones *et al.*, 2007). The accumulated ROS activates calcium channels, and the resulting Ca^{2+} influx further activates RHD2 (Takeda *et al.*, 2008). This positive feedback loop is thought to stabilize the tip-focused calcium gradient, which is required for the localization of vesicular transport to the tip. As for the mechanism by which ROP is initially localized to the future bulge-forming site, a basipetal auxin gradient in the root epidermis has been shown to act as a spatial clue (Fischer *et al.*, 2006). In addition, SUPERCENTIPEDE 1 (SCN1), a Rho GDP dissociation inhibitor (Rho-GDI), and FERONIA (FER), a receptor-like kinase (RLK) interacting with a ROP guanine nucleotide exchange factor (ROP-GEF), are thought to be involved in the ROP signaling (Carol *et al.*, 2005; Duan *et al.*, 2010). So far however, the molecular basis of the mechanism for establishing and sustaining the polarity of root hair tip growth is still unclear.

Signaling phospholipids are also known to function in plant cell morphogenesis (Fischer *et al.*, 2004; Heilmann, 2008; Thole and Nielsen, 2008; Xue *et al.*, 2009). Forward and reverse genetic studies in Arabidopsis have revealed the involvement of phosphatidic acid (PA), phosphoinositides, and their related genes in the regulation of root hair tip growth (Ohashi *et al.*, 2003; Anthony *et al.*, 2004; Bohme *et al.*, 2004; Vincent *et al.*, 2005; Preuss *et al.*, 2006; Kusano *et al.*, 2008; Lee *et al.*, 2008; Stenzel *et al.*, 2008; Thole *et al.*, 2008). In the following sections, we briefly overview phosphoinositide signaling in animals and fungi, especially its involvement in the regulation of cell architecture, then describe recent findings about phosphoinositide signaling and its interacting factors in root hair tip growth. Finally, we propose a hypothesis for the regulatory mechanism for establishing and sustaining the polarity of root hair tip growth focusing on phosphoinositide signaling.

Overview of Phosphoinositide Signaling in Eukaryotes

Phospholipids are a major constituent of membranes in eukaryotic cells. As well as their structural role, some of them act as signaling molecules involved in a wide variety of intracellular events. Phosphoinositides are typical signaling phospholipids that have been most intensively studied for the past decade (Di Paolo and De Camilli, 2006; Gamper and Shapiro, 2007; Krauss and Haucke, 2007; Sasaki *et al.*, 2009; Shewan *et al.*, 2011). Historically, phosphatidylinositol 4,5-bisphosphate (PtdIns(4,5)P_2) has been associated with the generation of two second messengers, diacylglycerol (DAG) and inositol 1,4,5-triphosphate (Ins(1,4,5)P_3), via phospholipase C (PLC)-mediated hydrolysis (Berridge and Irvine, 1984). At present however, phosphoinositides including PtdIns(4,5)P_2 are better known to act as signaling molecules by themselves. They promote subcellular events by recruiting their effector proteins to the membrane or membrane subdomain where they reside, as well as lead to cell-global responses accompanied by changes in gene expression as for cytosolic second messengers. Due to the subcellular function, phosphoinositides are involved in determining membrane identity and cell polarity as site-specific signals on membranes.

Phosphatidylinositol (PtdIns), the precursor of phosphoinositides, is synthesized primarily in the endoplasmic reticulum (ER) and then delivered to other membrane compartments by vesicular transport or PtdIns transfer proteins. Phosphorylation states at positions D3, D4, and D5 of the inositol moiety are altered by position-specific phosphoinositide kinases and phosphatases, resulting in seven phosphoinositide species (Figs 9.2 and 9.3; Fruman *et al.*, 1998; Anderson *et al.*, 1999; Rameh *et al.*, 1999). The spatiotemporal regulation of their metabolism allows them to serve as markers for cell compartment identity, called the PI code (Fig. 9.3; Bankaitis and Morris, 2003; Behnia and Munro, 2005; Kutateradze, 2010). It is notable in the metabolic system that strategic interconversion from one phosphoinositide to another is implicated in the mechanism for dynamic changes of membrane identity during membrane traffic and compartment maturation. In mammalian cells for example, PtdIns(4)P marking the Golgi complex and secretary vesicles is converted to PtdIns(4,5)P_2 on the plasma membrane, and PtdIns(3)P marking early endosomes is converted to PtdIns(3,5)P_2 on the late endosome membrane (Di Paolo and De Camilli, 2006; Kutateradze, 2010). Moreover, membrane subdomains may also be distinguished by the PI code. In mammalian epithelial cells, the coordinated functions of class I phosphoinositide 3-kinase (PI(3)K) and the phosphoinositide 3-phosphatase PTEN localize PtdIns(4,5)P_2 and PtdIns(3,4,5)P_3 to the apical and basolateral plasma membrane, respectively, leading to apicobasal cell polarity (Gassama-Diagne *et al.*, 2006; Comer and Parent, 2007; Martin-Belmonte *et al.*, 2007).

The PI code is basically deciphered by effector proteins, which recognize their specific phosphoinositides through conserved domain structures such as the pleckstrin homology (PH) domain or basic amino-acid clusters within unstructured regions (Table 9.1; Di Paolo and De Camilli, 2006; Kutateradze, 2010). However, the binding affinity of a single phosphoinositide-effector interaction is generally weak in fixing the effector protein on the membrane, and the effector can stably interact with its specific phosphoinositide only when it is coincidently bound to other ligands on the membrane. The coincident ligands may include other signaling phospholipids and membrane-associated proteins such as small GTPases (Wenk and De Camilli, 2004; Behnia and Munro, 2005). This coincidence detection mechanism increases the number of membrane recognition patterns, and hence allows highly specific and fine-tunable interactions between effector proteins and their specific membranes or membrane subdomains (Di Paolo and De Camilli 2006; Saarikangas *et al.* 2010).

Of the seven phosphoinositides, PtdIns(4,5)P_2 is the best-studied species and known to regulate various intracellular events including actin cytoskeletal reorganization and membrane traffic. PtdIns(4,5)P_2 plays a pivotal regulatory role in actin organization and dynamics through the physical interaction with

Table 9.1 Phospholipid-binding proteins. Representative interactions between signaling phospholipids and their physically interacting proteins in eukaryotes are listed with their binding modules and functions. Proteins which originated from plants are indicated in bold.

Phospholipid	Module	Protein name	Protein function	Ref
PtdIns(3)P	FYVE	EEA1	early endosomal antigen	j
	FYVE	HRS1	component of endosomal sorting complexes	j
	FYVE	SARA	Smad anchor for receptor activation, endosomal protein	j
	FYVE	PIKfyve	type III PIP kinase	j
	PHD	Pf1	PHD-containing transcription factor	i
	PX	SNX2, 3, 7, 13	sorting nexin, membrane traffic regulation	j
	–	Atg18	yeast autophagy related protein	p
	–	**SVB**	DUF538 domain protein, regulation of trichome size	y
	–	**ATG18**	Arabidopsis homolog of yeast Atg18, autophagy	y
PtdIns(4)P	A/ENTH	EpsinR	clathrin interacting protein	j
	PH	**AGD1**	Arf-GAP	u
	PH	FAPP1, 2	TGN-associated four-phosphate-adaptor protein	j
	PH	OSBP	oxysterol binding protein, steroid metabolism regulation	j
	–	**POL**	protein phosphatase 2C, stem cell maintenance	q
PtdIns(5)P	PHD	ING2	PHD-containing transcription factor	j
	PX	SNX13	sorting nexin, membrane traffic regulation	j
	–	**ATX**	trithorax transcription factor	i
$PtdIns(3,4)P_2$	PH	TAPP1, 2	tandem PH domain protein, protein recruitment	j
	PH	AKT/PKB	serine/threonine kinase	j
	PX	p47phox	NADPH oxidase regulatory subunit	j
$PtdIns(3,5)P_2$	A/ENTH	Ent3p, 5p	clathrin interacting protein	j
	GRAM	Myotubularin	phosphoinositide 3-phosphatase	j
	–	Atg18	yeast autophagy related protein	p
	–	**ATG18**	homolog of yeast autophagy 18G, autophagy related	y
	–	**Patellin1**	homolog of yeast Sec14, lipid transfer protein	y
	–	**PPR**	pentatricopeptide repeat protein, autophagy related	y
$PtdIns(4,5)P_2$	A/ENTH	AP180	clathrin assembly protein	j
	A/ENTH	Epsin	clathrin assembly protein	j
	A/ENTH	CALM	clathrin assembly protein	j
	A/ENTH	HIP1	huntingtin interacting protein, clathrin interaction	j
	C2	Synaptotagmin	regulator of SNARE	j

(*continued overleaf*)

Table 9.1 (*continued*)

Phospholipid	Module	Protein name	Protein function	Ref
	FERM	Ezrin	ERM family protein, actin-membrane linkage	j
	FERM	Radixin	ERM family protein, actin-membrane linkage	j
	FERM	Moesin	ERM family protein, actin-membrane linkage	j
	FERM	Talin	actin-interacting protein, actin-integrin linkage	j
	MORN	**PIP5K1**	type I PIPK	m
	PDZ	Syntenin	component of signal transduction complex	j
	PH	PLCδ1	PLC	j
	PH	Dynamin	GTPase, endocytosis	j
	PH	DBL	Rho-GEF	e
	PH	VAV	Rho-GEF	b
	PH	ARNO	Arf-GEF	d
	PH	ASAP1	Arf-GAP	c
	PTB	ARH	modulator of lipoprotein receptor functions	j
	PTB	Dab1	modulator of lipoprotein receptor functions	j
	PTB	SHC	component of signal transduction complex	j
	PX	PI3K II	class II PI(3) kinase	j
	PX	PLD1	PLD	n
	–	gelsolin	actin capping protein	r
	–	profilin	actin nucleotide exchanging protein	r
	–	ADF/cofilin	actin severing protein	r
	–	WASP	actin polymerization activator	r
	–	CAPS	Ca^{2+}-dependent activator protein for secretion	f
	–	**AtCP**	heterodimeric capping protein, actin binding	k
	–	**PCaP2**	Ca^{2+}-binding protein, regulation of root hair growth	w
PtdIns(3,4,5)P_3	PH	BTK	protein tyrosine kinase	j
	PH	AKT/PKB	serine/threonine kinase	j
	PH	PDK1	3' phosphoinositide dependent kinase, serine/threonine	a
	PH	DBL	Rho-GEF	e
	PH	VAV	Rho-GEF	b
	PH	ARNO	Arf-GEF	d
	PH	GRP1	Arf-GEF	j
	PTB	SHC	component of signal transduction complex	j
	PX	CISK	SGK, protein serine/threonine kinase	j
	DHR	DOCK180	RhoGEF	g

(*continued overleaf*)

Table 9.1 (*continued*)

Phospholipid	Module	Protein name	Protein function	Ref
PA	C2	PKCϵ	protein kinase C, serine/threonine kinase	h
	C2	Rab11-FIP I	class I Rab11 family interacting protein	h
	FRB	mTOR	serine/threonine kinase	h
	MORN	**PIP5K1**	type I PIP kinase	m
	PH	**AtPDK1**	PDK1 homolog, protein kinase	h
	PX	P47phox	NADPH oxidase regulatory subunit	h
	PX	PLD1	PLD	h
	Q2	Opi1	transcription factor, repressor	h
	TAPAS	PDE4A1	cAMP phosphodiesterase	h
	–	PIPK I	type I PIP kinase	h
	–	RafI	MAP triple kinase, serine/threonine kinase	h
	–	PP1Cγ	protein phosphatase	h
	–	AGAP1	Arf-GAP	h
	–	ARF	small GTPase	h
	–	SK1	sphingosin kinase	h
	–	NSF	N-ethylmaleimide-sensitive factor	h
	–	**ABI1**	protein phosphatase	h
	–	**PEPC**	phosphoenolpyruvate carboxylase	h
	–	**AtCP**	heterodimeric capping protein, actin binding	k
		CDPK	Ca^{2+}-dependent protein kinase	h
	–	**GAPDH**	glyceraldehyde 3-phosphate dehydrogenase	x
	–	**CHC**	clathrin heavy chain	x
	–	**CTR1**	MAP triple kinase, regulation of ethylene response	o
	–	**PTEN2**	phosphoinositide 3-phosphatase	s
	–	**MAP65-1**	microtubule associated protein, salt stress response	v
	–	**TGD4**	trigalactosyldiacylglycerol protein, lipid transfer	t

(a) Alessi *et al.* 1997; (b) Han *et al.*, 1998; (c) Kam *et al.*, 2000; (d) Macia *et al.*, 2000; (e) Russo *et al.*, 2001; (f) Grishanin *et al.*, 2004; (g) Cote *et al.*, 2005; (h) Testerink and Munnik, 2005; (i) Alvarez-Venegas *et al.*, 2006; (j) Di Paolo and De Camilli, 2006; (k) Huang *et al.*, 2006; (l) Kaadige *et al.*, 2006; (m) Im *et al.*, 2007; (n) Oude Weernink *et al.*, 2007; (o) Testerink *et al.*, 2007; (p) Obara *et al.*, 2008; (q) Gagne and Clark, 2010; (r) Saarikangas *et al.*, 2010; (s) Pribat *et al.*, 2012; (t) Wang *et al.*, 2012; (u) Yoo *et al.*, 2012; (v) Zhang *et al.*, 2012; (w) Kato *et al.*, 2013; (x) McLoughlin *et al.*, 2013; (y) Oxley *et al.*, 2013.

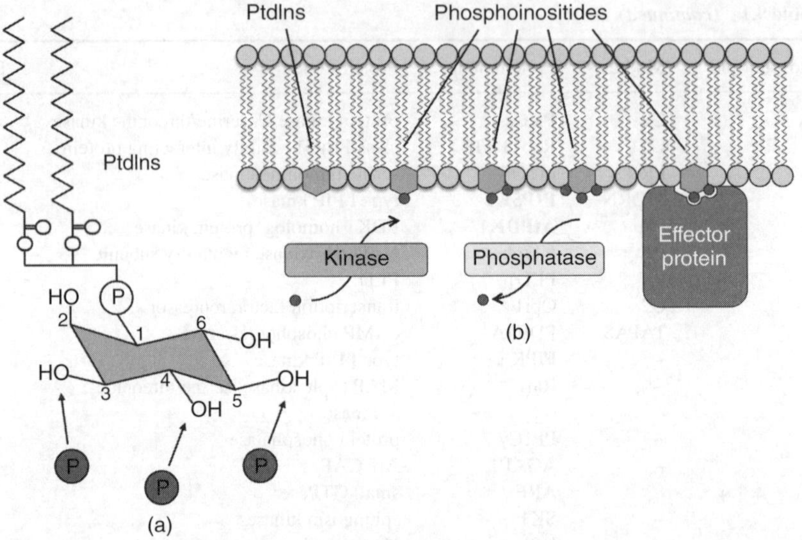

Figure 9.2 Structures of PtdIns and a dynamic image of phosphoinositides on biological membranes: (a) chemical structure and phosphorylation sites of PtdIns; and (b) phosphoinositide metabolism on biological membranes by kinases and phosphatases, and signaling interaction with their effector proteins.

most of the actin-biding proteins that regulate actin dynamics, including ERM (ezrin/radixin/moesin) family proteins, talin, WASP (Wiskott–Aldrich syndrome protein) family proteins, ADF/cofilin, profilin, gelsolin, and so on (Takenawa and Itoh, 2001; Saarikangas *et al.* 2010). PtdIns(4,5)P_2 is considered to be a positive signal for actin cytoskeleton formation because it typically activates actin-binding proteins, promoting the assembly of actin filaments, and suppresses those promoting disassembly. PtdIns(4,5)P_2 also has regulatory roles in membrane traffic via the binding to SNARE complex proteins, clathrin-associated proteins, CAPS, and so on (Itoh *et al.*, 2001; Olsen *et al.*, 2003; Grishanin *et al.*, 2004; Li *et al.*, 2006; Mayinger, 2012). In addition, PtdIns(4,5)P_2 regulates actin cytoskeleton and membrane traffic through regulatory proteins of small GTPases, for example, GTPase-activating proteins (GAPs) and guanine nucleotide exchange factors (GEFs; Paris *et al.*, 1997; Das *et al.*, 2000; Kam *et al.*, 2000; Russo *et al.*, 2001; Nie *et al.*, 2002). Another important class of PtdIns(4,5)P_2 effectors comprises plasma membrane-integrated proteins, including voltage-gated potassium and calcium channels and sensory transduction channels, which may also modulate cytoskeletal reorganization and membrane traffic (Runnels *et al.*, 2002; Wu *et al.*, 2002; Gamper *et al.*, 2004; Oliver *et al.*, 2004; Suh and Hille, 2005). Although PtdIns(4,5)P_2 localizes mainly to the plasma membrane, it is also known to function in the nucleus. PtdIns(4,5)P_2 and

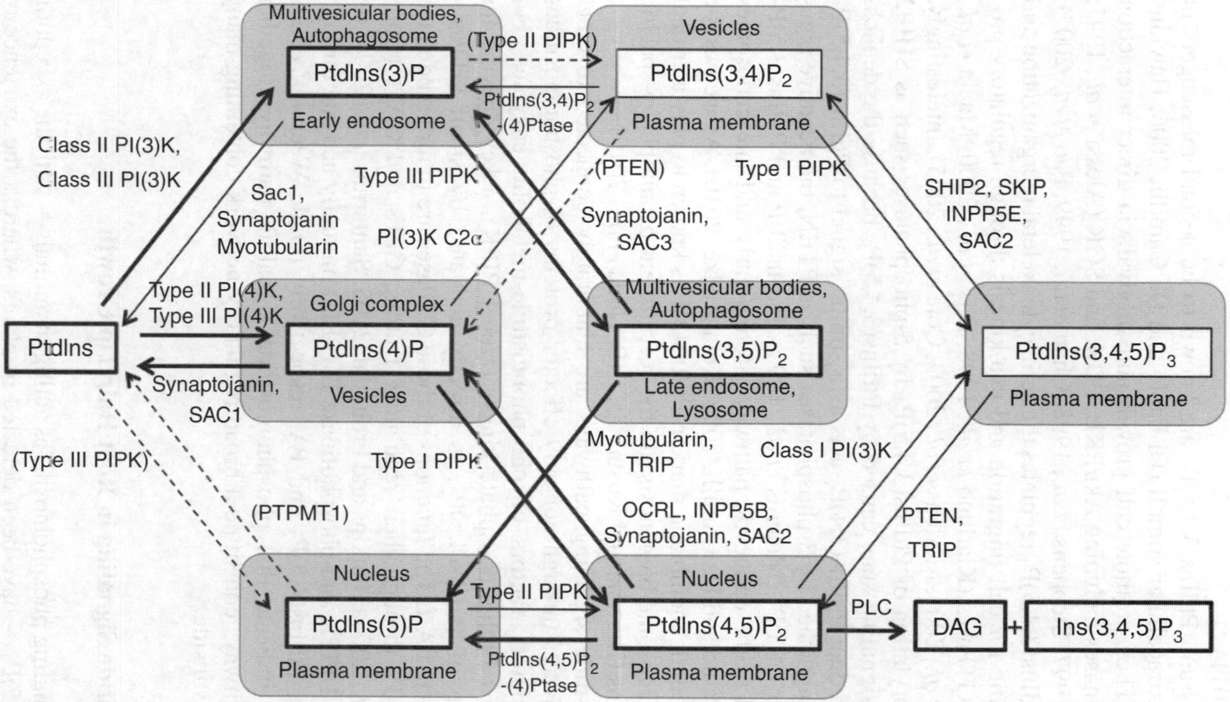

Figure 9.3 Metabolic pathways and intracellular localizations of phosphoinositides. Phosphoinositide metabolic reactions in animals are illustrated by arrows accompanied by enzyme names and boxes labeled with metabolite names. Dotted arrows and parenthesized enzyme names indicate reactions that have been detected *in vitro*, the significance of which in living cells remains unclear. Subcellular compartments where phosphoinositides mainly function as signaling molecules are shown in the rounded boxes encompassing phosphoinositide names. Thick arrows and boxes indicate reactions and metabolites, the existence of which in plant cells has been shown (Paolo and De Camilli, 2006; Sasaki *et al.*, 2009; Xue *et al.*, 2009; Barlow *et al.*, 2010; Kutateradze, 2010; Munnik and Vermeer, 2010; Shewan *et al.*, 2011; Boss and Im, 2012; Maffucci, 2013; Posor *et al.*, 2013).

PtdIns(5)P have been shown to regulate gene expression via pre-mRNA processing and chromatin remodeling in the nucleus (Barlow *et al.*, 2010; Keune *et al.*, 2011).

PtdIns(3,4)P_2 and PtdIns(3,4,5)P_3 are known to be second messengers in responses to extracellular stimuli (Di Paolo and De Camilli, 2006; Hawkins *et al.*, 2006). They promote cell growth and survival via direct interaction with protein kinases including Akt/PKB, PDK, and SGK (Alessi *et al.*, 1997; Stokoe *et al.*, 1997; Stephens *et al.*, 1998; Chan *et al.*, 1999; Pao *et al.*, 2007). In addition, PtdIns(3,4,5)P_3 regulates the actin cytoskeletal reorganization and membrane traffic for cell migration and cytokinesis through regulatory proteins of small GTPases (Klarlund *et al.*, 1998; Das *et al.*, 2000; Macia *et al.*, 2000; Russo *et al.*, 2001; Mertens *et al.*, 2003; Cote *et al.*, 2005). Interestingly, the dephosphorylation of PtdIns(3,4,5)P_3 by 5-phosphatases such as SHIP2 stabilizes the signaling state caused by PtdIns(3,4,5)P_3 because the dephosphorylation product PtdIns(3,4)P_2 acts as a similar signal to PtdIns(3,4,5)P_3 and is not extinguished by 3-phosphatases such as PTEN, the negative regulator of PtdIns(3,4,5)P_3 (Majerus *et al.*, 1999). PtdIns(3)P and PtdIns(3,5)P_2 are implicated in the endocytic pathway through early and late endosomes (Fig. 9.3; Di Paolo and De Camilli, 2006; Kutateradze, 2010). As well as the role as an endocytic membrane marker, PtdIns(3)P is known to play a critical role in autophagosome biogenesis via the recruitment of autophagy-specific effector proteins (Burman and Ktistakis, 2010; Noda *et al.*, 2010).

Phosphoinositide signaling pathways are elaborately connected to other signaling pathways through not only effector proteins of phosphoinositides but also regulatory factors of phosphoinositide-metabolic enzymes. For example, type I PtdIns phosphate kinase (type I PIPK), a key enzyme for the production of PtdIns(4,5)P_2, is directly activated by small GTPases including Rho and Arf. The functions of those GTPases are in turn modulated by PtdIns(4,5)P_2 through their regulatory factors, GEFs and GAPs (Oude Weernink *et al.*, 2004a; Jenkins and Frohman, 2005; Santarius *et al.*, 2006). In addition, type I PIPK and phospholipase D (PLD) mutually activate through their products, PtdIns(4,5)P_2 and PA, respectively (Oude Weernink *et al.*, 2007). These indicate that phosphoinositide signaling pathways together with other pathways constitute a broad regulatory network containing many feedback loops inside.

Phosphoinositide Signaling in Root Hair Tip Growth

The seven signaling phosphoinositides acting in animals – with the exception of PtdIns(3,4,5)P_3 – have been detected in plants, whereas the existence of PtdIns(3,4)P_2 in plants is still controversial (Zonia and Munnik, 2006; Munnik and Vermeer, 2010; Boss and Im, 2012). In addition, most of enzymatic activities metabolizing the phosphoinositides except PtdIns(3,4,5)P_3 and PtdIns(3,4)P_2 are conserved between animals and plants (Mueller-Roeber and Pical, 2002; Munnik and Vermeer, 2010; Boss and Im, 2012). These suggest that phosphoinositide signals in animals are basically analogous in plants, except for those of PtdIns(3,4,5)P_3 and PtdIns(3,4)P_2.

Genetic studies have shown that defects in Arabidopsis PtdIns transfer protein CAN OF WORMS 1/SEC FOURTEEN HOMOLOGS 1 (COW1/SFH1) cause short and distorted root hairs (Bohme *et al.*, 2004; Vincent *et al.*, 2005). COW1/SFH1 contains an N-terminal lipid-binding domain similar to the yeast PtdIns transfer protein Sec14 and a C-terminal domain similar to the *Lotus japonicus* nodulin Nij16. An SFH1-GFP fusion protein localized to the plasma membrane and cytoplasmic space in the apical region of growing root hairs, similar to the PtdIns(4,5)P_2 marker protein $PH_{PLC\delta 1}$-YFP (Vincent *et al.*, 2005). The loss of proper PtdIns(4,5)P_2 localization in *sfh1* mutant root hairs was closely linked to degeneration of the fine F-actin configuration at the tip. This fact, together with the PtdIns(4,5)P_2 function revealed in animal studies, strongly suggests that COW1/SFH1 generates phosphoinositide landmarks that are coupled to components of the actin cytoskeleton. Interestingly, during *Lotus* nodulogenesis, *LjPLP-IV* (the *Lotus* ortholog of *SFH1*) is suppressed, while its C-terminal nodulin-coding region is separately transcribed (Kapranov *et al.*, 2001). PtdIns transfer protein genes of the COW1/SFH1 class might serve as a genetic switch between the polarization and depolarization states in root hair tip growth.

The Arabidopsis type III PtdIns 4-kinase PI-4Kβ1 specifically bound to the GTP form of RabA4b in yeast and *in vitro*, and colocalized with it to the membrane fraction in the growing root hair tip (Preuss *et al.*, 2004, 2006). PI-4Kβ1 also physically interacted with the calcium ion sensor protein AtCBL1. These suggest that both Rab GTPase and calcium ion signals regulate the production of PtdIns(4)P through PI-4Kβ1 similar to the cases in mammal and yeast cells (Hendricks *et al.*, 1999; Bourne *et al.*, 2001; Zhao *et al.*, 2001; Huttner *et al.*, 2003; de Graaf *et al.*, 2004; Sciorra *et al.*, 2005). Mutant plants defective in both *PI-4Kβ1* and its closest relative *PI-4Kβ2* exhibited aberrant root hair morphologies accompanied by an altered structure of the RabA4b-labeled trans-Golgi network (TGN) compartments (Preuss *et al.*, 2006). This indicates that PI-4Kβ1/β2 activity is necessary for proper TGN organization and post-Golgi secretion in root hair tip growth. Supporting this idea, the Arabidopsis *ROOT HAIR DEFECTIVE 4* (*RHD4*) gene, a mutant of which results in short and distorted root hairs, has been found to encode a SAC1-like phosphoinositide phosphatase with substrate preference for PtdIns(4)P (Thole *et al.*, 2008). Similar to the case of PI-4Kβ1, a fusion protein YFP-RHD4 co-localized with RabA4b in wild-type root hairs, and the RabA4b localization was disturbed in *rhd4-1* mutant root hairs (Thole *et al.*, 2008). Moreover, the mutation significantly altered the distribution of PtdIns(4)P from a tip-localized plasma membrane domain to internal membrane compartments in root hairs (Thole *et al.*, 2008). These together indicate that both PI-4Kβ1 and RHD4 localize to RabA4b-labeled membranes in the growing root hair apical region and function to maintain proper PtdIns(4)P levels required for polarized secretion at the tip. PI-4Kβ1 and RHD4 may also affect root hair tip growth through the level of PtdIns(4,5)P_2 in the root hair apical region because PtdIns(4)P is the major substrate for the PtdIns(4,5)P_2 production in plants.

By analogy with other eukaryotic cell morphogenesis, PtdIns(4,5)P_2 is thought to be a pivotal regulator of plant cell architecture. The first evidence for its involvement in root hair tip growth is the immunofluorescence observation that PtdIns(4,5)P_2 is specifically localized to root hair bulges and growing root hairs in maize (Braun *et al.*, 1999). Also in Arabidopsis, PtdIns(4,5)P_2 has been detected at the plasma membrane and the cytoplasmic space in the growing root hair apices using the PI(4,5)P_2 marker protein $PH_{PLC\delta1}$-YFP (Vincent *et al.*, 2005). Plants have two classes of PtdIns(4)P 5-kinases, called type A PIP5Ks and type B PIP5Ks, the enzymatic activity of which corresponds to that of animal type I PIPKs. Type B PIP5Ks, but not type A PIP5Ks, contain N-terminal MORN repeats, which are though to bind to signaling phospholipids (Mueller-Roeber and Pical, 2002). The Arabidopsis type B PIP5K gene *PIP5K3* is expressed preferentially in root hair cells, and its defects caused significantly shorter root hairs than the wild type (Kusano *et al.*, 2008; Stenzel *et al.*, 2008). In contrast, the inducible overexpression of the gene caused longer root hairs (Kusano *et al.*, 2008). A PIP5K3-YFP fusion protein driven by the *PIP5K3* promoter could complement the short-root-hair phenotype of *pip5k3*, and localized to the plasma membrane of growing root hair apices (Kusano *et al.*, 2008; Stenzel *et al.*, 2008). This indicates that PtdIns(4,5)P_2 is produced by PIP5K3 on the plasma membrane of growing root hair apices. The fluorescence signal of PIP5K3-YFP was more intense in faster-growing root hairs and quickly disappeared when the growth ceased (Kusano *et al.*, 2008), suggesting that the amount of PIP5K3 on the apical plasma membrane is a quantitative factor of root hair growth rate. This idea is consistent with the fact that the level of PtdIns(4,5)P_2 on the plasma membrane of neural cells positively correlates with the size of the vesicle pool competent for exocytosis (Di Paolo *et al.*, 2004; Gong *et al.*, 2005; Milosevic *et al.*, 2005), because root hair growth also depends on active exocytosis focused to the apex.

PtdIns(4,5)P_2 is also known to be a regulatory signal for the growth of pollen tubes, another tip-growing cellular structure in plants. It specifically localizes to the apical plasma membrane of growing pollen tubes (Kost *et al.*, 1999). In tobacco and petunia, PLCs accumulated laterally on the plasma membrane in the subapical region of growing pollen tubes, and suppression of PLC activity led to PtdIns(4,5)P_2 delocalization and depolarized pollen tube growth (Dowd *et al.*, 2006; Helling *et al.*, 2006). Based on these results, it has been hypothesized that PLC restricts PI(4,5)P_2 to the pollen tube apex by digesting it in the subapical region (Dowd *et al.*, 2006; Helling *et al.*, 2006). RNA interference-induced suppression of an Arabidopsis type B PIP5K gene, *PIP5K6*, impaired pollen tube tip growth and inhibited clathrin-dependent endocytosis (Zhao *et al.*, 2010). In contrast, overexpression of the *PIP5K6* gene induced massive aggregation of the plasma membrane in pollen tube apices (Zhao *et al.*, 2010). The double mutation of two Arabidopsis A type PIP5K genes, *PIP5K10* and *PIP5K11*, both of which are expressed exclusively in pollens, made pollen tubes hypersensitive to Latrunculin B, an inhibitor of actin polymerization and organization (Ischebeck *et al.*, 2011). These support the idea that PtdIns(4,5)P_2 plays a regulatory role in pollen tube tip growth via actin cytoskeletal reorganization and membrane traffic. A genetic study has revealed that in *Physcomitrella patens*, an evolutional ancestor of land plants, PIP5Ks also play regulatory roles in developmental processes, particularly in cell tip growth (Saavedra *et al.*, 2011, 2012).

In addition to the PtdIns(4)P and PtdIns(4,5)P_2, the involvement of PtdIns(3)P in root hair tip growth has been shown in the context of ROS-mediated control of root hair growth. Transgenic plants expressing the PtdIns(3)P-binding FYVE domain in root hair cells exhibited shorter root hairs and a reduced root hair growth rate compared with the wild type (Lee *et al.*, 2008). LY294002, a specific inhibitor of PI(3)Ks, also affected root hair elongation. In the LY294002-treated root hair cells, membrane traffic at the final fusion stage of the late endosomes to the tonoplast was inhibited, and the intracellular ROS level decreased in a dose-dependent manner (Lee *et al.*, 2008). Proper levels of PtdIns(3)P are likely to be critical for root hair tip growth while it is still unclear how PtdIns(3)P and PI(3)Ks are involved in underlying mechanisms.

Signaling by Other Phospholipids in Root Hair Tip Growth

PA signaling also plays a regulatory role in cell morphogenesis. PA is produced from phosphatidylcholine (PC) by PLD or from DAG by DAG kinase (DGK) (Testerink and Munnik, 2005; Topham, 2006; Oude Weernink *et al.*, 2007; Li *et al.*, 2009; Jang *et al.*, 2012; Kolesnikov *et al.*, 2012). Plants have two types of PLDs: a C2 domain-containing type, which is unique to plants, and a PX-PH domain-containing type, which is common to most eukaryotes (Qin and Wang, 2002). Of those PLDs, the Arabidopsis PX-PH-type PLD PLDζ1 is involved in root hair cell patterning and morphogenesis (Ohashi *et al.*, 2003). Overexpression of PLDζ1 caused bulge formation from all root epidermal cell files and branched root hairs at a high frequency, while reduced PLDζ1 expression disrupted the planar polarity of bulge formation and the tip-growth polarity of root hair elongation. These findings indicate the involvement of PLDζ1 in the establishment and sustentation of root hair tip growth. Notably, PIP5K and PLD possibly constitute a positive feedback loop similar to those in animals (Oude Weernink *et al.*, 2007) because many plant PLDs, including PLDζ1, are activated by PtdIns(4,5)P_2 (Qin and Wang, 2002), and PA activates PIP5K1 through the MORN domain which is conserved in type B PIP5Ks (Im *et al.*, 2007).

DAG and Ins(1,4,5)P_3 are crucial second messengers produced via the PLC-mediated hydrolysis of PtdIns(4,5)P_2 in animals. So far however, orthologs of their effector proteins, such as protein kinase C (PKC) and Ins(1,4,5)P_3 receptor Ca^{2+} channels, have not been identified in land plants (Meijer and Munnik, 2003; Wheeler and Brownlee, 2008). Since DAG is rapidly converted to PA by DGK, PA likely acts as a second messenger downstream of PLC rather than DAG (Meijer and Munnik, 2003; Arisz *et al.*, 2009), while the possibility remains that DAG functions as a signaling molecule by itself (Dong *et al.*, 2012). As for Ins(1,4,5)P_3 function in plant cells, high-affinity Ins(1,4,5)P_3-binding sites have been detected in internal membranes and Ins(1,4,5)P_3 could trigger the release of calcium ions from internal stores (Alexsandre *et al.*, 1990; Tang *et al.*, 2007). In an experiment using caged probes, the release of Ins(1,4,5)P_3, as well as that of PI(4,5)P_2, resulted in an increase of pollen tube cytosolic Ca^{2+} and reorientation of pollen tube growth (Monteiro *et al.*, 2005). These results suggest that Ins(1,4,5)P_3 is recognized in plant cells by different effectors from those in animal cells.

Signal Integration by Phosphoinositide Interactors

Phosphoinositides bind to effector proteins with various molecular functions (Table 9.1). Some effector proteins, such as actin-binding proteins and clathrin assembly proteins, act as output devices of signaling pathways to regulate particular intracellular events, for example, actin cytoskeletal reorganization and membrane traffic. They are generally conserved in eukaryotes, and hence expected to play the same roles in both animals and plants. On the other hand, some act as nodes of signaling pathways to integrate phosphoinositide signals with other signals, including those of protein phosphorylation, small-GTPases, ROS, and Ca^{2+}. This section describes phosphoinositide interacting proteins that integrate phosphoinositide signals into various signaling pathways for root hair tip growth.

AGC protein kinases constitute a large protein kinase family in both animals and plants, and many of them contain phospholipid-binding domains (Pearce *et al.*, 2010). The 3'-phosphoinositide-dependent kinase PDK1 is a member of AGC protein kinases. Although animal PDK1 acts downstream in PI(3)K signaling via the binding to 3'-phosphoinositide such as PtdIns(3,4,5)P_3 (Alessi, *et al.* 1997), the Arabidopsis PDK1 preferentially bound to PA *in vitro*, and was activated by PA and PtdIns(4,5)P_2 in Arabidopsis protoplasts (Anthony *et al.*, 2004). In addition, the PA-dependent PDK1 activity (but not the PtdIns(4,5)P_2-dependent activity) lead to the activation of AGC2-1, another member of AGC protein kinases, a defect of which resulted in a short-root-hair phenotype (Anthony *et al.*, 2004). Interestingly, OXIDATIVE SIGNAL-INDUCIBLE 1 (OXI1, a protein kinase mediating oxidative stress responses) is identical to AGC2-1 (Rentel *et al.*, 2004). Moreover, defects in the *INCOMPLETE ROOT HAIR ELONGATION* (*IRE*) gene, which encodes another AGC protein kinase, caused a similar short-root-hair phenotype to *agc2-1* mutants (Oyama *et al.*, 2002). These findings, together with the existence of numerous AGC kinases in plants (Garcia *et al.*, 2012), suggest that AGC protein kinase cascades integrate various signals including phospholipids and ROS in the regulation of root hair tip growth. It is noteworthy that PDK1 phosphorylates PIONOID (PID), the AGC protein kinase that modulates auxin polar transport via the phosphorylation of PINFORMED (PIN) auxin transporters (Zegzouti *et al.*, 2006; Huang *et al.*, 2010). The signaling cascade PKD1-PID-PINs might mediate phospholipid signals, including PA and PtdIns(4,5)P_2, to the auxin signal pathway for root hair morphogenesis.

ARFs are small GTPases playing critical roles in multiple steps of membrane traffic in both animal and plant cells (D'Souza-Schorey and Chavrier, 2006; Nielsen *et al.*, 2008). Mutants of the Arabidopsis class I ARF-GAP gene *ARF-GAP DOMAIN 1* (*AGD1*) showed wavy root hairs (Yoo *et al.*, 2008). AGD1 contains a phosphoinositide-binding PH domain, and specifically binds to PtdIns monophosphates *in vitro* (Yoo *et al.*, 2012). Mutations in *PIP5K3* and *COW1/SFH1* genes enhanced the root-hair phenotype of *agd1*, but the *rhd4-1* mutation was epistatic to *agd1* (Yoo *et al.*, 2012). These facts suggest a tight connection between phosphoinositide metabolism, especially that of PtdIns(4)P, and the regulation of membrane traffic in root hair tip growth, while it is still unclear which ARF species AGD1 specifically interacts with.

PCaP2 is an Arabidopsis calcium-binding protein that also binds to the Ca^{2+}/calmodulin complex and phosphoinositides including PtdIns bisphosphates and PtdIns(3,4,5)P_3 *in vitro* (Kato *et al.*, 2010). It is predominantly expressed in root hair cells, and uniformly localizes throughout the plasma membrane (Kato *et al.*, 2010). Transgenic plants overexpressing full-length or N-terminal-truncated PCaP2 exhibited branched and bulbous root hairs, while a knockdown mutant of the gene exhibited longer root hairs than the wild type (Kato *et al.*, 2013). Interestingly, overexpression of its N-terminal basic peptide (N23), which contains the binding ability for both the Ca^{2+}/calmodulin complex and phosphoinositides, completely suppressed root hair development even in the root-hypocotyl transit region where long root hairs actively develop in the wild type. Since PtdIns(4,5)P_2 is the most probable phosphoinositide that PCaP2 can specifically recognize in the plasma membrane, it is suggested that PCaP2 negatively regulates root hair tip growth via the binding of its N23 moiety to PtdIns(4,5)P_2 in the plasma membrane. Supporting this idea, the phenotype of N23-overexpressing plants was partly rescued by the expression of PIP5K3-YFP in root hair cells. The binding of the Ca^{2+}/calmodulin complex to PCaP2 suppresses the binding of phosphoinositides (Kato *et al.*, 2010), suggesting that PCaP2 integrates the Ca^{2+} and phosphoinositide signals in the regulation of root hair tip growth by physically masking PtdIns(4,5)P_2 under low Ca^{2+} concentration conditions.

A Hypothetical Mechanism for Sustaining the Tip Growth Polarity

During root hair tip growth, the polarity is strictly sustained which results in straight root hairs without branches. For the sustentation of this highly polarized cell growth, a mechanism called tip-growth LENS (localization-enhancing network, self-sustaining) has been hypothesized (Cole and Fowler, 2006). The tip-growth LENS not only localizes events underlying cell expansion, such as cytoskeletal reorganization and membrane traffic, but also sustains the localization of its own components at the tip against the diffusion concomitant with cell expansion. Basically the same mechanism is thought to function for pollen tube tip growth, while its polarity is dynamically changed according to pollen tube guidance (Cole and Fowler, 2006; Kroeger and Geitmann, 2012; Guan *et al.*, 2013). Most importantly, tip-growth LENS is assumed to contain a positive feedback mechanism for the self-sustentation, and various combinations of signaling factors possibly constitute positive feedback loops in root hair tip growth.

In eukaryotes, Rho-family GTPases play a pivotal regulatory role in establishing cell polarity (Etienne-Manneville and Hall, 2002; Ridley, 2006; Heasman and Ridley, 2008; Hall, 2012). The Rho family generally comprises Rho, Rac, and Cdc42 in animals, and Rho and Cdc42 in yeasts (Takai *et al.*, 2001). On the other hand, Rho-family GTPases in plants constitute a unique class of proteins that are most similar to Rac in their primary structures; these are referred to as ROP or Rac/ROP (Zheng and Yang, 2000; Berken *et al.*, 2005). In many cases that cell polarity is established, the Rho-family GTPase activity accumulates at a specific locus on the plasma membrane via positive feedback loop(s) involving itself and its interacting factors. For example, in budding yeast cells, Cdc42 activity accumulates at the tip of a growing bud via a dual positive feedback mechanism in which Cdc42 recruits itself and its positive regulator (the Rho-GEF protein Cdc24) by promoting actin-based transport and conformational changes in a scaffolding protein, respectively, and then Cdc24 activates Cdc42 at the tip (Wedlich-Soldner *et al.*, 2003; Irazoqui *et al.*, 2003). In migrating neutrophil cells, Rac activity accumulates at the leading edge via a positive feedback loop, in which Rac activates a PI(3) kinase and the kinase product PtdIns(3,4,5)P_3 recruits and activates Rho-GEF, which in turn activates Rac (Weiner *et al.*, 2002).

The localized accumulation of ROPs has also been observed in various plant cells, including growing root hairs and pollen tubes (Kost *et al.*, 1999; Li *et al.*, 1999; Molendijk *et al.*, 2001; Jones *et al.*, 2002; Craddock *et al.*, 2012; Yang and Lavagi, 2012). Using a specific antibody and fluorescence protein tags in Arabidopsis, ROPs have been shown to localize to the apical plasma membrane of growing root hairs and to the future bulge formation site of trichoblasts (Molendijk *et al.*, 2001; Jones *et al.*, 2002). ROP overexpression enhanced root hair elongation, and disturbed the tip-growth polarity of growing root hairs and the planar polarity of trichoblasts (Molendijk *et al.*, 2001; Jones *et al.*, 2002). In contrast, a *rop2* mutant exhibited shorter root hairs than the wild type (Kusano *et al.*, unpublished data). These results support the idea that ROP is a constituent of the tip-growth LENS in root hairs. Interestingly, PIP5K3 showed the same characteristics as ROP2 not only in its intracellular localization pattern in growing root hairs and trichoblasts, but also in its mutant and overexpressor phenotypes (Kusano *et al.*, 2008). These strongly suggest that PIP5K3 and ROP2 are very close to each other in a regulatory pathway for root hair tip growth. Also in tobacco pollen tubes, overexpression of Arabidopsis PIP5K10 or PIP5K11 resulted in severe tip swelling and altered actin fine structure similar to those caused by the overexpression of the tobacco Rho-family GTPase Nt-Rac5 (Klahre *et al.*, 2006; Ischebeck *et al.*, 2011). Although there is no evidence for the direct interaction between PIP5K and ROP in pollen tubes, it has been reported that the Arabidopsis Rac2/ROP7 physically associates with PIP5K activity in tobacco pollen tubes (Kost *et al.* 1999). In animal cells, type I PIPKs (which correspond to PIP5Ks of Arabidopsis) are positively regulated by Rho-family GTPases as direct interactors of Rho and Rac and as indirect downstream factors of Cdc42 (Hartwig *et al.*, 1995; Matsui *et al.*, 1999; Takai *et al.*, 2001; Oude Weernink *et al.*, 2004b). Considering all these facts, PIP5K3 is very likely to function as an effector of ROP2 in root hair tip growth.

It is noteworthy that PtdIns(4,5)P_2 has various potentialities to be a positive regulator of Rho-family GTPases in animal systems. First, many Rho-GEFs and other positive regulators of Rho-family GTPases have been shown to physically interact with PtdIns(4,5)P_2 (Viaud *et al.*, 2012). In addition, phosphoinositides including PtdIns(4,5)P_2 are assumed to function as a RhoGDI-displacement factor (Rho-GDF), which promotes the membrane re-association of Rho-family GTPases by destabilizing the cytosolic complex between GDP-bound Rho-family GTPases and Rho-GDI (Faure *et al.*, 1999). Moreover, it has been reported that C-terminal polybasic clusters of animal Rho-family GTPases contribute to the plasma membrane localization via the binding to PtdIns(4,5)P_2 (Heo *et al.*, 2006; Johnson *et al.*, 2012). These

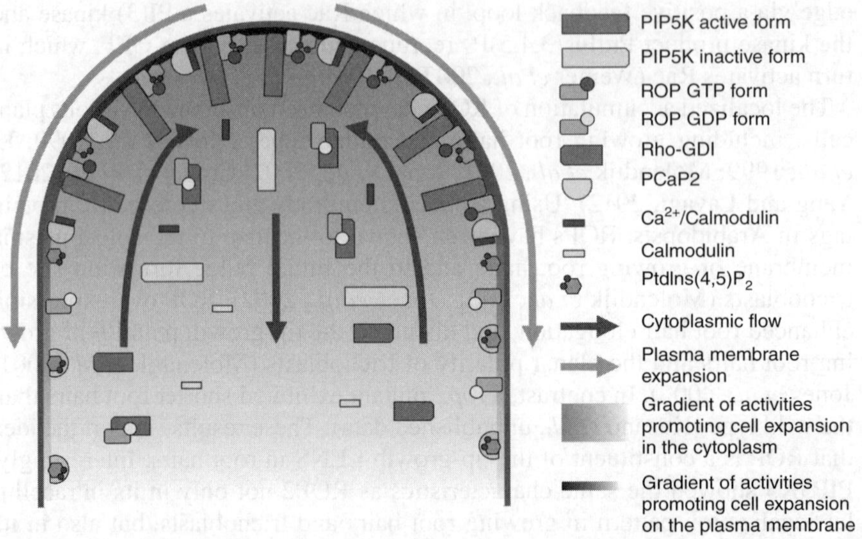

Figure 9.4 A hypothetical mechanism for sustaining the tip-growth polarity in Arabidopsis root hairs is schematically illustrated, focusing on the signaling roles of PIP5K, PtdIns(4,5)P_2, ROP, Rho-GDI, PCaP2, and calmodulin. The reverse fountain cytoplasmic flow and the plasma membrane expansion are depicted by arrows. The gradients of activities promoting cell expansion are illustrated by gradations in the cytoplasm and on the plasma membrane.

potentialities of PtdIns(4,5)P_2, in conjunction with the possibility of PIP5K activation by ROPs, suggest that ROPs, PIP5Ks, and PtdIns(4,5)P_2 can constitute a positive feedback loop for the tip-growth LENS in root hairs.

We propose a model of the tip-growth LENS for sustaining the polarity of growing root hairs, focusing on the roles of PtdIns(4,5)P_2 and ROP (Fig. 9.4). In the model, GTP-bound active ROP recruits and/or activates PIP5K at the apical plasma membrane, and PtdIns(4,5)P_2 produced by PIP5K recruits and/or activates ROP. This positive feedback mechanism results in the tip-focused accumulation of active ROP and PtdIns(4,5)P_2, both of which cooperatively stimulate exocytosis by promoting actin cytoskeletal reorganization and membrane traffic. The exocytosis brings not only materials for synthesizing the expanding plasma membrane and cell wall, but also membrane-associated signaling factors to the root hair apex. Those signaling factors, including Ca^{2+} channels and the NADPH oxidase RHD2, contribute to the robustness of the positive feedback mechanism. All these factors are however necessarily left behind the apex due to the membrane expansion concomitant with the tip growth. The laterally diffused ROP and PtdIns(4,5)P_2 signals might be cancelled actively in the subapical region by Rho-GAP and PLC, respectively, as in growing pollen tubes (Dowd *et al.*, 2006; Helling *et al.*, 2006; Hwang *et al.*, 2008). Moreover, PCaP2 might spatiotemporally modulate the PtdIns(4,5)P_2 signal by masking it under low Ca^{2+} concentration conditions (Kato *et al.*, 2010, 2013). In the subapical region, GDP-bound inactive ROP is withdrawn into the Rho-GDI complex in the cytoplasmic space, and brought back to the apex by the reverse fountain stream in the root hair cytoplasm (Miller *et al.*, 1997) together with other signaling factors released from the plasma membrane. In addition, membrane-integrated proteins may also be brought back to the apical plasma membrane by a constant retrograde membrane flow from the subapical region (Hepler *et al.*, 2001; Samaj *et al.*, 2005).

The model is thought to be applicable to the planar polarity establishment in root hair bulge formation to some extent, because the planar polarity establishment is also summed to require a positive feedback mechanism for making a single peak of signal activity on the trichoblast plasma membrane from a spatial clue by auxin (Fischer *et al.*, 2006). This idea is supported by the fact that fluorescence protein fusions of ROP2 and PIP5K3 localized to the future bulge-forming site on the trichoblast plasma membrane, and that both ROP2 and PIP5K3 overexpressors exhibited multiple root hairs from a single trichoblast (Jones *et al.*, 2002; Kusano *et al.*, 2008). Before the exocytosis for bulge formation starts, the positive feedback loop involving ROP2 and PIP5K3 is assumed to establish the planar polarity.

Concluding Remarks

Phosphoinositide signals are suitable to promote cell polarity due to the fact that they act as site-specific signals on membranes to localize intracellular events. In addition, their signaling pathways are connected to other pathways via a wide variety of phosphoinositide-interacting proteins, allowing robust and elaborate regulations of cell polarity. The latter characteristics and other complicated features of phosphoinositide signals prevent us from understanding their signaling pathways simply, however. PtdIns(4,5)P_2 localizes to the plasma membrane in the root hair apical region, and is considered to play a pivotal role in regulating actin cytoskeletal reorganization and membrane traffic, both of which are essential events for root hair tip growth. Moreover, PtdIns(4,5)P_2 together with PIP5K and ROP possibly act as a substantial component of the tip-growth LENS, the mechanism for sustaining the tip-growth polarity. At present however, our knowledge of the molecular basis of phosphoinositide signaling pathways and the regulatory network encompassing them are still fragmented. To elucidate, knowledge of the molecular basis of the tip-growth LENS for growing root hairs will lead to the understanding of a general mechanism through which plant cells establish and sustain their cell polarities.

References

Alessi, D.R., Deak, M., Casamayor, A., Caudwell, F.B., Morrice, N., Norman, D.G., Gaffney, P., Reese, C.B., MacDougall, C.N., Harison, D., Ashworth, A., and Bownes, M. (1997) 3-phosphoinositide-dependent protein kinase-1 (PDK1): structural and functional homology with the drosophila DSTPK61 kinase. *Current Biology*, **7**, 776–789.

Alexandre, J., Lassalles, J.F., and Kado, R.D. (1990) Opening of Ca^{2+} channels in isolated red beet vacuole membrane by inositol 1,4,5-trisphosphate. *Nature*, **343**, 567–570.

Alvarez-Venegas, R., Sadder, M., Hlavacka, A., Baluska, F., Xia, Y., Lu, G., Firsov, A., Sarath, G., Moriyama, H., Dubrovsky, J.G., and Avramova, Z. (2006) The *Arabidopsis* homolog of trithorax, ATX1, binds phosphatidylinositol 5-phosphate, and the two regulate a common set of target genes. *Proceedings of the National Academy of Science, USA*, **103**, 6049–6054.

Anderson, R.A., Boronenkov, I.V., Doughman, S.D., Kunz, J., and Loijens, J.C. (1999) Phosphatidylinositol phosphate kinase, multifaceted family of signaling enzymes. *Journal of Biological Chemistry*, **274**, 17794–17805.

Anthony, R.G., Henriques, R., Helfer, A., Mezaros, T., Rios, G., Testerink, C., Munnik, T., Deak, M., Koncz, C., and Bogre, L. (2004) A protein kinase target of a PDK1 signalling pathway is involved in root hair growth in *Arabidopsis*. *EMBO Journal*, **23**, 572–581.

Arisz, S.A., Testerink, C., and Munnik, T. (2009) Plant PA signaling via diacylglycerol kinase. *Biochimica et Biophysica Acta*, **1791**, 869–875.

Balkunde, R., Pesch, M., and Hulskamp, M. (2010) Trichome patterning in *Arabidopsis thaliana* from genetic to molecular models to molecular models. *Current Topics in Developmental Biology*, **91**, 299–321.

Bankaitis, V.A. and Morris, A.J. (2003) Lipids and the exocytotic machinery of eukaryotic cells. *Current Opinion in Cell Biology*, **15**, 389–395.

Barlow, C.A., Laishram, R.S., and Anderson, R.A. (2010) Nuclear phosphoinositides: a signaling enigma wrapped in a compartmental conundrum. *Trends in Cell Biology*, **20**, 25–35.

Behnia, R. and Munro, S. (2005) organelle identity and the signposts for membrane traffic. *Nature*, **438**, 597–604.

Berken, A., Thomas, C., and Wittinghofer, A. (2005) A new family of RhoGEFs activates the Rop molecular switch in plants. *Nature*, **436**, 1776–1180.

Berridge, M.J. and Irvine, R.F. (1984) Inositol triphosphate, a novel second messenger in cellular signal transduction. *Nature*, **312**, 315–321.

Bibikova, T.N. and Gilroy, S. (2002) Root hair development. *Journal of Plant Growth Regulation*, **21**, 383–415.

Bohme, K., Li, Y., Charlot, F., Grierson, C., Marrocco, K., Okada, K., Laloue, M., and Nogue, F. (2004) The *Arabidopsis COW1* gene encodes a phosphatidylinositol transfer protein essential for root hair tip growth. *Plant Journal*, **40**, 686–698.

Boss, W.F. and Im, Y.J. (2012) Phosphoinositide signaling. *Annual Review of Plant Biology*, **63**, 409–429.

Bourne, Y., Dannenberg, J., Pollmann, V., Marchot, P., and Pongs, O. (2001) Immunocytochemical localization and crystal structure of human frequenin (neuronal calcium sensor 1). *Journal of Biological Chemistry*, **276**, 11949–11955.

Braun, M., Baluska, F., von Witsch, M., and Menzel, D. (1999) Redistribution of actin, profiling and phosphatidylinositol-4,5-bisphosphate in growing and maturing root hairs. *Planta*, **209**, 435–443.

Burman, C. and Ktistakis, N.T. (2010) Regulation of autophagy by phosphatidylinositol 3-phosphate. *FEBS Letters*, **584**, 1302–1312.

Cardenas, L. (2009) New findings in the mechanisms regulating polar growth in root hair cells. *Plant Signal Behavior*, **4**, 4–8.

Carol, R.J. and Dolan, L. (2002) Building a hair: tip growth in *Arabidopsis thaliana* root hairs. *Philosophical Transactions of the Royal Society of London, B: Biological Sciences*, **357**, 815–1821.

Carol, R.J., Takeda, S., Linstead, P., Durrant, M.C., Kakesova, H., Derbyshire, P., Crea, S., Zarsky, V., and Dolan, L. (2005) A RhoGDP dissociation inhibitor spatially regulates growth in root hair cells. *Nature*, **438**, 1013–1016.

Chan, T.O., Rittenhouse, S.E., and Tsichlis, P.N. (1999) AKT/PBK and other D3 phosphoinositide-regulated kinase: kinase activation by phosphoinositide-dependent phosphorylation. *Annual Reviews of Biochemistry*, **68**, 965–1014.

Cole, R.A. and Fowler, J.E. (2006) Polarized growth: maintaining focus on the tip. *Current Opinion in Plant Biology*, **9**, 579–588.

Comer, F.I. and Parent, C.A. (2007) Phosphoinositides specify polarity during epithelial organ development. *Cell*, **128**, 239–240.

Cote, J.F., Motoyama, A.B., Bush, J.A., and Vuori, K. (2005) A novel and evolutionarily conserved PtdIns(3,4,5)P_3-binding domain is necessary for DOCK180 signalling. *Nature Cell Biology*, **7**, 797–807.

Craddock, C., Lavagi, I., and Yang, Z. (2012) New insights into Rho signaling from plant ROP/Rac GTPases. *Trends in Cell Biology*, **22**, 492–501.

Das, B., Shu, X.D., Day, G.J., Han, J., Krishna, U.M., Falck, J.R., and Broek, D. (2000) Control of intermolecular interactions between the pleckstrin homology and Db1 homology domains of Vav and Sos1 regulates Rac binding. *Journal of Biological Chemistry*, **275**, 15074–15081.

Datta, S., Kim, C.M., Pernas, M., Pires, N.D., Proust, H., Tam, T., Vijayakumar, P., and Dolan, L. (2011) Root hairs: development, growth and evolution at the plant-soil interface. *Plant Soil*, **346**, 1–14.

de Graaf, P., Zwart, W.T., van Dijken, R.A., Deneka, M., Schulz, T.K., Geijsen, N., Coffer, P.J., Gadella, B.M., Verkleij, A.J., van der Sluijs, P., and van Bergen en Henegouwen, P.M. (2004) Phosphatidylinositol 4-kinase beta is critical for functional association of rab11 with the Golgi complex. *Molecular Biology of the Cell*, **15**, 2038–2047.

Di Paolo, G. and De Camilli, P. (2006) Phosphoinositides in cell regulation and membrane dynamics. *Nature*, **443**, 651–675.

Di Paolo, G., Moskowitz, H.S., Gipson, K., Wenk, M.R., Voronov, S., Obayashi, M., Flavell, R., Fitzsimonds, R.M., Ryan, T.A., and De Camilli, P. (2004) Impaired PtdIns(4,5)P2 synthesis in nerve terminals produces defects in synaptic vesicle trafficking. *Nature*, **431**, 415–422.

Dolan, L. (2001) How and where to build a root hair. *Current Opinion in Plant Biology*, **4**, 550–554.

Dolan, L., Duckett, C.M., Grierson, C., Linstead, P., Schneider, K., Lawson, E., Dean, C., Poethig, S., and Roberts, K. (1994). Clonal relationships and cell patterning in the root epidermis of *Arabidopsis*. *Development*, **120**, 2465–2474.

Dong, W., Lv, H., Xia, G., and Wang, M. (2012) Does diacylglycerol serve as a signaling molecule in plants? *Plant Signal Behavior*, **7**, 472–475.

Dowd, P.E., Coursol, S., Skirpan, A.L., Kao, T.H., and Gilroy, S. (2006) Petunia phospholipase c1 is involved in pollen tube growth. *Plant Cell*, **18**, 1438–1453.

D'Souza-Schorey, C. and Chavrier, P. (2006) ARF proteins: roles in membrane traffic and beyond. *Nature Reviews Molecular Cell Biology*, **7**, 347–358.

Duan, Q., Kita, D., Li, C., Cheung, A.Y., and Wu, H.M. (2010) FERONIA receptor-like kinase regulates RHO GTPase signaling of root hair development. *Proceedings of the National Academy of Science, USA*, **107**, 17821–17826.

Etienne-Manneville, S. and Hall, A. (2002) RhoGTPases in cell biology. *Nature*, **420**, 629–635.

Faure, J., Vignais, P.V., and Dagher M-C. (1999) Phosphoinositide-dependent activation of RhoA involves partial opening of the RhoA/Rho-GDI complex. *European Journal of Biochemistry*, **262**, 879–889.

Fischer, U., Men, S., and Grebe, M. (2004) Lipid function in plant cell polarity. *Current Opinion in Plant Biology*, **7**, 670–676.

Fischer, U., Ikeda, Y., Ljung, K., Serralbo, O., Singh, M., Heidstra, R., Palme, K., Scheres, B., and Grebe, M. (2006). Vectorial information for *Arabidopsis* planar polarity is mediated by combined AUX1, EIN2, and GNOM activity. *Current Biology*, **16**, 2143–2149.

Foreman, J., Demidchik, V., Bothwell, J.H., Mylona, P., Miedema, H., Torres, M.A., Linstead, P., Costa, S., Brownlee, C., Jones, J.D., Davies, J.M., and Dolan, L. (2003) Reactive oxygen species produced by NADPH oxidase regulate plant cell growth. *Nature*, **422**, 442–446.

Fruman, D.A., Meyers, R.E., and Cantley, L.C. (1998) Phosphoinositide kinases. *Annual Review of Biochemistry*, **67**, 481–507.

Gagne, J.M. and Clark, S.E. (2010) The *Arabidopsis* stem cell factor POLTERGEIST is membrane localized and phospholipid stimulated. *Plant Cell*, **22**, 729–743.

Galway, M., Masucci, J., Lloyd, A., Walbot, V., Davis, R., and Schiefelbein, J. (1994). The *TTG* gene is required to specify epidermal cell fate and cell patterning in the *Arabidopsis* root. *Developmental Biology*, **166**, 740–754.

Gamper, N. and Shapiro, M.S. (2007) Target-specific PIP_2 signaling: How might it work? *Journal of Physiology*, **582**, 967–975.

Gamper, N., Reznikov, V., Yamada, Y., Yang, J., and Shapiro, M.S. (2004) Phophatidylinositol 4,5-bisphosphate signals underlie receptor-specific $G_{q/11}$-mediated modulation of N-type Ca^{2+} channels. *Journal of Neuroscience*, **24**, 10980–10992.

Garcia, A.V., Al-Yousif, M., and Hirt, H. (2012) Role of AGC kinases in plant growth and stress responses. *Cell and Molecular Life Sciences*, **69**, 3259–3267.

Gassama-Diagne, A., Yu, W., ter Beest, M., Martin-Belmonte, F., Kierbel, A., Engel, J., and Mostov, K. (2006) Phosphatidylinositol-3,4,5-trisphosphate regulates the formation of the basolateral plasma membrane in epithelial cells. *Nature Cell Biology*, **8**, 963–970.

Gilroy, S. and Jones, D.L. (2000) Through form to function: root hair development and nutrient uptake. *Trends in Plant Science*, **5**, 56–60.

Gong, L.W., Di Paolo, G., Diaz, E., Cestra, G., Diaz, M.E., Lindau, M., De Camilli, P., and Toomre, D. (2005) Phosphatidylinositol phosphate kinase type Ig regulates dynamics of large dense-core vesicle fusion. *Proceedings of the National Academy of Science, USA*, **102**, 5204–5209.

Grebe, M. (2012) The patterning of epidermal hairs in *Arabidopsis*– updated. *Current Opinion in Plant Biology*, **15**, 31–37.

Grishanin, R.N., Kowalchyk, J.A., Klenchin, V.A., Ann, K., Earles, C.A., Chapman, E.R., Gerona, R.R., and Martin, T.F. (2004) CAPS acts at a prefusion step in dens-core vesicle exocytosis as a PIP_2 binding protein. *Neuron*, **43**, 551–562.

Gu, F. and Nielsen, E. (2013) Targeting and regulation of cell wall synthesis during tip growth in plants. *Journal of Integrative Plant Biology*, **55**, 835–846.

Guan, Y., Guo, J., Li, H., and Yang, Z. (2013) Signaling in pollen tube growth: crosstalk, feedback, and missing links. *Molecular Plant*, **6**, 1053–1064.

Hall, A. (2012) Rho family GTPases. *Biochemical Society Transactions*, **40**, 1378–1382.

Han, J., Luby-Phelps, K., Das, B., Shu, X., Xia, Y., Mosteller, R.D., Krishna, U.M., Falck, J.R., White, M.A., and Broek, D. (1998) Role of substrates and products of PI 3-kinase in regulating activation of Rac-related guanosine triphosphatases by Vav. *Science*, **279**, 558–560.

Hartwig, J. H., Bokoch, G. M., Carpenter, C. L., Janmey, P. A., Taylor, L. A., Toker, A., and Stossel, T. P. (1995) Thrombin receptor ligation and activated Rac uncap actin filament barbed ends through phosphoinositide synthesis in permeabilized human platelets. *Cell*, **82**, 643–653.

Hawkins, P.T., Anderson, K.E., Davidson, K., and Stephens, L.R. (2006) Signalling through class I PI3Ks in mammalian cells. *Biochemical Society Transactions*, **34**, 647–662.

Heasman, S.J. and Ridley, A.J. (2008) Mammalian Rho GTPase: new insights into their functions from in vivo studies. *Nature Reviews Molecular Cell Biology*, **9**, 690–701.

Heilmann, I. (2008) Using genetic tools to understand plant phosphoinositide signaling. *Trends in Plant Science*, **14**, 171–179.

Helling, D., Possart, A., Cottier, S., Klahre, U., and Kost, B. (2006) Pollen tube tip growth depends on plasma membrane polarization mediated by tobacco PLC3 activity and endocytic membrane recycling. *Plant Cell*, **18**, 3519–3534.

Hendricks, K.B., Wang, B.Q., Schnieders, E.A., and Thorner, J. (1999) Yeast homologue of neuronal frequenin is a regulator of phosphatidylinositol-4-OH kinase. *Nature Cell Biology*, **1**, 234–241.

Heo, W.D., Inoue, T., Park, W.S., Kim, M.L., Park, B.O., Wandless, T.J., and Meyer, T. (2006) $PI(3,4,5)P_3$ and $PI(4,5)P_2$ lipids target proteins with polybasic clusters to the plasma membrane. *Science,* **314**, 1458–1461.

Hepler, P.K., Vidali, L., and Cheung, A.Y. (2001). Polarized cell growth in higher plants. *Annual Reviews of Cell and Developmental Biology*, **17**, 159–187.

Huang, F., Zago, M.K., Abas, L., van Marion, A., Galván-Ampudia, C.S., and Offringa, R. (2010). Phosphorylation of conserved PIN motifs directs *Arabidopsis* PIN1 polarity and auxin transport. *Plant Cell*, **22**, 1129–1142.

Huang, S., Gao, L., Blanchoin, L., and Staiger, C.J. (2006). Heterodimeric capping protein from *Arabidopsis* is regulated by phosphatidic acid. *Molecular Biology of the Cell*, **17**, 1946–1958.

Huttner, I.G., Strahl, T., Osawa, M., King, D.S., Ames, J.B., and Thorner, J. (2003) Molecular interactions of yeast frequenin (Frq1) with the phosphatidylinositol 4-kinase isoform, Pik1. *Journal of Biological Chemistry*, **278**, 4862–4874.

Hwang, J.U., Vernoud, V., Szumlanski, A., Nielsen, E., and Yang, Z. (2008) A tip- localized RhoGAP controls cell polarity by globally inhibiting Rho GTPase at the cell apex. *Current Biology*, **18**, 1907–1916.

Im, Y.J., Perera, I.Y., Brglez, I., Davis, A.J., Stevenson-Paulik, J., Phillippy, B.Q., Johannes, E., Allen, N.S., and Boss, W.F. (2007) Increasing plasma membrane phosphatudylinositol(4,5)bisphosphate biosynthesis increases phosphoinositide metabolism in *Nicotiana tabacum. Plant Cell*, **19**, 1603–1616.

Irazoqui, J.E., Gladfelter, A.S., and Lew, D.J. (2003) Scaffold-mediated symmetry breaking by Cdc42p. *Nature Cell Biology,* **5**, 1062–1070.

Ischebeck, T., Stenzel, I., Hempel, F., Jin, X., Mosblech, A. and Heilmann, I. (2011) Phosphatidylinositol-4,5-bisphosphate influences Nt-Rac5-mediated cell expansion in pollen tubes of Nicotiana tabacum. *Plant Journal*, **65**, 453–468.

Itoh, T., Koshiba, S., Kigawa, T., Kikuchi, A., Yokoyama, S., and Takenawa, T. (2001) Role of the ENTH domain in phosphatidylinositol-4,5,-bisphosphate binding and endocytosis. *Science*, **291**, 1047–1051.

Jang, J.-H., Lee, C.S., Hwang, D., and Ryu, S.H. (2012) Understanding of the roles of phospholipase D and phosphatidic acid through their binding partners. *Progress in Lipid Research*, **51**, 71–81.

Jenkins, G.M. and Frohman, M.A. (2005) Phospholipase D: a lipid centric review. *Cell and Molecular Life Sciences*, **62**, 2305–2316.

Johnson, J.L., Erickson, J.W. and Cerione, R.A. (2012) C-terminal di-arginine motif of Cdc42 protein is essential for binding to phosphatidylinositol 4,5-bisphosphate-containing membranes and inducing cellular transformation. *Journal of Biological Chemistry*, **287**, 5764–5774.

Jones, M.A., Shen, J.J., Fu, Y., Li, H., Yang, Z., and Grierson, C.S. (2002) The Arabidopsis Rop2 GTPase is a positive regulator of both root hair initiation and tip growth. *Plant Cell*, **14**, 763–776.

Jones, M.A., Raymond, M.J., Yang, Z., and Smirnoff, N. (2007) NADPH oxidase-dependent reactive oxygen species formation required for root hair growth depends on ROP GTPase. *Journal of Experimental Botany*, **58**, 1261–1270.

Jones, V.A.S. and Dolan, L. (2012) The evolution of root hairs and rhizoids. *Annals of Botany*, **110**, 205–212.

Kaadige, M.R. and Ayer, D.E. (2006) The polybasic region that follows the plant homeodomain zinc finger 1 of Pf1 is necessary and sufficient for specific phosphoinositide binding. *Journal of Biological Chemistry*, **281**, 28831–28836.

Kam, J.L., Miura, K., Jackson, T.R., Gruschus, J., Roller, P., Stauffer, S., Clark, J., Aneja, R., and Randazzo, P.A. (2000) Phosphoinositide-dependent activation of the ADP-ribosylation factor GTPase-activating protein ASAP1. Evidence for the pleckstrin homology domain functioning as an allosteric site. *Journal of Biological Chemistry*, **275**, 9653–9663.

Kapranov, P., Routt, S.M., Bankaitis, V.A., de Bruijn, F.J., and Szczglowski, K. (2001) Nodule-specific regulation of phosphatidylinositol transfer protein expression in *Lotus japonicus. Plant Cell*, **13**, 1369–1382.

Kato, M., Nagasaki-Takeuchi, N., Ide, Y. and Maeshima, M. (2010) An *Arabidopsis* hydrophilic Ca^{2+}-binding protein with a PEVK-rich domain, PCaP2, is associated with the plasma membrane and interacts with calmodulin and phosphatidylinositol phosphates. *Plant Cell Physiology*, **51**, 366–379.

Kato, M., Aoyama, T., and Maeshima, M. (2013) The Ca^{2+}-binding protein PCaP2 located on the plasma membrane is Involved in root hair development as a possible signal transducer. *Plant Journal*, **74**, 690–700.

Keune, W., Bultsma, Y., Sommer, L., Jones, D., and Divecha, N. (2011) Phosphoinositide signalling in the nucleus. *Advances in Enzyme Regulation*, **51**, 91–99.

Klahre, U., Becker, C., Schmitt, A.C., and Kost, B. (2006) Nt-RhoGDI2 regulates Rac/Rop signaling and polar cell growth in tobacco pollen tubes. *Plant Journal*, **46**, 1018–1031.

Klarlund, J.K., Rameh, L.E., Cantley, L.C., Buxton, J.M., Holik, J.J., Sakelis, C., Patki, V., Corvera, S., and Czech, M.P. (1998) Regulation of GRP1-catalyzed ADP rebosylation factor guanine nucleotide exchange by phosphatidylinositol 3,4,5-trisphosphate. *Journal of Biological Chemistry*, **273**, 1859–1862.

Kolesnikov, Y.S., Nokhrina, K.P., Kretynin, S.V., Volotovski, I.D., Martinec, J., Romanov, G.A., and Kravets, V.S. (2012) Molecular structure of phospholipase D and regulatory mechanisms of its activity in plant and animal cells. *Biochemistry*, **77**, 1–14.

Kost, B., Lemichez, E., Spielhofer, P., Hong, Y., Tolias, K., Carpenter, C., and Chua, N.-H. (1999) Rac homologues and compartmentalized phosphatidylinositol 4,5-bisphosphate act in a common pathway to regulate polar pollen tube growth. *Journal of Cell Biology*, **145**, 317–330.

Krauss, M. and Haucke, V. (2007) Phosphoinositides: Regulators of membrane traffic and protein function. *FEBS Letters*, **581**, 2105–2111.

Kroeger, J. and Geitmann, A. (2012) The pollen tube paradigm revisited. *Current Opinon in Plant Biology*, **15**, 618–624.

Kusano, H., Testerink, C., Vermeer, J.E., Tsuge, T., Shimada, H., Oka, A., Munnik, T. and Aoyama, T. (2008) The Arabidopsis Phosphatidylinositol Phosphate 5-Kinase PIP5K3 is a key regulator of root hair tip growth. *Plant Cell*, **20**, 367–380.

Kutateladze, T.G. (2010) Translation of the phophoinositide code by PI effectors. *Nature Chemical Biology*, **6**, 507–513.

Lee, Y.J. and Yang, Z. (2008) Tip growth: signaling in the apical dome. *Current Opinion in Plant Biology*, **11**, 662–671.

Lee, Y., Bak, G., Choi, Y., Chuang, WI., Cho, HT. and Lee, Y. (2008) Roles of phosphatidylinositol 3-kinase in root hair growth. *Plant Physiology*, **147**, 624–635.

Li, H., Lin, Y., Heath, R.M., Zhu, M.X., and Yang, Z. (1999) Control of pollen tube tip growth by a Rop GTPase-dependent pathway that leads to tip-localized calcium influx. *Plant Cell*, **11**, 1731–1742.

Li, L., Shin, O.-H., Rhee, J.-S., Arac, D., Rah, J.-C., Rizo, J., Sudhof, T., and Rosenmund, C. (2006) Phosphatidylinositol phosphates as co-activators of Ca^{2+} binding to C_2 domains of synaptotagmin 1. *Journal of Biological Chemistry*, **281**, 15845–15852.

Li, M., Hong, Y., and Wang, X. (2009) Phospholipase D- and Phosphatidic acid-mediated signaling in plants. *Biochimica et Biophysica Acta*, **1791**, 927–935.

Macia, E., Paris, S., and Chabre, M. (2000) Binding of the PH and polybasic C-terminal domains of ARNO to phosphoinositides and to acidic lipids. *Biochemistry*, **39**, 5893–5901.

Maffucci, T. (2012) An Introduction to Phosphoinositides. *Current Topics in Microbiology and Immunology*, **362**, 1–42.

Majerus, P.W., Kisseleva, M.V., and Norris, F.A. (1999) The role of phosphatases in inositol signaling reactions. *Journal of Biological Chemistry*, **274**, 10669–10672.

Martin-Belmonte, F., Gassama-Diagne, A., Datta, A., Yu, W., Rescher, U., Gerke, V., and Mostov, K. (2007) PTEN-mediated apical segregation of phosphoinositides controls epithelial morphogenesis through Cdc42. *Cell*, **128**, 383–397.

Matsui, T., Yonemura, S., Tsukita, S., and Tsukita, S. (1999) Activation of ERM proteins in vivo by Rho involves phosphatidylinositol 4-phosphate 5-kinase and not ROCK kinases. *Current Biology*, **9**, 1259–1262.

Mayinger, P. (2012) Phosphoinositides and vesicular membrane traffic. *Biochimica et Biophysica Acta*, **1821**, 1104–1113.

McLoughlin, F., Arisz, S.A., Dekker, H.L., Kramer, G., de Koster, C.G., Haring, M.A., Munnik, T., and Testerink, C. (2013) Identification of novel candidate phosphatidic acid-binding proteins involved in the salt-stress response of *Arabidopsis thaliana* roots. *Biochemical Journal*, **450**, 573–581.

Meijer, H.J.G. and Munnik, T. (2003) Phospholipid-based signaling in plants. *Annual Review of Plant Biology*, **54**, 265–306.

Mertens, A.E., Roovers, R.C., and Collard, J.G. (2003) Regulation of Tiam-Rac signaling. *FEBS Letters*, **546**, 11–16.

Miller, D.D., Norbert, C.A., de Ruijter, N.C.A., and Emons, A.M.C. (1997) From signal to form: aspects of the cytoskeleton-plasma membrane-cell wall continuum in root hair tips. *Journal of Experimental Botany*, **48**, 1881–1896.

Milosevic, I., Sørensen, J.B., Lang, T., Krauss, M., Nagy, G., Haucke, V., Jahn, R., and Neher, E. (2005) Plasmalemmal phosphatidylinositol-4,5-bisphosphate level regulates the releasable vesicle pool size in chromaffin cells. *Journal of Neuroscience*, **25**, 2557–2565.

Molendijk, A.J., Bischoff, F., Rajendrakumar, C.S., Friml, J., Braun, M., Gilroy, S., and Palme, K. (2001) *Arabidopsis thaliana* Rop GTPases are localized to tips of root hairs and control polar growth. *EMBO Journal*, **20**, 2799–2788.

Monteiro, D., Liu, Q., Lisboa, S., Sherer, G.E., Quader, H., and Malho, R. (2005) Phosphoinositides and phosphatidic acid regulate pollen tube growth and reorientation through modulation of $[Ca^{2+}]_c$ and membrane secretion. *Journal of Experimental Botany*, **56**, 1665–1674.

Mueller-Roeber, B. and Pical, C. (2002) Inositol phospholipid metabolism in *Arabidopsis*. Characterized and putative isoforms of inositol phospholipid kinase and phosphoinositide-specific phospholipase C. *Plant Physiology*, **130**, 22–46.

Munnik, T. and Vermeer, J.E. (2010) Osmotic stress-induced phosphoinositide and inositol phosphate signalling in plants. *Plant, Cell and Environment*, **33**, 655–69.

Nie, Z., Stanley, K.T., Stauffer, S., Jacques, K.M., Hirsch, D.S., Takei, J., and Randazzo, P.A. (2002) AGAP1, an endosome-associated, phosphoinositide-dependent ADP-ribosylation factor GTPase activating protein that affects actin cytoskeleton. *Journal of Biological Chemistry*, **277**, 48965–48975.

Nielsen, E., Cheung, A.Y., and Ueda, T. (2008) The regulatory RAB and ARF GTPases for vesicular trafficking. *Plant Physiology*, **147**, 1516–1526.

Noda, T., Matsunaga, K., Taguchi-atarashi, N., and Yoshimori, T. (2010) Regulation of membrane biogenesis in autophagy via PI3P dynamics. *Seminars in Cell and Developmental Biology*, **21**, 671–676.

Obara, K., Sekito, T., Niimi, K., and Ohsumi, Y. (2008) The Atg18–Atg2 complex is recruited to autophagic membranes via phosphatidylinositol 3-phosphate and exerts an essential function. *Journal of Biological Chemistry*, **283**, 23972–23980.

Ohashi, Y., Oka, A., Rodrigues-Pousada, R., Possenti, M., Ruberti, I., Morelli, G., and Aoyama, T. (2003) Modulation of phospholipid signaling by GLABRA2 in root-hair pattern formation. *Science*, **300**, 1427–1430.

Oliver, D., Lien, C.C., Soom, M., Baukrowitz, T., Jonas, P., and Fakler, B. (2004) Functional conversion between A-type and delayed rectifier K^+ channels by membrane lipids. *Science*, **304**, 265–270.

Olsen, H.L., Hoy, M., Zhang, W., Bertorello, A.M., Bokvist, K., Capito, K., Efanov, A.M., Meister, B., Thams, P., Yang, S.N., Rorsman, P., Berggren, P.O., and Gromada, J. (2003) Phosphatidylinositol 4-kinase serves as a metabolic sensor and regulates priming of secretory granules in pancreatic beta cells. *Proceedings of the National Academy of Science, USA*, **100**, 5187–5192.

Oude Weernink, P.A., Schmidt, M., and Jakobs, K.H. (2004a) Regulation and cellular roles of phosphoinositide 5-kinases. *European Journal of Pharmacology*, **500**, 87–99.

Oude Weernink, P.A., Meletiadis, K., Hommeltenberg, S., Hinz, M., Ishihara, H., Schmidt, M., and Jakobs, K.H. (2004b) Activation of type I phosphatidylinositol 4-phosphate 5-kinase isoforms by the Rho GTPases, RhoA, Rac1, and Cdc42. *Journal of Biological Chemistry*, **279**, 7840–7849.

Oude Weernink, P.A., Lopez de Jesus, M., and Schmidt, M. (2007) Phospholipase D signaling: orchestration by PIP_2 and small GTPases. *Naunyn-Schmiedeberg's Archives of Pharmacologyogy*, **374**, 399–411.

Oxley, D., Ktistakis, N., and Farmaki, T. (2013) Differential isolation and identification of PI(3)P and PI(3,5)P_2 binding proteins from *Arabidopsis thaliana* using an agarose-phosphatidylinositol-phosohate affinity chromatography. *Journal of Proteomics*, **91**, 580–594.

Oyama, T., Shimura, Y., and Okada, K. (2002) The *IRE* gene encodes a protein kinase homologue and modulates root hair growth in *Arabidopsis*. *Plant Journal*, **30**, 289–299.

Pao, A.C., McCormick, J.A., Li, H., Siu, J., Govaerts, C., Bhalla, V., Soundararajan, R., and Pearce, D. (2007) NH_2 terminus of serum and glucocorticoid-regulated kinase 1 binds to phosphoinositides and is essential for isoform-specific physiological functions. *American Journal of Physiology, Renal Physiology*, **292**, F1741–F1750.

Paris, S., Beraud-Dufour, S., Robineau, S., Bigay, J., Antonny, B., Chabre, M., and Chardin, P. (1997) Role of protein-phospholipid interactions in the activation of ARF1 by the guanine nucleotide exchange factor Arno. *Journal of Biological Chemistry*, **272**, 22221–22226.

Pearce, L.R., Komander, D., and Alessi, D.R. (2010) The nuts and bolts of AGC protein kinases. *Natre Reviews Molecular Cell Biology*, **11**, 9–22.

Peterson, R.L. and Farquhar, M.L. (1996) Root hairs: specialized tubular cells extending root surfaces. *Botanical Review*, **62**, 2–33.

Posor, Y., Eichhorn-Gruenig, M., Puchkov, D., Schoneberg, J., Ullrich, A., Lampe, A., Muller, R., Zarbakhsh, S., Gulluni, F., Hirsch, E., Krauss, M., Schultz, C., Schmoranzer, J., Noe, F., and Haucke, V. (2013) Spatiotemporal control of endocytosis by phosphatidylinositol-3,4-bisphosphate. *Nature*, **499**, 233–237.

Preuss, M.L., Serna, J., Falbel, T.G., Bednarek, S.Y. and Nielsen, E. (2004) The *Arabidopsis* Rab GTPase RabA4b localizes to the tips of growing root hair cells. *Plant Cell*, **16**, 1589–1603.

Preuss, M.L., Schmitz, A.J., Thole, J.M., Bonner, H.K.S., Otegui, M.S., and Nielsen, E. (2006) A role for the RabA4b effector protein PI-4Kβ1 in polarized expansion of root hair cells in *Arabidopsis thaliana*. *Journal of Cell Biology*, **172**, 991–998.

Pribat, A., Sormani, R., Rousseau-Gueutin, M., Julkowska, M.M., Testerink, C., Joubes, J., Castroviejo, M., Laguerre, M., Meyer, C., Germain, V., and Rothan, C. (2012) A novel class of PTEN protein in *Arabidopsis* displays unusual phosphoinositide phosphatase activity and efficiently binds phosphatidic acid. *Biochemical Journal*, **441**, 161–171.

Qin, C. and Wang, X. (2002) The Arabidopsis phospholipase D family. Characterization of a calcium-independent and phosphatidylcholine-selective PLDζ1 with distinct regulatory domains. *Plant Physiology*, **128**, 1057–1068.

Rameh, L.E. and Cantley, L.C. (1999) The role of phosphoinositide 3-kinase lipid products in cell function. *Journal of Biological Chemistry*, **274**, 8347–8350.

Rentel, M.C., Lecourieux, D., Ouaked, F., Usher, S.L., Petersen, L., Okamoto, H., Knight, H., Peck, S.C., Grierson, C.S., Hirt, H., and Knight, M.R. (2004) OXI1 kinase is necessary for oxidative burst-mediated signaling in *Arabidopsis*. *Nature*, **427**, 858–861.

Ridley, A.J. (2006) Rho GTPases and actin dynamics in membrane protrusions and vesicle trafficking. *Trends in Cell Biology*, **16**, 522–529.

Runnels, L.W., Yue, L., and Clapham, D.E. (2002) The TRPM7 channel is inactivated by PIP_2 hydrolysis. *Nature Cell Biology*, **4**, 329–336.

Russo, C., Gao, Y., Mancini, P., Vanni, C., Porotto, M., Falasca, M., Torrisi, M.R., Zheng, Y., and Eva, A. (2001) Modulation of oncogenic DBL activity by phosphoinositol phosphate binding to pleckstrin homology domain. *Journal of Biological Chemistry*, **276**, 19524–19531.

Ryan, E., Steer, M., and Dolan, L. (2001) Cell biology and genetics of root hair formation in *Arabidopsis thaliana*. *Protoplasma*, **215**, 140–149.

Saarikangas, J., Zhao, H., and Lappalainen, P. (2010) Regulation of the actin cytoskeleton-plasma membrane interplay by phosphoinositides. *Physiological Reviews*, **90**, 259–289.

Saavedra, L., Balbi, V., Lerche, J., Mikami, K., Heilmann, I., and Sommarin, M. (2011) PIPKs are essential for rhizoid elongation and caulonemal cell development in the moss *Physcomitrella patens*. *Plant Journal*, **67**, 635–647.

Saavedra, L., Mikami, K., Malho, R., and Sommarin, M. (2012) PIP kinases and their role in plants tip growing cells. *Plant Signaling and Behavior*, **7**, 1302–1305.

Samaj, J., Read, N.D., Volkmann, D., Menzel, D., and Baluska, F. (2005) The endocytic network in plants. *Trends in Cell Biology*, **15**, 425–433.

Samaj, J.J., Muller, J., Beck, M., Bohm, N. and Menzel, D. (2006) Vesicular trafficking, cytoskeleton and signaling in root hairs and pollen tubes. *Trends in Plant Science*, **11**, 594–600.

Santarius, M., Lee, C.H., and Anderson, R.A. (2006) Supervised membrane swimming: small G-protein lifeguards regulate PIPK signalling and monitor intracellular PtdIns(4,5)P_2 pools. *Biochemistry Journal*, **398**, 1–13.

Sasaki, T., Takasuga, S., Sasaki, J., Kofuji, S., Eguchi, S., Yamazaki, M., and Suzuki, A. (2009) Mammalian phosphoinositide kinases and phosphatases. *Progress in Lipid Research*, **48**, 307–343.

Schiefelbein, J.W. (2000). Constructing a plant cell. The genetic control of root hair development. *Plant Physiology*, **124**, 1525–1531.

Sciorra, V.A., Audhya, A., Parsons, A.B., Segev, N., Boone, C., and Emr, S.D. (2005). Synthetic genetic array analysis of the PtdIns 4-kinase Pik1p identifies components in a Golgi-specific Ypt31/rab- GTPase signaling pathway. *Molecular Biology of the Cell*, **16**, 776–793.

Shewan, A., Eastburn, D.J., and Mostov, K. (2011) Phosphoinositide in cell architecture. *Cold Spring Harbor Perspectectives in Biology*, doi: 10.1101/cshperspect.a004796.

Smith, L.G. and Oppenheimer, D.G. (2005) Spatial control of cell expansion by the plant cytoskeleton. *Annual Reviews in Cell and Developmental Biology*, **21**, 271–295.

Stenzel, I., Ischebeck, T., Konig, S., HoBubowska, A., Sporysz, M., Hause, B., and Heilmann, I. (2008) The type B phosphatidylinositol-4-phosphate 5-kinase 3 is essential for root hair formation in Arabidopsis thaliana. *Plant Cell*, **20**, 124–141.

Stephens, L.R., Anderson, K., Stokoe, D., Erdjument-Bromage, H., Painter, G.F., Holmes, A.B., Gaffney, P.R., Reese, C.B., McCormick, F., Tempst, P., Coadwell, J., and Hawkins, P.T. (1998) Protein kinase B kinases that mediate phosphatidylinositol 3,4,5-trisphosphate-dependent activation of protein kinase B. *Science*, **279**, 710–714.

Stokoe, D., Stephens, L.R., Copeland, T., Gaffney, P.R., Reese, C.B., Painter, G.F., Holmes, A.B., McCormick, F., and Hawkins, P.T. (1997) Dual role of phosphatidylinositol-3,4,5-trisphosphate in the activation of protein kinase B. *Science*, **277**, 567–570.

Suh, B.C. and Hille, B. (2005) Regulation of ion channels by phosphatidylinositol 4,5-bisphosphate. *Current Opinion in Neurobiology*, **15**, 370–378.

Takai, Y., Sasaki, T., and Matozaki, T. (2001) Small GTP-binding proteins. *Physiological Reviews*, **81**, 153–208.

Takeda, S., Gapper, C., Kaya, H., Bell, E., Kuchitsu, K., and Dolan, L. (2008) Local positive feedback regulation determines cell shape in root hair cells. *Science*, **319**, 1241–1244.

Takenawa, T and Itoh, T. (2001) Phosphoinositides, key molecules for regulation of actin cytoskeletal organization and membrane traffic form the plasma membrane. *Biochimica et Biophysica Acta*, **1533**, 190–206.

Tang, R.H., Han, S., Zheng, H., Cook, C.W., Choi, C.S., Woerner, T.E., Jackson, R.B., and Pei, Z.M. (2007) Coupling diurnal cytosolic Ca^{2+} oscillations to the CAS-IP_3 pathway in *Arabidopsis*. *Science*, **315**, 1423–1426.

Testerink, C. and Munnik, T. (2005) Phosphatidic acid: a multifunctional stress signaling lipid in plants. *Trends in Plant Science*, **10**, 368–375.

Testerink, C., Larsen, P.B., van der Does, D., van Himbergen, J.A., and Munnik, T. (2007) Phosphatidic acid binds to and inhibits the activity of *Arabidopsis* CTR1. *Journal of Experimental Botany*, **58**, 3905–3914.

Thole, J.M. and Nielsen, E. (2008) Phosphoinositides in plants: novel functions in membrane trafficking. *Current Opinion in Plant Biology*, **11**, 620–631.

Thole, J.M., Vermeer, J.E.M., Zhang, Y., Gadella Jr., T.W.J., and Nielsen, E. (2008) *ROOT HAIR DEFECTIVE4* encodes a phosphatidylinositol-4-phosphate phosphatase required for proper root hair development in Arabidopsis thaliana. *Plant Cell*, **20**, 381–395.

Tominaga-Wada, R., Ishida, T., and Wada, T. (2011) New insight into the mechanism of development of *Arabidopsis* root hairs and trichomes. *International Review of Cell and Molecular Biology*, **286**, 67–106.

Topham, M.K. (2006) Signaling roles of diacylglycerol kinases. *Journal of Cellular Biochemistry*, **97**, 474–484.

Viaud, J., Gaits-Iacovoni, F., and Payrastre, B. (2012) Regulation of the DH-PH tandem of guanine nucleotide exchanging factor for Rho GTPases by Phosphoinositides. *Advances in Biological Regulations*, **52**, 303–314.

Vincent, P., Chua, M., Nogue, F., Fairbrother, A., Mekeel, H., Xu, Y., Allen, N., Bibikova, T.N., Gilroy, S., and Bankaitis, V.A. (2005) A Sec 14p-nodulin domain phosphatidylinositol transfer protein polarizes membrane growth of *Arabidopsis thaliana* root hairs. *Journal of Cell Biology*, **168**, 801–812.

Wang, Z., Xu, C., and Benning, C. (2012) TGD4 involved in endoplasmic reticulum-to-chloroplast lipid trafficking is a phosphatidic acid binding protein. *Plant Journal*, **70**, 614–623.

Wedlich-Soldner, R., Altschuler, S., Wu, L., and Li, R. (2003) Spontaneous cell polarization through actomyosin-based delivery of the Cdc42 GTPase. *Science*, **299**, 1231–1235.

Weiner, O.D., Neilsen, P.O., Prestwich, G.D., Kirschner, M.W., Cantley, L.C., and Bourne, H.R. (2002). A PtdInsP(3)- and Rho GTPase-mediated positive feedback loop regulates neutrophil polarity. *Nature Cell Biology*, **4**, 509–513.

Wenk, M.R. and De Camilli, P. (2004) Protein–lipid interactions and phosphoinositide metabolism in membrane traffic: insights from vesicle recycling in nerveterminals. *Proceedings of the National Academy of Science, USA*, **101**, 8262–8269.

Wheeler, G.L. and Brownlee, C. (2008) Ca^{2+} signalling in plants and green algae-changing channels. *Trends in Plant Science*, **13**, 506–514.

Wu, L., Bauer, C.S., Zhen, X.G., Xie, C., and Yang, J. (2002) Dual regulation of voltage-gated calcium channels by PtdIns(4,5)P_2. *Nature*, **419**, 947–952.

Xue, H.-W., Chen, X, and Mei, Yu. (2009) Function and regulation of phospholipid signaling in plants. *Biochemical Journal*, **421**, 145–156.

Yang, Z. and Lavagi, I. (2012) Sppatial control of plasma membrane domains: ROP GTPase-bases symmetry breaking. *Current Opinion in Plant Biology*, **15**, 601–607.

Yoo, C.M., Wen, J., Motes, C.M., Sparks, J.A. and Blancaflor, E.B. (2008) A class I ADP ribosylation factor-GTPase activating protein is critical for maintaining directional root hair growth in *Arabidopsis*. *Plant Physiology*, **147**, 1659–1674.

Yoo, C.M., Quan, L., Cannon, A.E., Wen, J. and Blancaflor E.B. (2012) AGD1, a class1 ARF-GAP, acts in common signaling pathways with phosphoinositide metabolism and the actin cytoskeleton in controlling Arabidopsis root hair polarity. *Plant Journal*, **69**, 1064–1076.

Zegzouti, H., Anthony, R.G., Jahchan, N., Bogre, L., and Christensen, S.K. (2006) Phosphorylation and activation of PINOID by the phospholipid signaling kinase 3-phosphoinositide-dependent protein kinase 1 (PDK1) in *Arabidopsis*. *Proceedings of the National Academy of Science, USA*, **103**, 6404–6409.

Zhang, Q., Lin, F., Mao, T., Nie, J., Yan, M., Yuan, M., and Zhang, W. (2012) Phosphatidic acid regulates microtubule organization by interacting with MAP65-1 in response to salt stress in *Arabidopsis*. *Plant Cell*, **24**, 4555–4576.

Zhao, X., Varnai, P., Tuymetova, G., Balla, A., Toth, Z. E., Oker-Blom, C., Roder, J., Jeromin, A., and Balla, T. (2001) Interaction of neuronal calcium sensor-1 (NCS-1) with phosphatidylinositol 4-kinase b stimulates lipid kinase activity and affects membrane trafficking in COS-7 cells. *Journal of Biological Chemistry*, **276**, 40183–40189.

Zhao, Y., Yan, A., Feijó, J.A., Furutani, M., Takenawa, T., Hwang, I., Fu, Y., and Yang, Z. (2010) Phosphoinositides regulate clathrin-dependent endocytosis at the tip of pollen tubes in Arabidopsis and tobacco. *Plant Cell*, **22**, 4031–4044.

Zheng, Z.L. and Yang, Z. (2000) The Rop GTPase: an emerging signaling switch in plants. *Plant Molecular Biology*, **44**, 1–9.

Zonia, L. and Munnik, T. (2006) Cracking the green paradigm: functional coding of phosphoinositide signals in plant stress responses. *Subcellular Biochemistry*, **39**, 207–237.

10 Arabidopsis Trichome Morphogenesis and the Role of Microtubules in Controlling Trichome Branch Formation

M. David Marks

Department of Plant Biology, University of Minnesota, St Paul, MN

Introduction

Trichomes are hairs that form on the surface of most land plants. Depending on the plant species trichomes can be multicellular or unicellular, long- or short-lived, and either glandular (excreting secondary metabolites into an extracellular capsule) or non-glandular (Esau, 1977). Trichomes mainly function as defensive structures that can protect plants from both biotic and abiotic stresses. For example, they can deter herbaceous insects, reflect excess light, or modulate transpiration (Johnson, 1975). In Arabidopsis, trichomes are large unicellular structures containing a stalk and typically three–four branches. The development of leaf trichomes in Arabidopsis has long been used as a model to study the control of plant cell fate and cell differentiation (Haughn and Somerville, 1988; Marks *et al.*, 1991). An introduction to Arabidopsis trichome development is presented in this review along with some early findings concerning the identification of key transcriptional regulators that control trichome cell fate and differentiation. The remainder of the review will focus on more recent findings dealing with the role of microtubules in controlling the formation of branches on Arabidopsis trichomes.

Trichome development can be divided into several distinct developmental stages (Fig. 10.1; Szymanski *et al.*, 1998, 1999). An epidermal cell in a field of dividing cells commits to the trichome cell fate when it exits the normal cell cycle. The committed cell continues to complete abbreviated cycles lacking an M-phase resulting in the endoreduplication of the chromosomal DNA. Through this process a trichome cell accumulates between 32 and 64 C levels of chromosomal DNA (Hulskamp *et al.*, 1994). During this early step, considered to represent stage 1, the diameter of the incipient trichome increases relative to the surrounding cells. As endoreduplication proceeds, an early developing trichome progresses to stage 2 and expands out of the plane of the

Plant Cell Wall Patterning and Cell Shape, First Edition. Edited by Hiroo Fukuda.

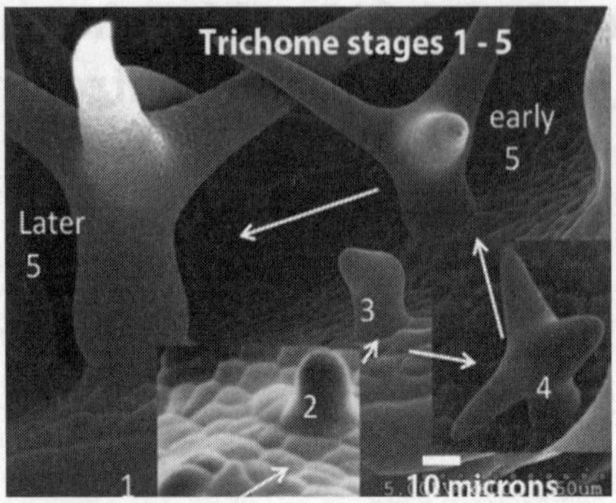

Figure 10.1 Stages of Arabidopsis trichome development. Trichomes proceed through distinct stages of development beginning with stage 1 shown in bottom middle. Sequential arrows show trichomes at different stages in development. All images are shown at the same scale. The white bar represents 10 μm.

epidermal surface. Leaf trichomes enter stage 3 when new foci of expansion develop on the upper dome of the expanding stage 2 trichomes. These focal points serve as the initiation sites of branches that continue to expand during stage 4. Stage 5 is characterized by a change in shape of the expanding branch tips from blunt to sharp. During this stage the trichomes rapidly expand in terms of both girth and length, and bumps or papillae emerge on the outer trichome surface. Expansion ceases during stage 6; however, the trichomes remain alive and continue to produce cell wall material. A mature trichome is composed of a single cell with a stalk approximately 100 μm in length and three–four branches up to 150 μm in length. The trichome surface is covered with papillae approximately 0.5–1 μm in diameter and the trichome cell wall can increase in thickness to over 1 μm (Marks *et al.*, 2008).

Transcriptional Control of Trichome Development

Much has been learned over the years about the regulatory mechanisms that govern the conversion of an epidermal cell into a trichome. This has recently been reviewed (Balkunde *et al.*, 2010; Grebe, 2012; Lin and Aoyama, 2012; Ryu *et al.*, 2013) and studied in some detail, and only the basic paradigm is presented here. The commitment to the trichome cell fate is marked by the upregulated expression of two transcription factors (TFs): *GL1* and *GL3* (Szymanski *et al.*, 2000). The AGI codes and complete names of all genes described in this review are given in Table 10.1. These TFs, along with the widely expressed scaffold protein TTG1, form a complex that activates additional transcription factors and regulatory genes (Morohashi and Grotewold, 2009). The early endoreduplication event is likely mediated by the upregulation of the *SIAMESE* (*SIM*) gene (a known target of the GL1/GL3/TTG1 complex; Morohashi and Grotewold, 2009). SIM has been shown to be an inhibitor of the CDK-cyclin D complex needed to drive mitotic cell cycles (Churchman *et al.*, 2006). The expression of other genes is also critical as mutations in *SIM* still allow trichome differentiation to initiate, but the resulting trichomes are multicellular (Walker *et al.*, 2000).

The quantity of GL1/GL3/TTG1 complex in developing trichomes and surrounding cells is modulated by an influence of negative factors including TRY and related proteins (Esch *et al.*, 2003, 2004; Wang *et al.*, 2008; Wester *et al.*, 2009). *TRY* is a transcriptional target of the complex and the TRY protein provides negative feedback on the activity of the complex by destabilizing the interaction between GL1 and GL3 (Esch *et al.*, 2003; Morohashi and Grotewold, 2009). TRY and related proteins move from early stage trichomes to neighboring uncommitted cells to prevent them from becoming trichomes (Wester *et al.*, 2009). TRY also limits the activity of the GL1/GL3/TTG1 complex in developing trichomes. This latter interaction influences trichome cell size, branch number, and the level of endoreduplication, all of which are enhanced in *try* mutants (Hulskamp *et al.*, 1994).

Mutations in other transcription factors also affect trichome morphogenesis. Strong mutant alleles of *GL2* result in trichomes that fail to fully expand out of the plane of the leaf surface (Koornneef *et al.*, 1982). Weaker alleles of *gl2* result in trichomes that are unbranched spikes. Likewise, mutations in *TTG2* also result in trichomes with reduced branching (Johnson *et al.*, 2002). In contrast, mutations in either *HDG11* or *NOK* result in trichomes with extra branches (Folkers *et al.*, 1997; Nakamura *et al.*, 2006; Jakoby *et al.*, 2008). Only a few downstream targets have been confirmed for the GL1/GL3/TTG1 complex and very little is known about the targets of the other transcription factors.

Analysis of Gene Expression in Developing Trichomes

Trichomes proceed through all six stages of development in less than 36 hours (see supplemental movie associated with Marks *et al.*, 2009). Because trichomes make up a small fraction of cells on a young leaf, it is not possible to use whole young shoot tissue to accurately analyze gene expression in developing trichomes. One proxy to circumvent this issue has been to use mutants exhibiting trichomes that are phenotypically arrested at early stages. One mutant that has proved valuable in this regard is an allelic variant of *GL3* called *gl3-shapeshifter* (*gl3-sst*) (Esch *et al.*, 2003). This allele contains

Table 10.1 ATH1 Affymetrix GeneChip expression values for genes described in the text.

AGI code[1]	Gene name	Gene abbreviation	Expression in early arrested trichomes[2]	Expression in intermediately arrested trichomes	Expression in wild-type mature trichomes	Expression in young shoots
AT1G27920	*GLABROUS1*	*GL1*	5067±430	455±482	88±41	Not detected
AT5G24520	*TRANSPARENT TESTA GLABRA1*	*TTG1*	572±39	842±66	805±12	863±77
AT5G04470	*SIAMESE*	*SIM*	6032±968	5209±1360	4477±2083	275±137
AT5G53200	*TRIPTYCHON*	*TRY*	351±94	260±140	634±204	200±101
AT1G79840	*GLABRA2*	*GL2*	7797±823	5798±1828	6756±2372	Not detected
AT2G37260	*TRANSPARENT TESTA GLABRA2*	*TTG2*	3640±784	2164±401	1388±525	Not detected
AT3G01140	*NOECK*	*NOK*	177±163	1005±480	167±40	Not detected
AT1G73360	*HOMEODOMAIN GLABROUS 11*	*HDG11*	225±27	492±80	494±318	Not detected
AT1G64690	*BRANCHLESS TRICHOME*	*BLT*	1588±131	133±82	Not detected	Not detected
AT4G14960	*Lefty1 (tubulin alpha 6)*	*TUA6*	1766±123	3627±1984	3656±2043	2877±1092
AT1G04820	*Lefty2 (tubulin alpha 4)*	*TUA4*	1288±168	2672±1282	2937±1663	1406±922
AT1G29990	*PREFOLDIN6*	*PFD6*	583±95	696±65	557±93	493±64
AT2G30410	*KIESEL*	*KIS*	1881±434	2535±644	2275±548	1469±421
AT5G17410	*GAMMA TUBULIN COMPLEX PROTEIN2 (SPIRAL3)*	*GCP2(SPR3)*	292±25	213±72	308±166	243±69
AT5G18580	*TONNEAU2*	*TON2*	2149±141	1936±275	1467±216	1006±147
AT2G34560	*FRAGILE FIBER2*	*FRA2*	292±62	774±180	793±143	1000±533
AT5G65930	*ZWICHEL*	*ZWI*	439±34	363±137	306±89	174±25
AT2G46600	*KCBP-INTERACTING CA2+ BINDING PROTEIN*	*KIC*	1254±309	5173±440	6224±1206	1.08E+04±4389
AT4G20780	*Calmodulin like 42*	*CML42*	567±64	1540±637	906±196	684±138

AT3G16630	*ATKINESIN-13A/KINESIN-13A; microtubule motor*	*KIN13A*	1164±66	1351±135	1024±194	763±201
AT2G02480	*STICHEL*	*STI*	321±50	191±48	Not detected	106±19
AT4G16340	*SPIKE1*	*SPK1*	415±53	270±96	224±44	278±62
AT5G62880	*RHO-RELATED PROTEIN FROM PLANTS 11*	*ROP11*	2775±462	824±218	398±23	417±48
AT3G53350	*ROP INTERACTING PROTEIN3/ MICROTUBULE DEPLETION DOMAIN1*	*RIP3/MIDD1*	296±55	129±13	76±23	66±20
AT3G44200	*NIMA (NEVER IN MITOSIS, GENE A)-RELATED 6*	*NEK6*	398±44	508±76	324±74	289±46
AT2G40070	*BASIC PROLINE-RICH PROTEIN1*	*BPP1*	361±16	Not detected	Not detected	Not detected

[1]Genes are listed in the order of their appearance in the text.

[2]The data represent three replicates for early arrested *gl3-sst nok* and intermediate arrested *gl3-sst*, five replicates for wild-type mature, and four replicates for young shoots. Data were normalized to an arithmetic mean of 1000 for all probes considered detected on each chip. Values are represented as the mean ± std. dev. values for probe sets derived from the indicated genes described in the first three columns (see Marks *et al.*, 2009; Gilding and Marks, 2010 for more analysis details).

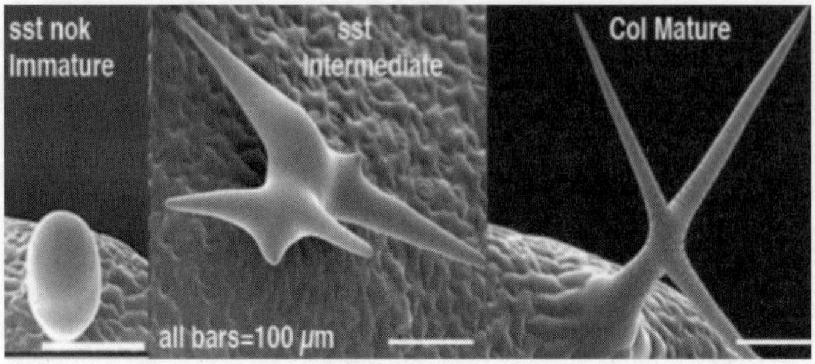

Figure 10.2 Genetically arrested trichomes compared to mature Columbia wild-type. Trichomes on *gl3-sst nok* (sst nok immature) double mutants arrest before branch initiation (stage 2), whereas *gl3-sst* (sst intermediate) trichomes form branches but lack surface papillae and exhibit blunt tips (stage 4). Mature trichomes (Col Mature) form surface papillae and have branches with sharp tips.

a point mutation resulting in an amino acid substitution in the GL3 peptide that reduces its affinity for GL1. The effect of this mutation is a phenotypic arrest of most trichomes at what appears to be stage 4 (Fig. 10.2). Combining the *gl3-sst* allele with the *nok* mutant results in an even more pronounced arrest in development (Gilding and Marks, 2010). Most of the trichomes on the double mutant lack branches and appear to arrest at stage 2 (Fig. 10.2). This demonstrates that some genes regulated by NOK are required to mediate the transition past stage 2, which was not predicted by its extra-branched trichome mutant phenotype.

To study the expression of genes in mature wild-type and arrested mutant trichomes, a technique for isolating intact trichomes from young shoots was developed (Marks *et al.*, 2008). Sufficient quantities of RNA were isolated from these trichomes to make probes for hybridization to the Arabidopsis ATH1 Affymetrix GeneChip. Data from these analyses were validated by the confirmation that many of the transcription factors described above were preferentially expressed in trichomes (Table 10.1; see also Marks *et al.*, 2008, 2009; Gilding and Marks, 2010). Analysis of genes expressed more highly in trichomes led to the identification of several new trichome mutants. Key in this review, the *BLT* gene was more highly expressed in the arrested trichomes than in either shoots or mature wild-type trichomes (Table 10.1). T-DNA insertions in this gene were found to eliminate trichome branch formation (Marks *et al.*, 2009). The reminder of this review will consider some of the cellular mechanisms and the roles of other genes such as *BLT* in regulating trichome branching. The genes to be discussed in this review are expressed in trichomes. However, as many of these genes are required for other aspects of plant development, they are also highly expressed in young shoots and other tissues.

Role of Microtubules in Trichome Branching

Microtubules (MTs) have long been implicated as playing a major role in the formation of branches during stage 3. This was shown in pharmacological studies on young seedlings treated with drugs or herbicides that alter the cytoskeleton (Mathur *et al.*, 1999; Szymanski *et al.*, 1999; Mathur and Chua, 2000). Seedlings were treated with agents that either depolymerize actin or microtubules. Treatments with either cytochalasin D or latrunculin B, which destabilize actin microfilaments (MFs), had little effect on the initiation of trichome branches. Trichomes on these MF-deleted plants exhibited normal trichome development up to stage 3. In stages 4 and 5 however, during which cell expansion accelerates, defects in trichome expansion became apparent. The treated trichomes displayed abnormal twisting and branches showed asymmetric expansion. These abnormal trichomes phenocopied those found on the distorted class of trichome mutants (Mathur *et al.* 1999; Szymanski *et al.* 1999). Subsequent studies revealed that the genes altered in the distorted mutants encode networks of interacting proteins that regulate MF formation. The function of these complexes is not discussed here, but see Yanagisawa *et al.* (2013) for a recent review. When early stage or initiating trichomes were treated with oryzalin, an MT destabilizing herbicide, stage 1 and 2 trichomes became bloated and did not form branches. These later results implicate dynamic MTs in early trichome expansion and branch formation.

Direct implication of MTs in trichome branch formation stems from studies of the subunits of MTs, which are composed of heterodimers of alpha and beta tubulin subunits. In Arabidopsis these subunits are encoded by multigene families; dominant mutations in two alpha tubulin genes have been shown to alter MT dynamics, resulting in a reduction in trichome branching (Abe *et al.*, 2004). The maturation of tubulin subunits is also important for MT dynamics. After the translation of tubulin subunits begins, a protein prefoldin complex containing a PFD6 protein binds to the nascent peptides. This binding promotes the transfer of the completed peptides to a chaperonin complex that mediates additional maturation steps. Mutations in *PFD6* have been reported to reduce trichome branch initiation (Gu *et al.*, 2008). A complex containing a protein called KIS completes the last step in tubulin subunit maturation. Mutations in *KIS* have a profound effect on trichome morphogenesis, affecting not just trichome branching but also general trichome outgrowth (Kirik *et al.*, 2002).

Mechanistically, plant cell expansion including trichome branch formation is passively driven by turgor pressure. Asymmetrical cell expansion is associated with the differential loosening of the cell wall (recently reviewed by Szymanski and Cosgrove, 2009). This can occur through chemical modifications of existing wall components or through the differential disposition of non-cellulosic components such as pectin or lignin or of cellulose in terms of orientation and abundance. The orientation of cellulose microfibers is largely controlled by the MTs that form tracks for guiding protein complexes containing cellulose synthases. Typically, cellulose microfibrils are oriented parallel to MTs and perpendicular to the orientation of cell expansion. For example, MTs form transverse cortical arrays in Arabidopsis hypocotyl cells during longitudinal cell expansion. Changes in MT orientation and abundance are associated with changes in asymmetric cell expansion as seen by the differential expansion of epidermal pavement cells. During pavement cell expansion, the indentations or necks are more MT rich than the expanding lobes (Fu *et al.*, 2005).

A model invoking the reorientation of MTs in stage 3 trichomes as the mechanism driving branch initiation has been presented (Szymanski, 2009). In this model, transverse and oblique cortical MTs rearrange into concentric arrays at the focal points of branch initiation sites. These arrays should mediate changes in the local deposition of cellulose microfibrils. These changes in microfibril orientation are predicted to create weak points in the cell wall. Subsequent passive expansion driven by turgor pressure at these weakened foci would result in the initiation of the branches. Images of MTs in stage 3 trichomes in which branch initiation is occurring support this notion (Fig. 10.3; see also Beilstein and Szymanski, 2004). The mechanisms controlling these changes

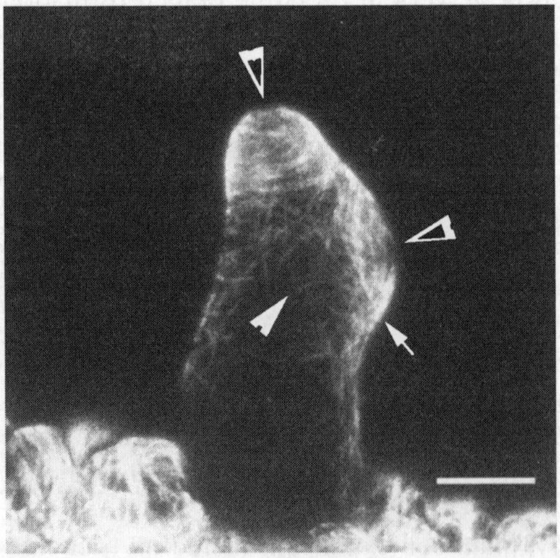

Figure 10.3 *In vivo* localization of MTs in a stage 3 trichome. Shown are initiating branches (open arrows), MTs beginning to form a concentric array at an emerging branch site (arrow with shaft) and MTs arranged randomly in regions of the cells not undergoing expansion (solid arrow). Scale bar: 10 μm. Image provided by Dr. Daniel B. Szymanski, Purdue University (for more details see Beilstein and Szymanski, 2004).

in MT dynamics are not known. A number of genes are currently being studied in a variety of systems that have roles in controlling various aspects of MT dynamics. As discussed in the following, mutational analyses show that many of these genes have roles in controlling trichome branch formation.

Molecular Processes Influencing MT Dynamics and Trichome Branching

Trichome branch initiation is reduced in mutants with alterations in genes required for the nucleation of MTs on existing cortical arrays. In general, the nucleation process involves the formation of MTs that project from gamma-tubulin ring complexes (γTuRC) attached to the sides of existing cortical MTs (Fishel and Dixit, 2013; Hashimoto, 2013). These new MTs typically grow out at a 40% angle relative to the mother MTs. A weak mutant allele of *GCP2*, which encodes a member of the γTuRC called *spr3*, results in a wider range of angles in MT outgrowth (Nakamura and Hashimoto, 2009). In trichomes, this change results in reduced trichome branching. A more extreme effect on trichome branching is seen in weak *ton2* mutants that also exhibit a deviation from the typical 40% angle of nascent MT outgrowth (Kirik *et al.*, 2012). Trichome branching is completely eliminated in these *ton2* mutants. TON2 is part of a phosphatase complex that localizes to γTuRC on cortical MTs (Kirik *et al.*, 2012). Mutations in *FRA2*, a subunit of the katanin complex which severs newly formed MTs at their minus end from the existing cortical MT, also reduces trichome branching (Burk *et al.*, 2001). MTs released by katanin can either be bundled to the mother MTs or they can treadmill away. This treadmilling involves the addition of tubulin subunits at the plus end at a faster rate than they are lost at the minus end.

It is not known if the concentric arrays that define branch foci are derived from existing cortical MTs, free treadmilling MTs that might be captured by complexes localized at these branch foci, or if the MTs are formed *de novo* at the foci. If MTs are formed *de novo* at the branch sites then the mechanistic cause of reduced branching in the above mutants might not be due to the observed alterations in existing cortical MTs, but instead due to a lack of proper organization of newly formed MTs.

The reorganization of MTs in concentric arrays at branch foci likely requires many different cellular activities. For example, mutational analyses have shown that the kinesin ZWI is required for branch formation (Oppenheimer *et al.*, 1997). *ZWI* encodes a MT motor protein that has MT bundling activity (Narasimhulu and Reddy, 1998). Mutations in *ZWI* have a profound effect on trichome morphogenesis, resulting in mostly unbranched or two-branched trichomes. ZWI is unusual in that in addition to its motor domain, it also has a calcium (Ca^{++}) calmodulin (CaM) binding domain. Studies have shown that MT binding is negatively regulated by CaM (Narasimhulu and Reddy, 1998).

The actual role of ZWI in trichome branching is still unknown. ZWI has been localized to the preprophase band, the mitotic spindle, and the phragmoplast, but ZWI has not been localized in developing trichomes (Bowser and Reddy, 1997). In addition to CaM, ZWI has been shown to interact with another Ca^{++} binding protein called KIC (Reddy *et al.*, 2004). This interaction also disrupts the ability of ZWI to bind MTs. The binding of KIC to ZWI requires a lower concentration of Ca^{++} than the binding of CaM to ZWI. While loss of function mutations in *KIC* do not affect trichome development, overexpression of *KIC* results in a reduction in branch formation possibly through its negative effect on ZWI MT binding. KIC also interacts with the Ca^{++} binding protein CML42 (Dobney *et al.*, 2009). Loss of function mutations in *CML42* result in extra-branched trichomes, which suggests that the inhibitory activity of KIC toward ZWI may be enhanced by CML42. These results highlight the likely role of the modulation of Ca^{++} concentrations during trichome morphogenesis. This is supported in the analysis of plants expressing the cameleon 3.6 Ca^{++} sensor (Monshausen *et al.*, 2008). Early-stage trichomes show higher levels of Ca^{++} in the tips of young branches than in the stalk (Fig. 10.4; Dorn and Marks, unpublished data).

Another kinesin, encoded by *KIN13A*, also plays a role in controlling trichome branching. Mutations in this gene result in extra-branched trichomes. Studies have shown that this kinesin co-localizes with the Golgi apparatus and

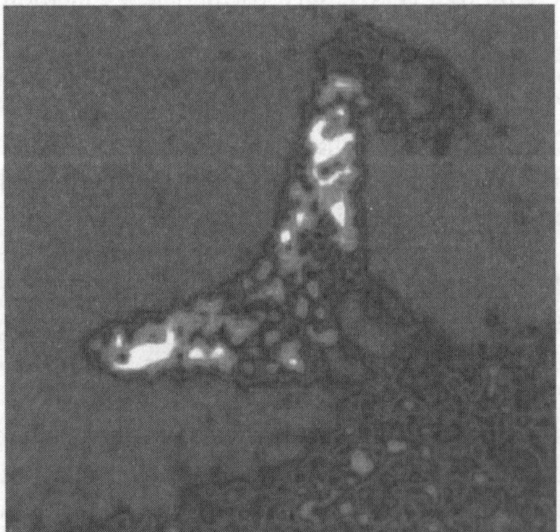

Figure 10.4 FRET analysis of stage 4 trichome expressing cameleon 3.6. The gradient of Ca^{++} from high to low is represented by the changes in color from pink to red to yellow to green to blue. Image captured and analyzed by Kevin Dorn (University of Minnesota) on a Leica TCS SP2 AOBS confocal microscope with the assistance of Elison B. Blancaflor at the Samuel Roberts Noble Foundation, Ardmore OK. (See plate section for the color version of this figure.)

mutants appear to exhibit aggregated Golgi complexes, suggesting that this kinesin plays a role in the dispersal of Golgi vesicles (Lu *et al.*, 2005). How this relates to the excessive trichome branching observed in the mutant is not known. More recent results suggest that KIN13A may play an additional role in modifying MT stability. This is discussed in more detail in the following paragraphs.

Two genes not known to play a direct role in MT dynamics are required for trichome branch formation. Loss of function mutations in either *STI* or *BLT* result in a complete loss of trichome branching and their overexpression results in extra branch formation (Fig. 10.5; Ilgenfritz *et al.*, 2003; Marks *et al.*, 2009; Kasili *et al.*, 2011). BLT does not contain any recognizable conserved motifs, but secondary structure analyses indicate that over two-thirds of the peptide should fold into a coiled-coil-type structure, known to mediate protein–protein interactions (Marks *et al.*, 2009). *STI* encodes a protein containing a central domain with a similarity to components of the Replication Factor C complex (RFC; Ilgenfritz *et al.*, 2003). In eukaryotes, this complex is composed of five related proteins (Cullmann *et al.*, 1995). During the initiation of replication the RFC loads PCNA onto DNA, and then DNA-bound PCNA loads DNA polymerase onto DNA at the replication forks (Podust and Fanning, 1997). A similar process is conserved in prokaryotes, in which the RFC complex is called the clamp loader. Overall, STI is more closely related to prokaryotic proteins involved in DNA replication than to the RFC counterparts found in eukaryotes (Khan *et al.*, 2007). *STI* appears to have a novel function as *sti* mutants do not exhibit defects in replication, and reach normal levels of endoreduplication (Ilgenfritz *et al.*, 2003). Furthermore, Arabidopsis contains RFC related genes more similar to those in other eukaryotes (Xia *et al.*, 2009, 2007; Liu *et al.*, 2010, 2012). STI may have a role in regulating MT functions, as Mathur and Chua (2000) showed that transient stabilization of MTs could induce limited branching in *sti* trichomes.

Close examination suggests that *blt* mutant trichomes are identical to those found on *sti* mutants (compare Fig. 10.5a–c). The double *sti blt* mutant trichomes are very similar to those of either single mutant, which implies that STI and BLT function in the same pathway (Marks *et al.*, 2009). This is supported by localization studies using GFP or RED tagged versions of *STI* or *BLT* constructs (Marks *et al.*, 2009). Transformants expressing these fused genes could partially rescue their respective mutants. Analysis of these transformants revealed that both proteins localize to the expanding tips of early staged (stages 2-4) trichomes (Fig. 10.5e and f). The interaction is also supported by yeast 2-hybrid analyses and by bimolecular complementation studies (Kasili *et al.*, 2011). The STI-BLT complex is a good candidate for either mediating or establishing a foundation for dynamic MT rearrangements at branch foci.

Rho of Plants (ROP) GTPases are also required for the initiation of trichome branching. In Arabidopsis, there are 11 ROPs (Vernoud *et al.*, 2003;

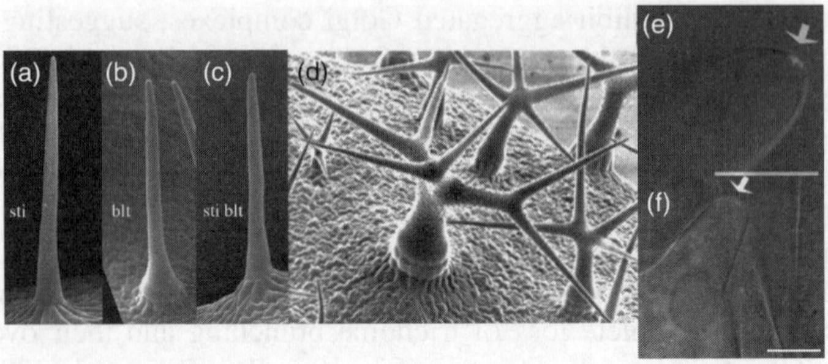

Figure 10.5 Analysis of *blt* and *sti* trichomes. (a–d) SEM images from of representative trichomes on *sti*, *blt*, *sti blt*, and a line overexpressing *STI*, respectively. (e, f) Localization of STI-Red and BLT-GFP at the tips of newly initiated branches (see Marks *et al.*, 2009 for more details).

Yang, 2008). When bound to GTP, they are active and modulate the activity of effector proteins. Through their endogenous GTPase activity, GTP is converted to GDP resulting in ROP deactivation. Reactivation requires ROP interaction with proteins called guanine exchange factors (GEFs) that replace the GDP with GTP. Mutations in *SPK1*, which encodes a GEF known to interact with and activate many different ROPs, result in a loss of trichome branch initiation (Qiu *et al.*, 2002; Basu *et al.*, 2008). A similar effect is observed in plants expressing the *C. difficile* toxin B protein (Singh *et al.*, 2012). This protein glucosylates ROPs and interferes with ROP activation. Plants expressing this toxin show a severe reduction in trichome branch formation. ROPs are known to modulate MT dynamics that could be important for trichome branch initiation. For example, ROP6 activates an effector called RIC (Lin *et al.*, 2013). This effector can recruit and activate katanins which, as discussed above, are required for wild-type levels of trichome branch initiation. Additionally, active ROP11 is known to interact with MIDD1/RIP3 (Oda and Fukuda, 2012). MIDD1 has been shown to interact with KIN13A in developing xylem cells and this interaction plays an important role in the formation of pits lacking secondary cell wall in xylem cells (Mucha *et al.*, 2010; Oda *et al.*, 2010). In these studies, it was shown that KIN13A acts as a plus-end MT destabilizer (Oda *et al.*, 2010). A complex containing ROP11, MIDD1, and KIN13A localizes to pits where the inhibition of MT formation prevents regional secondary cell wall deposition by cellulose synthase containing complexes (Oda and Fukuda, 2012). Given that *KIN13A* mutations result in extra-branched trichomes, KIN13A may function in trichomes to block excess branching by destabilizing MTs that mediate branch initiation.

Two proteins have been identified that have opposing roles in stimulating the formation of the concentric arrays of MTs that are proposed to promote branch initiation. Arrays similar to those seen in trichomes are associated with hypocotyl cell outgrowths that resemble early-stage trichomes (Motose *et al.*, 2011; Hamada *et al.*, 2013). The hypocotyl outgrowths are promoted on plants that either carry mutations in *NEK6* or that overexpress *BPP1* (Motose *et al.*, 2008, 2012; Sakai *et al.*, 2008; Hamada *et al.*, 2013). Mutations in *NEK6* also exhibit a reduction in trichome branching, whereas no mutant phenotypes were found to be associated with *BPP1* mutations, possibly due to gene redundancy (Motose *et al.*, 2008; Sakai *et al.*, 2008). In both cases the MTs in these plants are more resistant to the MT-destabilizing herbicide oryzalin and both proteins have been shown to interact with MTs. Both genes are expressed in early-stage trichomes where they could influence the formation of concentric arrays (Table 10.1). *NEK6* encodes a kinase with a coiled-coil domain that mediates homodimerization and interactions with related kinases (Motose *et al.*, 2011). BPP1 contains a basic domain that mediates MT binding (Hamada *et al.*, 2013). The mechanism by which these proteins influence MT dynamic is not known.

Conclusions

Most is known about the initiation and morphogenesis of trichomes at the TF level. Early work identified TFs required for controlling both early and late stages of trichome development. While the gene targets of some of the TFs are known, the targets for most of the TFs are completely unknown. This area of research remains largely unexplored. Additional studies that have focused on various aspects of MT function have begun to reveal a glimpse of the role of MTs in trichome morphogenesis. Structural analyses suggest that dynamic cortical MTs can form concentric arrays of MTs at trichome branch foci. These arrays could alter the orientation of deposited cellulose MFs in a way that creates a relative weakening in the cell wall at the focus of trichome branch initiation. In this review, most of the genes that were chosen for discussion were shown to have roles in influencing both MTs and trichome branch formation. For many of the studies, the primary focus was not on trichome development but on some other aspect of plant development. Given the ease with which trichomes can be analyzed, the study of these genes in trichomes will help facilitate a better understanding of how MT dynamics are regulated to promote various developmental events in plants.

References

Abe, T., Thitamadee, S., and Hashimoto, T. (2004) Microtubule defects and cell morphogenesis in the lefty1lefty2 tubulin mutant of Arabidopsis thaliana. *Plant and Cell Physiology*, **45**, 211–220.

Balkunde, R., Pesch, M., and Hulskamp, M. (2010) Trichome patterning in Arabidopsis thaliana from genetic to molecular models. *Current Topics in Developmental Biology*, **91**, 299–321.

Basu, D., Le, J., Zakharova, T., Mallery, E.L., and Szymanski, D.B. (2008) A SPIKE1 signaling complex controls actin-dependent cell morphogenesis through the heteromeric WAVE and ARP2/3 complexes. *Proceedings of the National Academy of Sciences of the United States of America*, **105**, 4044–4049.

Beilstein, M. and Szymanski, D. (2004) Cytoskeletal requirements during Arabidopsis trichome development, in *The Plant Cytoskeleton in Cell Differentiation and Development* (ed. P. Hussey), Blackwell, Oxford, pp. 265–289.

Bowser, J. and Reddy, A.S. (1997) Localization of a kinesin-like calmodulin-binding protein in dividing cells of Arabidopsis and tobacco. *The Plant Journal*, **12**, 1429–1437.

Burk, D.H., Liu, B., Zhong, R., Morrison, W.H., and Ye, Z.H. (2001) A katanin-like protein regulates normal cell wall biosynthesis and cell elongation. *The Plant Cell*, **13**, 807–828.

Churchman, M.L., Brown, M.L., Kato, N., Kirik, V., Hulskamp, M., Inze, D., De Veylder, L., Walker, J.D., Zheng, Z., Oppenheimer, D.G., Gwin, T., Churchman, J., and Larkin, J.C. (2006) SIAMESE, a plant-specific cell cycle regulator, controls endoreplication onset in Arabidopsis thaliana. *The Plant Cell*, **18**, 3145–3157.

Cullmann, G., Fien, K., Kobayashi, R., and Stillman, B. (1995) Characterization of the five replication factor C genes of Saccharomyces cerevisiae. *Molecular and Cellular Biology*, **15**, 4661–4671.

Dobney, S., Chiasson, D., Lam, P., Smith, S.P., and Snedden, W.A. (2009) The calmodulin-related calcium sensor CML42 plays a role in trichome branching. *The Journal of Biological Chemistry*, **284**, 31647–31657.

Esau, K. (1977) *Anatomy of Seed Plants*. John Wiley and Sons, New York.

Esch, J.J., Chen, M., Sanders, M., Hillestad, M., Ndkium, S., Idelkope, B., Neizer, J., and Marks, M.D. (2003) A contradictory GLABRA3 allele helps define gene interactions controlling trichome development in Arabidopsis. *Development*, **130**, 5885–5894.

Esch, J.J., Chen, M.A., Hillestad, M., and Marks, M.D. (2004) Comparison of TRY and the closely related At1g01380 gene in controlling Arabidopsis trichome patterning. *The Plant Journal*, **40**, 860–869.

Fishel, E.A. and Dixit, R. (2013) Role of nucleation in cortical microtubule array organization: variations on a theme. *The Plant Journal*, **75**, 270–277.

Folkers, U., Berger, J., and Hulskamp, M. (1997) Cell morphogenesis of trichomes in Arabidopsis: differential control of primary and secondary branching by branch initiation regulators and cell growth. *Development*, **124**, 3779–3786.

Fu, Y., Gu, Y., Zheng, Z., Wasteneys, G., and Yang, Z. (2005) Arabidopsis interdigitating cell growth requires two antagonistic pathways with opposing action on cell morphogenesis. *Cell*, **120**, 687–700.

Gilding, E.K. and Marks, M.D. (2010) Analysis of purified glabra3-shapeshifter trichomes reveals a role for NOECK in regulating early trichome morphogenic events. *The Plant Journal*, **64**, 304–317.

Grebe, M. (2012) The patterning of epidermal hairs in Arabidopsis - updated. *Current Opinion in Plant Biology*, **15**, 31–37.

Gu, Y., Deng, Z., Paredez, A.R., DeBolt, S., Wang, Z.Y., and Somerville, C. (2008) Prefoldin 6 is required for normal microtubule dynamics and organization in Arabidopsis. *Proceedings of the National Academy of Sciences of the United States of America*, **105**, 18064–18069.

Hamada, T., Tekeuchi, N.N., Kato, T., Fujiwara, M., Sonobe, S., Fukao, Y., and Hashimoto, T. (2013) Purification and characterization of novel microtubule-associated proteins from Arabidopsis cell suspension cultures. *Plant Physiology*, **163**, 1804–1816.

Hashimoto, T. (2013) A ring for all: gamma-tubulin-containing nucleation complexes in acentrosomal plant microtubule arrays. *Current Opinion in Plant Biology*, **16**, 698–703.

Haughn, G.W. and Somerville, C.R. (1988) Genetic-Control of Morphogenesis in Arabidopsis. *Developmental Genetics*, **9**, 73–89.

Hulskamp, M., Misra, S., and Jurgens, G. (1994) Genetic dissection of trichome cell development in Arabidopsis. *Cell*, **76**, 555–566.

Ilgenfritz, H., Bouyer, D., Schnittger, A., Mathur, J., Kirik, V., Schwab, B., Chua, N.H., Jurgens, G., and Hulskamp, M. (2003) The Arabidopsis STICHEL gene is a regulator of trichome branch number and encodes a novel protein. *Plant Physiology*, **131**, 643–655.

Jakoby, M.J., Falkenhan, D., Mader, M.T., Brininstool, G., Wischnitzki, E., Platz, N., Hudson, A., Hulskamp, M., Larkin, J., and Schnittger, A. (2008) Transcriptional profiling of mature Arabidopsis trichomes reveals that NOECK encodes the MIXTA-like transcriptional regulator MYB106. *Plant Physiology*, **148**, 1583–1602.

Johnson, C.S., Kolevski, B., and Smyth, D.R. (2002) TRANSPARENT TESTA GLABRA2, a trichome and seed coat development gene of Arabidopsis, encodes a WRKY transcription factor. *The Plant Cell*, **14**, 1359–1375.

Johnson, H.B. (1975) Plant pubescence: An ecological perspective. *Botanical Reviews*, **41**, 233–258.

Kasili, R., Huang, C.C., Walker, J.D., Simmons, L.A., Zhou, J., Faulk, C., Hulskamp, M., and Larkin, J.C. (2011) BRANCHLESS TRICHOMES links cell shape and cell cycle control in Arabidopsis trichomes. *Development*, **138**, 2379–2388.

Khan, H., Parks, N., Kozera, C., Curtis, B.A., Parsons, B.J., Bowman, S., and Archibald, J.M. (2007) Plastid genome sequence of the cryptophyte alga Rhodomonas salina CCMP1319: lateral transfer of putative DNA replication machinery and a test of chromist plastid phylogeny. *Molecular Biology and Evolution*, **24**, 1832–1842.

Kirik, A., Ehrhardt, D.W., and Kirik, V. (2012) TONNEAU2/FASS regulates the geometry of microtubule nucleation and cortical array organization in interphase Arabidopsis cells. *The Plant Cell*, **24**, 1158–1170.

Kirik, V., Grini, P.E., Mathur, J., Klinkhammer, I., Adler, K., Bechtold, N., Herzog, M., Bonneville, J.M., and Hulskamp, M. (2002) The Arabidopsis TUBULIN-FOLDING COFACTOR A gene is involved in the control of the alpha/beta-tubulin monomer balance. *The Plant Cell*, **14**, 2265–2276.

Koornneef, M., Dellaert, S.W.M., and van der Veen, J.H. (1982) EMS- and radiation-induced mutation frequencies at individual loci in Arabidopsis thaliana (L) Heynh. *Mutation Research*, **93**, 109–123.

Lin, D., Cao, L., Zhou, Z., Zhu, L., Ehrhardt, D., Yang, Z., and Fu, Y. (2013) Rho GTPase signaling activates microtubule severing to promote microtubule ordering in Arabidopsis. *Current Biology*, **23**, 290–297.

Lin, Q. and Aoyama, T. (2012) Pathways for epidermal cell differentiation via the homeobox gene GLABRA2: update on the roles of the classic regulator. *Journal of Integrative Plant Biology*, **54**, 729–737.

Liu, Q., Wang, J., Miki, D., Xia, R., Yu, W., He, J., Zheng, Z., Zhu, J.K., and Gong, Z. (2010) DNA replication factor C1 mediates genomic stability and transcriptional gene silencing in Arabidopsis. *The Plant Cell*, **22**, 2336–2352.

Liu, Y., Deng, Y., Li, G., and Zhao, J. (2012) Replication factor C1 (RFC1) is required for double-strand break repair during meiotic homologous recombination in Arabidopsis. *The Plant Journal*, **73**, 154–165.

Lu, L., Lee, Y.R., Pan, R., Maloof, J.N., and Liu, B. (2005) An internal motor kinesin is associated with the Golgi apparatus and plays a role in trichome morphogenesis in Arabidopsis. *Molecular Biology of the Cell*, **16**, 811–823.

Marks, M.D., Esch, J., Herman, P., Sivakumaran, S., and Oppenheimer, D. (1991) A model for cell-type determination and differentiation in plants. *Symposia of the Society for Experimental Biology*, **45**, 77–87.

Marks, M.D., Betancur, L., Gilding, E., Chen, F., Bauer, S., Wenger, J.P., Dixon, R.A., and Haigler, C.H. (2008) A new method for isolating large quantities of Arabidopsis trichomes for transcriptome, cell wall and other types of analyses. *The Plant Journal*, **56**, 483–492.

Marks, M.D., Wenger, J.P., Gilding, E., Jilk, R., and Dixon, R.A. (2009) Transcriptome analysis of Arabidopsis wild-type and gl3-sst sim trichomes identifies four additional genes required for trichome development. *Molecular Plant*, **2**, 803–822.

Mathur, J. and Chua, N.H. (2000) Microtubule stabilization leads to growth reorientation in Arabidopsis trichomes. *The Plant Cell*, **12**, 465–477.

Mathur, J., Spielhofer, P., Kost, B., and Chua, N. (1999) The actin cytoskeleton is required to elaborate and maintain spatial patterning during trichome cell morphogenesis in Arabidopsis thaliana. *Development*, **126**, 5559–5568.

Monshausen, G.B., Messerli, M.A., and Gilroy, S. (2008) Imaging of the Yellow Cameleon 3.6 indicator reveals that elevations in cytosolic Ca2+ follow oscillating increases in growth in root hairs of Arabidopsis. *Plant Physiology*, **147**, 1690–1698.

Morohashi, K. and Grotewold, E. (2009) A Systems Approach Reveals Regulatory Circuitry for Arabidopsis Trichome Initiation by the GL3 and GL1 Selectors. *PLoS Genetics*, **5**, e1000396.

Motose, H., Tominaga, R., Wada, T., Sugiyama, M., and Watanabe, Y. (2008) A NIMA-related protein kinase suppresses ectopic outgrowth of epidermal cells through its kinase activity and the association with microtubules. *The Plant Journal*, **54**, 829–844.

Motose, H., Hamada, T., Yoshimoto, K., Murata, T., Hasebe, M., Watanabe, Y., Hashimoto, T., Sakai, T., and Takahashi, T. (2011) NIMA-related kinases 6, 4, and 5 interact with each other to regulate microtubule organization during epidermal cell expansion in Arabidopsis thaliana. *The Plant Journal*, **67**, 993–1005.

Motose, H., Takatani, S., Ikeda, T., and Takahashi, T. (2012) NIMA-related kinases regulate directional cell growth and organ development through microtubule function in Arabidopsis thaliana. *Plant Signaling and Behavior*, **7**, 1552–1555.

Mucha, E., Hoefle, C., Huckelhoven, R., and Berken, A. (2010) RIP3 and AtKinesin-13A - a novel interaction linking Rho proteins of plants to microtubules. *European Journal of Cell Biology*, **89**, 906–916.

Nakamura, M. and Hashimoto, T. (2009) A mutation in the Arabidopsis gamma-tubulin-containing complex causes helical growth and abnormal microtubule branching. *Journal of Cell Science*, **122**, 2208–2217.

Nakamura, M., Katsumata, H., Abe, M., Yabe, N., Komeda, Y., Yamamoto, K.T., and Takahashi, T. (2006) Characterization of the class IV homeodomain-leucine zipper gene family in Arabidopsis. *Plant Physiology*, **141**, 1363–1375.

Narasimhulu, S.B. and Reddy, A.S. (1998) Characterization of microtubule binding domains in the Arabidopsis kinesin-like calmodulin binding protein. *The Plant Cell*, **10**, 957–965.

Oda, Y. and Fukuda, H. (2012) Initiation of cell wall pattern by a Rho- and microtubule-driven symmetry breaking. *Science*, **337**, 1333–1336.

Oda, Y., Iida, Y., Kondo, Y., and Fukuda, H. (2010) Wood cell-wall structure requires local 2D-microtubule disassembly by a novel plasma membrane-anchored protein. *Current Biology*, **20**, 1197–1202.

Oppenheimer, D.G., Pollock, M.A., Vacik, J., Szymanski, D.B., Ericson, B., Feldmann, K., and Marks, M.D. (1997) Essential role of a kinesin-like protein in Arabidopsis trichome morphogenesis. *Proceedings of the National Academy of Sciences of the United States of America*, **94**, 6261–6266.

Podust, V.N. and Fanning, E. (1997) Assembly of functional replication factor C expressed using recombinant baculoviruses. *The Journal of Biological Chemistry*, **272**, 6303–6310.

Qiu, J.L., Jilk, R., Marks, M.D., and Szymanski, D.B. (2002) The Arabidopsis SPIKE1 gene is required for normal cell shape control and tissue development. *The Plant Cell*, **14**, 101–118.

Reddy, V.S., Day, I.S., Thomas, T., and Reddy, A.S. (2004) KIC, a novel Ca2+ binding protein with one EF-hand motif, interacts with a microtubule motor protein and regulates trichome morphogenesis. *The Plant Cell*, **16**, 185–200.

Ryu, K.H., Zheng, X., Huang, L., and Schiefelbein, J. (2013) Computational modeling of epidermal cell fate determination systems. *Current Opinion in Plant Biology*, **16**, 5–10.

Sakai, T., Honing, H., Nishioka, M., Uehara, Y., Takahashi, M., Fujisawa, N., Saji, K., Seki, M., Shinozaki, K., Jones, M.A., Smirnoff, N., Okada, K., and Wasteneys, G.O. (2008) Armadillo repeat-containing kinesins and a NIMA-related kinase are required for epidermal-cell morphogenesis in Arabidopsis. *The Plant Journal*, **53**, 157–171.

Singh, M.K., Ren, F., Giesemann, T., Bosco, C.D., Pasternak, T.P., Blein, T., Ruperti, B., Schmidt, G., Aktories, K., Molendijk, A.J., and Palme, K. (2012) Modification of plant Rac/Rop GTPase signalling using bacterial toxin transgenes. *The Plant Journal*, **73**, 314–324.

Szymanski, D.B. (2009) Plant cells taking shape: new insights into cytoplasmic control. *Current Opinion in Plant Biology*, **12**, 735–744.

Szymanski, D.B. and Cosgrove, D.J. (2009) Dynamic coordination of cytoskeletal and cell wall systems during plant cell morphogenesis. *Current Biology*, **19**, R800–811.

Szymanski, D.B., Jilk, R.A., Pollock, S.M., and Marks, M.D. (1998) Control of GL2 expression in Arabidopsis leaves and trichomes. *Development*, **125**, 1161–1171.

Szymanski, D.B., Marks, M.D., and Wick, S.M. (1999) Organized F-actin is essential for normal trichome morphogenesis in Arabidopsis. *The Plant Cell*, **11**, 2331–2347.

Szymanski, D.B., Lloyd, A.M., and Marks, M.D. (2000) Progress in the molecular genetic analysis of trichome initiation and morphogenesis in Arabidopsis. *Trends in Plant Science*, **5**, 214–219.

Vernoud, V., Horton, A.C., Yang, Z., and Nielsen, E. (2003) Analysis of the small GTPase gene superfamily of Arabidopsis. *Plant Physiology*, **131**, 1191–1208.

Walker, J.D., Oppenheimer, D.G., Concienne, J., and Larkin, J.C. (2000) SIAMESE, a gene controlling the endoreduplication cell cycle in Arabidopsis thaliana trichomes. *Development*, **127**, 3931–3940.

Wang, S.C., Hubbard, L., Chang, Y., Guo, J.J., Schiefelbein, J., and Chen, J.G. (2008) Comprehensive analysis of single-repeat R3 MYB proteins in epidermal cell patterning and their transcriptional regulation in Arabidopsis. *BMC Plant Biology*, **8**, doi: 10.1186/1471-2229-8-81.

Wester, K., Digiuni, S., Geier, F., Timmer, J., Fleck, C., and Hulskamp, M. (2009) Functional diversity of R3 single-repeat genes in trichome development. *Development*, **136**, 1487–1496.

Xia, S., Xiao, L.T., Bi, D.L., and Zhu, Z.H. (2007) Arabidopsis replication factor C subunit 1 plays an important role in embryogenesis. *Journal of Plant Physiology and Molecular Biology*, **33**, 179–187.

Xia, S., Zhu, Z., Hao, L., Chen, J.G., Xiao, L., Zhang, Y., and Li, X. (2009) Negative regulation of systemic acquired resistance by replication factor C subunit3 in Arabidopsis. *Plant Physiology*, **150**, 2009–2017.

Yanagisawa, M., Zhang, C., and Szymanski, D.B. (2013) ARP2/3-dependent growth in the plant kingdom: SCARs for life. *Frontiers in Plant Science*, **4**, 166.

Yang, Z. (2008) Cell polarity signalling in Arabidopsis. *Annual Review of Cell and Developmental Biology*, **24**, 551–575.

11 Transfer Cells: Novel Cell Types with Unique Wall Ingrowth Architecture Designed for Optimized Nutrient Transport

David W. McCurdy

School of Environmental and Life Sciences, University of Newcastle, Callaghan NSW, Australia

Introduction

Transfer cells provide a fascinating example of how the cell wall profoundly influences cell function in plants. As the name implies, transfer cells (TCs) are involved in nutrient transport events where nutrients pass across a plasma membrane at apoplasmic/symplasmic junctions in nutrient acquisition or delivery pathways. To facilitate this process, TCs become anatomically modified by developing elaborate wall ingrowth structures which project into the cytoplasmic volume of the cell and serve as a scaffold to amplify plasma membrane surface area. This increase in surface area supports increased densities of plasma membrane nutrient transporters, thus providing a mechanistic basis for facilitating enhanced rates of nutrient transport across the plasma membrane.

The term 'transfer cell' was first used by Gunning and coworkers (Gunning *et al.*, 1968) to describe various phloem cells in leaf minor veins which contained inwardly directed projections of wall material, so-called 'wall ingrowths' (Fig. 11.1). These TCs were proposed to function in active uptake of solutes, principally sugars, from the cell wall space for subsequent transfer to sieve elements and thus distribution to sink tissues (Gunning *et al.*, 1968). Extensive surveys then undertaken by Gunning and Pate yielded the realization that numerous other cell types in the plant kingdom contained similar wall ingrowths; they therefore expanded the definition of 'transfer cell' to include all cell types containing a distinctive 'wall-membrane apparatus' consisting of wall ingrowths ensheathed by an amplified plasma membrane (Gunning and Pate, 1969, 1974). They also noted that mitochondria and endomembrane structures were typically abundant in TCs, often accumulating adjacent to the wall ingrowths themselves and therefore defining a feature of

Plant Cell Wall Patterning and Cell Shape, First Edition. Edited by Hiroo Fukuda.

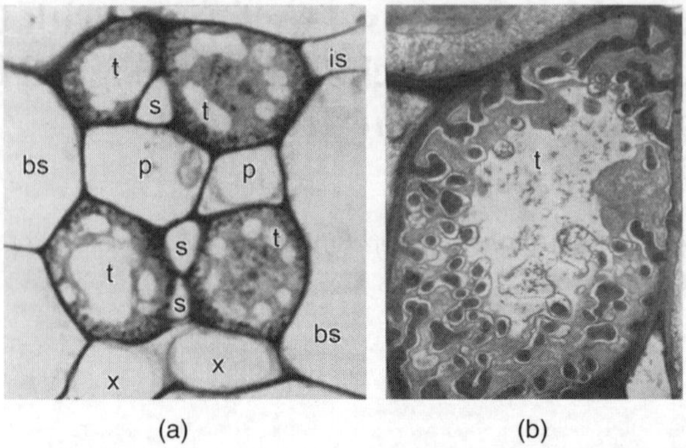

Figure 11.1 TCs in *Pisum arvense* minor vein. Images from Gunning *et al.* (1968) report the first comprehensive account of the structure of TCs in a plant tissue. (a) Transverse section of minor vein showing the position of cytoplasmically dense TCs (t) adjacent to phloem parenchyma (p) cells and sieve elements (s) of the phloem. Also shown are tracheary elements (x), two bundle sheath (bs) cells and an intercellular space (is). (b) Higher magnification view of a TC (t) showing numerous and distinctive protruberances of the wall structure, defined as 'wall ingrowths'. The wall ingrowths in this TC essentially occur around the full surface of the cell, a feature subsequently identified as a characteristic of companion TCs in pea minor veins. Images reproduced from Gunning *et al.* (1968) *Journal of Cell Biology* 37:C7-C12, with permission (The Rockefeller University Press).

these cells consistent with a role in membrane transport events (Gunning and Pate, 1974).

The seminal reviews by Gunning and Pate (1969, 1974) provided the first detailed survey of sites where TC development occur in plants and algae. Indeed, these reviews essentially defined the field of TC biology and posed questions about mechanism and function that continue to shape current research in this field. The more recent review by Offler *et al.* (2003) compiled the most extensive survey to date of published accounts of the existence of TCs, noting that they occur in all plant taxonomic groups and in algae and fungi. TCs are strategically positioned at various interfaces where transport demand is high, such as the maternal/filial interface in developing seed (Thompson *et al.*, 2001), symplasmic/apoplasmic interfaces involved in nutrient loading and unloading of vascular pipelines, and at interfaces where exchange between these pipelines occurs, such as xylem nodal tissue (Gunning and Pate, 1969), secretory organs such as floral nectaries (Kronestedt and Robards, 1987) and salt glands (Gunning and Pate, 1969), and at nutrient acquisition interfaces such as cortical/epidermal cells of roots (Schikora and Schmidt, 2002) and leaf epidermal cells of aquatic plants (Rascio *et al.*, 1994). Development of TCs may also occur as a consequence of symbiotic associations and plant/parasite interactions (Offler *et al.*, 2003) or, in the case of nematode infections, giant cells with TC-like wall ingrowths develop in root vascular tissue (Jones and Gunning, 1976). A unique characteristic of TCs is that they come into existence by *trans*-differentiation of an existing differentiated cell type. Epidermal cells or phloem parenchyma cells may therefore *trans*-differentiate to become epidermal TCs or phloem parenchyma TCs, respectively. The name of the subsequent TC maintains the identity of the original cell type.

The architecture and building of wall ingrowths is a fascinating example of localized wall deposition in plant cells and, in many instances, probably involves molecular mechanisms not previously documented in plant biology. Recent advances in imaging TCs, primarily using scanning electron microscopy (SEM) or field-emission SEM (FESEM), have revealed that wall ingrowth architecture can be classified within two major categories: reticulate and flange. The deposition of flange ingrowths resembles localized secondary wall deposition in many respects, whereas the deposition of reticulate wall ingrowths probably involves mechanisms unique to plant cell wall biology (McCurdy *et al.*, 2008). In a few known cases TCs can deposit both types of wall ingrowths, a process that is fascinating to contemplate from a mechanistic perspective.

This chapter focuses primarily on describing wall ingrowth architecture in TCs and the numerous and unique variations that make these functionally important cell types so absorbing to study. Consideration is also given to other aspects of TC biology, such as the signals that induce their *trans*-differentiation, changes in gene expression accompanying this *trans*-differentiation, and speculation about cellular mechanisms involved in building wall ingrowths, particularly those of the reticulate class. Consideration is also given to describing the features of phloem parenchyma TCs in minor veins of the model species *Arabidopsis thaliana* (Arabidopsis) as a way forward to develop genetic approaches to study TC biology.

Reticulate Wall Ingrowth Morphology

Reticulate wall ingrowths are by far the most common wall ingrowth morphology seen in the plant kingdom (Gunning and Pate, 1969; Offler *et al.*, 2003). The overwhelming conclusion from early structural studies using mostly light and transmission electron microscopy (TEM) was that wall ingrowths of the reticulate type are very diverse in architecture and extent, but ultimately can be defined by the initial formation of discrete papillate ingrowths arising from the parent wall (Gunning and Pate, 1969). The distribution of wall ingrowths can be more or less uniform around the cell, as seen in companion TCs in minor veins of pea leaves, and are defined as 'type A' by Gunning and Pate. More commonly, the wall ingrowths are highly localized to specific regions of the cell reflecting the assumed direction of nutrient flow, as seen in phloem parenchyma TCs in minor veins of *Arabidopsis*. Such TCs are defined as 'type B' (Gunning and Pate, 1969; Pate and Gunning, 1969).

The complex and diverse architecture of wall ingrowths was appreciated from early TEM and, to a lesser extent, SEM studies (Gunning and Pate, 1969), but the full complexity and amazing architecture of these wall structures was more fully realized by SEM analysis of 'freeze-fractured' or 'dry-cleaved' plant tissues (Talbot *et al.*, 2001, 2002). For freeze-fracturing, more accurately described as 'fracturing frozen tissue', excised tissue pieces were snap-frozen in liquid nitrogen and subsequently fractured on a liquid nitrogen-cooled block before being extracted extensively with detergent and viewed by SEM. The dry-cleaving procedure, developed to view the cytoplasmic face of epidermal cells of *Vicia faba* cotyledons, involved adhering detergent-extracted critical-point-dried cotyledons to support stubs and then tearing away the bulk of the cotyledon tissue to leave behind a thin paradermal strip of mostly epidermal tissue, thus exposing the cytoplasmic face of the outer periclinal wall of the epidermal TCs (Talbot *et al.*, 2001).

Viewing sheets of either abaxial or adaxial epidermal cells in this way enabled a detailed ontology of reticulate wall ingrowth formation to be constructed. This process is initiated by the formation of small papillate projections of wall material arising from the cytoplasmic face of the outer periclinal parent wall (Fig. 11.2a, b). These projections form at discrete loci (Fig. 11.2b) and as they develop they become more variable in morphology; some form bifurcated structures while others appear to make connections via thin lateral projections (Fig. 11.2c, d). The culmination of this activity is the formation of a network of interconnected papillae which ultimately creates a flattened, fenestrated sheet which is connected to but not continuous with the underlying parent wall (Fig. 11.2e–g). The fenestrae become occluded, or 'filled-in', by apparent growth of small protrusions extending laterally from the inner surface of the fenestrae (Talbot *et al.*, 2001). Concurrently, new papillae form on the surface of the fenestrated layer (Fig. 11.2f, g), which then generate a new round of bifurcation and lateral fusion to build a second, interconnected, fenestrated layer. Continuation of this cyclic process results in construction of a multilayered, fenestrated ingrowth wall network (Fig. 11.2h).

Variation in reticulate wall ingrowth morphology is seen amongst different TCs. In xylem parenchyma TCs in *Tradescantia virginiana* and nodes of pea (*Pisum sativum*), the ingrowth network is composed of mostly smooth tubular ingrowths. These create a very open meshwork, presumably enabling maximum surface area of the ensheathing plasma membrane (Fig. 11.3). In contrast, wall ingrowth networks in xylem parenchyma TCs in nodes of basil (*Ocimum basilicum*) and nucellar projection TCs of wheat (*Triticum aestivum*) are broader and more flattened in appearance, giving the impression of a more closed or compressed ingrowth network with consequently less amplification of plasma membrane surface area (Talbot *et al.*, 2002). These different morphologies all derive from initial deposition of papillae at discrete loci on the parent wall. This process may be restricted to single papillae with limited branching or fusion, as seen in xylem parenchyma TCs in roots of *V. faba* (Talbot *et al.*, 2002), or more typically involves many rounds of papillate deposition followed by branching and lateral fusion to build a multilayered wall ingrowth network, often extending some distance into the cytoplasmic volume

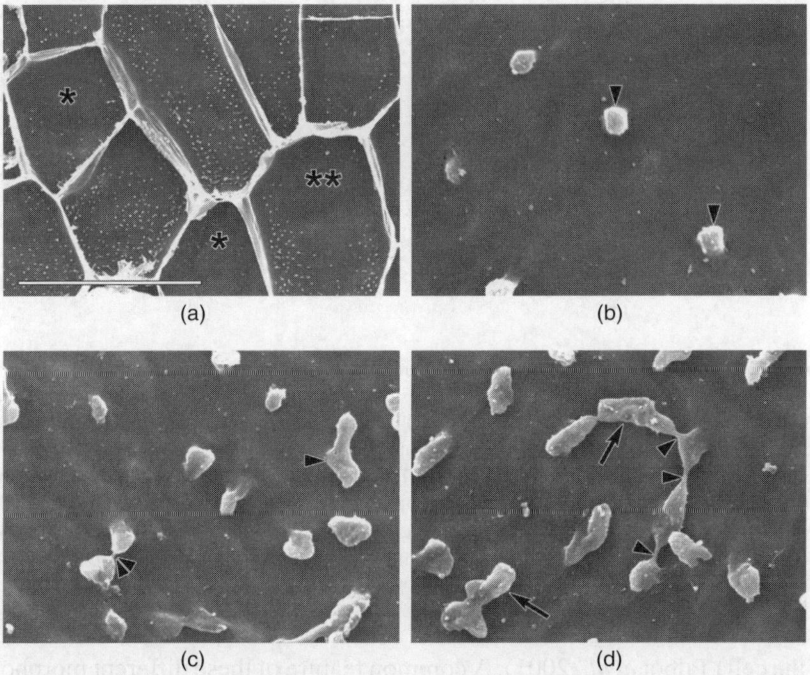

Figure 11.2 Morphology of reticulate wall ingrowth development on the outer periclinal wall of epidermal TCs of *Vicia faba* cotyledons. Material was prepared by a 'dry-cleave' technique and imaged by SEM. (a) Low magnification view of the cytoplasmic face of the outer periclinal wall of several epidermal cells, showing asynchronous deposition of papillate wall ingrowths (small dots) and spatial heterogeneity of the emerging papillae within individual cells. Epidermal cells with few wall ingrowths are indicated by a single asterisk; the double asterisk indicates a cell with asynchronous development of wall ingrowths across the face of the cell. The sheared anticlinal walls of the epidermal cells are seen as white borders around each cell. (b) Higher-magnification view of individual papillate wall ingrowths (arrowheads) deposited at discrete but apparently random loci on the surface of the underlying parent wall. (c) Individual papillae develop various morphologies, including those that bifurcate (arrowhead) or join together via slender lateral connections (double arrowheads). (d) As the number of papillate wall ingrowths increases, those with lateral appendages (arrows) become more common as well as those with slender connections between each other (arrowheads).

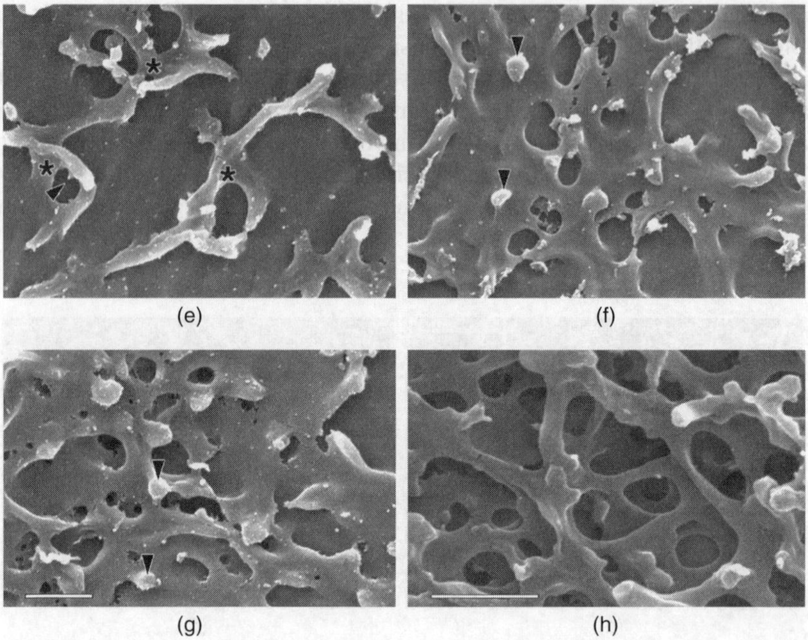

Figure 11.2 (e) The fusion of lateral connections and appendages form local areas of raised wall ingrowth material (asterisks) which can form fenestrae. Small protrusions are seen emerging laterally from the edges of some fenestrae (arrowhead). (f) Continued deposition results in the formation of a flattened, fenestrated layer of wall material. New papillate wall ingrowths (arrowheads) form on the cytoplasmic face of this first fenestrated layer. (g) A wall ingrowth labyrinth comprising two layers of fenestrated wall material with new papillate projections (arrowheads), presumably initiating a third fenestrated layer. (h) Multi-layered wall ingrowth network comprising of at least three layers. Scale bars: 20 μm (a); 1 μm (b–g); 1 μm (h). Images reproduced from Talbot *et al.* (2001) *Protoplasma* 215:191–203, with permission (Springer).

of the cell (Talbot *et al.*, 2001). A common feature of these different morphologies appears to be the presence of small finger-like or rounded protrusions projecting laterally from the edges of the wall ingrowths and fenestrae. In open networks these finger-like protrusions are slender and distinctive (Fig. 11.3f), whereas in the more flattened, closed networks these structures pepper the surface of the ingrowths and are more numerous. Talbot *et al.* (2002) suggested that the different ingrowth morphologies may be due to degrees of filling in of the fenestrae by these surface projections.

The deposition of reticulate wall ingrowths appears to be preceded in many instances by construction of a uniform layer of wall material. The presence of this wall layer is most clearly seen in TCs where highly polarized deposition of wall ingrowths occurs, such as epidermal TCs of *V. faba* cotyledons

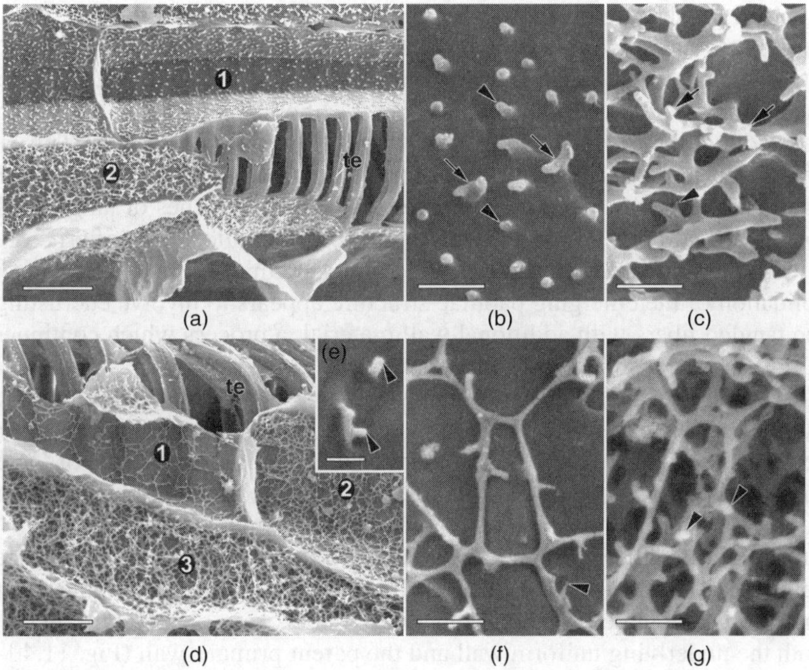

Figure 11.3 Reticulate wall ingrowths in xylem parenchyma TCs of longitudinally fractured stem nodes of (a–c) *Pisum sativum* and (d–g) *Tradescantia virginiana*. (a, d) Low-magnification views of TCs positioned adjacent to underlying tracheary elements (te); (b, c, e–g) higher-magnification views of regions of these TCs. (a) Two TCs (1 and 2) show different stages of reticulate wall ingrowth development. (b) Numerous single papillae (arrowheads) or bifurcating wall ingrowths (arrows) in TC 1 in (a) are deposited at discrete loci on the cytoplasmic face of the parent wall. (c) At a later stage of development, the wall ingrowths in TC 2 in (a) are branched and intertwined to form a complex open network. New papillae (arrows) emerge from this network and a small lateral protrusion is seen (arrowhead). (d) Three TCs (1, 2 and 3) at different stages of reticulate ingrowth development. (e) Small papillate or branched ingrowths (arrowheads) deposited on the cytoplasmic surface of the parent wall. (f) Fused papillae form an interconnected, flattened network on the surface of the wall. Small lateral protrusions (arrowhead) are seen. (g) A complex three-dimensional network of interconnected tubular wall ingrowths is seen, with new papillate projections (arrowheads) emerging from the most recently deposited strands of this labyrinth. Scale bars: 10 μm (a, d); 1 μm (b, c, f and g); 0.5 μm (e). Figure reproduced from Talbot *et al.* (2002) *Protoplasma* 219:197–209, with permission (Springer).

(Farley *et al.*, 2000; Talbot *et al.*, 2007b) and nucellar projection TCs in developing wheat grains (Wang *et al.*, 1994). The existence of a uniform wall layer in TCs where non-polarized wall ingrowth deposition occurs is more equivocal, but hints that it exists in such TCs can be seen in some instances (Wimmers and Turgeon, 1991). Irrespective, polarized deposition of wall ingrowths appears to be preceded invariantly by polarized deposition of this new wall layer. FESEM imaging shows that cellulose microfibrils in the thin primary wall of adaxial epidermal cells in *V. faba* cotyledons, prior to ingrowth deposition, are randomly organized with only minimal local alignment. In contrast, the uniform wall layer (deposited within 2–4 hours of TC induction) is composed of highly ordered cellulose microfibrils (Talbot *et al.*, 2007b). This layer of ordered microfibrils forms a platform for subsequent deposition of small papillate ingrowths, which are initiated as localized patches of tangled cellulosic material (Fig. 11.4a–c). Unlike the linear cellulose microfibrils of the underlying uniform wall, these tangled fibers are thicker and coalesced. Consolidation of the emerging papillae structure appears to involve encrusting of the tangled fibers with additional wall material, a process which continues to build a more consolidated papillate structure (Fig. 11.4b, c).

Viewing these nascent papillae in transverse view by TEM showed that the earliest appearance of the papillate structures corresponded to discrete semi-rounded patches of material that accumulated on the surface of the uniform wall layer (Fig. 11.4d). This material appeared to be globular in nature but, as the ingrowth developed a more elongated finger-like morphology, the substructure of the ingrowth appeared more linear, similar in appearance to cellulose microfibrils. At all stages, however, heavy metal staining showed that the accumulated material and linear fibers were clearly different from both the underlying uniform wall and the parent primary wall (Fig. 11.4d–f). A common feature of TEM views was the presence of an electron translucent layer encasing the emerging wall ingrowths but interior to the plasma membrane (Fig. 11.4d–f). This layer is suspected to be caused by, at least in part, fixation artifacts (Gunning and Pate, 1969; Talbot *et al.*, 2007b). Imaging mature ingrowth labyrinths in epidermal TCs of *V. faba* cotyledons by FESEM showed that the wall ingrowth surfaces were smooth and solid-looking, presumably reflecting substantial deposition of hemicellulosic and wall matrix materials in these ingrowths (Talbot *et al.*, 2007b).

Flange Wall Ingrowth Morphology

Flange wall ingrowths are characterized by deposition of curvilinear, rib-shaped projections of wall material. Similar to reticulate ingrowths, the flange category encompasses many variations and degrees of elaboration. Simple flange ingrowths, seen in xylem parenchyma TCs of vegetative nodes in monocotyledons such as wheat, maize (*Zea mays*), and rice (*Oryza sativa*), are typically smooth-surfaced structures deposited in regularly spaced, mostly parallel arrays transverse to the long axis of the cell (Fig. 11.5). These distinctive cell wall ridges are initiated as long thin ribs which then develop into thicker, often forked, bands (Fig. 11.5b, d). In many respects, flange wall ingrowths are similar to secondary wall thickenings, albeit generally thinner

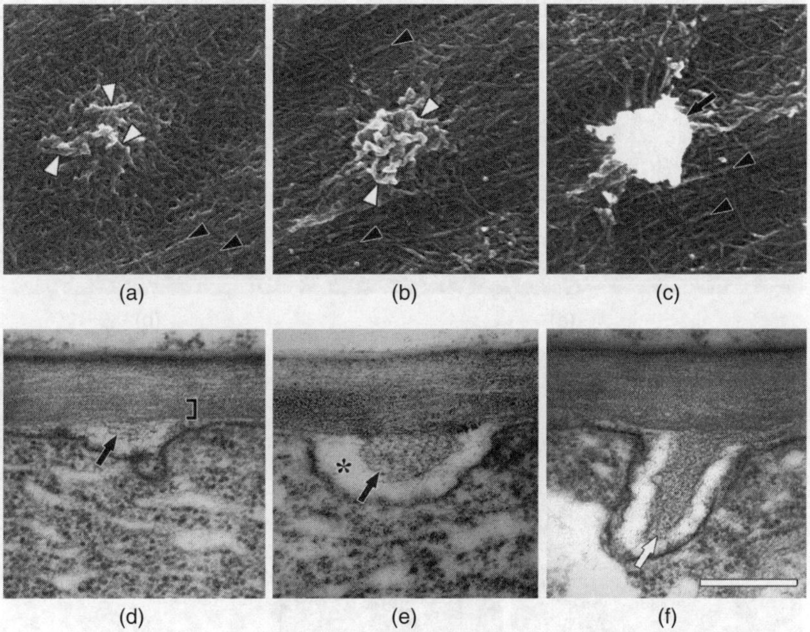

Figure 11.4 Structure of developing papillate wall ingrowths in epidermal TCs of *Vicia faba* cotyledons. (a–c) Initial stages of reticulate wall ingrowth deposition observed by FESEM. (d–f) Equivalent stages of wall ingrowth deposition observed in transverse sections by TEM. (a) The first stage of wall ingrowth formation is the localized deposition of raised patches of disorganized, presumably cellulosic, material (white arrowheads). These patches are deposited onto or emerge from a layer of parallel cellulose microfibrils (black arrowheads) which constitute the new uniformly thickened wall layer seen in TCs prior to reticulate ingrowth deposition. (b) An emerging wall ingrowth structure appears to be consolidated by accumulation of thickened structures, forming a localized mass of entangled, coalesced wall material (white arrows) deposited over parallel microfibrils of the underlying wall (black arrowheads). (c) At a later stage of ingrowth formation, the typical papillate morphology is achieved by further encrusting of additional wall material (black arrow). The substructure of the papillate ingrowth is not visible due to excessive electron charging in the FESEM beam. Parallel microfibrils of the underlying parent wall (arrowheads) remain visible. (d) Earliest stage of papillate ingrowth formation in transverse section appears as a localized accumulation of wall material (arrow) deposited between the plasma membrane and the uniformly thickened wall layer (bracket) deposited over the surface of the parent primary wall. (e) The developing ingrowth appears as a rounded mass (arrow) of apparently globular material adjacent to the underlying wall. A thick layer of electron translucent material (asterisk) covers the emerging papillae. (f) As the ingrowth develops a more elongated papillate morphology, its substructure appears to be composed of linear microfibrils (white arrow). The ingrowth remains surrounded by a layer of electron translucent material. Scale bar: 500 nm (a–f). Images reproduced from Talbot *et al.* (2007b) *Plant Cell and Physiology* 48:147–158, with permission (Springer).

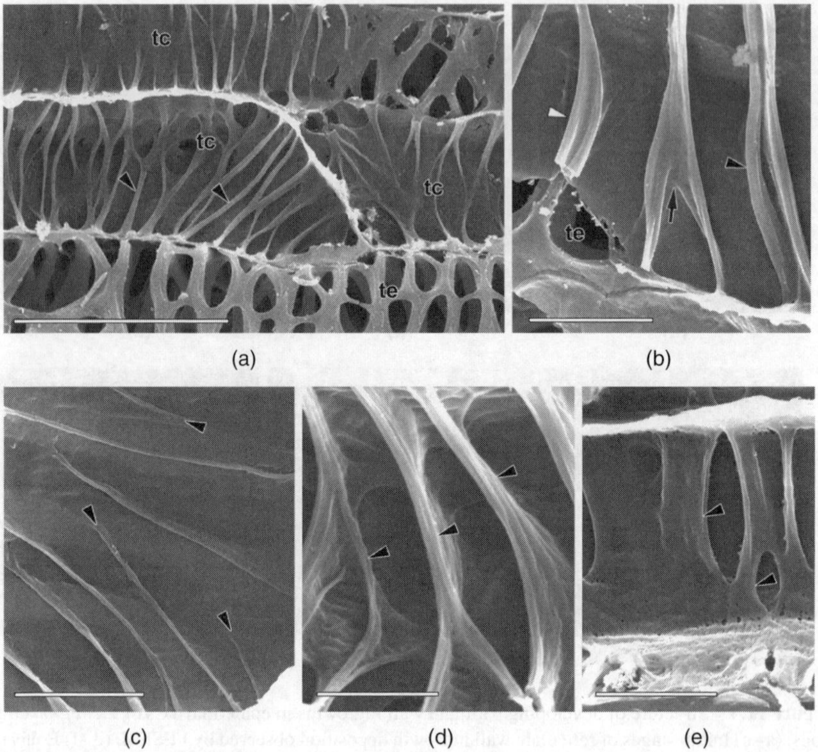

Figure 11.5 Flange wall ingrowths in xylem parenchyma TCs of longitudinally fractured nodes of (a, b) *Triticum aestivum*, (c, d) *Zea mays*, and (e) *Oryza sativa*. (a) The flange ingrowths in the TCs (tc) are roughly parallel, long slender thickenings (arrowheads) which bifurcate and fuse linearly with neighboring ingrowths. The ingrowths are thinner than secondary wall thickenings of the adjacent tracheary element (te). (b) Higher-magnification view showing the Y-shaped structure of individual ingrowths (arrowheads) and a bifurcating ingrowth (arrow). An adjacent tracheary element (te) is seen under the transfer cell. A flange ingrowth (white arrow) directly overlies a secondary wall thickening of the tracheary element (te). (c) In a young TC, slender ribs of wall material (arrowheads) appear to emerge from the underlying parent wall. (d) In an older TC, these ribs have developed into thin ridges of wall ingrowths (arrowheads) which often bifurcate. (e) Flange wall ingrowths in *O. sativa* are deposited as broad, flattened plates (arrowheads). Scale bars: 10 μm (a); 5 μm (b–d); 2 μm (e). Figure reproduced from Talbot *et al.* (2002) *Protoplasma* 219:197–209, with permission (Springer).

and more numerous (Fig. 11.5a). The positioning of flange ingrowths in nodes is often seen juxtaposed against a secondary wall thickening of a neighboring tracheary element (Fig. 11.5b), implying that the deposition of both elements could be spatially coordinated.

An example of the elaboration of flange wall ingrowth morphology is seen in basal endosperm TCs of *Z. mays* kernals. The flange ingrowths are deposited in an anticlinal orientation relative to the interface of the basal endosperm TCs and placental cavity and extend along the full length of the cell, but become progressively more cross-linked and fused towards the base (Fig. 11.6). The cross-linking between individual flanges appears to be initiated by thin lateral finger-like protrusions which emerge from neighboring flanges (Fig. 11.6b) to eventually form a dense interconnected network of wall ingrowth material that fills much of the cell volume (Fig. 11.6a, c). The extent of cross-linking varies along the length of the flange ingrowths, such that the upper portion of a basal endosperm TC consists of densely packed but essentially individual flanges, whereas the basal (lower) region consists of a dense, highly cross-linked network of wall material (Fig. 11.6a, c). Higher magnification views suggest that the cross-linked network appears to emerge directly from the underlying outer periclinal wall of basal endosperm TCs (Fig. 11.6c), reminiscent of a fenestrated reticulate wall ingrowth network.

Interestingly, this cross-linked network structure does not occur in the adjacent inner layer of basal endosperm TCs in maize, suggesting that the mechanism for building the cross-linked lower region is specific to the most basal layer of TCs in this tissue. Indeed, a recent study by Monjardino *et al.* (2013) claims that these TCs do indeed deposit reticulate-type wall ingrowths along their basal periclinal face. These authors used TEM and confocal microscopy of Calcofluor White-stained sections to conclude that reticulate ingrowths are located exclusively along the outer (basal) periclinal wall and adjacent regions of the anticlinal wall, whereas flange ingrowths form on all remaining faces of the cell. Interestingly, unlike other examples of TCs displaying polarized deposition of reticulate wall ingrowths, Monjardino *et al.* (2013) did not detect the presence of a distinctive uniform wall layer on the cytoplasmic face of the outer periclinal wall. If the formation of such a layer is a prerequisite for polarized reticulate wall ingrowth deposition (see the previous section), then the conclusions by Monjardino *et al.* (2013) are controversial. Since most of their observations were based on TEM, where nascent flange ingrowths can appear reticulate-like in transverse section (Davis *et al.*, 1990; Kang *et al.*, 2009), further analysis (preferably by freeze-fracture SEM; see Talbot *et al.*, 2002) is required to validate their interesting claim.

The use of FESEM in combination with polysaccharide extraction using acetic/nitric acid treatment enabled the organization of cellulose microfibrils in nascent and mature flange wall ingrowths to be observed (Talbot *et al.*, 2007a). In xylem TCs of wheat nodes, mature flange ingrowths are composed of acid-resistant cellulose microfibrils oriented parallel to one another along the full length of the ingrowth (Fig. 11.7a, b). The use of detergent extraction enabled similar views of linear flange ingrowths in basal endosperm TCs of corn (Talbot *et al.*, 2007a). Under these conditions, mature flange ingrowths were also seen to be composed of linear, parallel cellulose microfibrils (Fig. 11.7e, f). Nascent flange ingrowths in these cells were seen as localized

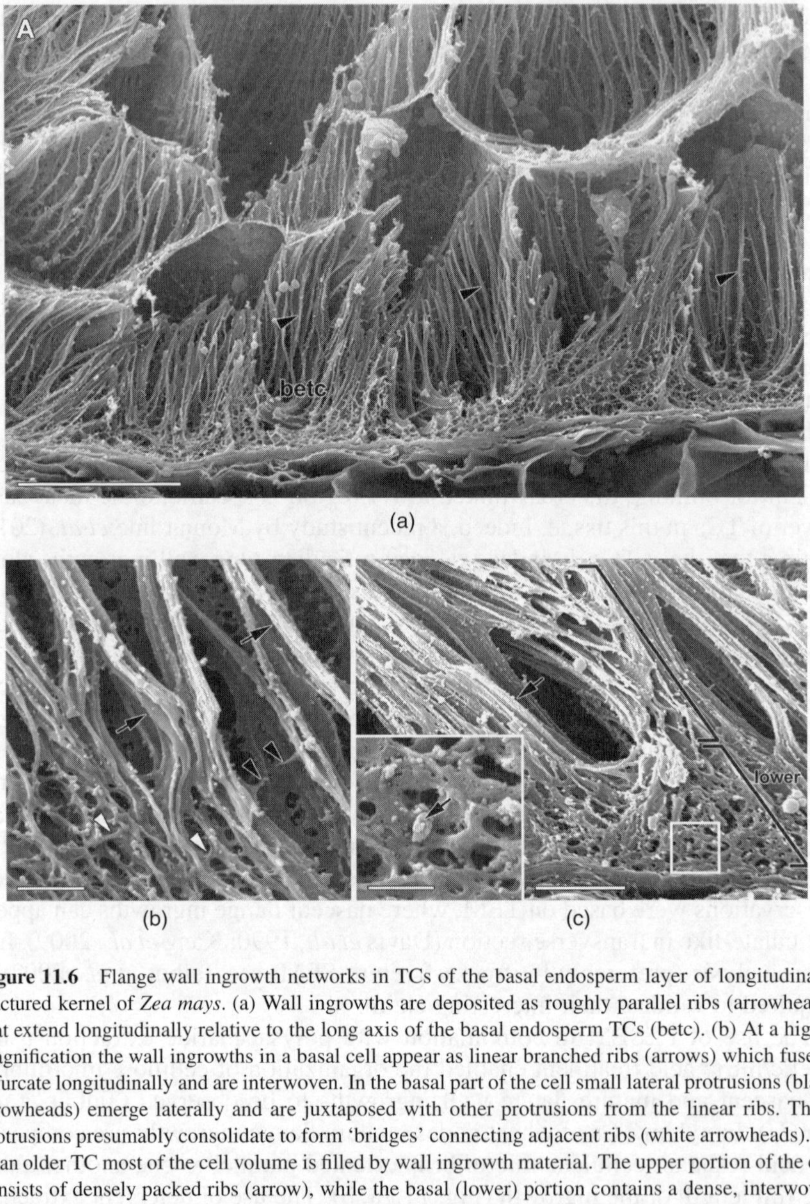

Figure 11.6 Flange wall ingrowth networks in TCs of the basal endosperm layer of longitudinally fractured kernel of *Zea mays*. (a) Wall ingrowths are deposited as roughly parallel ribs (arrowheads) that extend longitudinally relative to the long axis of the basal endosperm TCs (betc). (b) At a higher magnification the wall ingrowths in a basal cell appear as linear branched ribs (arrows) which fuse or bifurcate longitudinally and are interwoven. In the basal part of the cell small lateral protrusions (black arrowheads) emerge laterally and are juxtaposed with other protrusions from the linear ribs. These protrusions presumably consolidate to form 'bridges' connecting adjacent ribs (white arrowheads). (c) In an older TC most of the cell volume is filled by wall ingrowth material. The upper portion of the cell consists of densely packed ribs (arrow), while the basal (lower) portion contains a dense, interwoven network of wall material. The high-magnification view of the boxed area in (c) shows this network, which also appears to contain examples of small papillate projections emerging in places from the surface of the network (arrow). Scale bars: 20 μm (a); 2 μm (b); 5 μm (c); 1 μm (inset in c). Figure reproduced from Talbot *et al.* (2002) *Protoplasma* 219:197–209, with permission (Springer).

groupings of linear cellulose microfibrils organized in a somewhat criss-cross pattern across the cytoplasmic face of the parent cell wall (Fig. 11.7c, d). These criss-crossed assemblies of cellulose microfibrils appear to merge to initiate a more substantial nascent wall ingrowth, which becomes consolidated along

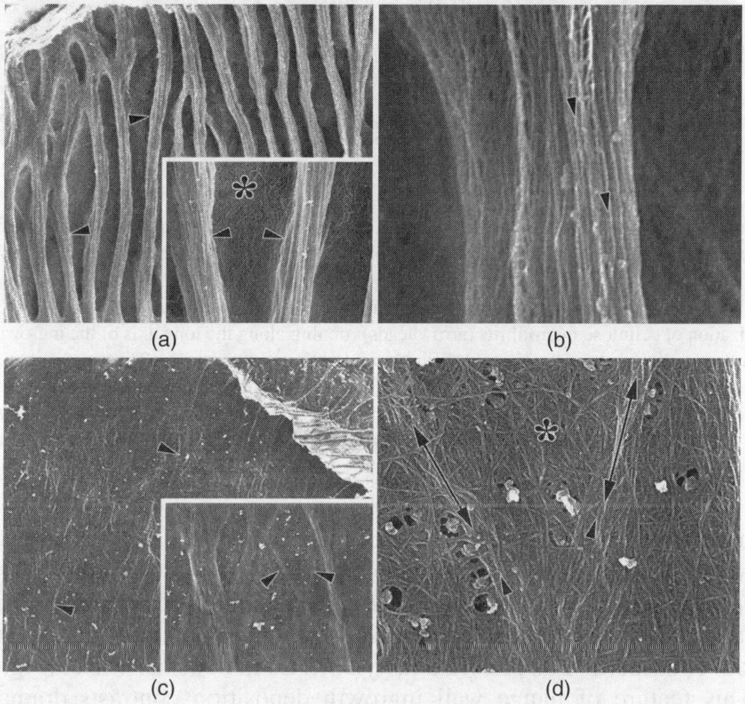

Figure 11.7 Cellulose microfibril orientation of flange wall ingrowths and underlying parent wall. (a, b) Xylem TCs in nodes of wheat (*Triticum aestivum*). (c–f) Basal endosperm TCs of corn (*Zea mays*). FESEM was used to observe microfibrils or bundles of microfibrils in each tissue. (a) A single TC showing the linear arrangement of flange ingrowth projections (arrowheads) and their fusion with neighboring projections in places. The inset in (a) shows a high magnification view of two flange ingrowths showing their linear substructure (arrowheads) and the random organization of microfibrils in the underlying wall between the ingrowths (asterisk). (b) High-magnification view of a flange ingrowth showing the longitudinal arrangement of parallel cellulose microfibrils (arrowheads). These structures survived extraction in boiling acetic/nitric acid indicating they are composed of crystalline cellulose. (c) An early stage of flange wall ingrowth deposition in an endosperm TC interior to the most basal layer. Emerging wall ingrowths are seen traversing the surface of the inner wall (arrowheads). The inset in (c) shows that the emerging ingrowths (arrowheads) are deposited in a criss-cross pattern. (d) High-magnification view of the surface of the inner wall of the cell shown in (c). The emerging wall ingrowths appear to arise from local ordering of cellulose microfibrils into bundles (arrowheads) roughly parallel to each other, and which define the axis of the emerging wall ingrowths (double-headed arrows). These nascent ingrowths appear to form over the more randomly arranged microfibrils of the underlying wall (asterisk).

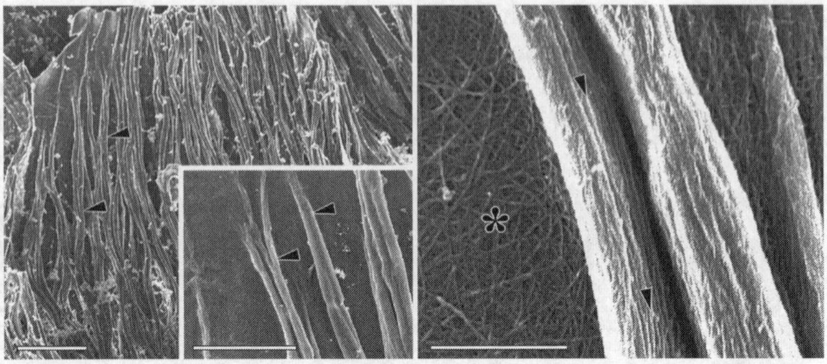

Figure 11.7 (e) A more mature stage of flange wall ingrowth deposition in a basal endosperm TC of a corn kernel. The flange ingrowths (arrowheads) are thicker and show increased lateral fusion or joining with neighboring ingrowths. The inset in (e) shows the tapered ends of the flange ingrowths (arrowheads). (f) High-magnification view of flange ingrowths in a basal endosperm TC showing the parallel organization of cellulose microfibrils (arrowheads) running along the long axis of the ingrowths. Note the random organization of the microfibrils in the underlying wall (asterisk). The endosperm tissue was extracted with detergent. Scale bars: 10 μm (a, c, e; bar shown in e); 3 μm (insets in a, c, e; bar shown in inset e); 500 nm (b, d, f; bar shown in f). Figure reproduced from Talbot *et al.* (2007a) *Functional Plant Biology* 34:307–313, with permission (CSIRO Publishing).

its length to form a mature flange ingrowth composed of linearly organized cellulose microfibrils (Fig. 11.7e). Significantly, the FESEM observations showed that the parent cell wall between the emerging wall ingrowths is composed of more randomly organized cellulose microfibrils (Fig. 11.7d, f), and this situation is maintained in regions containing mature flange ingrowths in both endosperm TCs of corn and xylem TCs in nodes of wheat (Fig. 11.7a, f). This feature of flange wall ingrowth deposition contrasts dramatically with that of reticulate wall ingrowths, which are deposited as essentially tangled lumps of presumably non-crystalline cellulose microfibrils over or onto a wall layer composed of parallel microfibrils. This contrasting structural organization has implications for mechanisms by which the two different wall ingrowth types are deposited.

Chemical Composition of Wall Ingrowths

Early histochemical analysis of wall ingrowths noted that they contained cellulose microfibrils and were abundant in pectic polysaccharides, characteristics consistent with the ingrowth wall itself facilitating diffusion of solutes either to or from the ensheathing plasma membrane (Gunning and Pate, 1974). TEM observations noted that cellulose microfibrils in reticulate wall ingrowths appeared to be loosely packed and arranged in apparent whorls compared to the underlying parent wall (Farley *et al.*, 2000; Talbot *et al.*, 2007b). The first quantitative analysis of wall ingrowth composition was undertaken by DeWitt *et al.* (1999) who compared a maize cell line which develops a TC-like wall morphology with that of a control line which does not. Their study did not detect quantitative differences in mass or composition of the three major polysaccharide fractions tested, namely hemicelluloses, pectin, or cellulose. Based on this outcome, they concluded that substantial changes in wall morphology may not require gross changes in wall composition, but rather subtle spatiotemporal changes in minor but key regulatory components of wall architecture (DeWitt *et al.*, 1999).

This conclusion was broadly supported by subsequent immunocytochemical studies. Dahiya and Brewin (2006) used a panel of antibody probes to analyze wall ingrowth composition of TCs in pea nodules, finding that xyloglucan and pectin antigens were uniformly distributed throughout the ingrowths (similar to that seen in the adjacent primary wall) and that arabinogalactans uniformly labeled the plasma membrane surrounding the reticulate wall ingrowths. Antibodies against callose strongly labeled the base of nascent ingrowths and localized to discrete domains or apparent channels seen throughout more developed wall ingrowths. These authors interpreted their observations as evidence for the existence of callose-containing channels, not unlike plasmodesmata, that extend throughout the ingrowths and serve to reduce the diffusional resistance of nutrients in the apoplasm (Dahiya and Brewin, 2006).

However, this distribution of callose is not seen in other reticulate-type wall ingrowths. Vaughn *et al.* (2007) used a similar panel of antibodies and affinity probes and concluded that wall ingrowths of epidermal TCs of *V. faba* cotyledons were fundamentally primary wall-like in composition but with notable distinctions. Both the primary wall and the uniformly deposited new wall layer reacted strongly with antibodies against esterified pectin, xyloglucan and other hemicelluloses, and a cellulase-gold affinity probe used as a marker for cellulose. The electron opaque inner region of the wall ingrowths reacted similarly to these probes, except those reacting against highly de-esterified pectins (Vaughn *et al.*, 2007). In contrast to the results of Dahiya and Brewin (2006), anti-callose reactivity was only seen in the electron translucent outer layer of mature wall ingrowths. Vaughn and co-workers interpreted the presence of callose in this region as providing a spreading effect to assist penetration of the ingrowths into the cytoplasm, analogous in some respects to how callose is proposed to assist the spreading properties of developing cell plates. In the study by Vaughn *et al.* (2007) the plasma-membrane–wall interface was labeled strongly with antibodies against arabinogalactan protein, similar to that observed by Dahiya and Brewin (2006). In *V. faba* cotyledons, arabinogalactan labeling was prominent in nascent wall ingrowths and associated specifically with regions of plasma membrane ensheathing the wall ingrowths (Vaughn *et al.*, 2007).

Arabinogalactan proteins have been implicated in coordinating cell wall assembly, possibly by maintaining solubility of wall polymers (Roy *et al.*, 1998). In this context, arabinogalactan proteins might be performing the role of the unidentified minor but key regulatory component of wall architecture suggested by DeWitt *et al.* (1999). The observation that β-D-glucosyl Yariv reagent, a synthetic phenyl-glycoside used to broadly inhibit arabinogalactan protein function, caused significant disruption to wall ingrowth deposition in *V. faba* cotyledon TCs supports this possibility (Vaughn *et al.*, 2007).

Other studies have used similar approaches to investigate the compositional profile of wall ingrowths in TCs. Ligrone *et al.* (2011) used a combination of cytochemical and immunocytochemical procedures to analyze the wall ingrowth apparatus of leaf TCs in *Elodea canadensis*. The leaf lamina of this aquatic angiosperm, like many others, consists of only two cell layers with cells of the abaxial (lower) layer developing as TCs. Cytochemical analysis showed that the outer periclinal plasma membrane of abaxial cells was enriched in light-dependent ATPase activity, while wall ingrowths in these cells were compositionally divergent (at least in part) from the parent primary wall. Specifically, while both the wall ingrowths and underlying primary wall cross-reacted with antibodies directed against glucuronoarabinoxylan and rhamnogalacturonan I, the wall ingrowths lacked cross-reactivity with antibodies against esterified (JIM7 antibody) or de-esterified (JIM5 antibody) pectins (Ligrone *et al.*, 2011) compared to the primary wall.

This observation contrasts with wall ingrowth composition of TCs in both pea nodules (Dahiya and Brewin, 2006) and *V. faba* cotyledons (Vaughn *et al.*, 2007), where the electron opaque inner region of the wall ingrowths are strongly reactive with JIM7 antibody, indicating the presence of esterified pectins. De-esterification of pectins reduces cell wall extensibility (Röckel *et al.*, 2008), so high levels of esterified pectins in these wall ingrowths is presumed to play a role in maintaining flexibility of the developing ingrowths as they branch and fuse to form fenestrated layers. The elevated expression of a pectin methylesterase inhibitor during wall ingrowth deposition (Dibley *et al.*, 2009) is consistent with this proposition, since pectin methylesterases have a major role in pectin remodeling (Pelloux *et al.*, 2007). An unknown consideration here is why the wall ingrowths in abaxial leaf epidermal cells of *Elodea* lack appreciable levels of esterified pectins (Ligrone *et al.*, 2011). One possibility is that these wall ingrowths are primarily designed to maximize plasma membrane-localized ATPase activity to enhance acidification of the cell wall space and thus promote pH-dependent conversion of HCO_3^- to CO_2. This priority may be incompatible with maintaining flexible wall ingrowth structures via inhibiting pectin methylesterase activity.

Boughanmi *et al.* (2010) examined the effect of salinity stress on the distribution of wall ingrowth components in companion TCs in minor veins of the glycophyte *Medicago sativa* (cultivar Gabès). Immunogold labeling showed that the wall ingrowths in these type A TCs were composed of cellulose, acidic pectins, and type I rhamnogalacturonans. Similar to epidermal TCs of *V. faba* cotyledons, these wall ingrowths showed high amounts of methyl-esterified pectins as well as arabinogalactan and arabinogalactan-like proteins associated predominantly with the wall ingrowth surface (Boughanmi *et al.*, 2010). Intriguingly, in salt-stressed plants the wall ingrowths are characterized by higher levels of de-esterified pectins, enhanced levels of fucosylated xyloglucans, and an increased abundance of arabinogalactan proteins. They proposed that the increased levels of de-esterified pectins, presumably caused by salt-induced stimulation of pectin methylesterases (Micheli, 2001), may aid in sequestering Na^+ ions in the wall ingrowth matrix, and that increasing fucosylated xyloglycans with their increased cross-linking of cellulose microfibrils may provide a mechanism for stabilizing wall rigidity to better accommodate resistance to high internal turgor pressure associated with salt stress. Equally, salt-induced accumulation of arabinogalactan proteins (Lamport *et al.*, 2006) may stabilize the plasma membrane as well as provide a so-called 'plasticizer' effect to loosen the pectin network in response to increases in turgor pressure (Boughanmi *et al.*, 2010). Collectively, these results demonstrate that wall composition of TCs responds to environmental perturbations, in this case salt stress, and that this response is central to the capacity of glycophytes to tolerate the stress-induced recirculation of Na^+ by phloem.

Mechanisms of Reticulate Wall Ingrowth Deposition: A Unique Type of Localized Wall Deposition in Plants

The deposition of a papillate wall ingrowth as the first step in constructing a reticulate ingrowth wall network represents a unique form of localized wall deposition in plants. The 'primary-wall-like' (Vaughn *et al.*, 2007) composition and cylindrical shape of papillate wall ingrowths implies that their construction involves both cellulose synthesis (Talbot *et al.*, 2007b) and targeted secretion of cell wall vesicles. However, unlike a process such as tip growth in pollen tubes where cell wall vesicles are targeted to a single apical domain (Cheung and Wu, 2008), in the case of papillate wall ingrowths, targeted or localized secretion of vesicles presumably occurs at numerous discrete loci across the face of the parent wall. The mechanistic details of such a process are unknown in plant biology.

The deposition of tangled lumpy fibers of presumed cellulosic material, which marks the first visible indication of papillate wall ingrowth deposition (Fig. 11.4), implies aberrant cellulose biosynthesis, possibly caused by either reduced rates of cellulose production or restricted movement of the cellulose synthase rosette complexes within the plasma membrane

(McCurdy *et al.*, 2008). Abnormal cellulose deposition reminiscent of these thicker fibers is seen when cellulose biosynthesis is disrupted by mutation (Sato *et al.*, 2001; Sugimoto *et al.*, 2003; Fujita *et al.*, 2013) or drugs (Sugimoto *et al.*, 2001; Vaughn *et al.*, 2007), or when the CesA composition of primary wall cellulose synthase complexes are altered genetically (Desprez *et al.*, 2007; Persson *et al.*, 2007). It is therefore feasible that the transition from ordered synthesis of microfibrils, that make up the uniform wall layer (Fig. 11.4a–c), to aberrant cellulose biosynthesis, that marks deposition of emerging papillate structures, coincides with changes in expression of primary wall *CESAs*. Supporting this possibility, Desprez *et al.* (2007) argue that competition between the three CesA6-related isoforms for the third invariant component of rosette complexes provides a mechanism for altering cellulose microfibril synthesis at different stages of cell growth. Alternatively, or additionally, post-translational phosphorylation of CesAs causing disrupted mobility of rosette complexes (Chen *et al.*, 2010) may be sufficient to cause production of tangled fibers which mark the initiation of papillate wall ingrowth deposition.

Deposition of cellulose microfibrils and targeted secretion of cell wall vesicles invokes a role for the cytoskeleton in deposition of reticulate wall ingrowths; however, surprisingly few studies in this area have been published. Immunofluorescence microscopy showed that microtubules become redistributed and polarized towards sites of ingrowth deposition in both abaxial epidermal TCs of *V. faba* cotyledons (Bulbert *et al.*, 1998) and differentiating placental TCs of *Lilium longiflorum* pistils (Singh *et al.*, 1999), but it was unclear from these studies whether this redistribution preceded wall ingrowth deposition or otherwise. A more recent investigation using adaxial epidermal TCs in *V. faba* cotyledons yielded the unexpected result that disruption of the microtubule cytoskeleton with either oryzalin or taxol had no detectable effect on the construction of a reticulate wall ingrowth network (H.M. Zhang *et al.*, unpublished data). This result seems surprising at first consideration, given the well-documented link between cortical microtubules and cellulose microfibril deposition in plant cells. This link invariably accounts for linear deposition of crystalline cellulose microfibrils however, whereas the deposition of lumpy and tangled non-crystalline microfibrils presumably occurs independently of any influence by microtubules.

The highly localized deposition of papillate wall ingrowths implies some form of targeted secretion of cell wall vesicles to deliver pectins and other matrix components specifically to the wall ingrowths. The actin cytoskeleton is a likely candidate in such a process, given its role in controlling plant cell morphogenesis (Smith and Oppenheimer, 2005). Again however, how the actin cytoskeleton may coordinate multi-loci-targeted secretion raises questions about mechanisms not previously considered in plant biology. For instance, is actin filament assembly directed to individual ingrowth loci, thereby providing a specific vesicle delivery pathway to each ingrowth? Thin transverse bundles of actin are suggested to be involved in close-range delivery of CesA-containing rosette complexes to sites of secondary wall deposition (Wrightman and Turner, 2008), but the organizational complexity required for this process appears to be substantially less than that required for actin involvement in targeted delivery to dozens, if not hundreds, of emerging papillate wall ingrowths to a single wall surface in a TC (see Fig. 11.2a). Alternatively, exocytosis of vesicles may be restricted to discrete domains of the plasma membrane by localized depolymerization of a dense cortical actin meshwork (Offler *et al.*, 2003). Local regulation of rapid growth and severing activity of actin filament dynamics by actin-binding proteins, possibly regulated by Ca^{2+}, might provide a mechanism for such outcomes (Staiger *et al.*, 2009; Blanchoin *et al.*, 2010).

A role for actin in wall ingrowth deposition can be implied by the finding that AtFH6, a plasma-membrane-anchored formin, is upregulated in giant feeding cells induced by root knot nematode infection (Favery *et al.*, 2004). These giant feeding cells develop wall ingrowths reminiscent of those in TCs, but are deposited on all walls of the feeding cells. Formins stimulate *de novo* polymerization of actin filaments at cell membranes (van Gisbergen and Bezanilla, 2013), and AtFH6 was found to be uniformly distributed around these giant cells, consistent from a positional sense at least, with a role in building the wall ingrowth-like structures. However, determining whether AtFH6-anchored actin filaments are distinctly involved in wall ingrowth building, rather than vesicle trafficking required for extensive isotrophic expansion of these giant cells, will require more detailed genetic and structural analyses.

Mechanisms of Flange Wall Ingrowth Deposition: Similarities to Secondary Wall Thickenings

A striking feature of flange ingrowths in many instances is their similarity in general morphology to localized bands of secondary wall thickenings (Talbot *et al.*, 2007a). In xylem elements, secondary wall deposition is preceded by clustering of cortical microtubules into transverse bands which predict and subsequently direct localized wall deposition (Falconer and Seagull, 1985; Oda *et al.*, 2005; Wightman and Turner, 2008). Real-time confocal imaging of microtubule dynamics in *trans*-differentiating xylem elements showed that microtubules are initially bundled directly under developing wall thickenings and then become displaced laterally as the thickenings develop (Oda *et al.*, 2005; Wightman and Turner, 2008). Furthermore, the localization of cellulose synthase complexes to these developing secondary wall thickenings was dependent on the underlying bundles of microtubules (Wightman and Turner, 2008).

A role for microtubules in directing deposition of flange wall ingrowths appears to be broadly similar to that of secondary wall thickenings in xylem elements. Talbot *et al.* (2007a) showed that in xylem TCs of wheat nodes, bundles of microtubules were frequently positioned directly beneath or to the sides of the ingrowths. The authors interpreted these observations as evidence that microtubules move from positions predominantly underneath developing flange ingrowths to more lateral positions flanking mature ingrowths. Given that self-assembly of cellulose can drive local microfibril order in the absence of microtubules (Baskin, 2001), and that microtubules are required to concentrate cellulose synthase enzymes to sites of secondary wall thickenings (Gardiner *et al.*, 2003; Wightman and Turner, 2008), Talbot *et al.* (2007b) proposed a model for regulation of flange wall ingrowth deposition in TCs. In this model the initial clustering of cortical microtubules is required to establish specific domains in the plasma membrane that allow access by secretory vesicles to exocytose their contents into the cell wall space and deliver CesA complexes to the plasma membrane. Once inserted, the CesA complexes synthesize microfibrils aligned with the bundled microtubules and secreted matrix polysaccharides self-assemble with the microfibrils. Once linear domains of wall ingrowth deposition with parallel microfibrils have been established, self-assembly of both microfibrils and matrix polysaccharides further consolidate the linear flange ingrowths. This last step is independent of microtubules, which by this stage have been displaced to the sides of the ingrowths, although these microtubules might have a continuing role in confining CesA complexes to the edges of the flange ingrowths (see Wightman and Turner, 2008). Given this apparently similar role for microtubules in directing flange wall ingrowth deposition and secondary wall thickenings, it will be fascinating to unravel the influence of microtubule-associated proteins in coordinating microtubule dynamics associated with flange wall ingrowth deposition (see Chapter 7 and Oda and Fukuda, 2012). An unresolved issue related to this question is how the deposition of cellulose microfibrils transitions from random to parallel as new flange ingrowths are laid down. Local ordering of microfibrils required to initiate formation of a flange ingrowth might involve recruitment of random microfibrils of the underlying wall (Fig. 11.7). Alternatively, and probably more realistically, the parallel microfibrils presumably represent a newly deposited population being directed by association with underlying bundles of cortical microtubules.

No studies analyzing a role for the actin cytoskeleton in regulating flange wall ingrowth deposition have been published. Taking a lead again from secondary wall biology, early studies demonstrated a role for the actin cytoskeleton in directing transverse alignments of microtubules essential for the establishment of secondary wall thickenings (Kobayashi *et al.*, 1988). Broadly similar roles for the actin cytoskeleton were established for cell shaping of developing mesophyll of wheat (Wernicke and Jung, 1992). As discussed previously, Wightman and Turner (2008) provide evidence suggesting that a subpopulation of transversely oriented actin filaments target delivery of cellulose synthase complexes directly to sites of secondary wall synthesis. Discovering similar roles for actin in directing deposition of flange wall ingrowths most likely requires development of experimental systems in which flange wall ingrowth deposition can be manipulated genetically.

An intriguing aspect of TC biology is the observation that in some cells papillate wall ingrowth projections can be deposited over the surface of secondary wall thickenings, as seen in xylem parenchyma TCs in roots of *V. faba* (Talbot *et al.*, 2002), or indeed over flange wall ingrowths, as seen in seed coat TCs in cotton (Pugh *et al.*, 2010). The cellular mechanisms required for sequential construction of different types of wall ingrowths in a single TC are fascinating to consider, and certainly beyond current molecular understanding of how plant cell wall deposition may be regulated.

Signals Inducing Transfer Cell Differentiation

The *trans*-differentiation of TC development occurs across specific developmental windows where enhanced rates of plasma membrane transport of nutrients at symplasmic/apoplasmic boundaries are required for normal plant development. This process appears to be strongly correlated with the availability of solutes for transport, particularly sugars (see Offler *et al.*, 2003). Examples include the differentiation of epidermal TCs in *V. faba* cotyledons coinciding with commencement of the linear phase of seed fill (Bonnemain *et al.*, 1991), and differentiation of companion TCs in leaf minor veins of pea coinciding with photoassimilate supply (Gunning and Pate, 1974; Henry and Steer, 1980).

Alternatively, TC development can be induced in response to external stresses, either biotic in the case of nematode attack (Jones and Gunning, 1976) and invasion by pathogens or symbiotic organisms (Offler *et al.*, 2003), or abiotic such as nutrient deficiency (Schikora and Schmidt, 2002), salt stress (Boughanmi *et al.*, 2003), or exposure to high light (Amiard *et al.*, 2007) or cold (Maeda *et al.*, 2006). Ethylene has been implicated directly in regulating the *trans*-differentiation of rhizodermal TCs in tomato roots in response to iron deficiency (Schikora and Schmidt, 2002), and indirectly via the finding that genes encoding ethylene metabolism and signaling are upregulated in barley aleurone TCs (Thiel *et al.*, 2008). Furthermore, many genes selectively induced in adaxial epidermal TCs of *V. faba* cotyledons carry auxin- and ethylene-response elements in promoter regions of orthologous genes (Dibley *et al.*, 2009). Indeed, these authors have since shown that wall ingrowth deposition in adaxial epidermal TCs is preceded by a rapid burst in ethylene production driven by auxin and involving an epidermal-specific ethylene signaling cascade, including post-translational upregulation of an ortholog of the EIN3 transcription factor (VfEIN3-1), and differential expression of three ethylene response factor genes (*VfERF1-3*) (Zhou *et al.*, 2010). Progression of this ethylene signaling pathway through VfEIN3-1 was regulated by intracellular glucose concentrations sensed through a hexokinase-dependent pathway (Andriunas *et al.*, 2011).

More recent experimentation using this system established the involvement of reactive oxygen species (ROS) signaling downstream of ethylene and directing polarized deposition of wall ingrowths (Andriunas *et al.*, 2012). In this case, ethylene signaling elicited epidermal-specific expression of two *V. faba* respiratory burst oxidase homologues (*VfrbohA* and *VfrbohC*), which coincided with production of extracellular ROS (H_2O_2) restricted to the outer periclinal walls of epidermal cells (Andriunas *et al.*, 2012; Xia *et al.*, 2012). The extracellular H_2O_2 is speculated to provide the positional cue to direct targeted secretion of cell wall vesicles to the outer periclinal wall required for polarized deposition of the uniform wall layer which precedes, and is essential for, subsequent papillate wall ingrowth deposition in these TCs (Andriunas *et al.*, 2012). This conclusion was reached because delocalization of the polarized ROS signal by exposing cotyledons to excess H_2O_2 resulted in loss of deposition of the polarized uniform wall layer and subsequent reallocation of wall building to other faces of the epidermal cells (Andriunas *et al.*, 2012). Overall, the body of research by Andriunas and colleagues provides the most detailed analysis to date of signaling pathways initiating plant TC development. A key question is the extent to which these inductive signaling pathways may be conserved throughout the plant kingdom, or how different elements may differentially influence TC development in different developmental scenarios and locations (Andriunas *et al.*, 2013).

The Transfer Cell Transcriptome

Transcript profiling experiments using *V. faba* cotyledons (Dibley *et al.*, 2009) and barley endosperm (Thiel *et al.*, 2008) revealed that *trans*-differentiation of TCs is accompanied by differential expression of approximately 600–800 genes. This process presumably results in generation of a TC-specific transcriptome which is responsible for generating the unique morphological and functional characteristics of TCs. Several genes nominated as TC-specific have been identified in barley (Doan *et al.*, 1996) and particularly in basal endosperm TCs of maize. Most of these genes in maize code for small hydrophilic proteins possibly involved in cell defense, such as four defensin-like genes *BETL1–4* (Hueros *et al.*, 1995), and a novel cell-wall-related protein MEG1 (Gutiérrez-Marcos *et al.*, 2004). Significantly,

Gómez *et al.* (2002) identified *ZmMRP-1* (*Zea mays* MYB-related protein-1) as a transcription factor expressed exclusively in endosperm TCs in maize kernals. This transcription factor can *trans*-activate *BETL* and *MEG1* promoters (Gutiérrez-Marcos *et al.*, 2004; Barrero *et al.*, 2006) and has been shown to act as a key regulator of the differentiation of TCs in the basal endosperm layer in maize (Gómez *et al.*, 2009). Furthermore, when expressed in Arabidopsis and other species, the *ZmMRP-1* promoter is induced at sites of active transport between source and sink tissues, including vascular connections between developing organs and the main plant vasculature (Barrero *et al.*, 2009). Whether TCs are present at these vascular junctions in Arabidopsis remains to be established, however.

An intriguing recent development is the discovery of a role for two-component signaling pathways in TC development. Muñiz and colleagues identified two atypical type-A response regulators, *ZmTCRR-1* and *ZmTCRR-2*, which are specifically expressed in the endosperm TC layer in maize (Muñiz *et al.*, 2006, 2010). Interestingly, unlike mRNA derived from either gene, both protein products were detected in endosperm tissue some distance from the TC layer, suggesting the operation of a non-cell autonomous signaling pathway (Muñiz *et al.*, 2006, 2010). Evidence of causal relationships between these genes and regulation of TC development awaits further experimentation; however, a role for two-component signaling in TC development, at least in endosperm tissue, is supported from recent 454 transcriptome sequencing in barley (Thiel *et al.*, 2012). Sequencing of transcripts captured by laser microdissection of endosperm TCs across early stages of development identified up to 40 genes likely involved in two-component signaling, including histidine kinases, histidine-containing phosphotransfer proteins, and three classes of downstream response regulators (Thiel *et al.*, 2012). The number of putative two-component signaling genes expressed in this tissue, possibly the entire barley gene set for this biological function, implies a significant role for this pathway in regulating the differentiation of endosperm TCs. Endosperm TCs only constitute a small cohort of cell types other than mostly xylem vascular tissue however, which deposit flange wall ingrowths. In contrast, TCs with reticulate wall ingrowth morphologies are significantly more common throughout the plant kingdom (Offler *et al.*, 2003; Andriunas *et al.*, 2013); it will therefore be informative to discover if expression of two-component signaling genes is equally widespread in TCs of the reticulate type.

Comprehensive transcript profiling of barley endosperm TCs identified numerous ethylene biosynthesis and signaling genes which were differentially transcribed early in their development (Thiel *et al.*, 2012), consistent with a conserved role for ethylene signaling in TC induction (Zhou *et al.*, 2010; Andriunas *et al.*, 2011, 2012). Cell cycle genes were also upregulated early in endosperm TC development, consistent with findings by Dibley *et al.* (2009), and indicating that *trans*-differentiation of TCs involves re-initiation of cell division. Numerous other genes involved in cellular processes expected to contribute to wall ingrowth formation, such as vesicle trafficking, cell wall biosynthesis, cell expansion, and shape control, were identified as differentially expressed (Thiel *et al.*, 2012). Genes associated with amino acid catabolism, carbohydrate oxidation, mitochondrial activity, and starch turnover were also upregulated during endosperm TC development, consistent with high metabolic demand accompanying wall ingrowth deposition (Thiel *et al.*, 2012; see also Dibley *et al.*, 2009). The challenge from this and related studies is to establish TC-specificity for these cohorts of genes and demonstrate causal relationships with wall ingrowth deposition.

Phloem Parenchyma Transfer Cell in *Arabidopsis thaliana*: A Genetic Model to Study Transfer Cell Biology

Progress in unraveling the molecular details of TC development will undoubtedly be enhanced by development of genetic approaches to study this process. In Arabidopsis, TCs with reticulate wall ingrowths form in phloem parenchyma (PP) cells in minor veins of leaves (Haritatos *et al.*, 2000) and sepals (Chen *et al.*, 2012). Ingrowth deposition in these cells is polarized to cell surfaces primarily adjacent to sieve elements (SEs) and, to a lesser extent, companion cells (CCs) (Haritatos *et al.*, 2000; Maeda *et al.*, 2006; Amiard *et al.*, 2007). This anatomy implies an active role in apoplasmic phloem loading whereby sucrose delivered symplasmically from adjacent bundle sheath cells (Haritatos *et al.*, 2000) is effluxed across the plasma membrane of PP TCs into the apoplasm and subsequently loaded into cells of the SE/CC complex via a sucrose/H^+ co-transporter (SUC2) in the plasma membrane of CCs (Truernit and Sauer 1995; Gottwald *et al.*, 2000). The efflux mechanism was recently attributed to AtSWEET11 and 12, two novel plasma-membrane-localized sucrose uniporters that are expressed specifically in PP TCs in leaf tissue (Chen *et al.*, 2012). A *atsweet11 atsweet12* double mutant displayed phenotypes consistent with defective unloading of sugars from source leaves, thus demonstrating a central role for PP TCs in apoplasmic (active) phloem loading in Arabidopsis (Chen *et al.*, 2012). Intriguingly however, asymmetrically branched plasmodesmata occur with reasonable frequency between PP TCs and adjacent CCs in Arabidopsis veins (Haritatos *et al.*, 2000), prompting the suggestion that passive symplasmic phloem loading of sugars in Arabidopsis may occur under certain physiological circumstances (Slewinski and Braun, 2010).

A key element for using Arabidopsis as a genetic model for TC studies is the observation that wall ingrowth deposition in PP TCs is responsive to stresses such as high light, cold, or exposure to methyl jasmonate (Maeda *et al.*, 2006; Amiard *et al.*, 2007). Edwards *et al.* (2010) used a bioinformatics approach to identify genes commonly upregulated in leaf tissue in response to these stresses and identified *GIGANTEA* (*GI*) as a component regulating TC development. This regulation may be indirect however, as *GI* mutants typically show increased anti-oxidant capacity probably resulting from constitutive overexpression of genes encoding superoxide dismutase and ascorbate peroxidase (Cao *et al.*, 2006). Consequently, Edwards *et al.* (2010) proposed that *GI* may be regulating wall ingrowth deposition downstream of stress signals such as ROS, a known polarizing signal for wall ingrowth deposition (Andriunas *et al.*, 2012; Xia *et al.*, 2012).

A role for jasmonic acid in signaling TC development in Arabididopsis PP TCs has also been implied from recent genetic analyses. Oxidative stress triggered by exposure to high light causes increased wall ingrowth deposition, similar to that triggered by exposure to methyl jasmonate alone (Amiard *et al.*, 2007). More importantly, the high-light response was reduced significantly in the jasmonate-deficient double mutant *fad7-1 fad8-1* (Amiard *et al.*, 2007) but increased significantly in *npq1-2 lut2-1*, a double mutant lacking zeaxanthin, and its isomer lutein, photoprotective agents expected to suppress production of jasmonates (Demmig-Adams *et al.*, 2013). The absence of this suppression may cause higher levels of jasmonic acid upon exposure to high light, hence the observed increase in wall ingrowth deposition compared to controls (Demmig-Adams *et al.*, 2013). A consideration from these studies is whether jasmonic acid and other signals such as ROS are acting as true inductive signals of TC *trans*-differentiation, or rather as downstream modulators of wall ingrowth deposition. For example, ethylene acts as an inductive signal for the *trans*-differentiation of epidermal TCs in *V. faba* cotyledons, but is also required to maintain ongoing deposition of wall ingrowths (Andriunas *et al.*, 2011, 2012).

Low temperature also acts as an abiotic stress causing increased levels of wall ingrowth deposition in PP TCs (Maeda *et al.*, 2006, 2008). Transfer of wild-type plants to cold resulted in substantially increased levels of polarized wall ingrowth deposition in PP TCs within three days of transfer (Maeda *et al.*, 2008). Subjecting the vitamin-E-deficient mutant *vte2* to low temperature caused greatly increased levels of abnormal and non-polarized deposition of wall ingrowths in PP TCs, including excess synthesis of callose throughout the wall ingrowths and elsewhere in these cells (Maeda *et al.*, 2006, 2008). Not surprisingly, this low-temperature phenotype in *vte2* caused dramatically reduced export of sugars from the leaf, detectable within 6 hours after transfer to low temperature, and consequently increased levels of soluble sugars in leaves of cold-treated plants (Maeda *et al.*, 2006). A screen for suppressor mutations of the *vte2* phenotype identified genes involved in lipid metabolism such as *fad2*, an endoplasmic reticulum (ER)-localized fatty acid desaturase and *tgd1* (*trigalactosyldiacylglycerol1*), a component of the ER-to-plastid lipid ATP-binding cassette transporter (Song *et al.*, 2010). This result suggests the low-temperature phenotype of *vte2* is due to disrupted ER-to-plastid lipid trafficking. How this process specifically affects wall ingrowth deposition in PP TCs in *vte2* is not known, but movement of lipohilic compounds between plastids and the ER has previously been documented (see Song *et al.*, 2010).

The ability to use Arabidopsis as a genetic model for TC studies has been facilitated by high-throughput fluorescence staining of PP TCs in leaf tissue. Edwards *et al.* (2010) used Calcofluor White to stain the cellulose-rich wall ingrowths in these cells. Recently, Arun Chinnappa *et al.* (2013) developed Aniline Blue as a method to detect the callose component of wall ingrowths in PP TCs. High-throughput staining of leaf tissue in T-DNA insertional mutants enabled identification of two NAC-domain and two MYB-related genes as putative transcriptional regulators of TC development in PP TCs (Arun Chinnappa *et al.*, 2013). This approach, combined with analysis of mutants such as *vte2* and the *fad7-1 fad8-1* double mutant, provide experimental systems to dissect the genetic control of wall ingrowth deposition in PP TCs in Arabidopsis.

Conclusions

From their first documentation by Gunning and Pate plant biologists have been fascinated by TCs due to their elaborate and diverse wall ingrowth architectures and their capacity to *trans*-differentiate in response to nutrient transport demands arising from developmental programs or in response to stress. Significant progress has been made in defining signaling pathways that induce the formation of TCs and the transcriptional programs that accompany this process. Progress is also being made in identifying potential TC-identity genes that presumably act to regulate these downstream transcriptional programs, although the number of candidates is low compared to the panel of transcription factors identified as 'master' genes regulating localized secondary wall deposition (Zhong *et al.*, 2010). Given the unique aspects of wall ingrowth morphology, a major focus of future research will be to unravel the cellular mechanisms responsible for this process. The unique aspects of reticulate wall ingrowth development imply the operation of molecular mechanisms not previously described in plant biology; the discovery of such mechanisms will therefore provide new understanding of localized wall deposition in plant cells. Access to genetic systems to study TC biology, analogous to those used to provide exquisite molecular detail of how cells build secondary wall thickenings (Oda *et al.*, 2010), will be needed to achieve such advances and thus provide a deeper understanding of these fascinating and functionally important cell types.

Acknowledgements

I thank Dr David Collings for assistance with figure production and acknowledge funding from ARC Discovery Project grants (DP0556217, DP0664626, DP110100770).

References

Amiard, V., Demmig-Adams, B., Mueh, K.E., Turgeon, R., Combs, A.F., and Adams, W.W. III,. (2007) Role of light and jasmonic acid signaling in regulating foliar phloem cell wall ingrowth development. *New Phytologist*, **173**, 722–731.

Andriunas, F.A., Zhang, H.M., Weber, H., McCurdy, D.W., Offler, C.E., and Patrick, J.W. (2011) Glucose and ethylene signalling pathways converge to regulate *trans*-differentiation of epidermal transfer cells in *Vicia narbonensis* cotyledons. *Plant Journal*, **68**, 987–998.

Andriunas, F.A., Zhang, H.M., Xia, X., Offler, C.E., McCurdy, D.W., and Patrick, J.W. (2012) Reactive oxygen species form part of a regulatory pathway initiating *trans*-differentiation of epidermal transfer cells in *Vicia faba* cotyledons. *Journal of Experimental Botany*, **63**, 3617–3630.

Andriunas, F.A., Zhang, H.M., Xia, X., Patrick, J.W., and Offler, C.E. (2013) Intersection of transfer cells with phloem biology - broad evolutionary trends, function, and induction. *Frontiers in Plant Science*, **4**, 221, doi: 10.3389/fpls.2013.00221.

Arun Chinnappa, K.S., Nguyen, T.T.S., Hou, J., Wu, Y., and McCurdy, D.W. (2013) Phloem parenchyma transfer cells in *Arabidopsis* – an experimental system to identify transcriptional regulators of wall ingrowth formation. *Frontiers in Plant Science*, **4**, 102, doi: 10.3389/fpls.2013.00102.

Barrero, C., Muniz, L.M., Gomez, E., Hueros, G., and Royo, J. (2006) Molecular dissection of the interaction between the transcriptional activator *ZmMRP-1* and the promoter of *BETL-1*. *Plant Molecular Biology*, **62**, 655–668.

Barrero, C., Royo, J., Grijota-Martinez, C., Faye, C., Paul, W., Sanz, S., Steinbiss, H.-H., and Hueros, G. (2009) The promoter of *ZmMRP-1*, a maize transfer cell-specific transcriptional activator, is induced at solute exchange surfaces and responds to transport demands. *Planta*, **229**, 235–247.

Baskin, T.I. (2001) On the alignment of cellulose microfibrils by cortical microtubules, a review and a model. *Protoplasma*, **215**, 150–171.

Blanchoin, L., Boujemaa-Paterski, R., Henty, J.L., Khurana, P., and Staiger, C.J. (2010) Actin dynamics in plant cells, a team effort from multiple proteins orchestrates this very fast-paced game. *Current Opinions in Plant Biology*, **13**, 714–723.

Bonnemain, J.L., Bourquin, S., Renault, S., Offler, C.E., and Fisher, D.G. (1991) Transfer cells, structure and physiology, in *Recent Advances in Phloem Transport and Assimilate Compartmentation* (eds J.L. Bonnemain, S. Delrot, W.L. Lucas, and L. Dainty), Ouest Edition, Presses Académiques, Nantes, Paris, pp. 74–83.

Boughanmi, N., Michonneau, P., Verdus, M.-C., Piton, F., Ferjani, E., Bizid, E., and Fleurat-Lessard, P. (2003) Structural changes induced by NaCl in companion and transfer cells of *Medicago sativa* blades. *Protoplasma*, **220**, 179–187.

Boughanmi, N., Thibault, F., Decou, R., Fleurat-Lessard, P., Béré, E., Costa, G., and Lhernould, S. (2010) NaCl effect on the distribution of wall ingrowth polymers and arabinogalactan proteins in type A transfer cells of *Medicago sativa* Gabès leaves. *Protoplasma*, **242**, 69–80.

Bulbert, M.W., Offler, C.E., and McCurdy, D.W. (1998) Polarized microtubule deposition coincides with wall ingrowth formation in transfer cells of *Vicia faba* L. cotyledons. *Protoplasma*, **201**, 8–16.

Cao, S., Jiang, S., and Zhang, R. (2006) The role of *GIGANTEA* gene in mediating the oxidative stress response in *Arabidopsis*. *Plant Growth Regulator*, **48**, 261–270.

Chen, S., Ehrhardt, D.W., and Somerville, C.R. (2010) Mutations of cellulose synthase (CESA1) phosphorylation sites modulate anisotropic cell expansion and bidirectional mobility of cellulose synthase. *Proceedings of the National Academy of Science USA*, **107**, 17188–17193.

Chen, L.-Q., Qu, X.-Q., Hou, B.-H., Sosso, D., Osorio, S., Fernie, A.R., and Frommer, W.B. (2012) Sucrose efflux mediated by SWEET proteins as a key step for phloem transport. *Science*, **335**, 207–211.

Cheung, A.Y., Wu, H.M. (2008) Structural and signaling networks for the polar cell growth machinery in pollen tubes. *Annual Review of Plant Biology*, **59**, 547–572.

Dahiya, P. and Brewin, N.J. (2006) Immunogold localization of callose and other cell wall components in pea nodule transfer cells. *Protoplasma*, **214**, 210–218.

Davis, R.W., Smith, J. D., and Cobb, B.G. (1990) A light and electron microscope investigation of the transfer cell region of maize caryopses. *Canadian Journal of Botany*, **68**, 471–479.

Demmig-Adams, B., Cohu, C.M., Amiard, V., Zadelhoff, G., Veldink, G.A., Muller, O., and Adams, W.W.III. (2013) Emerging trade-offs – impact of photoprotectants (PsbS, xanthophylls, and vitamin E) on oxylipins as regulators of development and defense. *New Phytologist*, **197**, 720–729.

Desprez, T., Juraniec, M., Crowell, E.F., Jouy, H., Pochylova, Z., Parcy, F., Höfte H, Gonneau, M., and Vernhettes, S. (2007) Organization of cellulose synthase complexes involved in primary cell wall synthesis in *Arabidopsis thaliana*. *Proceedings of the National Academy of Science USA*, **104**, 15572–15577.

DeWitt, G., Richards, J., Mohnen, D., and Jones, A.M. (1999) Comparative compositional analysis of walls with two different morphologies, archetypical versus transfer-cell-like. *Protoplasma*, **209**, 238–245.

Dibley, S.J., Zhou, Y., Andriunas, F.A., Talbot, M.J., Offler, C.E., Patrick, J.W., and McCurdy, D.W. (2009) Early gene expression programs accompanying *trans*-differentiation of epidermal cells of *Vicia faba* cotyledons into transfer cells. *New Phytologist*, **182**, 863–877.

Doan, D.N., Linnestad, C., and Olsen, O.A. (1996) Isolation of molecular markers from the barley endosperm coenocyte and the surrounding nucellus cell layers. *Plant Molecular Biology*, **31**, 877–886.

Edwards, J., Martin, A.P., Andriunas, F., Offler, C.E., Patrick, J.W., and McCurdy, D.W. (2010) *GIGANTEA* is a component of a regulatory pathway determining wall ingrowth deposition in phloem parenchyma transfer cells of *Arabidopsis thaliana*. *Plant Journal*, **63**, 651–661.

Falconer, M.M. and Seagull, R.W. (1985) Immunofluorescent and calcofluor white staining of developing tracheary elements in *Zinnia elegans* L. suspension cultures. *Protoplasma*, **125**, 190–198.

Farley, S.J., Patrick, J.W., and Offler, C.E. (2000) Functional transfer cells differentiate in cultured cotyledons of *Vicia faba* L. seeds. *Protoplasma*, **214**, 102–117.

Favery, B., Chelysheva, L.A., Lebris, M., Jammes, F., Marmagne, A., de Almeida-Engler, J., Lecomte, P., Vaury, C., Arkowitz, R.A., and Abada, P. (2004) *Arabidopsis* formin AtFH6 is a plasma membrane–associated protein upregulated in giant cells induced by parasitic nematodes. *Plant Cell*, **16**, 2529–2540.

Fujita, M., Himmelspach, R., Ward, J., Whittington, A., Hasenbein, N., Liu, C., Truong, T.T., Galway, M.E., Mansfield, S.D., Hocart, C.H., and Wasteneys, G.O. (2013) The *anisotropy1* D604N mutation in the Arabidopsis cellulose synthase1 catalytic domain reduces cell wall crystallinity and the velocity of cellulose synthase complexes. *Plant Physiology*, **162**, 74–85.

Gardiner, J.C., Taylor, N.G., and Turner, S.R. (2003) Control of cellulose synthase complex localization in developing xylem. *Plant Cell*, **15**, 1740–1748.

Gómez, E., Royo, J., Guo, Y., Thompson, R., and Hueros, G. (2002) Establishment of cereal endosperm expression domains: Identification and properties of a maize transfer cell-specific transcription factor, *ZmMRP-1*. *Plant Cell*, **14**, 599–610.

Gómez, E., Royo, J., Muñiz, L.M., Sellam, O., Paul,W., Gerentes, D., Barrero, C., López, M., Perez, P., Hueros, G. (2009) The maize transcription factor myb-related protein-1 is a key regulator of the differentiation of transfer cells. *Plant Cell*, **21**, 2022–2035.

Gottwald, J.R., Krysan, P.J., Young, J.C., Evert, R.F., and Sussman, M.R. (2000) Genetic evidence for the *in planta* role of phloem-specific plasma membrane sucrose transporters. *Proceedings of the National Academy of Science USA*, **97**, 13979–13984.

Gunning, B.E.S. and Pate, J.S. (1969) 'Transfer Cells' plant cells with wall ingrowths in relation to short distance transport of solutes — their occurrence, structure, and development. *Protoplasma*, **68**, 107–133.

Gunning, B.E.S. and Pate, J.S. (1974) Transfer cells, in *Dynamic Aspects of Plant Ultrastructure* (ed. A.W. Robards), McGraw-Hill, London, pp. 441–479.

Gunning, B.E.S., Pate, J.S., and Briarty, L.G. (1968) Specialized 'transfer cells' in minor veins of leaves and their possible significance in phloem translocation. *Journal of Cell Science*, **37**, C7–C12.

Gutierrez-Marcos, J.F., Costa, L.M., Biderre-Petit, C., Khbaya, B., O'Sullivan, D.M., Wormald, M., Perez, P., and Dickinson, H.G. (2004) *maternally expressed gene1* is a novel maize endosperm transfer cell-specific gene with a maternal parent-of-origin pattern of expression. *Plant Cell*, **16**, 1288–1301.

Haritatos, E., Medville, R., and Turgeon, R. (2000) Minor vein structure and sugar transport in *Arabidopsis thaliana*. *Planta*, **211**, 105–111.

Henry, Y. and Steer, M.W. (1980) A re-examination of the induction of phloem transfer cell development in pea leaves (*Pisum sativum*). *Plant and Cell Environment*, **3**, 377–380.

Hueros, G., Varotto, S., Salamini, F., and Thompson, R.D. (1995) Molecular characterization of *BET1*, a gene expressed in the endosperm transfer cells of maize. *Plant Cell*, **7**, 747–757.

Jones, M.G.K. and Gunning, B.E.S. (1976) Transfer cells and nematode induced giant cells in *Helianthemum*. *Protoplasma*, **87**, 273–279.

Kang, B.-H., Xiong, Y., Williams, D.S., Pozueta-Romero, D., and Chourey, P.S. (2009) *Miniature1*-encoded cell wall invertase is essential for assembly and function of wall-in-growth in the maize endosperm transfer cell. *Plant Physiology*, **151**, 1366–1376.

Kobayashi, H., Fukuda, H., and Shibaoka, H. (1988) Interrelation between the spatial disposition of actin filaments and microtubules during the differentiation of tracheary elements in cultured *Zinnia* cells. *Protoplasma*, **143**, 29–37.

Kronestedt, E.C. and Robards, A.W. (1987) Sugar secretion from the nectary of *Strelitzia*, an ultrastructural and physiological study. *Protoplasma*, **137**, 168–182.

Lamport, D.T.A., Kieliszewski, M.J., and Showalter, A.M. (2006) Salt stress upregulates periplasmic arabinogalactan proteins: using salt stress to analyse AGP function. *New Phytologist*, **169**, 479–492.

Ligrone, R., Vaughn, K.C., and Rascio, N. (2011) A cytochemical and immunocytochemical analysis of the wall labyrinth apparatus in leaf transfer cells in *Elodea canadensis*. *Annals of Botany*, **107**, 717–722.

Maeda, H., Song, W., Sage, T.L., and DellaPenna, D. (2006) Tocopherols play a crucial role in low-temperature adaptation and phloem loading in *Arabidopsis*. *Plant Cell*, **18**, 2710–2732.

Maeda, H., Sage, T.L., Isaac, G., Welti, R., and DellaPenna, D. (2008) Tocopherols modulate extraplastidic polyunsaturated fatty acid metabolism in *Arabidopsis* at low temperature. *Plant Cell*, **20**, 452–470.

McCurdy, D.W., Patrick, J.W., and Offler, C.E. (2008) Wall ingrowth formation in transfer cells: novel examples of localized wall deposition in plant cells. *Current Opinion in Plant Biology*, **11**, 653–661.

Micheli, F. (2001) Pectin methyl esterases, cell wall enzymes with important roles in plant physiology. *Trends in Plant Science*, **6**, 414–419.

Monjardino, P., Rocha, S., Tavares, A.C., Fernandes, R., Sampaio, P., Salema, R., and da Câmara Machado, A. (2013) Development of flange and reticulate wall ingrowths in maize (*Zea mays* L.) endosperm transfer cells. *Protoplasma*, **250**, 495–503.

Muñiz, L.M., Royo, J., Gomez, E., Barrero, C., Bergareche, D., and Hueros, G. (2006) The maize transfer cell-specific type-A response regulator *ZmTCRR-1* appears to be involved in intercellular signalling. *Plant Journal*, **48**, 17–27.

Muñiz, L.M., Royo, J., Gómez, E., Baudot, G., Paul, W., and Hueros, G. (2010) Atypical response regulators expressed in the maize endosperm transfer cells link canonical two component systems and seed biology. *BMC Plant Biology*, **10**, 84.

Oda, Y. and Fukuda, H. (2012) Secondary cell wall patterning during xylem differentiation. *Current Opinion in Plant Biology*, **15**, 38–44.

Oda, Y., Mimura, T., and Hasezawa, S. (2005) Regulation of secondary wall development by cortical microtubules during tracheary element differentiation in *Arabidopsis* cell suspensions. *Plant Physiology*, **137**, 1027–1036.

Oda, Y., Iida, Y., Kondo, Y., and Fukuda, H. (2010) Wood cell-wall structure requires local 2D-microtubule disassembly by a novel plasma membrane-anchored protein. *Current Biology*, **20**, 1197–1202.

Offler, C.E., McCurdy, D.W., Patrick, J.W., and Talbot, M.J. (2003) Transfer cells: cells specialized for a special purpose. *Annual Review of Plant Biology*, **54**, 431–454.

Pate, J.S. and Gunning, B.E.S. (1969) Vascular transfer cells in angiosperm leaves. A taxonomic and morphological survey. *Protoplasma*, **68**, 135–156.

Pelloux, J., Rustérucci, C., and Mellerowicz, E.J. (2007) New insights into pectin methylesterase structure and function. *Trends in Plant Science*, **12**, 267–277.

Persson, S., Paredez, A., Carroll, A., Palsdottir, H., Doblin, M., Poindexter, P., Khitrov, N., Auer, M., and Somerville, C.R. (2007) Genetic evidence for three unique components in primary cell-wall cellulose synthase complexes in *Arabidopsis*. *Proceedings of the National Academy of Science USA*, **104**, 15566–15571.

Pugh, D.A., Offler, C.E., Talbot, M.J., and Ruan, Y.-L. (2010) Evidence for the role of transfer cells in the evolutionary increase in seed and fiber biomass yield in cotton. *Molecular Plant*, **3**, 1075–1086.

Rascio, N., Mariani, P., Dalla Vecchia, F., Zanchin, A., Pool, A., and Larcher, W. (1994) Ultrastructural and photosynthetic features of leaves and stems of *Elodea canadensis*. *Plant Physiology*, **144**, 314–323.

Röckel, N., Wolf, S., Kost, B., Rausch, T., and Greiner, S. (2008) Elaborate spatial patterning of cell-wall PME and PMEI at the pollen tube tip involves PMEI endocytosis, and reflects the distribution of esterified and de-esterified pectins. *Plant Journal*, **53**, 133–143.

Roy, S., Jauh, G.-Y., Hepler, P.K., and Lord, E.M. (1998) Effects of Yariv phenylglycoside on cell wall assembly in the lily pollen tube. *Planta*, **204**, 450–458.

Sato, S., Kato, T., Kakegawa, K., Ishii, T., Liu, Y.G., Awano, T., Takabe, K., Nishiyama, Y., Kuga, S., Nakamura, Y., Tabata, S., and Shibata, D. (2001) Role of the putative membrane-bound endo-1,4-β-glucanase KORRIGAN in cell elongation and cellulose synthesis in *Arabidopsis thaliana*. *Plant and Cell Physiology*, **42**, 251–263.

Schikora, A. and Schmidt, W. (2002) Formation of transfer cells and H^+-ATPase expression in tomato roots under P and Fe deficiency. *Planta*, **215**, 304–311.

Singh, S., Lazzaro, M.D. and Walles, B. (1999) Microtubule organization in the differentiating transfer cells of the placenta in *Lilium* spp. *Protoplasma*, **207**, 75–83.

Slewinski, T.L. and Braun, D.M. (2010) Current perspectives on the regulation of whole-plant carbohydrate partitioning. *Plant Science*, **178**, 341–349.

Smith, L.G. and Oppenheimer, D.G. (2005) Spatial control of cell expansion by the plant cytoskeleton. *Annual Review in Cell Developmental Biology*, **21**, 271–295.

Song, W., Maeda, H., and DellPenna, D. (2010) Mutations of the ER to plastid lipid transporters *TGD1*, *2*, *3* and *4* and the ER oleate desaturase *FAD2* suppress the low temperature-induced phenotype of Arabidopsis tocopherol-deficient mutant *vte2*. *Plant Journal*, **62**, 1004–1018.

Staiger, C.J., Sheahan, M.B., Khurana, P., Wang, X., McCurdy, D.W., and Blanchoin, L. (2009) Actin filament dynamics are dominated by rapid growth and severing activity in the *Arabidopsis* cortical array. *Journal of Cell Biology*, **184**, 269–280.

Sugimoto, K., Williamson, R.E., and Wasteneys, G.O. (2001) Wall architecture in the cellulose-deficient *rsw1* mutant of *Arabidopsis thaliana*: microfibrils but not microtubules lose their transverse alignment before microfibrils become unrecognizable in the mitotic and elongation zones of roots. *Protoplasma*, **215**, 172–183.

Sugimoto, K., Himmelspach, R., Williamson, R.E., and Wasteneys, G.O. (2003) Mutation or drug-dependent microtubule disruption causes radial swelling without altering parallel cellulose microfibril deposition in *Arabidopsis* root cells. *Plant Cell*, **15**, 1414–1429.

Talbot, M.J., Franceschi, V.R., McCurdy, D.W., and Offler, C.E. (2001) Wall ingrowth architecture in epidermal transfer cells of *Vicia faba* cotyledons. *Protoplasma*, **215**, 191–203.

Talbot, M.J., Offler, C.E., and McCurdy, D.W. (2002) Transfer cell architecture: a contribution towards understanding localized wall deposition. *Protoplasma*, **219**, 197–209.

Talbot, M.J., Wasteneys, G.O., McCurdy, D.W., and Offler, C.E. (2007a) Deposition patterns of cellulose microfibrils in flange wall ingrowths of transfer cells indicate clear parallels with those of secondary wall thickenings. *Functional Plant Biology*, **34**, 307–313.

Talbot, M.J., Wasteneys, G.O., Offler, C.E., and McCurdy, D.W. (2007b) Cellulose synthesis is required for deposition of reticulate wall ingrowths in transfer cells. *Plant and Cell Physiology*, **48**, 147–158.

Thiel. J., Weier, D., Sreenivasulu, N., Strickert, M., Weichert, N., Melzer, M., Czauderna, T., Wobus, U., Weber, H., and Weschke, W. (2008) Different hormonal regulation of cellular differentiation and function in nucellar projection and endosperm transfer cells: a microdissection-based transcriptome study of young barley grains. *Plant Physiology*, **148**, 1436–1452.

Thiel, J., Riewe, D., Rutten, T., Melzer, M., Friedel, S., Bollenbeck, F., Weschke, W., and Weber, H. (2012) Differentiation of endosperm transfer cells of barley: a comprehensive analysis at the micro-scale. *Plant Journal*, **71**, 639–655.

Thompson, R.D., Hueros, G., Becker, H.-A., and Maitz, M. (2001) Development and function of seed transfer cells. *Plant Science*, **160**, 775–783.

Truernit, E. and Sauer, N. (1995) The promoter of the *Arabidopsis thaliana* SUC2 sucrose-H^+ symporter gene directs expression of β-glucuronidase to the phloem: evidence for phloem loading and unloading by SUC2. *Planta*, **196**, 564–570.

van Gisbergen, P.A.C. and Bezanilla, M. (2013) Plant formins: membrane anchors for actin polymerization. *Trends in Cell Biology*, **23**, 227–233.

Vaughn, K.C., Talbot, M.J., Offler, C.E., and McCurdy, D.W. (2007) Wall ingrowths in epidermal transfer cells of *Vicia faba* cotyledons are modified primary walls marked by localized accumulations of arabinogalactan proteins. *Plant and Cell Physiology*, **48**, 159–168.

Wang, H.-L., Offler, C.E., and Patrick, J.W. (1994) Nucellar projection transfer cells in the developing wheat grain. *Protoplasma*, **182**, 39–52.

Wernicke, W. and Jung, G. (1992) Role of cytoskeleton in cell shaping of developing mesophyll of wheat (*Triticum aestivum* L.). *European Journal of Cell Biology*, **57**, 88–94.

Wightman, R. and Turner, S.R. (2008) The roles of the cytoskeleton during cellulose deposition at the secondary cell wall. *Plant Journal*, **54**, 794–805.

Wimmers, L.E. and Turgeon, R. (1991) Transfer cells and solute uptake in minor veins of *Pisum sativum* leaves. *Planta*, **186**, 2–12.

Xia, X., Zhang, H.M., Andriunas, F.A., Offler, C.E., and Patrick, J.W. (2012) Extracellular hydrogen peroxide, produced through a respiratory burst oxidase/superoxide dismutase pathway, directs ingrowth wall formation in epidermal transfer cells of *Vicia faba* cotyledons. *Plant Signaling and Behavior*, **7**, 1–4.

Zhong, R., Lee, C., Ye, Z. H. (2010) Evolutionary conservation of the transcriptional network regulating secondary cell wall biosynthesis. *Trends in Plant Science*, **15**, 625–632.

Zhou, Y., Andriunas, F., Offler, C.E., McCurdy, D.W., and Patrick, J.W. (2010) An epidermal-specific ethylene signal cascade regulates *trans* differentiation of transfer cells in *Vicia faba* cotyledons. *New Phytologist*, **185**, 931–943.

Section 3
Developmental Regulations of Cell Shape

12 Regulation of Guard Cell Formation by Integration of Transcriptional and Signaling Regulation

Chin-Min Kimmy Ho[1] and Dominique C. Bergmann[1,2]

[1]*Department of Biology, Stanford University, Stanford, CA*
[2]*Howard Hughes Medical Institute, Stanford University, Stanford, CA*

Introduction

To survive the rigors of life on land, among other adaptations plants have an epidermis and cuticle that serve as a water and gas impermeable barrier. This protection against desiccation comes at a cost, however. Plants need regulated intake of atmospheric carbon dioxide (CO_2) to serve as a raw material for photosynthetic carbon fixing and they need an ultimate sink for water vapor at the end of the transpiration stream. In their role as epidermal valves, stomata are solutions to this problem. In nearly all known extant and fossil plants, stomata consist of two sister cells of epidermal origin that flank a pore they can open and close by a turgor-driven mechanism. These two stomatal guard cells are the end product of a dedicated, yet flexible, developmental program.

Stomatal development in Arabidopsis is an interesting system because the final product – a pair of symmetric kidney-shaped guard cells – is only attained through a multistage process. The journey begins when some cells are selected from a field of cells possessing (what appears to be) equivalent potential. These lineage founder cells are dispersed in the young epidermis and each can divide a variable number of times before committing to their final stomatal fate. Divisions within a lineage and communication within and between lineages are required to generate physiologically optimized numbers and patterns of stomata. Asymmetric divisions in the stomatal lineage share properties with stem-cell-like divisions in other systems and are under the control of master regulator transcription factors whose expression and activities are, in turn, responsive to signals from a variety of sources.

Plant Cell Wall Patterning and Cell Shape, First Edition. Edited by Hiroo Fukuda.

Basic Stomatal Development Pathway in Arabidopsis

Stomata develop on the surfaces of most photosynthetic aerial organs (e.g., leaves, stems, and carpels). Stomatal distribution follows some organ-specific guidelines but, in nearly all cases, is well described by a 'one-cell-spacing' rule that stipulates that two mature stomata (a stoma being defined as two sister guard cells and the pore between them) are always separated by at least one intervening non-stomatal cell. Stomata are formed as the organ develops; in leaves, stomatal lineage divisions produce the majority of the cells in the epidermis (Geisler *et al.*, 2000). It is not obvious what triggers some cells in the epidermis to become stomatal precursors; thus far, no overt morphological or molecular signs unambiguously mark these cells. Retrospective analysis through serial imaging allows one to define Meristemoid Mother Cells (MMCs) as cells that divide asymmetrically to produce meristemoids and stomatal lineage ground cells (SLGCs; Fig. 12.1), both of which have predictable behaviors. Meristemoids can divide again asymmetrically, recreating themselves and producing an SLGC. Meristemoids may repeat this so-called 'amplifying' division another one–three times before they differentiate into the next precursor stage, the guard mother cell (GMC). GMCs will divide once, symmetrically, to produce the two immature guard cells (GCs). Guard cells develop their characteristic morphology and structure by production of a thickened cell wall between them, which is later digested to form the stomatal pore (Yang and Sack, 1995). Rearrangement of cytoskeletal elements, reinforcement of the cell wall and expression of ion channels is required for full differentiation of the guard cells (Bergmann and Sack, 2007; Negi *et al.*, 2013). The flexible division pattern of the meristemoid has been likened to stem-cell divisions in animal systems. It is not quite analogous however, in that *both* daughters of the MMC division have the capability to divide again repeatedly. The SLGC can undergo an asymmetric 'spacing' division that produces another meristemoid/SLGC pair. This is similar to the division of the MMC with one important addition. To enforce a one-cell-spacing rule, the spacing divisions must be oriented such that the new meristemoid is not contacting the previous meristemoid (Fig. 12.1a). Division orientation is not just in relationship to the cells that share the same lineage, but relative to meristemoids, GMCs, and GCs produced by other lineages. Cell–cell signaling is therefore paramount to this process.

Arabidopsis is a good model for stomatal pattern in many dicot species. Stomatal production in monocots (especially the grasses) and in basal plant species differs at a morphological level and the reader is referred to excellent recent discussions of these other species (e.g., Vatén and Bergmann, 2012; Facette and Smith, 2012; Rudall and Knowles, 2013). In this chapter, we

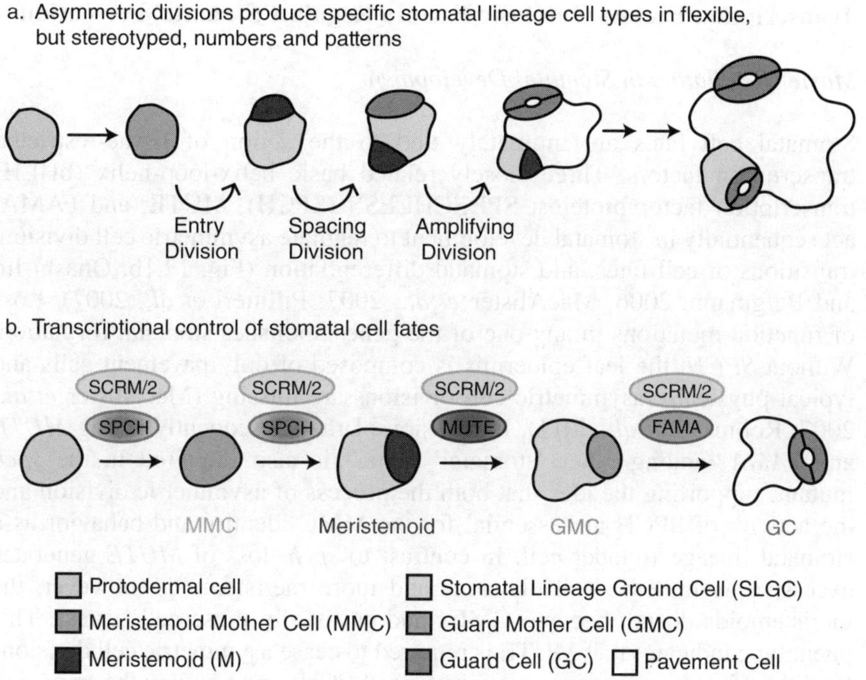

Figure 12.1 Diagram of stomatal lineage cell divisions and fates. (a) In the typical progression through differentiation, a protodermal cell (gray) becomes a meristemoid mother cell (MMC; pink) when it acquires the ability to divide asymmetrically. An MMC's first asymmetric cell division, also called an entry division, creates a smaller meristemoid (red) and larger stomatal lineage ground cell (SLGC; beige). Both meristemoids and SLGCs can divide again asymmetrically. SLGCs can either differentiate into a pavement cell or undergo another asymmetric 'spacing' cell division to form another meristemoid/SLGC pair with the meristemoid spaced away from committed stomatal precursor cells. 'Amplifying division' refers to the divisions made by meristemoids to produce another meristemoid/SLGC pair. After amplification, a meristemoid transitions into a guard mother cell (GMC; orange) and eventually differentiates into paired guard cells (GCs; green). (b) Stomatal lineage progression is mediated by heterodimers of three closely related and cell-type specifically expressed bHLHs, SPEECHLESS(SPCH), MUTE, and FAMA with partners SCRM and/or SCRM2. These transcription factor complexes are necessary and sufficient to determinate the meristemoid, GMC, and GC fates during stomatal development. (See plate section for the color version of this figure.)

focus on the molecular and genetic underpinnings of stomatal development as identified in Arabidopsis. Interestingly, despite differences in morphology, homologues of many of the genes required for Arabidopsis stomatal development are found in other species (Vatén and Bergmann, 2012) and there is even some functional evidence for their conservation (Liu *et al.*, 2009).

Transcription Factors: Intrinsic Factors Regulate Stomatal Formation

Master Regulators in Stomatal Development

Stomatal cell fates are intimately tied to the action of tissue-restricted transcription factors. Three closely related basic helix-loop-helix (bHLH) transcription factor proteins, SPEECHLESS (SPCH), MUTE, and FAMA, act sequentially in stomatal development to regulate asymmetric cell division, transitions of cell fates, and stomatal differentiation (Fig. 12.1b; Ohashi-Ito and Bergmann, 2006; MacAlister *et al.*, 2007; Pillitteri *et al.*, 2007). Loss of function mutations in any one of the genes abolishes stomatal formation. Without *SPCH*, the leaf epidermis is composed of only pavement cells and typical physically asymmetric cell divisions are missing (MacAlister *et al.*, 2007; Robinson *et al.*, 2011). Expression of the subsequently acting *MUTE* and *FAMA* (among other 'stomatal' genes) is also impaired in the *spch* mutant, supporting the idea that both the process of asymmetric division and the activity of SPCH are essential for the MMC identity and behavior as a stomatal lineage founder cell. In contrast to *spch*, loss of *MUTE* generates excessive asymmetric cell divisions and more meristemoids; however, the meristemoids do not become GMCs and eventually those cells arrest. This phenotype indicates that MUTE is required to cease asymmetric cell divisions (and therefore is somewhat antagonistic to SPCH) and to allow the transition to guard mother cell fate (MacAlister *et al.*, 2007; Pillitteri *et al.*, 2007). The *fama* mutant also fails to make any recognizable stomata but, instead of arresting cells as single-celled GMCs, *fama* mutants have 'caterpillar-like' tumors that represent additional rounds of symmetric cell divisions occurring in the place of the guard cells (Ohashi-Ito and Bergmann, 2006). These 'tumors' express GMC fate markers, suggesting that FAMA functions in the termination of the symmetric cell division in GMCs and in the differentiation of GCs. Overall, the sequentially combined actions of these three bHLH transcription factors initiate and regulate the asymmetric cell division in MMCs and meristemoids and subsequently promote GMC fate, followed by termination and differentiation of stomatal guard cells.

Two Additional bHLHs are Required Partners for Activities of Master Regulators

bHLH transcription factors typically dimerize in homo- or hetero-combinations. A forward genetic screen identified a dominant mutant, *scrm-D*, with an important role in stomatal development revealed by the phenotype of the epidermis nearly entirely composed of GCs. *scrm-D* was found to result from a point mutation leading to a single amino acid change in the ICE1 protein (Kanaoka *et al.*, 2008). *ICE1* was originally recognized as a master regulator in cold stress (Chinnusamy *et al.*, 2003), and *ICE1/SCRM* (hereafter referred to as *SCRM*) encodes a bHLH transcription factor expressed throughout stomatal development (Kanaoka *et al.*, 2008). Its paralog *SCRM2* is also specifically detected in stomatal lineage cells (Kanaoka *et al.*, 2008). Although loss of either *SCRM* or *SCRM2* alone has little phenotypic consequence, *scrm;scrm2* double mutants fail to initiate the stomatal lineage and resemble *spch*. *scrm;scrm2/+* mutants exhibit lineage defects comparable to *mute* (Kanaoka *et al.*, 2008). Their expression and mutant phenotypes suggest *SCRM* and *SCRM2* function together with *SPCH*, *MUTE* and *FAMA* and their proteins structures suggest this could be via physical interaction. The hypothesis is supported by yeast-two-hybrid and split-GFP (green fluorescent protein) experiments that demonstrate that none of SPCH, MUTE and FAMA strongly homodimerize, but each interacts with SCRM or its 'activated' form scrm-D (Kanaoka *et al.*, 2008).

MYB Domain Proteins Balance between Cell Differentiation and Cell Division in GMCs

Although the bHLHs play key roles in stomatal formation, other transcriptional factors add other layers of regulation to this developmental process. The balance between developmental programming and cell cycle progression is highly coordinated during stomatal development. This is best understood at a mechanistic level during the transition from the GMC to GC stages. Plants with mutations in *FOUR LIPS* (*FLP*), which encodes an R2R3 MYB-type protein, produce adjacent pairs of stomata (Payne *et al.*, 2000). In addition, double mutants of *FLP* and its homolog *MYB88* produce clusters of putative GMCs and GCs, suggesting the function of FLP and MYB88 is to promote the GMC to differentiate fully into a GC and to restrict it to one symmetric cell division (Lai *et al.*, 2005). Although MYB transcription factors are often found to partner with bHLH proteins in plants, neither FLP nor MYB88 physically interacts with FAMA in split-GFP assays (Ohashi-Ito and Bergmann, 2006). In addition, additive phenotypes result from combining *fama* and *flp* mutations, indicating they are unlikely to be exclusive partners of each other.

How do FLP/MYB88 regulate cell cycle progress to terminate symmetric cell division and promote differentiation? FLP/MYB88 competes with E2F-DP, a transcriptional activator of the cell cycle, for a binding site in the promoter of CYCLIN-DEPENDENT KINASE B1;1 (CDKB1;1), thereby stopping the cell cycle and promoting GMC differentiation (Boudolf *et al.*, 2004; Xie *et al.*, 2010). This idea is also supported by observation of triple mutants of plant A-type cyclins, *cyca2;134* and *cyca2;234*. Compared to WT, the triple mutants have a high proportion of single guard cells suggesting that CYCA2s promote symmetric GMC cell divisions to produce paired GCs (Vanneste *et al.*, 2011). In addition, CYCA2;3 expression is significantly higher in *flp;myb88* double mutants than in WT, indicating that FLP and MYB88 are required to suppress the expression of CYCA2;3 and to stop cell cycles (Vanneste *et al.*, 2011).

Other Transcription Factors Regulate Aspects of Guard Cell Formation and Behavior

While the bHLHs may have major roles in stomatal progression, they are likely just the 'tip of the iceberg' in terms of total regulation of stomatal behaviors. Quantitative effects on stomatal divisions are seen with the MADS-box gene, *AGAMOUS-LIKE 16* (*AGL16*), known to be expressed in guard cells (Alvarez-Buylla *et al.*, 2000). Interestingly, AGL16 mRNA is a unique target of miR824, which enables AGL16 sequence-specific degradation (Kutter *et al.*, 2007). Expression of miR824-resistant *AGL16* mRNA (but not wild-type *AGL16* mRNA) results in an elevated frequency of higher-order stomatal complexes. This phenotype comes about from extra rounds of entry and amplifying divisions. In contrast, in plants lacking *AGL16* or in lines overexpressing miR824 there is a decrease in the proportion of higher-order stomatal complexes (Kutter *et al.*, 2007). Another role for transcription factors is in ensuring the production of functional GCs. Recently, the plant-specific Dof-type transcription factor STOMATAL CARPENTER 1 (SCAP1) was shown to regulate guard cell differentiation (Negi *et al.*, 2013). *SCAP1* is specifically expressed in mature guard cells but not in GMCs. Loss of *SCAP1* results in irregularly shaped stomata lacking the ability to regulate stomatal movement in response to environmental stimuli, such as CO_2-induced stomatal closing and light-induced stomatal opening. SCAP1 modulates the expression of key components in stomatal movement and morphogenesis including the MYB60 transcription factor (see 'Molecular and Developmental Adaptations to Drought Stress'), pectin methylesterases for building the thickened cell walls found at the pores, and potassium channel proteins to regulate pore apertures (Negi *et al.*, 2013).

The Signaling Cascade: Receiving Extracellular Signals by Membrane-bound Receptors

Receptor-like Kinases (RLKs) of the ERECTA Family Negatively Regulate Stomatal Patterning

Arabidopsis Landsberg *erecta* is a popular laboratory ecotype due to its erect growth habit; much of this morphology (including a compact inflorescence, short petioles, and blunt siliques) is due to inactivation of a leucine-rich repeat receptor-like Ser/Thr Kinase (LRR-RLK) that is the product of the *ERECTA (ER)* gene (Torii *et al.*, 1996). The reduced organ sizes seen in *er* mutants are due to lessened cell proliferation activity (Shpak *et al.*, 2003). Mutations in two *ERECTA-LIKE* genes, *ERL1* and *ERL2* (Shpak *et al.*, 2004) exacerbate the *er* phenotype, but in unique ways (Shpak *et al.*, 2004). Promoter swap experiments suggest the functional differences among the ER family (ERf) members are mainly due to their intricate expression patterns rather than their protein structures (Shpak *et al.*, 2004).

Not only functioning in plant growth, ERf members also regulate the balance between cell proliferation and differentiation during stomatal development (Shpak *et al.*, 2005). Severe increases in stomatal density and stomata mis-patterning into clusters are only found in *er;erl1;erl2* triple mutants, implying these genes function redundantly as negative regulators of stomatal fate and proliferation (Shpak *et al.*, 2005). Phenotypic analyses of *er* single mutants and combinations of double mutants also reveal specificities in their stomatal patterning roles (Fig. 12.2a), however. *ER* is responsible for restricting entry division in stomatal lineage. *ERL1* inhibits GMC differentiation and promotes the correct orientation of meristemoid-making divisions. *ERL2* appears to control the amplifying cell division in SLGCs (Shpak *et al.*, 2005). Transcriptional reporters show that all *ERf* members are present in the protodermal cells of emerging leaf primordial; later, however, *ER* expression is diminished before epidermal cell differentiation while *ERL1* and *ERL2* expression remains high in the stomatal-lineage cells (Shpak *et al.*, 2005). Taken together, expression and genetic data suggest that, rather than regulating protodermal asymmetric cell divisions and stomatal lineage initiation (as for ER), ERL1 and ERL2 more likely function in perceiving the positional cues for stomatal patterning in lineage cells at later stages.

A Receptor-like Protein (RLP) Coordinates with ERf to Control Stomatal Development

TOO MANY MOUTHS (TMM), one of the first genetically defined stomatal development regulators (Yang and Sack, 1995), appears to partner with ERf receptors. Different from ERf members, phenotypes of loss of *TMM* have only been described in the stomatal lineage, and this leucine-rich repeat-containing receptor-like protein (LRR-RLP) is expressed in small epidermal cells with high proliferative activities (early stages of the stomatal lineage), but is not detectable in protoderm or mature GCs (Nadeau and Sack, 2002). Based on its membrane localization, cell type expression pattern, mutant phenotype and protein structure, TMM is likely to be a component of a receptor complex needed to sense positional cues. TMM lacks an intracellular kinase domain implying that it either: (1) transduces a signal by associating with other

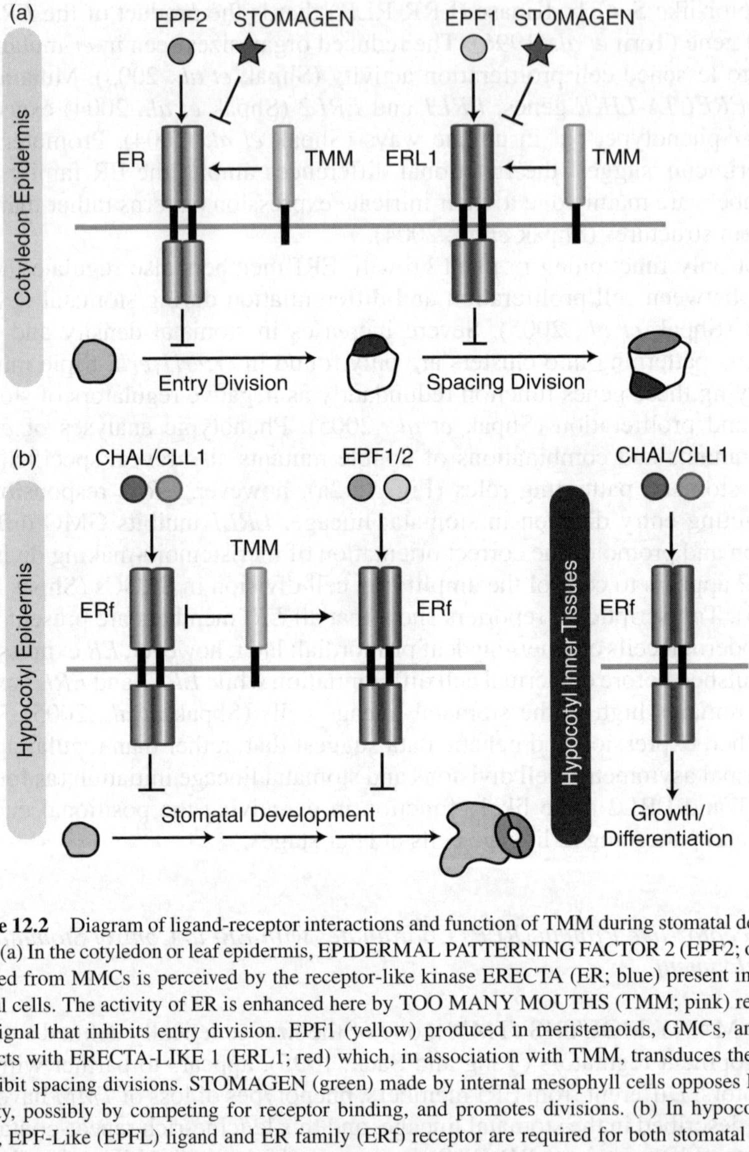

Figure 12.2 Diagram of ligand-receptor interactions and function of TMM during stomatal development. (a) In the cotyledon or leaf epidermis, EPIDERMAL PATTERNING FACTOR 2 (EPF2; orange) secreted from MMCs is perceived by the receptor-like kinase ERECTA (ER; blue) present in protodermal cells. The activity of ER is enhanced here by TOO MANY MOUTHS (TMM; pink) resulting in a signal that inhibits entry division. EPF1 (yellow) produced in meristemoids, GMCs, and GCs interacts with ERECTA-LIKE 1 (ERL1; red) which, in association with TMM, transduces the signal to inhibit spacing divisions. STOMAGEN (green) made by internal mesophyll cells opposes EPF1/2 activity, possibly by competing for receptor binding, and promotes divisions. (b) In hypocotyls or stems, EPF-Like (EPFL) ligand and ER family (ERf) receptor are required for both stomatal pattern and stem growth. CHALLAH (CHAL; dark green) and its family members (e.g. CHALLAH-Like 1, CLL1; light green), whose normal roles are in stem growth, could 'escape' from internal tissues and negatively regulate stomatal development. TMM is only expressed in stomatal lineage cells. When TMM is present, CHAL/CLL1 signaling through ERf receptors is dampened while EPF1/2 signaling is permitted. In the absence of TMM however, a CHAL/CLL1-ERf module transmits information to prevent stomatal production; consequently, a 'no stomata' phenotype is found in *tmm* hypocotyls. Arrows indicate positive influences in signaling process and T-bars represent negative influences. (See plate section for the color version of this figure.)

proteins possessing the kinase domain such as *ERf* (Nadeau and Sack, 2002; Lee *et al.*, 2012); or (2) that its role may be to modulate signaling by sequestering ligands or receptors in inactive complexes.

Partnerships between LRR-RLKs and LRR-RLPs are not unique to epidermal development. In the shoot apical meristem (SAM) the RLK CLAVATA1 (CLV1) and the RLP CLAVATA2 (CLV2) regulate the balance between stem cell populations in the SAM central domain and cell differentiation in flanking regions (Jeong *et al.*, 1999). Because stomatal development happens in the epidermis of many different organs, the relationship between the ERf RLKs and TMM RLP may not always be the same, an idea supported by numerous genetic studies. For example, *TMM* promotes or inhibits stomatal formation depending on organ type (Fig. 12.2). Similar to phenotypes found in *ERf* mutants (Shpak *et al.*, 2005) in *tmm* leaves, MMCs and SLGCs undergo asymmetric cell division more frequently and meristemoids differentiate into GCs more rapidly, resulting in excessive stomatal clusters (Geisler *et al.*, 2000). However, in *tmm* stems and hypocotyls, asymmetric cell divisions appear rarely and these divisions do not yield cells that become stomata (Yang and Sack, 1995; Bhave *et al.*, 2009). This indicates that *ERf* and *TMM* act cooperatively to inhibit stomatal formation in leaves, but that *TMM* and *ERL1* work antagonistically in hypocotyls and stems since *erl1*;*tmm* double mutants release the 'no-stomata' phenotype found in *tmm* hypocotyl (Shpak *et al.*, 2005). These data also suggest that TMM modulates the compartmentalized and distinct functions of the ERf receptors in stomatal signaling. Determining how TMM might perform this role requires an examination of the signals TMM and the ERf perceive.

Members of EPIDERMAL PATTERNING FACTOR LIKE Family Provide Positional Cues in Stomatal Patterning

The EPIDERMAL PATTERNING FACTOR LIKE (EPFL) family in Arabidopsis encodes 11 plant-specific cysteine-rich secreted peptides. Their bioactive forms are approximately 50 amino acids in length with a cysteine knot-like conformation created by intramolecular disulfide bonds (Hara *et al.*, 2009). Among the 11, EPF1, EPF2, STOMAGEN/EPFL9, CHALLAH (CHAL)/EPFL6, CHALLAH-LIKE 1 (CLL1)/EPFL5, and CHALLAH-LIKE2 (CLL2)/EPFL4 can play roles in controlling stomatal density and patterning in the plant epidermis (Fig. 12.2; Hara *et al.*, 2007, 2009; Hunt and Gray, 2009; Abrash and Bergmann, 2010; Kondo *et al.*, 2010; Sugano *et al.*, 2010; Abrash *et al.*, 2011; Niwa *et al.*, 2013). To become activated, signaling peptides are often processed by extracellular proteases; a candidate protease, STOMATAL DENSITY AND DISTRIBUTION-1 (SDD1), has been known for many years to regulate stomatal pattern (Berger and Altmann, 2000;

Von Groll *et al.*, 2002) and it was speculated that SDD1 would promote EPF processing. Genetic analyses suggest that EPFs function independently of this factor (Hara *et al.*, 2007, 2009; Hunt and Gray, 2009) however, so the processing factors are still at large.

Negative Regulators: EPF1 and EPF2

Expression patterns and genetic analyses of EPF1 and EPF2 have dissected their roles in ERf signaling at distinct phases during stomatal development. EPF2 is expressed in the protodermal cells undergoing asymmetric cell divisions at early stages of the stomatal lineage (Hara *et al.*, 2009). In the *epf2* loss-of function mutants, protodermal cells acquire MMC identity and divide asymmetrically more frequently. This results in hyperproliferation of meristemoids and increased numbers of both GCs and pavement cells (the products of meristemoid divisions), although stomatal pattern is minimally affected (Hara *et al.*, 2009; Hunt and Gray, 2009). Overexpression of EPF2 or *in vitro* application of bioactive EPF2 peptides suppresses the stomatal entry divisions, resulting in an epidermis devoid of stomata and composed of pavement cells only (Hara *et al.*, 2009; Lee *et al.*, 2012). These data suggest a model that EPF2 is secreted from MMCs to neighboring cells to keep the neighbors from acquiring stomata lineage cell fate.

In contrast to the role of EPF2 in inhibiting protodermal cells from becoming MMCs, EPF1 likely serves as a positional signal to provide spatial information about existing stomata and stomatal precursor cells. EPF1 is expressed later than EPF2 in meristemoids, GMCs, and young GCs (Hara *et al.*, 2007). Loss-of-function mutations in EPF1 produce clustered stomata (phenotypically resembling *tmm*), suggesting the newly formed stomata are placed incorrectly relative to established stomata (Hara *et al.*, 2007). Overexpression of EPF1 or *in vitro* application of bioactive EPF1 peptides does not affect MMC formation, but instead represses the transition from meristemoids to GMCs. These halted cells subsequently transdifferentiate into pavement cells (Hara *et al.*, 2007; Lee *et al.*, 2012). Taken together, EPF2 and EPF1 appear to be the major local regulators of divisions and patterns. It is interesting to consider that these peptides are produced by the same cell types that they act upon, so mechanisms to ensure a signal affects the neighbor cells, rather than a cell's own development, are required.

Overexpression phenotypes of EPF1 and EPF2 are dependent on *TMM* and *ERf*, suggesting both EPF1 and EPF2 work upstream of LRR-RLK signaling in stomatal development (Hara *et al.*, 2007, 2009). Interesting observations were found in the *tmm*;*epf1* and *tmm*;*epf2* double mutants, however. Loss of EPF1 and EPF2 affect stomatal clustering and density, respectively; however, in a *tmm* background, careful analysis of *tmm*;*epf2* double mutants reveals that *tmm;epf2* leaves produce clustered and arrested meristemoid cells similar to the *epf2* single mutant and reduce stomatal cluster sizes relative to *tmm* (Hunt and Gray, 2009). These results suggest that EPF2 can activate ERf signaling in either a TMM-dependent or TMM-independent manner.

Although overexpression of either EPF1 or EPF2 can suppress stomatal production, promoter swap experiments reveal functional differences between these two signaling peptides (Hara *et al.*, 2009). In the *epf1* mutant background, expression of *EPF1p::EPF2* but not *EPF2p::EPF1* partially rescues the clustered stomata phenotype (Hara *et al.*, 2009). This suggests that the specificity of EPF1 in regulating stomata clustering is due to its distinct expression pattern. On the other hand, in *epf2*, expression of *EPF1p::EPF2* but not *EPF2p::EPF1* partially rescues the elevated stomatal density, suggesting that there are intrinsic protein differences between these two peptides. These data provide a hint that the divergent functions of EPF1 and EPF2 may be due to their specific receptor targets. Given that *ER* and *EPF2* share expression in the MMCs and both *epf2* and *er* mutants overproliferate meristemoids, a reasonable hypothesis is that *EPF2* is a specific ligand for *ER* (Shpak *et al.*, 2005; Hara *et al.*, 2009; Hunt and Gray, 2009).

Such specificity hypotheses were tested in elegant biochemical studies using co-immunoprecipitation (co-IP) assays and biosensor chips; they identified ER–EPF2 and ERL1–EPF1 as ligand-receptor pairs (Lee *et al.*, 2012). Blocking ER signaling using a kinase-dead ER receptor results in seedlings resistant to EPF2 but not EPF1. Conversely, seedlings expressing kinase-dead ERL1 were resistant to EPF1 treatment. These *in vitro* and *in vivo* data show the specific functions of the ER–EPF2 pair in entry divisions and the ERL1–EPF1 pair in spacing divisions (Lee *et al.*, 2012). In addition, ER and ERL1 can homodimerize or heterodimerize with each other. The receptor-like protein TMM can associate with ER or ERL1 but not with itself, suggesting a model whereby ERf members are primary receptors for EPFs and TMM is a signaling modulator affecting productive ERf signaling complex formation (Lee *et al.*, 2012).

Positive Regulator: STOMAGEN

In contrast to the negative roles of EPF1 and EPF2 in stomatal development, STOMAGEN is a positively acting signaling peptide (Kondo *et al.*, 2010; Sugano *et al.*, 2010). As a member of the EPF family (EPFL9), it is remarkable that this peptide has differential effects on stomatal production. Biochemical analysis elucidated a variable loop of approximately 20 amino acids that confers the functional specificities between STOMAGEN and EPF1/2, whereas the conserved cysteine residues form scaffolds structurally required for all their activities (Ohki *et al.*, 2011). Overexpression or *in vitro* application of bioactive STOMAGEN on leaves increases stomatal density as well as the size of stomatal clusters in a dose-dependent manner (Kondo *et al.*, 2010; Sugano *et al.*, 2010). In contrast, RNAi-knockdown of *STOMAGEN* reduces the stomata number (Sugano *et al.*, 2010). Interestingly, STOMAGEN is expressed broadly in the mesophyll but not in the epidermis or stomatal lineage cells, another way it differs from EPF1 and EPF2. STOMAGEN may therefore act as a signaling molecule connecting mesophyll behavior to the epidermis (Kondo *et al.*, 2010; Sugano *et al.*, 2010). This interlayer communication may have physiological relevance as the mesophyll cells contain the bulk of the photosynthetic machinery and stomata are the portals for CO_2 entry. Co-application of chemically synthesized STOMAGEN and EPF2 in seedlings shows that EPF2 functions antagonistically to STOMAGEN (Ohki *et al.*, 2011). The treatment of synthesized EPF2 reduces the efficacy of STOMAGEN, suggesting STOMAGEN competes with EPF2 for binding to receptors. The competing mechanism of EPFLs provides a way to fine-tune the stomata formation in the response to physiological changes during development.

CHALLAH and its Relatives: Dissecting Overlapping but Distinct Functions of TMM

The behaviors of an EPFL subclass add to the complexity in peptide singling during development. Mutations in *CHAL/EPFL6* were identified by their ability to suppress *tmm* phenotypes in a tissue-specific manner, restoring stomata to *tmm* hypocotyls and stems while having no effects in leaves (Abrash and Bergmann, 2010). Fitting with this, *CHAL* is present in hypocotyls and stems at higher levels than in leaves. Here, similar to STOMAGEN, CHAL is restricted to internal layers rather than expressed in stomatal lineage cells (Abrash and Bergmann, 2010). Overexpression of CHAL produced little effect in a WT background; however, when overexpressed in a *tmm* background, CHAL eliminated stomata in leaf epidermis (Abrash and Bergmann, 2010). Recall that loss of *TMM* renders overexpression of EPF1 or EPF2 incapable of blocking stomatal formation. Different ligands of the same family therefore display opposite genetic interactions with their putative receptor, TMM. How might this work at a molecular level?

From phenotypes of *CHAL* family mutants in combination with, and compared to, *ERf* and *TMM* mutants, a plausible model for activity begins to emerge. First, reappearance of hypocotyl stomata in *chal;tmm* resembles *erl1;tmm* (Shpak *et al.*, 2005; Abrash and Bergmann, 2010). Second, although neither *chal* single nor *chal* family multiple mutants have stomatal phenotypes when *TMM* is present, *chal;cll2* and *chal;cll1;cll2* plants are short and resemble *er* mutants morphologically (Abrash and Bergmann, 2010; Abrash *et al.*, 2011; Uchida *et al.*, 2012). Moreover, the ability of overexpressed

CHAL, CLL1/EPFL5, or CLL2/EPFL4 to suppress stomatal formation in a *tmm* background is dependent on the presence of functional ERf receptors (Abrash *et al.*, 2011). Taken together with their internal expression patterns, these phenotypes indicate that CHAL family ligands are likely perceived by ERf to promote stem growth, and that their involvement in stomatal formation is 'accidental'. An attractive model is that TMM is a stomatal lineage signaling specificity factor that allows the same receptors (ERfs) to distinguish between EPFL signals for stomatal development (EPF1/2) and those for growth (CHAL family) by forming complexes with the ERfs that have different affinities for different EPFL subclasses (Fig. 12.2). When TMM is present, CHAL family signaling through ERf receptors would be repressed while EPF1/2 signaling is promoted. In the absence of TMM however, the CHAL family members expressed in the stem interior may inappropriately activate the ERfs to prevent stomatal production, resulting in a 'no stomata' phenotype in *tmm* hypocotyls.

YDA MAPK Pathway: Transducing Extracellular Information to Intracellular Transcription Factors

Mitogen-activated protein kinases (MAPKs) are part of an evolutionary conserved signaling cascade found in nearly all eukaryotes. MAPK cascades play important roles in biological activities such as immune responses, response to abiotic and biotic stresses, and growth and development processes (Rodriguez *et al.*, 2010). In transducing environmental and developmental cues into intracellular responses, the typical signaling/phosphorylation relay starts from signals at the plasma membrane activating a MAPK kinase kinase (MAPKKK), which activates a MAPK kinase (MAPKK) and then a MAPK to regulate the activities of target proteins by phosphorylation. The genome of *A. thaliana* encodes approximately 60 MAPKKKs, 10 MAPKKs, and 20 MAPKs. Of those whose functions are known, most are associated with plant immunity and abiotic stress response (Rodriguez *et al.*, 2010). In some cases, modules consisting of common MAPKKKs, MAPKKs, and MAPKs are involved in both development (Fig. 12.3; Lukowitz *et al.*, 2004; Meng *et al.*, 2012) and immune responses (Rodriguez *et al.*, 2010). The sharing of components suggests the system could flexibly modulate plant growth in response to environmental stimuli, but also presents problems of signaling specificity.

YDA MAPK Signaling in Stomatal Development

The connection of MAPK signaling to stomatal development came from loss-of-function mutations in *YODA(YDA)*, a MAPKKK gene, that resulted

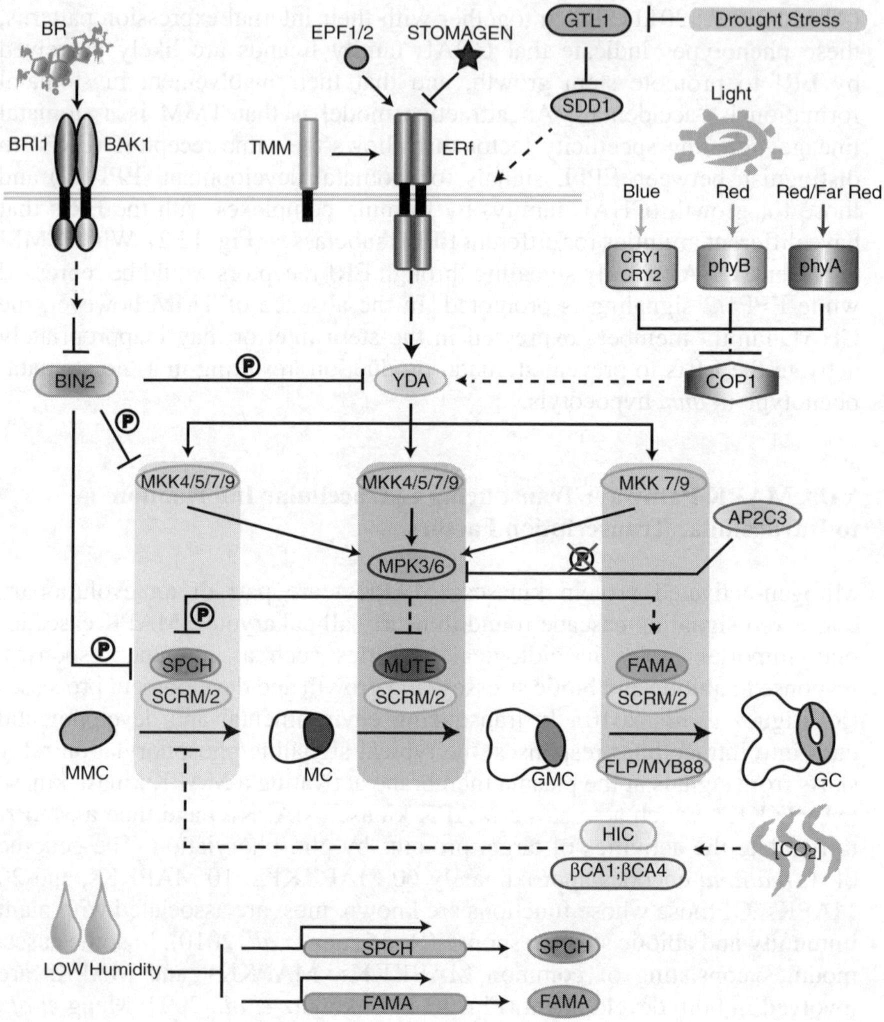
GTL1
Drought Stress
BR
EPF1/2
STOMAGEN
SDD1
Light
BRI1
BAK1
TMM
ERf
Blue
Red
Red/Far Red
CRY1
CRY2
phyB
phyA
BIN2
P
YDA
COP1
P
MKK4/5/7/9
MKK4/5/7/9
MKK 7/9
AP2C3
MPK3/6
P
P
P
SPCH
MUTE
FAMA
SCRM/2
SCRM/2
SCRM/2
MMC
MC
GMC
FLP/MYB88
GC
HIC
[CO2]
βCA1;βCA4
LOW Humidity
SPCH
SPCH
FAMA
FAMA

in an epidermis composed nearly entirely of stomata (Bergmann *et al.*, 2004). This phenotype closely resembles the *ERf* triple mutants (Shpak *et al.*, 2005). Deletion of an internal inhibitory domain created a constitutively active YDA version (CA-YDA) that could repress stomatal production in WT as well as in *sdd1* and *tmm* mutants (Bergmann *et al.*, 2004). These genetic results suggest that YDA is the key player in transducing a signal from receptors to an intracellular signaling cascade. Later, MAPKK4 and MAPKK5 (MKK4/5) and MAPK3 and MAPK6 (MPK3/6) were identified as downstream components of YDA signaling cascade (Wang *et al.*, 2007). Downregulation of MKK4/5 or MPK3/6 by RNA interference (RNAi) and/or mutations resulted in massive proliferation of stomata, much like in *yda* (Wang *et al.*, 2007). Broad expression of an activated homologue of MKK4/5 eliminated stomata (Wang *et al.*, 2007). Activated versions of two additional MAPKKs, MKK7/9 and MKK4/5, were subsequently shown to be able to eliminate stomata if expressed only in protodermal cells and MMCs (Lampard *et al.*, 2009).

←

Figure 12.3 Hormonal and environmental signals integrated into signal transduction pathways during stomatal development. Core MAPK and Brassinosteroid (BR) signaling cascades exhibit extensive cross-talk (phosphorylation of pathway components). In the canonical stomatal pathway, ligand receptor complexes transduce negative or positive signals (depending on specific ligand) to a downstream MAPK cascade through currently unidentified intermediates. The specific MAPK members demonstrated to affect stomatal development are indicated, along with their point and direction of action. In the case of MMC-meristemoid transitions, SPCH has been identified as the key downstream phosphorylated target. Active MPKs can be dephosphorylated and inactivated by AP2C3. BR signaling influences stomatal development through several regulatory interactions. The main cytoplasmic effector of BR signaling, the GSK3-like kinase BIN2, can deactivate YDA and MKK4/5 by phosphorylation, resulting in de-repression of SPCH and consequently promoting stomatal formation. However, BIN2 can also phosphorylate and inhibit SPCH directly. BIN2 may therefore fine-tune stomatal development in different tissues depending on systemic conditions. Environmental stimuli also play an important role in controlling the number of stomata. Drought inhibits the expression of transcription factor GLT1 (which targets and represses SDD1). SDD1 acts upstream of ERf and inhibits stomatal production. Signals from light including blue, red, and far-red light perceived by CRY1/2, phyB, phyA, respectively, all converge into the ubiquitin E3 ligase COP1, a central repressor in light signaling. Genetically, COP1 acts upstream of YDA. CO_2 concentration is inversely correlated with stomatal numbers. Enzymes (HIC, a 3-ketoacyl-CoA synthase) and β-CARBONIC ANHYDRASE 1 (βCA1) and βCA4 are produced in GCs and can non-autonomously repress stomatal initiation. Under low humidity conditions, *SPCH* and *FAMA* are *de novo* cytosine methylated and transcriptionally repressed. Arrows represent positive regulations and T-bars indicate negative regulations. Solid lines represent genetically or biochemically confirmed interactions. Dashed lines indicate indirect interactions. BAK1, BRI1-ASSOCIATED RECEPTOR KINASE 1; BIN2, BR INSENSITIVE 2; BRI1, BR INSENSITIVE 1; COP1, CONSTITUTIVE PHOTOMORPHOGENIC 1; CRY1/2, CRYPTOCHROME1/2; EPF1/2, EPIDERMAL PATTERNING FACTOR1/2; ERf, ERECTA family; FLP, FOUR LIPS; GTL1, GT-2 LIKE 1; HIC, HIGH CARBON DIOXIDE; MAPKKK, MAP KINASE KINASE KINASE; phyA, phytochrome A; phyB, phytochrome B; SDD1, STOMATAL DENSITY AND DISTRIBUTION 1; TMM, TOO MANY MOUTHS; YDA, YODA.

Genetic results and similarity of phenotypes lead to the hypothesis that extracellular signals perceived at the membrane by TMM and ERf were transduced by YDA (as MAPKKK), MKK4/5/7/9, and MPK3/6, ultimately leading to phosphorylation of downstream targets that regulate asymmetric cell divisions and cell fate specification. It is interesting to note that the components in this signaling module negatively regulate stomatal formation. We will readdress this negative activity in 'Activity of SPCH is Modulated by Phosphorylation'.

MAPK Phosphatases

Signal transduction in MAPK cascades proceeds via sequential phosphorylation events, so the active status of MAPKs can be reversed by regulated dephosphorylation (Canagarajah *et al.*, 1997). Recent studies identified a PP2C type phosphatase, ARABIDOPSIS PROTEIN PHOSPHATASE 2C (AP2C3) expressed specifically in stomatal lineage cells (Umbrasaite *et al.*, 2010). Overexpression of AP2C3 mimics loss of *YDA* or *MPK3/6* mutants, in that excessive and clustered stomata are produced. *In vitro*, AP2C3 can dephosphorylate MPK6 (Umbrasaite *et al.*, 2010). It was expected that loss of AP2C3 would resemble overactive MAPK signaling; no aberrant phenotypes were seen in AP2C3 mutants however, suggesting that there may be redundancy with related phosphatases. Taken together, AP2C3 family enzymes may downregulate MAPK signaling activity by dephosphorylating MPK6, subsequently de-repressing stomatal initiation.

Bridging BR Signaling Through the YDA MAPK Cascade

MAPK cascades are not the only signal transduction systems that regulate stomatal development; this is not surprising, as stomatal production must be coordinated with overall plant growth. Brassinosteroid (BR) is a plant steroid hormone with effects on cell growth, development, and physiological processes (reviewed by Wang *et al.*, 2012). Recent studies show that intermediates in the BR signaling pathway modulate stomatal density through their interactions with several components of the MAPK pathway (Kim *et al.*, 2012; Khan *et al.*, 2013). The GLYCOGEN SYNTHASE KINASE 3 (GSK3)-like kinase, BR INSENSITIVE2 (BIN2), serves as the major link between BR perception at the plasma membrane mediated by receptor kinases BRASSINOSTEROID INSENSITIVE1 (BRI1) and BRI1-ASSOCIATED KINASE1 (BAK1) (and a number of membrane-associated kinases) and downstream transcription factor effectors such as BRASSINOZOLE RESISTANT1 (BZR1; reviewed by Kim and Wang, 2010). Overexpression of BIN2 or downregulation of BIN2 and its related kinases results in clusters of or reduced numbers of stomata, depending on organ and growth conditions tested (Gudesblat *et al.*, 2012; Kim *et al.*, 2012). A quadruple mutant of the phosphatases that oppose BIN2 results in a cotyledon epidermis full of stomata (Kim *et al.*, 2012). An *in vitro* kinase assay supports the hypothesis that BIN2 inhibits YDA kinase activity through phosphorylation (Kim *et al.*, 2012). BR can therefore regulate stomatal development by suppressing BIN2, leading to de-repression of YDA. Another study shows that BIN2 can phosphorylate MAPK kinases, MKK4 and MKK5, *in vitro* (Khan *et al.*, 2013). The BIN2-mediated phosphorylation of MKK4 reduces MKK4 activity against its substrate MPK6 *in vitro*. Conversely, overexpression of MKK4 and MKK5 rescues the stomatal defects in BR-deficient plants (Khan *et al.*, 2013). Taken together, BR-pathway/MAPK crosstalk has potential roles bridging developmental and physiological pathways to allow coordinated responses to systemic needs of the plant.

Activity of SPCH is Modulated by Phosphorylation

Both MAPK and BR signaling cascades have multiple targets. What is the specific mechanism by which each pathway affects stomata? Multiple lines of evidence indicate that SPCH is a common target and that phosphorylation of SPCH leads to its inactivation. In the epidermis, the *spch* loss-of-function phenotype is indistinguishable from that produced by CA-YDA (Bergmann *et al.*, 2004). *SPCH* is epistatic to *YDA* as indicated by the stomataless (*spch*) phenotype, rather than stomatal overprolifersation (*yda*) phenotype observed in *yda;spch* double mutants (MacAlister *et al.*, 2007). Although SPCH, MUTE, and FAMA are closely related, protein sequences alignments reveals a unique 93 amino acid sequence in SPCH with 10 serine/threonine followed by proline (S/TP) sites. S/TP is the minimal site for MAPK phosphorylation; moreover, five are PxS/TP, which is considered a high-stringency MAPK phosphorylation target site. An *in vitro* kinase assay confirms that MPK3/6 phosphorylate the SPCH in this domain, so it has been termed the MAPK target domain (MPKTD; Lampard *et al.*, 2008). Deletion of the MPKTD abolishes phosphorylation of SPCH *in vitro* and *in vivo,* results in hyperaccumulation of SPCH and the production of excessive meristemoids and stomatal clusters (Lampard *et al.*, 2008). SPCH variants with point mutations in individual PxS/TP sites can rescue *spch,* but lead to dose-dependent increases in SPCH accumulation and meristemoid production. The more sites eliminated, the more meristemoids produced (Lampard *et al.*, 2008). The results indicate that MAPK signaling is upstream of SPCH and can regulate the initiation of the stomatal lineage. Because SPCH is expressed in a limited number of cells, its expression serves as a way of granting specificity to the MAPK cascade.

The same sites that mediate MAPK phosphorylation can also be target sites of other kinases such as the cyclin-dependent kinases (CDKs, for cell cycle regulation) and BIN2. Given the recent association of BR signaling with stomatal development and the MAPK cascade, it was exciting to find that BIN2 could also phosphorylate SPCH directly (Gudesblat *et al.*, 2012). *In vitro* kinase assays and mass spectrometry analysis of samples treated with BR reveal that BIN2 can phosphorylate some of the same MPKTD residues targeted by MPK3/6, but also five residues near the amino terminus of SPCH (Gudesblat *et al.*, 2012). Elimination of these amino-terminal BIN2-target sites resulted in a increased number of epidermal cells but not mature stomata, suggesting that BIN2-dependent phosphorylation could regulate early stomatal lineage proliferation (Gudesblat *et al.*, 2012). Together, these results point to SPCH as an integration point between signaling pathways. By controlling the phosphorylation status of SPCH, thereby regulating activity of the key factor in creating the stomata lineage, BR and MAPKs can fine-tune development to coordinate epidermal cell division and stomata formation in line with systematic demands.

Potential Roles for MAPK Signaling in Late Stomatal Development and Potential of Stomatal Development in Understanding Signaling Specificity

Identification of SPCH as a downstream target of MAPKs and BIN2 provides some cell-type specificity to the signals. This however leaves the question open of whether MAPKs and BR signals influence later stages and, if so, what the downstream targets would be. Also, since MAPK signaling networks feature in a broad range of biologic processes from cell division to regulating developmental pathways and responses to abiotic and biotic stresses, modulating activities of components in the YDA MAPK module often results in pleiotropic effects. For example, *yda* mutant embryos are malformed and plants are dwarfed and sterile (Lukowitz *et al.*, 2004), and *mpk3;mpk6* mutants are embryo lethal (Wang *et al.*, 2007).

To define the roles of MAPK signaling at later stages, cell-type specific manipulation of the pathway was therefore needed. As a test case, the expression of constitutively active MAPKKs was driven with stomatal-stage-specific promoters (Lampard *et al.*, 2009). Briefly, the SPCH promoter drives expression in cells undergoing asymmetric cell divisions to form meristemoids (MacAlister *et al.*, 2007). The MUTE promoter drives expression during the transition from meristemoids to GMCs (MacAlister *et al.*, 2007; Pillitteri *et al.*, 2007) and the FAMA promoter drives expression during the transition from GMC to GC (Ohashi-Ito and Bergmann, 2006). This strategy limits activities of MAPK components to a particular stomatal development stage, thus reducing phenotypes associated with other developmental events.

As expected, when expressed in the earliest stage of stomatal lineage, YDA, MKK4/5, and MAKK7/9 are negative regulator progression because their CA (constitutively activated) versions suppress stomata formation, importantly, without introducing phenotypes outside of the epidermis (Lampard *et al.*, 2009). Surprisingly however, when expressed with FAMApro, CA-YDA and CA-MKK7/9 do not halt stomatal production but rather induce the formation of excess stomata. FAMA-promoter-driven expression of CA-MKK4 and CA-MKK5 had no effect at this stage, however (Lampard *et al.*, 2009). The protein sequences of MKK4/5 and MKK7/9 pairs are quite similar, so why do MKK7/9 but not MKK4/5 influence stomata formation at the final stage? Analysis of the small differences in these proteins' sequences, stability, subcellular localization, and quantitative ability to phosphorylate targets may reveal keys to specificity of MAPK signaling with wide-ranging implications for preventing erroneous signaling cross-talk.

Modulation of Stomatal Development by Environmental Stimuli

Plants have developed sophisticated ways to sense signals from the outside and to alter their growth in response to environmental challenges. Modulating stomatal behavior (opening and closing) and stomatal development are some of the ways plants react to environmental stresses. Although there have been extensive studies on environmental control of mature stomatal behavior (Kim *et al.*, 2010), little is known about how environmental cues regulate stomatal production. Recent pivotal studies provide a hint about how environmental factors such as light, CO_2 levels, drought stress, and low humidity affect stomatal development.

Effects of Light Quality and Quantity on Stomatal Development

Stomatal density increases upon elevation of light intensity. Light quality also matters, as a high red to far-red ratio also promotes stomata formation although this has little effect on patterning (Boccalandro *et al.*, 2009; Casson *et al.*, 2009). phyB, the red-light-absorbing phytochrome, is the dominant mediator of red-light-induced stomata production; *phyB* mutants have a low stomatal density and low stomatal index (SI, the ratio of stomata to total epidermal cell number) even when exposed to high-intensity red light (Boccalandro *et al.*, 2009; Casson *et al.*, 2009). Genetically downstream of *phyB*, *PHYTOCHROME-INTERACTING FACTOR4 (PIF4)*, a bHLH transcription factor gene, is needed to increase stomatal numbers and SI in response to high-intensity red-light conditions. It is attractive to consider a model in which PIF4 interacts with stomatal bHLH proteins such as SPCH, MUTE, FAMA, or SCRM1/2 to regulate stomata production and that the PIF4–bHLH interaction is disrupted in the presence of phyB (Casson *et al.*, 2009).

Stomatal development is also inhibited in mutants of the blue-light receptors, cryptochrome 1 and 2 (*cry1*;*cry2*), and the far-red-light receptor *phyA* (Kang *et al.*, 2009). The downstream signaling of these receptors converges on the repression of the E3 ubiquitin ligase CONSTITUTIVE PHOTOMORPHOGENIC 1 (COP1) (Kang *et al.*, 2009; reviewed by Lau and Deng, 2012). In turn, COP1 is genetically upstream of *YDA* as the stomatal clusters in *COP1* mutants can be suppressed by *CA-YDA* (Kang *et al.*, 2009). Further biochemical experiments are needed to demonstrate how COP1 suppresses stomata formation through the YDA pathway; one might speculate that COP1 degrades a stomatal-inhibiting factor (such as EPF1/2) upstream of *YDA*, or that COP1 acts on YDA itself. Mutations in ubiquitin-conjugating enzyme (E2) variant protein, CONSTITUTIVE PHOTOMORPHOGENIC 10 (COP10), also result in increased number of stomatal clusters in cotyledons and hypocotyls (Delgado *et al.*, 2012a, 2012b). A detailed time-course study of *cop10* mutants suggests that COP10 regulates the frequency of entry and amplifying divisions during stomatal development and suppresses premature acquisition of stomatal cell fate (Delgado *et al.*, 2012b).

Molecular and Developmental Adaptations to Drought Stress

To balance between CO_2 uptake and water loss during gas exchange, plants regulate stomatal activities and numbers in response to environmental changes (Gray *et al.*, 2000; Hashimoto *et al.*, 2006). Water-use efficiency (WUE) is a term that encapsulates this balance; plants have high WUE if they lose little water per amount of carbon fixed in photosynthesis. In drought conditions, water loss is critical and the immediate response is to reduce stomatal pore size. A key regulator in this process is the R2R3-type MYB protein, MYB60 (Cominelli *et al.*, 2005). MYB60 is specifically expressed in guard cells and its expression is negatively regulated under drought conditions. The *myb60* null mutant has constitutive reduction in stomatal opening and increased drought tolerance. Some stress-response genes are identified as targets of MYB60, suggesting that MYB60 links physiological responses in GCs to environmental stimuli (Cominelli *et al.*, 2005). Overexpression lines of two MYB60 variants are also more susceptible to water stress than WT (Oh *et al.*, 2011).

A second long-term response to drought is to decrease stomatal formation. Loss-of-function mutations in the *GT-2 LIKE1* (*GTL1*) transcription factor increase tolerance to water deficiency and enhance WUE without sacrificing biomass (Yoo *et al.*, 2010). *gtl1* mutants have lower transpiration rates and an approximately 25% reduction in abaxial stomata numbers (Yoo *et al.*, 2010). GTL1 is expressed in abaxial epidermal nuclei and its expression is suppressed in water-deficit conditions. A molecular model for GTL action was uncovered when it was found that *gtl1* mutants had increased SDD1 expression and chromatin immunoprecipitation studies showed that GTL1 bound to a GT3 box (GGTAAA) in the *SDD1* promoter. Taken together with the role of *SDD1* as a negative regulator of stomatal production, the model posits that dehydration inhibits GLT1 expression thus releasing GTL1 repression of the *SDD1* promoter. Elevated *SDD1* subsequently inhibits stomata formation and reduces transpiration (Yoo *et al.*, 2010).

Physiological and Developmental Responses to Altering CO_2 Concentrations

An inverse correlation between stomatal density and/or stomatal index and atmospheric CO_2 was identified in many species over geological timescales and under laboratory conditions (Woodward, 1987). As with drought response, elevated CO_2 results in changes in stomatal opening and stomatal density, and the former has received more attention (reviewed by Kim *et al.*, 2010). There are two groups of proteins identified as potential mediators of the developmental response to CO_2, however.

Plants doubly mutant for genes encoding CO_2-binding proteins, β-carbonic anhydrase 1 (β*CA1*) and β*CA4*, have impaired CO_2-mediated stomatal movement and elevated stomata numbers (Hu *et al.*, 2010). These enzymes are expressed in both mesophyll cells and GCs; however, GC-specific expression of β*CA1* and β*CA4* is sufficient to restore the CO_2 responsiveness (a cell-autonomous role) of stomata closure and to create plants with normal stomatal density (Hu *et al.*, 2010). Since the guard cells are the final differentiated cell type in stomata lineage, the signals produced in guard cells by βCA1 and βCA4 must affect the initial cells in other stomata lineages though non-cell-autonomous mechanisms.

Another gene hinting at non-autonomous regulation of response to CO_2 is *HIGH CARBON DIOXIDE* (*HIC*), whose mutants exhibit increased stomatal density under elevated CO_2 conditions, suggesting that HIC is a condition-sensitive negative regulator of stomatal development (Gray *et al.*, 2000). The *HIC* gene encodes a putative 3-ketoacyl-CoA synthase, an enzyme involved in the synthesis of very-long-chain fatty acids (VLCFAs) in cuticular wax. Interestingly, the expression of HIC is restricted to guard cells; as with the β-carbonic anhydrases, activity at the end of the stomatal development pathway affects stomatal initiation elsewhere. How alterations in VLCFAs could affect stomatal development is unknown. It is possible that differential permeability of the GC cuticle to moisture would send a signal to inhibit new stomata, or that VLCFAs are biochemical precursors to lipid-based signaling molecules.

Potential for Epigenetic Reprogramming by Low Humidity

So far, we have presented evidence that environmental cues impinge on developmental processes through transcriptional regulation (GTL1 repression of SDD1) or through post-translational modification (BIN2 phosphorylation of SPCH and MAPK components). In addition to these two types of modulation, it has been reported that low humidity regulates stomatal development epigenetically (Tricker *et al.*, 2012). Arabidopsis grown in low-humidity conditions has a reduced SI. This is correlated with *de novo* DNA methylation on *SPCH* and *FAMA*. Consistent with this methylation status, both *SPCH* and *FAMA* become transcriptionally repressed. The low-humidity-induced responses are abolished in mutants involved in *de novo* DNA methylation, in GC methylation maintenance, and in mutants involved in siRNA (short-interfering non-coding RNA) production in the RNA-directed DNA methylation pathway (Tricker *et al.*, 2012). These results suggest that these two stomatal genes are negatively regulated both transcriptionally and post-transcriptionally in response to low humidity stress.

Conclusion

The creation of stomata in species, organ, and environmentally appropriate numbers and patterns arises from a distinctive cell lineage informed by signals from cells within the lineage, from other tissues, and from the environment. In this chapter we have presented the core genes and genetic pathways (Table 12.1) that specify which cells in the epidermis can enter the lineage and how transitions through this lineage are regulated. It is striking how a few key transcription factors expressed in a limited number of cells are subject to multiple levels of regulation from transcriptional to post-translational and how this is mediated by pathways that report both developmental and environmental information. In the future, we expect to see large-scale approaches applied to defining the entire transcriptional networks defining cell types and transitions between stages. Moreover, understanding how those core networks are modified under different conditions will provide intriguing basic biological insights as well as the potential to engineer plants with better WUE or the ability to withstand future climate pressures.

Table 12.1 Genes involved in stomatal development

Symbol	AGI code	Full name	Gene description	References
Transcriptional regulation and fate determination				
SPCH	AT5G53210	*SPEECHLESS*	bHLH protein	MacAlister *et al.*, 2007; Pillitteri *et al.*, 2007
MUTE	AT3G06120	*MUTE*	bHLH protein	MacAlister *et al.*, 2007; Pillitteri *et al.*, 2007
FAMA	AT3G24140	*FAMA*	bHLH protein	Ohashi-Ito and Bergmann, 2006
SCRM	AT3G26744	*SCRM/ICE1*	bHLH protein	Kanaoka *et al.*, 2008
SCRM2	AT1G12860	*SCRM2*	bHLH protein	Kanaoka *et al.*, 2008
FLP	AT1G14350	*FOUR LIPS*	R2R3 MYB protein	Lai *et al.*, 2005; Xie *et al.*, 2010
MYB88	AT2G02820	*MYB88*	R2R3 MYB protein	Lai *et al.*, 2005; Xie *et al.*, 2010
CDKB1;1	AT3G54180	*CYCLIN DEPENDENT KINASE B1;1*	Cyclin-dependent kinase	Boudolf *et al.*, 2004; Xie *et al.*, 2010
CYCA2;1	AT5G25380	*CYCLIN A2;1*	Cyclin	Vanneste *et al.*, 2011
CYCA2;2	AT5G11300	*CYCLIN A2;2*	Cyclin	Vanneste *et al.*, 2011
CYCA2;3	AT1G15570	*CYCLIN A2;3*	Cyclin	Vanneste *et al.*, 2011
CYCA2;4	AT1G80370	*CYCLIN A2;4*	Cyclin	Vanneste *et al.*, 2011
AGL16	AT3G57230	*AGAMOUS-LIKE 16*	MADS-box protein	Alvarez-Buylla *et al.*, 2000; Kutter *et al.*, 2007
miR824	AT4G24415	*MicroRNA 824*	MicroRNA	Kutter *et al.*, 2007
SACP1	AT5G65590	*STOMATAL CARPENTER 1*	Dof protein	Negi *et al.*, 2013
Spacing and patterning				
ER	AT2G26330	*ERECTA*	Leucine-rich repeat receptor kinase	Shpak *et al.*, 2005
ERL1	AT5G62230	*ERECTA-Like 1*	Leucine-rich repeat receptor kinase	Shpak *et al.*, 2005
ERL2	AT5G07180	*ERECTA-Like2*	Leucine-rich repeat receptor kinase	Shpak *et al.*, 2005
TMM	AT1G80080	*TOO MANY MOUTHS*	Leucine-rich repeat receptor like protein	Yang and Sack, 1995; Geisler *et al.*, 2000; Nadeau and Sack, 2002; Bhave *et al.*, 2009

(*continued overleaf*)

Table 12.1 (*continued*)

Symbol	AGI code	Full name	Gene description	References
SDD1	AT1G04110	*STOMATAL DENSITY AND DISTRIBUTION-1*	Subtilisin-like protease	Berger and Altmann, 2000; Von Groll *et al.*, 2002
EPF1	AT2G20875	*EPIDERMAL PATTERNING FACTOR1*	Cystein-rich secreted peptide	Hara *et al.*, 2007
EPF2	AT1G34245	*EPIDERMAL PATTERNING FACTOR2*	Cystein-rich secreted peptide	Hara *et al.*, 2009; Hunt and Gray, 2009
STOMAGEN	AT4G12970	*STOMAGEN/EPF-LIKE9*	Cystein-rich secreted peptide	Kondo *et al.*, 2010; Sugano *et al.*, 2010
CHAL	AT2G30370	*CHALLAH/EPF-LIKE6*	Cystein-rich secreted peptide	Abrash and Bergmann, 2010
CLL1	AT3G22820	*CHALLAH-LIKE1/EPF-LIKE5*	Cystein-rich secreted peptide	Abrash *et al.*, 2011; Niwa *et al.*, 2013
CLL2	AT4G14723	*CHALLAH-LIKE2/EPF-LIKE4*	Cystein-rich secreted peptide	Abrash *et al.*, 2011
YDA	AT1G63700	*YODA*	MAPKKK	Bergmann *et al.*, 2004; Lampard *et al.*, 2008
MKK4	AT1G51660	*MAP KINASE KINASE 4*	MAPKK	Wang *et al.*, 2007; Lampard *et al.*, 2009
MKK5	AT3G21220	*MAP KINASE KINASE 5*	MAPKK	Wang *et al.*, 2007; Lampard *et al.*, 2009
MKK7	AT1G18350	*MAP KINASE KINASE 7*	MAPKK	Lampard *et al.*, 2009
MKK9	AT1G73500	*MAP KINASE KINASE 9*	MAPKK	Lampard *et al.*, 2009
MPK3	AT3G45640	*MAP KINASE 3*	MAPK	Wang *et al.*, 2007
MPK6	AT2G43790	*MAP KINASE 6*	MAPK	Wang *et al.*, 2007
AP2C3	AT2G40180	*Arabidopsis protein phosphatase 2C*	Phosphatase	Umbrasaite *et al.*, 2010

Hormonal and environmental signaling				
BIN2	AT4G18710	*BRASSINOSTEROID INSENSITIVE 2*	GSK3-like Kinase	Kim *et al.*, 2012; Khan *et al.*, 2013; Gudesblat *et al.*, 2012
CRY1	AT4G08920	*CRYPTOCHROME 1*	Blue light receptor	Kang *et al.*, 2009
CRY2	AT1G04400	*CRYPTOCHROME 2*	Blue light receptor	Kang *et al.*, 2009
PHYA	AT1G09570	*PHYTOCHROME A*	Red/Far-red light receptor	Kang *et al.*, 2009
PHYB	AT2G18790	*PHYTOCHROME B*	Red/Far-red light receptor	Boccalandro *et al.*, 2009; Casson *et al.*, 2009; Kang *et al.*, 2009
COP1	AT2G32950	*CONSTITUTIVE PHOTOMORPHOGENIC 1*	Ubiquitin ligase	Kang *et al.*, 2009
COP10	AT3G13550	*CONSTITUTIVE PHOTOMORPHOGENIC 10*	Ubiquitin ligase	Delgado *et al.*, 2012b, a
PIF4	AT2G43010	*PHYTOCHROME-INTERACTING FACTOR 4*	bHLH protein	Casson *et al.*, 2009
GTL1	AT1G33240	*Arabidopsis thaliana GT-2 Like 1*	GT2 protein	Yoo *et al.*, 2010
MYB60	AT1G08810	*MYB60*	MYB protein	Cominelli *et al.*, 2005; Oh *et al.*, 2011
HIC	AT2G46720	*HIGH CARBON DIOXIDE*	3-ketoacyl-CoA synthase	Gray *et al.*, 2000
β*CA1*	AT3G01500	β carbonic anhydrase 1	CO2-binding protein	Hu *et al.*, 2010
β*CA4*	AT1G70410	β carbonic anhydrase 4	CO2-binding protein	Hu *et al.*, 2010

References

Abrash, E.B. and Bergmann, D.C. (2010) Regional specification of stomatal production by the putative ligand CHALLAH. *Development*, **137**, 447–455.

Abrash, E.B., Davies, K.A., and Bergmann, D.C. (2011) Generation of signaling specificity in Arabidopsis by spatially restricted buffering of ligand-receptor interactions. *Plant Cell*, **23**, 2864–2879.

Alvarez-Buylla, E.R., Liljegren, S.J., Pelaz, S., Gold, S.E., Burgeff, C., Ditta, G.S., Vergara-Silva, F., and Yanofsky, M.F. (2000) MADS-box gene evolution beyond flowers, expression in pollen, endosperm, guard cells, roots and trichomes. *Plant Journal*, **24**, 457–466.

Berger, D. and Altmann, T. (2000) A subtilisin-like serine protease involved in the regulation of stomatal density and distribution in Arabidopsis thaliana. *Genes and Development*, **14**, 1119–1131.

Bergmann, D.C. and Sack, F.D. (2007) Stomatal development. *Annual Review of Plant Biology*, **58**, 163–181.

Bergmann, D.C., Lukowitz, W., and Somerville, C.R. (2004) Stomatal development and pattern controlled by a MAPKK kinase. *Science*, **304**, 1494–1497.

Bhave, N.S., Veley, K.M., Nadeau, J.A., Lucas, J.R., Bhave, S.L., and Sack, F.D. (2009) TOO MANY MOUTHS promotes cell fate progression in stomatal development of Arabidopsis stems. *Planta*, **229**, 357–367.

Boccalandro, H.E., Rugnone, M.L., Moreno, J.E., Ploschuk, E.L., Serna, L., Yanovsky, M.J., and Casal, J.J. (2009) Phytochrome B enhances photosynthesis at the expense of water-use efficiency in Arabidopsis. *Plant Physiology*, **150**, 1083–1092.

Boudolf, V., Barrôco, R., Engler, J. de A., Verkest, A., Beeckman, T., Naudts, M., Inzé, D., and De Veylder, L. (2004) B1-type cyclin-dependent kinases are essential for the formation of stomatal complexes in Arabidopsis thaliana. *Plant Cell*, **16**, 945–955.

Canagarajah, B.J., Khokhlatchev, A., Cobb, M.H., and Goldsmith, E.J. (1997) Activation mechanism of the MAP kinase ERK2 by dual phosphorylation. *Cell*, **90**, 859–869.

Casson, S.A., Franklin, K.A., Gray, J.E., Grierson, C.S., Whitelam, G.C., and Hetherington, A.M. (2009) phytochrome B and PIF4 regulate stomatal development in response to light quantity. *Current Biology*, **19**, 229–234.

Chinnusamy, V., Ohta, M., Kanrar, S., Lee, B.-H., Hong, X., Agarwal, M., and Zhu, J.-K. (2003) ICE1, a regulator of cold-induced transcriptome and freezing tolerance in Arabidopsis. *Genes and Development*, **17**, 1043–1054.

Cominelli, E., Galbiati, M., Vavasseur, A., Conti, L., Sala, T., Vuylsteke, M., Leonhardt, N., Dellaporta, S.L., and Tonelli, C. (2005) A guard-cell-specific MYB transcription factor regulates stomatal movements and plant drought tolerance. *Current Biology*, **15**, 1196–1200.

Delgado, D., Ballesteros, I., Mena, M., and Fenoll, C. (2012a) Roles of constitutive photomorphogenic 10 in Arabidopsis stomata development. *Plant Signaling and Behavior*, **7**, 990–993.

Delgado, D., Ballesteros, I., Torres-Contreras, J., Mena, M., and Fenoll, C. (2012b) Dynamic analysis of epidermal cell divisions identifies specific roles for COP10 in Arabidopsis stomatal lineage development. *Planta*, **236**, 447–461.

Facette, M.R. and Smith, L.G. (2012) Division polarity in developing stomata. *Current Opinion in Plant Biology*, **15**, 585–592.

Geisler, M., Nadeau, J., and Sack, F.D. (2000) Oriented asymmetric divisions that generate the stomatal spacing pattern in arabidopsis are disrupted by the too many mouths mutation. *Plant Cell*, **12**, 2075–2086.

Gray, J.E., Holroyd, G.H., van der Lee, F.M., Bahrami, A.R., Sijmons, P.C., Woodward, F.I., Schuch, W., and Hetherington, A.M. (2000) The HIC signalling pathway links CO2 perception to stomatal development. *Nature*, **408**, 713–716.

Groll, Von, U., Berger, D., and Altmann, T. (2002) The subtilisin-like serine protease SDD1 mediates cell-to-cell signaling during Arabidopsis stomatal development. *Plant Cell*, **14**, 1527–1539.

Gudesblat, G.E., Schneider-Pizoń, J., Betti, C., Mayerhofer, J., Vanhoutte, I., van Dongen, W., Boeren, S., Zhiponova, M., de Vries, S., Jonak, C., and Russinova, E. (2012) SPEECHLESS integrates brassinosteroid and stomata signalling pathways. *Nature Cell Biology*, **14**, 548–554.

Hara, K., Kajita, R., Torii, K.U., Bergmann, D.C., and Kakimoto, T. (2007) The secretory peptide gene EPF1 enforces the stomatal one-cell-spacing rule. *Genes and Development*, **21**, 1720–1725.

Hara, K., Yokoo, T., Kajita, R., Onishi, T., Yahata, S., Peterson, K.M., Torii, K.U., and Kakimoto, T. (2009) Epidermal cell density is autoregulated via a secretory peptide, EPIDERMAL PATTERNING FACTOR 2 in Arabidopsis leaves. *Plant and Cell Physiology*, **50**, 1019–1031.

Hashimoto, M., Negi, J., Young, J., Israelsson, M., Schroeder, J.I., and Iba, K. (2006) Arabidopsis HT1 kinase controls stomatal movements in response to CO2. *Nature Cell Biology*, **8**, 391–397.

Hu, H., Boisson-Dernier, A., Israelsson-Nordström, M., Böhmer, M., Xue, S., Ries, A., Godoski, J., Kuhn, J.M., and Schroeder, J.I. (2010) Carbonic anhydrases are upstream regulators of CO2-controlled stomatal movements in guard cells. *Nature Cell Biology*, **12**, 87–93.

Hunt, L. and Gray, J.E. (2009) The signaling peptide EPF2 controls asymmetric cell divisions during stomatal development. *Current Biology*, **19**, 864–869.

Jeong, S., Trotochaud, A.E., and Clark, S.E. (1999) The Arabidopsis CLAVATA2 gene encodes a receptor-like protein required for the stability of the CLAVATA1 receptor-like kinase. *Plant Cell*, **11**, 1925–1934.

Kanaoka, M.M., Pillitteri, L.J., Fujii, H., Yoshida, Y., Bogenschutz, N.L., Takabayashi, J., Zhu, J.-K., and Torii, K.U. (2008) SCREAM/ICE1 and SCREAM2 specify three cell-state transitional steps leading to arabidopsis stomatal differentiation. *Plant Cell*, **20**, 1775–1785.

Kang, C.-Y., Lian, H.-L., Wang, F.-F., Huang, J.-R., and Yang, H.-Q. (2009) Cryptochromes, phytochromes, and COP1 regulate light-controlled stomatal development in Arabidopsis. *Plant Cell*, **21**, 2624–2641.

Khan, M., Rozhon, W., Bigeard, J., Pflieger, D., Husar, S., Pitzschke, A., Teige, M., Jonak, C., Hirt, H., and Poppenberger, B. (2013) Brassinosteroid-regulated GSK3/Shaggy-like kinases phosphorylate mitogen-activated protein (MAP) kinase kinases, which control stomata development in Arabidopsis thaliana. *Journal of Biological Chemistry*, **288**, 7519–7527.

Kim, T.-H., Böhmer, M., Hu, H., Nishimura, N., and Schroeder, J.I. (2010) Guard cell signal transduction network, advances in understanding abscisic acid, CO2, and Ca2+ signaling. *Annual Review of Plant Biology*, **61**, 561–591.

Kim, T.-W. and Wang Z.Y. (2010) Brassinosteroid signal transduction from receptor kinases to transcription factors. *Annual Review of Plant Biology*, **61**, 681–704.

Kim, T.-W., Michniewicz, M., Bergmann, D.C., and Wang, Z.-Y. (2012) Brassinosteroid regulates stomatal development by GSK3-mediated inhibition of a MAPK pathway. *Nature*, **482**, 419–422.

Kondo, T., Kajita, R., Miyazaki, A., Hokoyama, M., Nakamura-Miura, T., Mizuno, S., Masuda, Y., Irie, K., Tanaka, Y., Takada, S., Kakimoto, T., and Sakagami, Y. (2010) Stomatal density is controlled by a mesophyll-derived signaling molecule. *Plant and Cell Physiology*, **51**, 1–8.

Kutter, C., Schöb, H., Stadler, M., Meins, F., and Si-Ammour, A. (2007) MicroRNA-mediated regulation of stomatal development in Arabidopsis. *Plant Cell*, **19**, 2417–2429.

Lai, L.B., Nadeau, J.A., Lucas, J., Lee, E.-K., Nakagawa, T., Zhao, L., Geisler, M., and Sack, F.D. (2005) The Arabidopsis R2R3 MYB proteins FOUR LIPS and MYB88 restrict divisions late in the stomatal cell lineage. *Plant Cell*, **17**, 2754–2767.

Lampard, G.R., MacAlister, C.A., and Bergmann, D.C. (2008) Arabidopsis stomatal initiation is controlled by MAPK-mediated regulation of the bHLH SPEECHLESS. *Science*, **322**, 1113–1116.

Lampard, G.R., Lukowitz, W., Ellis, B.E., and Bergmann, D.C. (2009) Novel and expanded roles for MAPK signaling in Arabidopsis stomatal cell fate revealed by cell type-specific manipulations. *Plant Cell*, **21**, 3506–3517.

Lau, O.S. and Deng, X.W. (2012) The photomorphogenic repressors COP1 and DET1, 20 years later. *Trends in Plant Science*, **17**, 584–593.

Lee, J.S., Kuroha, T., Hnilova, M., Khatayevich, D., Kanaoka, M.M., McAbee, J.M., Sarikaya, M., Tamerler, C., and Torii, K.U. (2012) Direct interaction of ligand-receptor pairs specifying stomatal patterning. *Genes and Development*, **26**, 126–136.

Liu, T., Ohashi-Ito, K., and Bergmann, D.C. (2009) Orthologs of Arabidopsis thaliana stomatal bHLH genes and regulation of stomatal development in grasses. *Development*, **136**, 2265–2276.

Lukowitz, W., Roeder, A., Parmenter, D., and Somerville, C. (2004) A MAPKK kinase gene regulates extra-embryonic cell fate in Arabidopsis. *Cell*, **116**, 109–119.

MacAlister, C.A., Ohashi-Ito, K., and Bergmann, D.C. (2007) Transcription factor control of asymmetric cell divisions that establish the stomatal lineage. *Nature*, **445**, 537–540.

Meng, X., Wang, H., He, Y., Liu, Y., Walker, J.C., Torii, K.U., and Zhang, S. (2012) A MAPK cascade downstream of ERECTA receptor-like protein kinase regulates Arabidopsis inflorescence architecture by promoting localized cell proliferation. *Plant Cell*, **24**, 4948–4960.

Nadeau, J.A. and Sack, F.D. (2002) Control of stomatal distribution on the Arabidopsis leaf surface. *Science*, **296**, 1697–1700.

Negi, J., Moriwaki, K., Konishi, M., Yokoyama, R., Nakano, T., Kusumi, K., Hashimoto-Sugimoto, M., Schroeder, J.I., Nishitani, K., Yanagisawa, S., and Iba, K. (2013) A Dof transcription factor, SCAP1, is essential for the development of functional stomata in Arabidopsis. *Current Biology*, **23**, 479–484.

Niwa, T., Kondo, T., Nishizawa, M., Kajita, R., Kakimoto, T., and Ishiguro, S. (2013) EPIDERMAL PATTERNING FACTOR LIKE5 peptide represses stomatal development by inhibiting meristemoid maintenance in Arabidopsis thaliana. *Bioscience, Biotechnology, and Biochemistry*, **77**, 1287–1295.

Oh, J.E., Kwon, Y., Kim, J.H., Noh, H., Hong, S.-W., and Lee, H. (2011) A dual role for MYB60 in stomatal regulation and root growth of Arabidopsis thaliana under drought stress. *Plant Molecular Biology*, **77**, 91–103.

Ohashi-Ito, K. and Bergmann, D.C. (2006) Arabidopsis FAMA controls the final proliferation/differentiation switch during stomatal development. *Plant Cell*, **18**, 2493–2505.

Ohki, S., Takeuchi, M., and Mori, M. (2011) The NMR structure of stomagen reveals the basis of stomatal density regulation by plant peptide hormones. *Nature Communications*, **2**, 512.

Payne, C.T., Zhang, F., and Lloyd, A.M. (2000) GL3 encodes a bHLH protein that regulates trichome development in arabidopsis through interaction with GL1 and TTG1. *Genetics*, **156**, 1349–1362.

Pillitteri, L.J., Sloan, D.B., Bogenschutz, N.L., and Torii, K.U. (2007) Termination of asymmetric cell division and differentiation of stomata. *Nature*, **445**, 501–505.

Robinson, S., Barbier de Reuille, P., Chan, J., Bergmann, D., Prusinkiewicz, P., and Coen, E. (2011) Generation of spatial patterns through cell polarity switching. *Science*, **333**, 1436–1440.

Rodriguez, M.C.S., Petersen, M., and Mundy, J. (2010) Mitogen-activated protein kinase signaling in plants. *Annual Review of Plant Biology*, **61**, 621–649.

Rudall, P.J. and Knowles, E.V.W. (2013) Ultrastructure of stomatal development in early-divergent angiosperms reveals contrasting patterning and pre-patterning. *Annals of Botany*, **112**(6), 1031–1043.

Shpak, E.D., Lakeman, M.B., and Torii, K.U. (2003) Dominant-negative receptor uncovers redundancy in the Arabidopsis ERECTA Leucine-rich repeat receptor-like kinase signaling pathway that regulates organ shape. *Plant Cell*, **15**, 1095–1110.

Shpak, E.D., Berthiaume, C.T., Hill, E.J., and Torii, K.U. (2004) Synergistic interaction of three ERECTA-family receptor-like kinases controls Arabidopsis organ growth and flower development by promoting cell proliferation. *Development*, **131**, 1491–1501.

Shpak, E.D., McAbee, J.M., Pillitteri, L.J., and Torii, K.U. (2005) Stomatal patterning and differentiation by synergistic interactions of receptor kinases. *Science*, **309**, 290–293.

Sugano, S.S., Shimada, T., Imai, Y., Okawa, K., Tamai, A., Mori, M., and Hara-Nishimura, I. (2010) Stomagen positively regulates stomatal density in Arabidopsis. *Nature*, **463**, 241–244.

Torii, K.U., Mitsukawa, N., Oosumi, T., Matsuura, Y., Yokoyama, R., Whittier, R.F., and Komeda, Y. (1996) The Arabidopsis ERECTA gene encodes a putative receptor protein kinase with extracellular leucine-rich repeats. *Plant Cell*, **8**, 735–746.

Tricker, P.J., Gibbings, J.G., Rodríguez López, C.M., Hadley, P., and Wilkinson, M.J. (2012) Low relative humidity triggers RNA-directed de novo DNA methylation and suppression of genes controlling stomatal development. *Journal of Experimental Botany*, **63**, 3799–3813.

Uchida N., Lee J.S., Horst R.J., Lai H.H., Kajita R., Kakimoto T., Tasaka M., and Torii, K.U. (2012) Regulation of inflorescence architecture by intertissue layer ligand-receptor communication between endodermis and phloem. *Proceedings of the National Academy of Science, USA*, **109**, 6337–6342.

Umbrasaite, J., Schweighofer, A., Kazanaviciute, V., Magyar, Z., Ayatollahi, Z., Unterwurzacher, V., Choopayak, C., Boniecka, J., Murray, J.A.H., Bogre, L., and Meskiene, I. (2010) MAPK phosphatase AP2C3 induces ectopic proliferation of epidermal cells leading to stomata development in Arabidopsis. *PLoS ONE*, **5**, e15357.

Vanneste, S., Coppens, F., Lee, E., Donner, T.J., Xie, Z., Van Isterdael, G., Dhondt, S., De Winter, F., De Rybel, B., Vuylsteke, M., De Veylder, L., Friml, J., Inzé, D., Grotewold, E., Scarpella, E., Sack, F., Beemster, G.T.S., and Beeckman, T. (2011) Developmental regulation of CYCA2s contributes to tissue-specific proliferation in Arabidopsis. *EMBO Journal*, **30**, 3430–3441.

Vatén, A. and Bergmann, D.C. (2012) Mechanisms of stomatal development, an evolutionary view. *Evodevo*, **3**, 11.

Wang, H., Ngwenyama, N., Liu, Y., Walker, J.C., and Zhang, S. (2007) Stomatal development and patterning are regulated by environmentally responsive mitogen-activated protein kinases in Arabidopsis. *Plant Cell*, **19**, 63–73.

Wang, Z.-Y., Bai, M.-Y., Oh, E., and Zhu, J.-Y. (2012) Brassinosteroid signaling network and regulation of photomorphogenesis. *Annual Review of Genetics*, **46**, 701–724.

Woodward, F.I. (1987) Stomatal numbers are sensitive to increases in CO2 from pre-industrial levels. *Nature*, **327**, 617–618.

Xie, Z., Lee, E., Lucas, J.R., Morohashi, K., Li, D., Murray, J.A.H., Sack, F.D., and Grotewold, E. (2010) Regulation of cell proliferation in the stomatal lineage by the Arabidopsis MYB FOUR LIPS via direct targeting of core cell cycle genes. *Plant Cell*, **22**, 2306–2321.

Yang, M. and Sack, F.D. (1995) The too many mouths and four lips mutations affect stomatal production in Arabidopsis. *Plant Cell*, **7**, 2227–2239.

Yoo, C.Y., Pence, H.E., Jin, J.B., Miura, K., Gosney, M.J., Hasegawa, P.M., and Mickelbart, M.V. (2010) The Arabidopsis GTL1 transcription factor regulates water use efficiency and drought tolerance by modulating stomatal density via transrepression of SDD1. *Plant Cell*, **22**, 4128–4141.

13 Transcriptional Regulation of Biosynthesis of Cell Wall Components during Xylem Differentiation

Ruiqin Zhong and Zheng-Hua Ye

Department of Plant Biology, University of Georgia, Athens, GA

Introduction

Xylem is a fluid-conducting tissue that is responsible for the transport of water and minerals from roots to other parts of the plant body. The conducting cells in the xylem are tracheary elements, which are further divided into two types: tracheids and vessels. Tracheids were the first type of tracheary elements to evolve, which are found in seedless vascular plants and gymnosperms. Vessels evolved later than tracheids and are found in angiosperms as well as in the division of Gnetophyte that belongs to gymnosperms (Raven *et al.*, 1999). The evolution of xylem together with phloem as fluid-conducting cells solved the problem of long-distance fluid transport, which is considered to be a pivotal step for vascular plants to appear in terrestrial habitats during the Silurian period, around 430 million years ago (Raven *et al.*, 1999). The prominent feature of xylem tracheary elements is the presence of thickened secondary walls in addition to the primary walls (Fig. 13.1). The deposition of secondary walls reinforces the walls of tracheary elements, which not only confers tracheary elements with the ability to withstand the negative pressure generated during transpiration but also makes them strong mechanical tissues to support the whole plant body. The evolution of the ability for first vascular plants to deposit secondary walls in xylem tracheary elements was therefore crucial for their conquest of land.

Secondary walls are mainly composed of three types of polymers, including cellulose, hemicelluloses (xylan and glucomannan), and lignin, the biosynthesis of which requires multiple enzymes (Mellerowicz *et al.*, 2001). How the first vascular plants evolved to acquire these secondary wall biosynthetic genes to make secondary walls is an important issue in plant biology. Genomic analyses have revealed that the moss *Physcomitrella patens*, a non-vascular plant, possesses homologs of the biosynthetic genes for cellulose, xylan, and lignin although the moss plants do not make secondary walls (Roberts

Plant Cell Wall Patterning and Cell Shape, First Edition. Edited by Hiroo Fukuda.

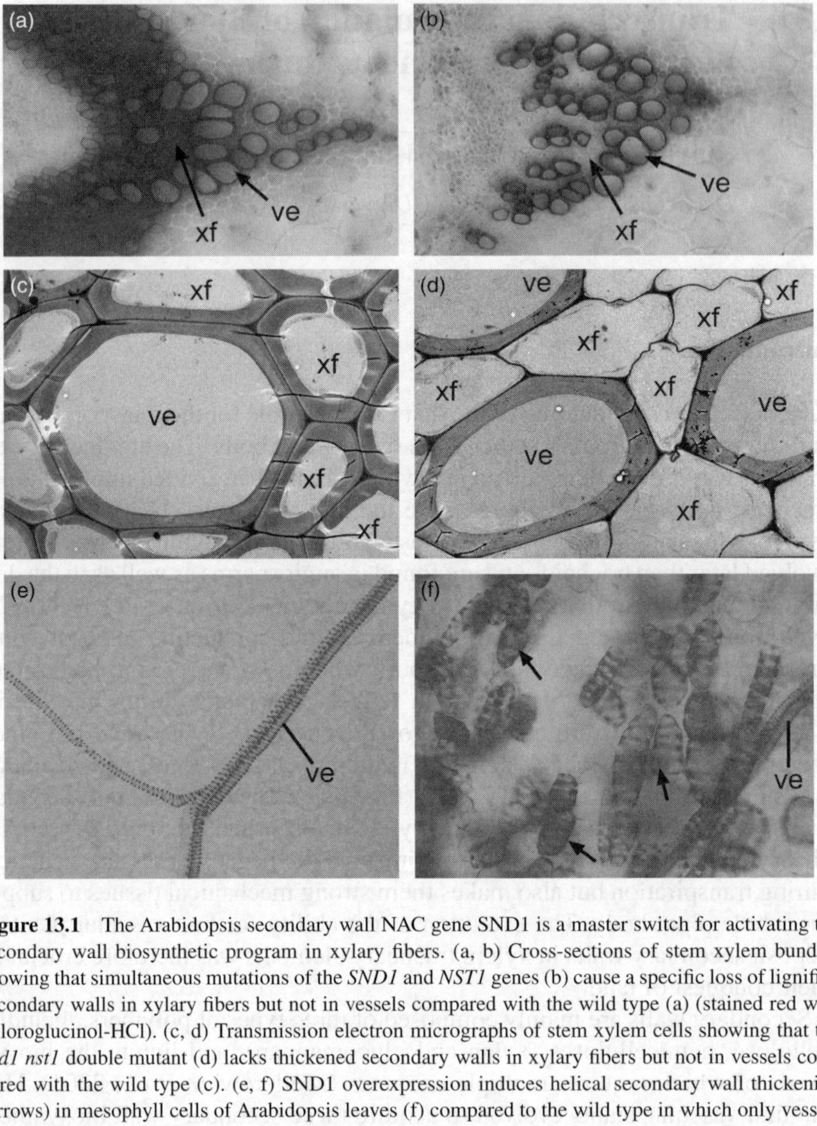

Figure 13.1 The Arabidopsis secondary wall NAC gene SND1 is a master switch for activating the secondary wall biosynthetic program in xylary fibers. (a, b) Cross-sections of stem xylem bundles showing that simultaneous mutations of the *SND1* and *NST1* genes (b) cause a specific loss of lignified secondary walls in xylary fibers but not in vessels compared with the wild type (a) (stained red with phloroglucinol-HCl). (c, d) Transmission electron micrographs of stem xylem cells showing that the *snd1 nst1* double mutant (d) lacks thickened secondary walls in xylary fibers but not in vessels compared with the wild type (c). (e, f) SND1 overexpression induces helical secondary wall thickening (arrows) in mesophyll cells of Arabidopsis leaves (f) compared to the wild type in which only vessels in leaf veins have helical secondary wall thickening (e). (See plate section for the color version of this figure.)

and Bushoven, 2007; Rensing *et al.*, 2008; Xu *et al.*, 2009). One plausible hypothesis is that instead of constructing new secondary wall biosynthetic genes *de novo*, the first-evolved vascular plants might have co-opted these pre-existing ancestral cell wall genes, via gene duplication and diversification, to carry out new functions (e.g., biosynthesis of secondary walls for tracheary elements). This hypothesis is further substantiated by recent findings that the moss *Physcomitrella patens* is able to produce 1,4-β-linked xylan, the same kind of polymer found in secondary walls of vascular plants (Kulkarni *et al.*, 2012). Another important question is how the first-evolved vascular plants recruited these cell wall biosynthetic genes to be coordinately expressed in cells destined to become tracheary elements. The coordinated expression of secondary wall biosynthetic genes is crucial for the proper synthesis, secretion, and assembly of secondary wall components.

In the past decade, molecular and genetic analyses of xylem development have uncovered master transcriptional regulators that control the expression of secondary wall biosynthetic genes temporarily and spatially, which provides a glimpse into how vascular plants evolved to coordinately regulate secondary wall biosynthetic genes for the construction of tracheary elements. A number of previous review articles have summarized the important players involved in the transcriptional regulation of secondary wall biosynthesis (Demura and Fukuda, 2007; Zhong and Ye, 2007, 2009; Caño-Delgado *et al.*, 2010; Demura and Ye, 2010; Ohashi-Ito and Fukuda, 2010; Zhong *et al.*, 2010a; Wang and Dixon, 2012; Zhang *et al.*, 2011; Schuetz *et al.*, 2013). In this chapter, we focus on the molecular mechanism underlying the transcriptional control of the biosynthesis of cell wall components during xylem differentiation.

Xylem Cell Wall Composition and Biosynthesis

To understand how the biosynthesis of cell wall components is transcriptionally regulated during xylem differentiation, it is important to first comprehend briefly what components xylem cell walls are made of and what genes are responsible for the biosynthesis of these wall components. Xylems from seedless vascular plants and gymnosperms primarily consist of tracheids, and those from angiosperms contain tracheary elements, fibers, and parenchyma cells (Fig. 13.1a). Although primary cell wall remnant still remains the bulk of components in mature xylems are secondary walls, which are the major constituent of biomass produced by vascular plants. Secondary walls are mainly composed of cellulose, hemicelluloses (xylan and glucomannan), and lignin. Cellulose and hemicelluloses form the main load-bearing network and lignin impregnates to the network to strengthen secondary walls (Plomion *et al.*, 2001).

Cellulose consists of linear chains of β-1,4-linked glucosyl residues, and it is estimated that 36 of these chains are glued together via hydrogen bonding to produce a microfibril with a diameter of about 5 nm (McCann *et al.*, 1990). Cellulose biosynthesis occurs at the plasma membrane, in which the cellulose synthase complexes are visualized via transmission electron microscopy as rosettes. Each rosette is seen to be made of six particles and each particle is thought to be an aggregate of six cellulose synthase subunits (Somerville, 2006). The genes encoding cellulose synthases have been identified and three functionally non-redundant cellulose synthase subunits, which belong to glycosyltransferase family GT2, are required for the normal assembly of the rosette complexes and cellulose biosynthesis (Endler and Persson, 2011). In Arabidopsis, the three cellulose synthase genes involved in secondary wall biosynthesis during xylem differentiation are CesA4, CesA7, and CesA8, and mutation of any of them is sufficient to cause a severe reduction in secondary wall thickening and concomitantly a collapsed xylem phenotype due to the inability of vessels to withstand the negative pressure generated during transpiration (Taylor *et al.*, 2004).

The major hemicelluloses in the xylem are xylan and glucomannan (Timell, 1967). Xylan is made of a linear chain of β-1,4-linked xylosyl residues, some of which may be substituted with α-1,2-linked glucuronic acid, α-1,2-linked methylglucuronic acid, α-1,2/α-1,3-linked arabinose, and acetylated at C-2 and C-3 (Scheller and Ulvskov, 2010). Xylan from dicots typically contains glucuronic acid side chains and is therefore named glucuronoxylan. On the other hand, xylan from gymnosperms and grass often comprises glucuronic acid and arabinose as side chains, and hence is called glucuronoarabinoxylan. In addition to the xylosyl backbone, xylan from gymnosperms and dicots has a unique tetrasaccharide sequence located at the reducing end, which is composed of β-D-Xyl-(1→3)-α-l-Rha-(1→2)-α-D- Gal*p*A-(1→4)-D-Xyl*p* (Shimizu *et al.*, 1976; Johansson and Samuelson, 1977; Andersson *et al.*, 1983; Pena *et al.*, 2007; Lee *et al.*, 2009). The biosynthesis of xylan requires a number of genes encoding glycosyltransferases, acetyltransferases, and methyltransferases. Recent studies in Arabidopsis and other species have revealed that the elongation of the xylan backbone entails two families of glycosyltransferases, GT43 and GT47 (Brown *et al.*, 2007, 2009; Lee *et al.*, 2007a, 2011; Pena *et al.*, 2007; Wu *et al.*, 2009, 2010). In Arabidopsis, two functionally non-redundant groups of GT43 members function cooperatively as xylosyltransferases in catalyzing the xylosyl transfer during xylan backbone synthesis (Lee *et al.*, 2010, 2012c, 2012d). The addition of glucuronic acid side chains in xylan is catalyzed by xylan glucuronyltransferases belonging to GT8 (Mortimer *et al.*, 2010; Lee *et al.*, 2012a; Rennie *et al.*, 2012), and the addition of arabinose side chains is mediated by arabinoxylan arabinosyltransferase belonging to GT61 (Anders *et al.*, 2012). The methylation of glucuronic acid side chains in xylan is carried out by glucuronoxylan methyltransferases belonging to a group of DUF579-domain-containing proteins (Lee *et al.*, 2012b; Urbanowicz *et al.*, 2012) and the acetylation of xylan is mediated by xylan acetyltransferases whose strong candidates are a group of DUF231-domain-containing proteins (Xiong *et al.*, 2013; Yuan *et al.*, 2013). The biosynthesis of the xylan reducing end tetrasaccharide sequence requires glycosyltransferases from families GT47 and GT8 (Zhong *et al.*, 2005; Brown *et al.*, 2007; Lee *et al.*, 2007b; Pena *et al.*, 2007). The biosynthesis of glucomannan, the other major hemicellulose in the xylem, is mediated by glucomannan synthase belonging to the cellulose synthase-like family A (Suzuki *et al.*, 2006; Liepman *et al.*, 2007).

Lignin is a complex aromatic polymer resulting from oxidative polymerization of three monolignols called *p*-coumaryl, coniferyl, and sinapyl alcohols (Boerjan *et al.*, 2003). The biosynthesis of monolignols starts from the common phenylpropanoid pathway leading to the production of hydroxycinnamoyl-CoA esters, which are further routed into the lignin branch pathway to generate monolignols. The ten enzymes catalyzing monolignol biosynthesis are phenylalanine ammonia lyase (PAL), cinnamate 4-hydroxylase (C4H), 4-commarate CoA ligase (4CL), hydroxycinnamoyl CoA:shikimate hydroxycinnamoyl transferase (HCT), *p*-coumaroyl shikimate 3′-hydroxylase (C3′H), caffeoyl CoA *O*-methyltransferase (CCoAOMT), ferulate 5-hydroxylase (F5H), caffeic acid *O*-methyltransferase (COMT), cinnamoyl CoA reductase (CCR), and cinnamyl alcohol dehydrogenase (CAD). The genes encoding monolignol biosynthetic enzymes have all been identified and functionally characterized. Monolignols synthesized in the cytosol are believed to be transported through the plasma membrane via transporters into cell walls, where they are condensed into lignin via oxidative polymerization catalyzed by oxidases, including laccases and peroxidases (Bonawitz and Chapple, 2010).

Coordinated Expression of Secondary Wall Biosynthetic Genes during Xylem Differentiation

Since the biosynthesis of secondary wall components during xylem differentiation requires a suite of enzymes, it is conceivable that genes encoding these enzymes are coordinately expressed during secondary wall biosynthesis. Early transcriptome profiling study of xylem differentiation demonstrated that the expression of genes encoding secondary wall biosynthetic enzymes is temporarily and spatially associated with xylem development (Demura *et al.*, 2002; Oh *et al.*, 2003; Aspeborg *et al.*, 2005; Brown *et al.*, 2005; Ehlting *et al.*, 2005; Kubo *et al.*, 2005; Zhao *et al.*, 2005). One of the best examples is the study of secondary wall biosynthetic genes during xylem development in hybrid aspen (Aspeborg *et al.*, 2005). By transcriptome profiling of cryosectioned-developing xylem and subsequent bioinformatic analysis, Aspeborg *et al.* (2005) identified 25 xylem-specific glycosyltransferases belong to families GT2, GT8, GT14, GT31, GT43, GT47, and GT61. These glycosyltransferases include cellulose synthases, glucomannan synthase, and those required for xylan biosynthesis. These transcriptome profiling studies clearly demonstrated that genes involved in the biosynthesis of xylem cell wall components are coordinately expressed, which indicates that there must be transcriptional switches coordinately activating their expression during xylem development.

Promoter Analysis of Xylem-specific Expression

Since the expression of a gene is driven by its promoter sequence, analysis of promoters of secondary wall biosynthetic genes to identify *cis*-elements directing xylem-specific expression is one of the valuable approaches to dissect how the coordinated expression of secondary wall biosynthetic genes is achieved. It has been demonstrated through GUS reporter gene analysis that the promoters of genes for the biosynthesis of cellulose, xylan, and lignin are able to drive the GUS reporter gene expression in xylem (Hatton *et al.*, 1995; Lacombe *et al.*, 2000; Wu *et al.*, 2000; Pena *et al.*, 2007), indicating that the promoters contain *cis*-elements mediating xylem-specific expression. One attractive hypothesis is that xylem-specific secondary wall biosynthetic genes may share common *cis*-elements that direct their coordinated expression during xylem development. Bioinformatic analyses of secondary wall biosynthetic genes have revealed several putative *cis*-elements potentially involved in regulation of their expression in xylem cells. For example, analysis of promoter sequences of a set of xylem-specific genes led to identification of a putative *cis*-element, ACAAAGAA, that is present in 13 out of 52 xylem-specific genes (Ko *et al.*, 2006). Creux *et al.* (2008) conducted a comparative analysis of promoters of cellulose synthase genes from Arabidopsis, *Populus*, and *Eucalyptus* and identified a number of putative *cis*-elements conserved in their promoters. Similarly, Shi *et al.* (2009) recognized five core *cis*-elements present in the promoters of 18 monolignol biosynthetic genes from *Populus*. These analyses led to the hypothesis that these putative *cis*-elements could serve as regulatory elements responsible for the coordinated expression of secondary wall biosynthetic genes, but their functional significance awaits further investigation.

The first functionally characterized *cis*-element responsible for xylem-specific gene expression is the AC element identified from the promoters of lignin biosynthetic genes. Promoter deletion analysis of the bean phenylalanine ammonia-lyase gene 2 (PAL2) defined the minimal sequence required for xylem-specific expression to be within a 254-bp promoter region, indicating that this region contains all *cis*-elements necessary for driving xylem-specific expression (Leyva *et al.*, 1992; Hatton *et al.*, 1995). Further footprinting and electrophoretic mobility shift assay revealed that the 254-bp promoter sequence contains three AC elements (AC-I, ACCTACC; AC-II, ACCAACC; AC-III, ACCTAAC) and a G-box. Mutation of any one of the three AC elements in the 254-bp promoter sequence reduces xylem-associated expression of the GUS reporter gene. Simultaneous mutations of two or three AC elements abolishes xylem-specific expression (Hatton *et al.*, 1995). The functional significance of the AC elements was further substantiated by the finding that the AC-II heptamer is sufficient to drive the xylem-specific expression of GUS reporter gene (Séguin *et al.*, 1997). Promoter deletion analysis also identified a 50-bp region and an 80-bp region in promoters of *Eucalyptus* lignin biosynthetic genes *EgCCR* and *EgCAD2*, respectively, necessary and sufficient for xylem-specific expression (Lacombe *et al.*, 2000; Lauvergeat *et al.*, 2002). Further *in vivo* footprinting analysis of the 80-bp *EgCAD2* promoter sequence revealed four *cis*-elements that are associated with the binding of transcription factors. Two of them correspond to the previously identified AC-I element, ACCTACC, and the other two have the identical sequence CTGGTT (Rahantamalala *et al.*, 2010). The AC elements are present in the promoters of 7 out of the 10 monolignol biosynthetic genes in Arabidopsis (Raes *et al.*, 2003). Together, the promoter deletion and *cis*-element mapping analyses indicate that the 7-bp AC element ACC(A/T)A(A/C)C could function as a *cis* regulatory element coordinating the expression of monolignol biosynthetic genes during xylem differentiation.

Transcriptional Switches Activating Secondary Wall Deposition

Early Work on Xylem-specific MYBs and Lignification

The identification of the lignin pathway-specific AC regulatory element prompted a search for transcription factors that bind and activate the AC element. The AC element is similar to the binding sequence CC(T/A)ACC recognized by the maize MYB protein P (Grotewold *et al.*, 1994), indicating that the transcription factors that activate the AC element during xylem development may be MYB proteins (Hatton *et al.*, 1995). The first direct proof of the involvement of xylem-specific MYB transcription factors in activating lignin biosynthetic genes through the AC element came from the study of two tree MYBs, EgMYB2 and PtMYB4, isolated from developing wood of *Eucalyptus* and pine, respectively (Patzlaff *et al.*, 2003a; Goicoechea *et al.*, 2005). Both EgMYB2 and PtMYB4 are able to bind to the AC elements and activate transcription in an AC element-dependent manner. When overexpressed in tobacco, they effectively induced the expression of lignin biosynthetic genes and caused ectopic deposition of lignin or an alteration of lignin profile. The studies of EgMYB2 and PtMYB4 indicate that xylem-specific MYBs bind to the AC elements and thereby activate the lignin biosynthetic pathway during xylem development in tree species.

NAC Master Switches Activating the Entire Secondary Wall Biosynthetic Program

Secondary wall deposition during xylem development requires the coordinated activation of the biosynthetic pathways for all wall components including cellulose, hemicellulose, and lignin. The coordinated activation of the entire secondary wall biosynthetic program was demonstrated to be controlled by a group of secondary wall NAC master switches (collectively called SWNs; Fig. 13.2). Kubo *et al.* (2005) identified a subfamily of 7 NAC domain genes, namely *VASCULAR-RELATED NAC-DOMAIN PROTEIN1* (*VND1*), *VND2*, *VND3*, *VND4*, *VND5*, *VND6*, and *VND7*, that are upregulated during xylem vessel element differentiation in Arabidopsis *in vitro* cultured cells. These VND genes are close homologs of a zinnia NAC domain gene that is also upregulated during the zinnia *in vitro* xylem vessel element formation (Demura *et al.*, 2002). Detailed study of VND6 and VND7 revealed that they are specifically expressed in xylem vessels of Arabidopsis, and their overexpression is sufficient to cause ectopic deposition of secondary walls in cells that are normally parenchymatous. Furthermore, dominant repression of VND6 and VND7 results in a loss of secondary wall deposition in xylem vessels (Kubo *et al.*, 2005; Yamaguchi *et al.*, 2008, 2010a). These findings provide direct evidence that VND6 and VND7 are master switches capable of activating the entire secondary wall biosynthetic program during vessel differentiation in Arabidopsis.

The two secondary-wall-containing cell types in dicot xylem are vessels and fibers (Fig. 13.1; Zhong *et al.*, 2001). While the secondary wall biosynthesis in vessels is controlled by VND6 and VND7, in xylem fibers it is regulated by the VND close homologs, NAC SECONDARY WALL THICKENING PROMOTING FACTOR1 (NST1) and SECONDARY WALL-ASSOCIATED NAC DOMAIN PROTEIN1 (SND1, also called NST3; Zhong *et al.*, 2006, 2007b; Ko *et al.*, 2007; Mitsuda *et al.*, 2007). Both *SND1* and *NST1* are specifically expressed in xylem fibers and interfascicular fibers in Arabidopsis stems. When overexpressed, they are able to induce the expression of biosynthetic genes for cellulose, xylan, and lignin and concomitantly lead to ectopic secondary wall thickening in cells that are normally parenchymatous (Fig. 13.1e, f). Although single T-DNA knockout mutation of either *SND1* or *NST1* does not cause apparent defects in secondary wall thickening, simultaneous T-DNA mutations or repression of both *SND1* and *NST1* results in a complete loss of secondary wall thickening in both xylem fibers and interfascicular fibers. Consistent with their specific expression in fiber cells, mutations of *SND1* and *NST1* have no effect on secondary wall deposition in vessels (Fig. 13.1a–d). As a result, the mutant plants grow normally as the wild type but have a pendent stem phenotype due to lack of fibers for structural support (Zhong *et al.*, 2006, 2007b; Mitsuda *et al.*, 2007). These findings unequivocally demonstrate

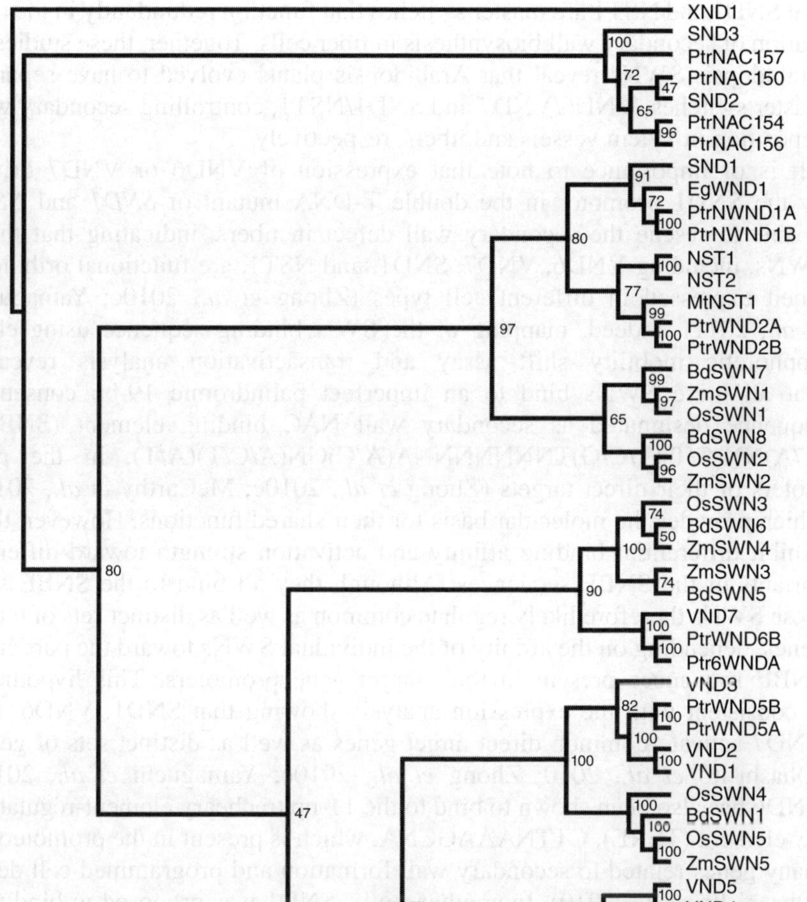

Figure 13.2 Phylogenetic relationship of NAC transcription factors that are known to be involved in regulating secondary wall biosynthesis. The NAC transcription factors are from *Arabidopsis thaliana*, poplar (*Populus trichocarpa*; Ptr), *Eucalyptus gunnii* (Eg), *Medicago truncatula* (Mt), *Brachypodium distachyon* (Bd), maize (*Zea mays*; Zm), and rice (*Oryza sativa*; Os). The phylogenetic tree was constructed with the neighbor-joining algorithm and displayed using the TREEVIW program. Bootstrap values are shown in percentages at the nodes.

that SND1 and NST1 are master switches that function redundantly in the regulation of secondary wall biosynthesis in fiber cells. Together, these studies of Arabidopsis SWNs reveal that Arabidopsis plants evolved to have separate master switches, VND6/VND7 and SND1/NST1, controlling secondary wall deposition in xylem vessels and fibers, respectively.

It is of importance to note that expression of VND6 or VND7 driven by the SND1 promoter in the double T-DNA mutant of *SND1* and *NST1* is able to rescue the secondary wall defect in fibers, indicating that these SWNs, including VND6, VND7, SND1, and NST1, are functional orthologs albeit expressed in different cell types (Zhong *et al.*, 2010c; Yamaguchi *et al.*, 2011). Indeed, mapping of the SWN binding sequence using electrophoretic mobility shift assay and transactivation analysis revealed that all these SWNs bind to an imperfect palindromic 19-bp consensus sequence designated as secondary wall NAC binding element (SNBE), (T/A)NN(C/T)(T/C/G)TNNNNNNNA(A/C)GN(A/C/T)(A/T), in the promoters of their direct targets (Zhong *et al.*, 2010c; McCarthy *et al.*, 2011), which provides the molecular basis for their shared functions. However, they exhibit differential binding affinity and activation strength toward different variants of the SNBE sequences. Although they all bind to the SNBE site, these SWNs therefore likely regulate common as well as distinct sets of target genes, depending on the affinity of the individual SWNs toward the particular SNBE sequences present in their target gene promoters. This hypothesis is consistent with the expression analysis showing that SND1, VND6, and VND7 activate common direct target genes as well as distinct sets of genes (Ohashi-Ito *et al.*, 2010; Zhong *et al.*, 2010c; Yamaguchi *et al.*, 2011). VND6 has also been shown to bind to the 11-bp tracheary element-regulating *cis*-element (TERE), CTTNAAAGCNA, which is present in the promoters of many genes related to secondary wall formation and programmed cell death (Ohashi-Ito *et al.*, 2010). In another study, SND1 was proposed to bind to a 14-bp core motif TACXTTXXXXATGA but, as the authors pointed out, most of the genes containing this core motif are not regulated by SWNs (Wang *et al.*, 2011).

Discovery of SWNs as master switches activating the entire secondary wall biosynthetic program in Arabidopsis raises a question of whether the use of SWNs as secondary wall master switches during xylem development is evolutionarily conserved. SWN homologs are found in diverse taxa of vascular plants, including seedless vascular plants, gymnosperms, and angiosperms (Fig. 13.2; Zhong *et al.*, 2010a). SWN homologs from dicot tree species (poplar and *Eucalyptus*), a herbaceous dicot (*Medicago truncatula*), and monocots (rice, maize, *Brachypodium distachyon*) have proven to be functional orthologs of Arabidopsis SWNs in regulating secondary wall biosynthesis. In poplar, 6 pairs of duplicated SWN genes, namely PtrWNDs (for wood-associated NAC domain transcription factors) /PtrVNSs (for

VND-, NST/SND-, and SMB-related proteins), are expressed in developing wood (secondary xylem). When overexpressed in Arabidopsis and poplar, they are capable of activating the expression of secondary wall biosynthetic genes and inducing ectopic secondary wall deposition (Zhong *et al.*, 2010b; Zhong and Ye, 2010; Ohtani *et al.*, 2011). In contrast, dominant repression of PtrWND functions leads to a significant reduction in secondary wall thickening in poplar wood (Zhong *et al.*, 2011b). Similarly, SWN homologs from *Eucalyptus*, rice, maize, and *Brachypodium* are able to activate the entire secondary wall biosynthetic program, leading to ectopic deposition of cellulose, xylan, and lignin when overexpressed (Zhong *et al.*, 2010a, 2011a; Valdivia *et al.*, 2013). The *Medicago* SWN homolog MtNST1 was identified through mutant screening for altered patterns of lignin autofluorescence in stems. Mutation of the *MtNST1* gene causes a reduction in the contents of lignin and cell wall polysaccharides (Zhao *et al.*, 2010a). Taken together, these studies have demonstrated that SWN homologs from both dicot woody species and monocots perform the same functions as Arabidopsis SWNs in regulating secondary wall biosynthesis. As SWN homologs from poplar, *Eucalyptus*, rice, and maize are all able to complement the secondary wall defects in fibers of the Arabidopsis *snd1 nst1* double mutants (Zhong *et al.*, 2011a, b), they must activate the same target genes as the Arabidopsis SWNs. Indeed, it has been shown that the Arabidopsis SWNs and their orthologs from poplar, rice, and maize all bind to and activate the same DNA binding element SNBE, indicating the evolutionary conservation of these secondary wall NAC master switches in the recognition of the DNA binding element in their target genes.

The expression level or activity of SWNs may be regulated either positively or negatively by several factors. It has been shown that two Arabidopsis ASSYMETRIC LEAVES2 (AS2) family proteins, ASL19 and ASL20, are positive regulators of *VND6* and *VND7* expression (Soyano *et al.*, 2008). Both *ASL19* and *ASL20* are expressed in developing vessels and their overexpression is able to induce ectopic deposition of secondary walls in non-vascular cells. Since the *ASL20* expression is activated by VND7 and vice versa, it has been proposed that ASL19 and ASL20 are involved in a positive feedback loop for *VND6* and *VND7* expression (Soyano *et al.*, 2008). SWNs themselves may also play a positive feedback role in regulating their expression as SWN overexpression increases their own expression (Wang *et al.*, 2011; Zhong *et al.*, 2011b). In contrast to the positive regulation, a VND-INTERACTING2 (VNI2) protein has been shown to be a negative regulator of VND function in Arabidopsis (Yamaguchi *et al.*, 2010b). VNI2 is able to interact with VND proteins and its interaction represses the expression of VND7-regulated genes and VND7-induced vessel development. VNI2 is therefore considered to be a transcriptional repressor for VND function. Repression of SWN function may also be mediated by a dominant negative form of SWNs (Li *et al.*, 2012b). In poplar, alternative splicing of a wood-associated NAC domain transcription factor, PtrWND1B/PtrSND1-A2, leads to synthesis of a truncated form of the protein that contains only the NAC dimerization domain without the DNA-binding and transcriptional activation domains. This truncated NAC protein, called PtrSND1-A2^{IR}, likely forms a heterodimer with other PtrWNDs, thus creating non-functional dominant negative transcriptional complexes. Although the functional significance of the alternative spliced form of PtrWND during xylem development is unclear, one proposed role for the transcriptional repression of SWN functions is to maintain transcriptional homeostasis in the SWN-regulated network (Li *et al.*, 2012b). Transcriptional repression of secondary wall deposition during xylem development has also been demonstrated for an Arabidopsis NAC domain protein, XND1 (Zhao *et al.*, 2008), but it is currently unknown whether it exerts its transcriptional repression through the SWN-regulated pathway.

MYB Master Switches Activating the Entire Secondary Wall Biosynthetic Program

The study of SWNs has unequivocally demonstrated that they are master switches regulating secondary wall biosynthesis during xylem development. The next important issue is how SWNs activate downstream genes, thus leading to the secondary wall deposition. Analysis of SWN-regulated genes in Arabidopsis has revealed that a transcriptional network encompassing SWNs and a suite of their downstream transcription factors governs the activation of the secondary wall biosynthetic program (Zhong and Ye, 2007). Among the SWN-regulated downstream transcription factors, two Arabidopsis MYB transcription factors – MYB46 and MYB83 – have been found to act as second-level master switches controlling secondary wall biosynthesis (Zhong *et al.*, 2007a; McCarthy *et al.*, 2009). Both MYB46 and MYB83 are direct targets of SWNs and are expressed in vessels and fibers during xylem development. Overexpression of either MYB46 or MYB83 is sufficient to induce the expression of biosynthetic genes for cellulose, xylan, and lignin and consequently result in ectopic deposition of secondary walls. Although T-DNA mutation of either MYB46 or MYB83 does not cause significant defect in secondary wall thickening, simultaneous mutations of MYB46 and MYB83 lead to a loss of secondary wall thickening in vessels and concomitantly a growth arrest at the seedling stage. The essential role of MYB46 and MYB83 in regulating secondary wall biosynthesis was verified by their dominant repression and RNA interference inhibition, showing a severe defect in secondary wall thickening in both vessels and fibers during xylem development (McCarthy *et al.*, 2009). These findings demonstrate that Arabidopsis evolved to have two levels of master switches, SWNs and their direct targets MYB46/MYB83, to regulate the biosynthesis of secondary wall components during xylem development.

The existence of second-level MYB master switches in regulating secondary wall biosynthesis has also been demonstrated in a number of other species, including gymnosperms (pine), dicot woody species (poplar and *Eucalyptus*), and monocots (rice and maize) (Fig. 13.3). A pine-wood-associated MYB, PtMYB4, is a close homolog of MYB46 and MYB83; when overexpressed, it is able to induce ectopic deposition of secondary walls (Patzlaff *et al.*, 2003a; Zhong and Ye, 2009). Furthermore, expression of PtMYB4 in the *myb46 myb83* double mutant is sufficient to rescue the defective secondary wall phenotype, confirming that these pine MYBs are functional orthologs of MYB46/MYB83 (Zhong *et al.*, 2010a). Similarly, MYB46/MYB83 homologs from poplar (PtrMYB2, PtrMYB3, PtrMYB20, and PtrMYB21), eucalyptus (EgMYB2), rice (OsMYB46), and maize (ZmMYB46) are capable of activating the entire secondary wall biosynthetic program and complement the *myb46 myb83* double mutant phenotypes (Goicoechea *et al.*, 2005; Zhong and Ye, 2009; Zhong *et al.*, 2010a, 2013; McCarthy *et al.*, 2010). The available lines of evidence indicate that as for the first-level NAC master switches, the operation of second-level MYB master switches in regulating the secondary wall biosynthetic program is also evolutionarily conserved in vascular plants.

MYB Transcription Factors Specifically Regulating Lignin Biosynthesis

Some of the SWN-regulated downstream transcription factors have been found to be transcriptional activators specific to the lignin biosynthetic pathway. Of note are two closely related Arabidopsis MYB genes, *MYB58* and *MYB63*, which are direct targets of MYB46 and MYB83 (Zhou *et al.*, 2009). *MYB58* and *MYB63* are specifically expressed in vessels and fibers during xylem development and their dominant repression or RNA interference inhibition results in a reduction in lignin content and secondary wall thickening. Overexpression of MYB58 and MYB63 induces the expression of genes specific to the lignin biosynthetic pathway and, as a result, it leads to ectopic deposition of only lignin but not cellulose or xylan. Direct target analysis further revealed that they directly activate the expression of lignin biosynthetic pathway genes through binding to the AC elements in the lignin pathway gene promoters. Together, these findings demonstrate that MYB58 and MYB63 are specific transcriptional switches of lignin biosynthesis. A poplar-wood-associated MYB58/MYB63 close homolog PtrMYB28 (Fig. 13.3) is also able to activate the promoter of a lignin biosynthetic gene in the AC element-dependent manner indicating that, similar to MYB58/MYB63, PtrMYB28 is likely involved in the regulation of lignin biosynthesis during wood formation in poplar (Zhong and Ye, 2009).

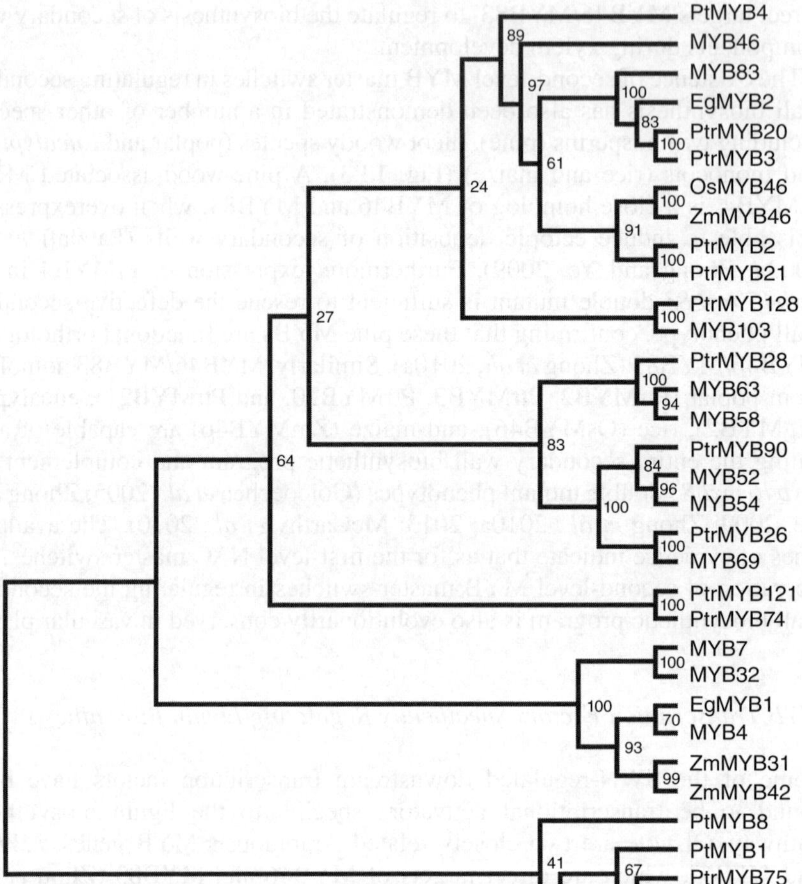

Figure 13.3 Phylogenetic relationship of MYB transcription factors that are known to be involved in regulating secondary wall biosynthesis. The MYB transcription factors are from *Arabidopsis thaliana*, pine (*Pinus taeda*; Pt), *Eucalyptus gunnii* (Eg), poplar (*Populus trichocarpa*; Ptr), maize (*Zea mays*; Zm), and rice (*Oryza sativa*; Os). Bootstrap values are shown in percentages at the nodes.

In addition to MYB58/MYB63, another Arabidopsis SWN-regulated transcription factor MYB85 has been shown to be a specific activator of lignin biosynthesis since it only induces the expression of lignin pathway genes and causes ectopic lignin deposition when overexpressed (Zhong *et al.*, 2008).

It has been demonstrated that a pine-wood-associated MYB85 close homolog PtMYB1 activates the lignin pathway genes via binding to the AC element and, when overexpressed, it is able to cause ectopic lignin deposition (Patzlaff *et al.*, 2003b; Bomal *et al.*, 2008). Another pine MYB, PtMYB4, and a *Eucalyptus* MYB, EgMYB2, were previously thought to be specific regulators of lignin biosynthesis (Patzlaff *et al.*, 2003a; Goicoechea *et al.*, 2005), but subsequent studies have revealed that they are MYB46/MYB83 functional orthologs regulating the entire secondary wall biosynthetic program (Zhong *et al.*, 2010a). In contrast to the activation of the whole lignin biosynthetic pathway by MYB58/MYB63/MYB85, another SWN-regulated MYB transcription factor, MYB103, appears to be specific in regulating *F5H* expression. As a result, MYB103 mutation causes a specific reduction in syringyl lignin (Ohman *et al.*, 2013). It was claimed in one report that SND1 directly activated the *Medicago F5H* expression (Zhao *et al.*, 2010b). However, direct target study and transactivation analysis showed that SND1 does not directly activate Arabidopsis *F5H* expression, indicating that F5H is unlikely a direct target of SND1 (Zhong *et al.*, 2010c; Ohman *et al.* 2013). In poplar, several other PtrWND-regulated wood-associated transcription factors, PtrMYB90 and PtrMYB26, appear to specifically activate the promoters of lignin biosynthetic genes, indicating their potential roles in regulating lignin biosynthesis (Zhong *et al.*, 2011b). As well as MYBs, a tobacco LIM family protein, NtLIM1, has been proposed to be a regulator of lignin biosynthesis as it is able to activate the AC element-driven GUS reporter gene expression and its antisense inhibition leads to a reduction in lignin content in tobacco stems (Kawaoka *et al.*, 2000).

In contrast to the transcriptional activators of lignin biosynthesis, a number of transcription factors that belong to the same subfamily of MYBs may act as transcriptional repressors of lignin biosynthesis. Three closely related Arabidopsis MYBs – MYB4, MYB7, and MYB32 – are direct targets of MYB46/MYB83 (Zhong and Ye, 2012); they are proposed to be repressors of lignin biosynthesis as mutations of *MYB4* and *MYB32* induce the expression of a few of lignin biosynthetic pathway genes (Jin *et al.*, 2000; Preston *et al.*, 2004). Several MYB4 homologs from maize (ZmMYB31 and ZmMYB42; Sonbol *et al.*, 2009; Fornalé *et al.*, 2010), switchgrass (PvMYB4; Shen *et al.*, 2012), and *Eucalyptus* (EgMYB1; Legay *et al.*, 2007, 2010) have also been demonstrated to downregulate lignin pathway genes when overexpressed, further confirming that MYB4 and its close homologs act as transcriptional repressors of lignin biosynthesis. EgMYB1 causes a reduction in the expression of genes for the biosynthesis of cellulose and xylan in addition to those for lignin biosynthesis when heterologously overexpressed in poplar and Arabidopsis (Legay *et al.*, 2010). It was therefore proposed that EgMYB1 may be a repressor for the entire secondary wall biosynthetic program. These MYB repressors have been shown to bind to the AC element (Fornalé *et al.*, 2010;

Legay *et al.*, 2010; Shen *et al.*, 2012); their transcriptional repression of lignin pathway genes is therefore likely mediated through their competitive binding to the AC element. With the exception of EgMYB1, whose expression is wood-associated, it has not been determined whether other MYB repressors are expressed in developing xylem; their *in vivo* functional significance in regulating lignin biosynthesis during xylem development therefore awaits further investigation.

Downstream Transcription Factors Regulated by NAC and MYB Master Switches

In addition to the MYBs that are specific to the lignin biosynthetic pathway, a number of other transcription factors have been demonstrated to be regulated by the secondary wall NAC and MYB master switches, indicating the complexity of the transcriptional network regulating secondary wall biosynthesis (Fig. 13.4; Zhong *et al.*, 2008, 2010c, 2011b; Ko *et al.*, 2009; Ohashi-Ito *et al.*,

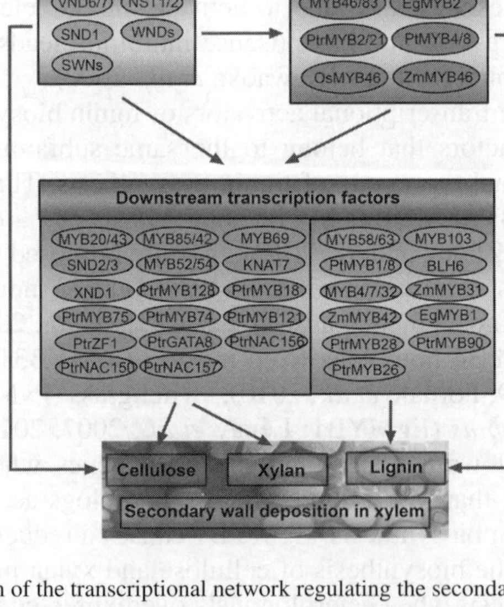

Figure 13.4 Diagram of the transcriptional network regulating the secondary wall biosynthetic program. The regulatory structure in this network is a multileveled feed-forward loop, in which the first-level NAC master switches and the second-level MYB master switches regulate downstream transcription factors. They all together activate the secondary wall biosynthetic program, leading to deposition of secondary walls in xylem.

2010; Yamaguchi *et al.*, 2011). Some of these downstream transcription factors, including SND2, SND3, MYB52, MYB54, MYB69, MYB42, MYB43, MYB20, KNAT7, and the above-described lignin pathway-specific MYBs, have been shown to be specifically expressed in secondary-wall-forming cells in Arabidopsis (Zhong *et al.*, 2007a, 2008; McCarthy *et al.*, 2009; Zhou *et al.*, 2009). Among the SWN-upregulated transcription factors, MYB46, MYB83, SND3, MYB103, KNAT7, and XND1 are direct targets of SWNs (Zhong *et al.*, 2008, 2010c). As discussed above, MYB46 and MYB83 are the second-level master switches regulating the entire secondary wall biosynthetic program (Zhong *et al.*, 2007a; Ko *et al.*, 2009; McCarthy *et al.*, 2009), MYB103 appears to be specific in regulating *F5H* expression (Ohman *et al.*, 2013), and XND1 is proposed to be transcriptional repressor of secondary wall biosynthesis (Zhao *et al.*, 2008). Dominant repression of SND2 or SND3 causes a reduction in secondary wall thickening and its overexpression leads to increased secondary wall thickening in fiber cells, suggesting its involvement in regulating secondary wall biosynthesis (Zhong *et al.*, 2008; Hussey *et al.*, 2011). Mutation of *KNAT7* results in a collapsed vessel phenotype, which is most likely due to a reduced vessel wall thickness (Brown *et al.*, 2005). However, KNAT7 is proposed to be a transcriptional repressor of secondary wall biosynthesis and it may mediate its transcriptional repression via interacting with MYB75 and OVATE FAMILY PROTEIN4 (OFP4) (Bhargava *et al.*, 2010, 2013; Li *et al.*, 2011, 2012a). Both MYB75 and OFP4 are also proposed to be transcriptional repressors of secondary wall biosynthesis as their mutations cause increased secondary wall thickening in fiber cells. However, since MYB75 and OFP4 were shown not to be expressed in interfascicular fiber cells (Bhargava *et al.*, 2010; Li *et al.*, 2011), their proposed functions in secondary wall biosynthesis in fiber cells are questionable. The functional roles of other SWN-upregulated transcription factors in regulating the secondary wall biosynthetic program requires further investigation, which will undoubtedly enrich our understanding of the transcriptional network controlling secondary wall biosynthesis.

In poplar, PtrWNDs upregulate the expression of close homologs of all the Arabidopsis SWN-regulated transcription factors. In addition, they induce the expression of 13 other transcription factors, whose Arabidopsis homologs are not known to be activated by SWNs (Zhong *et al.*, 2011b). Transactivation analysis has revealed that, in addition to the MYB46/MYB83 functional orthologs (PtrMYB2, PtrMYB3, PtrMYB20 and PtrMYB21), a number of these PtrWND-regulated transcription factors, including PtrNAC150, PtrNAC156, PtrNAC157, PtrMYB18, PtrMYB74, PtrMYB75, PtrMYB121, PtrMYB128, PtrZF1, and PtrGATA8, are able to activate the promoter activities of the biosynthetic genes for cellulose, xylan, and lignin. This is in contrast to the fact that, in Arabidopsis, only MYB46 and MYB83 are known to be capable of activating the biosynthetic genes for all three major wall components. These findings indicate that the transcriptional network regulating wood formation in tree species is much more complicated than that in herbaceous Arabidopsis.

Arabidopsis MYB46 and its Eucalyptus ortholog EgMYB2 have also been shown to upregulate a group of transcription factors such as MYB4, MYB7, MYB32, MYB42, MYB43, MYB52, MYB54, MYB58, MYB63, MYB85, KNAT7, BLH6, HB5, and HB15 (Zhong *et al.*, 2007a; Ko *et al.*, 2009; McCarthy *et al.*, 2009; Zhong and Ye, 2012; Cassan-Wang *et al.*, 2013). Among them, MYB4, MYB7, MYB32, MYB43, MYB52, MYB54, MYB58, MYB63, KNAT7, and BLH6 are direct targets of MYB46 (Zhong and Ye, 2012). The functional roles of some of these MYB46-regulated transcription factors have been characterized. As discussed above, MYB4, MYB7, and MYB32 are proposed to be transcriptional repressors of lignin biosynthesis (Jin *et al.*, 2000; Preston *et al.*, 2004), MYB58, MYB63, and MYB85 are transcriptional activators of lignin biosynthesis (Zhong *et al.*, 2008; Zhou *et al.*, 2009), and KNAT7 is proposed to be a transcriptional repressor of the whole secondary wall biosynthetic program (Li *et al.*, 2012a). Phenotypic analysis of mutants for six MYB46/EgMYB2-upregulated genes, including BLH6, a zinc finger transcription factor (At3g46620) MYB52, a MYB-like transcription factor (At3g11280) HB5, and HB15, showed their possible involvement in regulating lignin biosynthesis (Cassan-Wang *et al.*, 2013). The functional roles of many additional MYB46-regulated transcription factors remain to be characterized and further studies of their functions will help dissect the complexity of the transcriptional regulation of secondary wall biosynthesis.

Feed-forward Loop Regulatory Network Controlling the Expression of Secondary Wall Biosynthetic Genes

The discovery of SWNs together with their downstream transcription factors provides an unprecedented opportunity to unravel the transcriptional regulatory network controlling secondary wall biosynthesis. It is now well established that in Arabidopsis, both SWNs and their direct targets MYB46/MYB83 act as master switches capable of turning on the entire secondary wall biosynthetic program. The next challenging task is to find out how SWNs and a suite of their downstream transcription factors activate the expression of the biosynthetic genes for cellulose, xylan, and lignin, leading to the deposition of secondary walls. The available lines of evidence indicate that the transcriptional network regulating secondary wall biosynthesis is not a simple linear regulatory cascade; instead, vascular plants evolved a complex feed-forward loop regulatory mechanism, in which top-level transcription factors directly activate second-level transcription factors and together they regulate their downstream targets, which ultimately lead to activation of the secondary wall biosynthetic program (Fig. 13.4). In this transcriptional network, the SWN master regulators initiate the transcriptional hierarchy. They directly activate the expression of the second-level master regulators *MYB46/*MYB83 and an array of other transcription factors, some of which are also directly activated by MYB46/MYB83 (Zhong *et al.*, 2008; Zhong and Ye, 2012). MYB46/MYB83 directly regulate the expression of *MYB58/MYB63* and a set of other downstream transcription factors, including *MYB4*, *MYB7*, and *MYB32* (Zhong and Ye, 2012).

It has been demonstrated that both MYB46/MYB83 and their direct targets MYB58/MYB63 directly activate the expression of lignin biosynthetic genes. Mapping of the DNA binding element revealed that MYB46 and MYB83 bind to the 7-bp secondary wall MYB-responsive element (SMRE), ACC(A/T)A(A/C)(T/C) (Zhong and Ye, 2012). The SMRE motif includes eight variants of DNA-binding sequences (SMRE1, ACCAAAT; SMRE2, ACCAACT; SMRE3, ACCAAAC; SMRE4, ACCAACC; SMRE5, ACCTAAT; SMRE6, ACCTACT; SMRE7, ACCTAAC; and SMRE8, ACCTACC), which encompass three previously characterized AC elements, AC-I (ACCTACC; SMRE8), ACII (ACCAACC; SMRE4) and AC-III (ACCTAAC; SMRE7) that are bound and activated by MYB58/MYB63. MYB46 was shown in another study to bind to the M46R motif (Kim *et al.*, 2012), which is the same as the SMRE site except for the inclusion of one additional nucleotide. MYB46 orthologs from poplar (PtrMYB2, PtrMYB3, PtrMYB20, and PtrMYB21), *Eucalyptus* (EgMYB2) and pine (PtMYB4) have also been demonstrated to bind to and activate the SMRE sequences (Zhong *et al.*, 2013). These findings indicate that the activation of lignin biosynthetic pathway is more complicated than previously described; the core *cis* element for the coordinated regulation of lignin biosynthetic genes includes not only the three AC elements but also the other five SMRE variants.

In addition, some genes implicated in monolignol transport and polymerization are also directly activated by these MYB transcription factors. It has been suggested that ABC transporters play a role in monolignol transport (Miao and Liu, 2010; Alejandro *et al.*, 2012), and peroxidases and laccases are involved in monolignol polymerization (Berthet *et al.*, 2011; Herrero *et al.*, 2013). The laccase 4 gene has been shown to be a direct target of MYB58/MYB63 and an ABC transporter gene and two laccase genes (laccase 11 and laccase 12) have been demonstrated to be direct targets of MYB46/MYB83 (Zhou *et al.*, 2009; Zhong and Ye, 2012). The genes of a few ABC transporters, two peroxidases (peroxidase 72 and peroxidase 12), and a laccase (laccase 11) are also directly activated by SWNs, which bind to and activate the SNBE motif present in the promoters of their direct target genes (Zhong *et al.*, 2010c). Together, these findings demonstrate that transcription factors at different levels in the regulatory network function cooperatively to activate secondary wall biosynthesis.

It should be pointed out that MYB46 has been shown to bind to the promoters of the cellulose synthase genes, *CesA4*, *CesA7*, and *CesA8*, and mutations of the MYB46 binding sites in these promoters lead to the inability of cellulose synthase genes to complement the *cesa* mutant phenotypes (Kim *et al.*, 2013a, b), indicating that these MYB46 binding sites are important for expression of the cellulose synthase genes. Although the expression of CesA genes were shown to be induced directly by MYB46 in one report (Kim *et al.*, 2013b), global analysis of direct targets of MYB46 showed no direct activation of CesA genes by MYB46 (Zhong and Ye, 2012) indicating that the direct activation, if any, of CesA genes by MYB46 is minimal.

Concluding Remarks

Much progress has been made in the past decade on understanding how the expression of secondary wall biosynthetic genes is coordinately regulated. It is now well established that a transcriptional network encompassing SWN and MYB master switches together with their downstream transcription factors regulates the secondary wall biosynthetic program during xylem development. The discovery of a battery of downstream transcription factors regulated by SWN and MYB master switches provides unprecedented tools to further investigate how they work synergistically to turn on the secondary wall biosynthetic genes, leading to patterned deposition of secondary walls in xylem cells. Since evolution of the ability of vascular plants to make secondary-wall-reinforced xylem-conducting cells was pivotal for their conquest of land, a complete understanding of the molecular mechanisms controlling xylem cell wall deposition will help decipher the evolution of vascular plants.

Acknowledgements

The work in the authors' lab is supported by funding from the US Department of Agriculture National Institute of Food and Agriculture (AFRI Plant Biology program #2010-65116-20468) and the National Science Foundation (ISO-1051900).

References

Alejandro, S., Lee, Y., Tohge, T., Sudre, D., Osorio, S., Park, J., Bovet, L., Lee, Y., Geldner, N., Fernie, A.R., and Martinoia, E. (2012) AtABCG29 is a monolignol transporter involved in lignin biosynthesis. *Current Biology*, **22**, 1207–1212.

Anders, N., Wilkinson, M.D., Lovegrove, A., Freeman, J., Tryfona, T., Pellny, T.K., Weimar, T., Mortimer, J.C., Stott, K., Baker, J.M., Defoin-Platel, M., Shewry, P.R., Dupree, P., and Mitchell, R.A. (2012) Glycosyl transferases in family 61 mediate arabinofuranosyl transfer onto xylan in grasses. *Proceedings of the National Academy of Science, USA*, **109**, 989–993.

Andersson, S.-I., Samuelson, O., Ishihara, M., and Shimizu, K. (1983) Structure of the reducing end-groups in spruce xylan. *Carbohydrate Research*, **111**, 283–288.

Aspeborg, H., Schrader, J., Coutinho, P.M., Stam, M., Kallas, A., Djerbi, S., Nilsson, P., Denman, S., Amini, B., Sterky, F., Master, E., Sandberg, G., Mellerowicz, E., Sundberg, B., Henrissat, B., and Teeri, T.T. (2005) Carbohydrate-active enzymes involved in the secondary cell wall biogenesis in hybrid aspen. *Plant Physiology*, **137**, 983–997.

Berthet, S., Demont-Caulet, N., Pollet, B., Bidzinski, P., Cézard, L., Le Bris, P., Borrega, N., Hervé, J., Blondet, E., Balzergue, S., Lapierre, C., and Jouanin, L. (2011) Disruption of LACCASE4 and 17 results in tissue-specific alterations to lignification of *Arabidopsis thaliana* stems. *Plant Cell*, **23**, 1124–1137.

Bhargava, A., Mansfield, S.D., Hall, H.C., Douglas, C.J., and Ellis, B.E. (2010) MYB75 functions in regulation of secondary cell wall formation in the Arabidopsis inflorescence stem. *Plant Physiology*, **154**, 1428–1438.

Bhargava, A., Ahad, A., Wang, S., Mansfield, S.D., Haughn, G.W., Douglas, C.J., and Ellis, B.E. (2013) The interacting MYB75 and KNAT7 transcription factors modulate secondary cell wall deposition both in stems and seed coat in Arabidopsis. *Planta*, **237**, 1199–1211.

Boerjan, W., Ralph, J., and Baucher, M. (2003) Lignin biosynthesis. *Annual Review of Plant Biology*, **54**, 519–546.

Bomal, C., Bedon, F., Caron, S., Mansfield, S.D., Levasseur, C., Cooke, J.E., Blais, S., Tremblay, L., Morency, M.J., Pavy, N., Grima-Pettenati, J., Séguin, A., and Mackay, J. (2008) Involvement of *Pinus taeda* MYB1 and MYB8 in phenylpropanoid metabolism and secondary cell wall biogenesis, a comparative in planta analysis. *Journal of Experimental Botany*, **59**, 3925–3939.

Bonawitz, N.D. and Chapple, C. (2010) The genetics of lignin biosynthesis, connecting genotype to phenotype. *Annual Review of Genetics*, **44**, 337–363.

Brown, D.M., Zeef, L.A.H., Ellis, J., Goodacreb, R., and Turner, S.R. (2005) Identification of novel genes in Arabidopsis involved in secondary cell wall formation using expression profiling and reverse genetics. *Plant Cell*, **17**, 2281–2295.

Brown, D.M., Goubet, F., Wong, V.W., Goodacre, R., Stephens, E., Dupree, P., and Turner, S.R. (2007) Comparison of five xylan synthesis mutants reveals new insight into the mechanisms of xylan synthesis. *Plant Journal*, **52**, 1154–1168.

Brown, D.M., Zhang, Z., Stephens, E., Dupree, P., and Turner, S.R. (2009) Characterization of IRX10 and IRX10-like reveals an essential role in glucuronoxylan biosynthesis in Arabidopsis. *Plant Journal*, **57**, 732–746.

Caño-Delgado, A., Lee, J.Y., and Demura, T. (2010) Regulatory mechanisms for specification and patterning of plant vascular tissues. *Annual Review of Cell and Developmental Biology*, **26**, 605–637.

Cassan-Wang, H., Goué, N., Saidi, M.N., Legay, S., Sivadon, P., Goffner, D., and Grima-Pettenati, J. (2013) Identification of novel transcription factors regulating secondary cell wall formation in Arabidopsis. *Frontiers in Plant Science*, **4**, 189.

Creux, N.M., Ranik, M., Berger, D.K., and Myburg, A.A. (2008) Comparative analysis of orthologous cellulose synthase promoters from *Arabidopsis*, *Populus* and *Eucalyptus*, evidence of conserved regulatory elements in angiosperms. *New Phytologist*, **179**, 722–737.

Demura, T. and Fukuda, H. (2007) Transcriptional regulation in wood formation. *Trends in Plant Science*, **12**, 64–70.

Demura, T. and Ye, Z.H. (2010) Regulation of plant biomass production. *Current Opinion in Plant Biology*, **13**, 299–304.

Demura, T., Tashiro, G., Horiguchi, G., Kishimoto, N., Kubo, M., Matsuoka, N., Minami, A., Nagata-Hiwatashi, M., Nakamura, K., Okamura, Y., Sassa, N., Suzuki, S., Yazaki, J., Kikuchi, S., and Fukuda, H. (2002) Visualization by comprehensive microarray analysis of gene expression programs during transdifferentiation of mesophyll cells into xylem cells. *Proceedings of the National Academy of Science, USA*, **99**, 15794–15799.

Ehlting, J., Mattheus, N., Aeschliman, D.S., Li, E., Hamberger, B., Cullis, I.F., Zhuang, J., Kaneda, M., Mansfield, S.D., Samuels, L., Ritland, K., Ellis, B.E., Bohlmann, J., and Douglas, C.J. (2005) Global transcript profiling of primary stems from *Arabidopsis thaliana* identifies candidate genes for missing links in lignin biosynthesis and transcriptional regulators of fiber differentiation. *Plant Journal*, **42**, 618–640.

Endler, A. and Persson, S. (2011) Cellulose synthases and synthesis in Arabidopsis. *Molecular Plant*, **4**, 199–211.

Fornalé, S., Shi, X., Chai, C., Encina, A., Irar, S., Capellades, M., Fuguet, E., Torres, J.L., Rovira, P., Puigdomènech, P., Rigau, J., Grotewold, E., Gray, J., and Caparrós-Ruiz, D. (2010) ZmMYB31 directly represses maize lignin genes and redirects the phenylpropanoid metabolic flux. *Plant Journal*, **64**, 633–644.

Goicoechea, M., Lacombe, E., Legay, S., Mihaljevic, S., Rech, P., Jauneau, A., Lapierre, C., Pollet, B., Verhaegen, D., Chaubet-Gigot, N., and Grima-Pettenati, J. (2005) EgMYB2, a new transcriptional activator from Eucalyptus xylem, regulates secondary cell wall formation and lignin biosynthesis. *Plant Journal*, **43**, 553–567.

Grotewold, E., Drummond, B.J., Bowen, B., and Peterson, T. (1994) The myb-homologous P gene controls phlobaphene pigmentation in maize floral organs by directly activating a flavonoid biosynthetic gene subset. *Cell*, **76**, 543–553.

Hatton, D., Sablowski, R., Yung, M.H., Smith, C., Schuch, W., and Bevan, M. (1995) Two classes of *cis* sequences contribute to tissue-specific expression of a *PAL2* promoter in transgenic tobacco. *Plant Journal*, **7**, 859–876.

Herrero, J., Esteban-Carrasco, A., and Zapata, J.M. (2013) Looking for *Arabidopsis thaliana* peroxidases involved in lignin biosynthesis. *Plant Physiology and Biochemistry*, **67**, 77–86.

Hussey, S.G., Mizrachi, E., Spokevicius, A.V., Bossinger, G., Berger, D.K., and Myburg, A.A. (2011) SND2, a NAC transcription factor gene, regulates genes involved in secondary cell wall development in Arabidopsis fibres and increases fibre cell area in Eucalyptus. *BMC Plant Biology*, **11**, 173.

Jin, H., Cominelli, E., Bailey, P., Parr, A., Mehrtens, F., Jones, J., Tonelli, C., Weisshaar, B., and Martin, C. (2000) Transcriptional repression by AtMYB4 controls production of UV-protecting sunscreens in *Arabidopsis*. *EMBO Journal*, **19**, 6150–6161.

Johansson, M.H. and Samuelson, O. (1977) Reducing end groups in birch xylan and their alkaline degradation. *Wood Science and Technology*, **11**, 251–263.

Kawaoka, A., Kaothien, P., Yoshida, K., Endo, S., Yamada, K., and Ebinuma, H. (2000) Functional analysis of tobacco LIM protein Ntlim1 involved in lignin biosynthesis. *Plant Journal*, **22**, 289–301.

Kim, W.C., Ko, J.H., and Han, K.H. (2012) Identification of a cis-acting regulatory motif recognized by MYB46, a master transcriptional regulator of secondary wall biosynthesis. *Plant Molecular Biology*, **78**, 489–501.

Kim, W.C., Kim, J.Y., Ko, J.H., Kim, J., and Han, K.H. (2013a) Transcription factor MYB46 is an obligate component of the transcriptional regulatory complex for functional expression of secondary wall-associated cellulose synthases in *Arabidopsis thaliana*. *Journal of Plant Physiology*, **170**, 1374–1378.

Kim, W.C., Ko, J.H., Kim, J.Y., Kim, J.M., Bae, H.J., and Han, K.H. (2013b) MYB46 directly regulates the gene expression of secondary wall-associated cellulose synthases in Arabidopsis. *Plant Journal*, **73**, 26–36.

Ko, J.H., Beers, E.P., and Han, K.H. (2006) Global comparative transcriptome analysis identifies gene network regulating secondary xylem development in *Arabidopsis thaliana*. *Molecular Genetics and Genomics*, **276**, 517–531.

Ko, J.H., Yang, S.H., Park, A.H., Lerouxel, O., and Han, K.H. (2007) ANAC012, a member of the plant-specific NAC transcription factor family, negatively regulates xylary fiber development in *Arabidopsis thaliana*. *Plant Journal*, **50**, 1035–1048.

Ko, J.H., Kim, W.C., and Han, K.H. (2009) Ectopic expression of MYB46 identifies transcriptional regulatory genes involved in secondary wall biosynthesis in Arabidopsis. *Plant Journal*, **60**, 649–65.

Kubo, M., Udagawa, M., Nishikubo, N., Horiguchi, G., Yamaguchi, M., Ito, J., Mimura, T., Fukuda, H., Demura, T. (2005) Transcription switches for protoxylem and metaxylem vessel formation. *Genes and Development*, **19**, 1855–1860.

Kulkarni, A.R., Peña, M.J., Avci, U., Mazumder, K., Urbanowicz, B.R., Pattathil, S., Yin, Y., O'Neill, M.A., Roberts, A.W., Hahn, M.G., Xu, Y., Darvill, A.G., and York, W.S. (2012) The ability of land plants to synthesize glucuronoxylans predates the evolution of tracheophytes. *Glycobiology*, **22**, 439–451.

Lacombe, E., Doorsselaere, J.V., Boerjan, W., Boudet, A.M., and Grima-Pettenati, J. (2000) Characterization of *cis*-elements required for vascular expression of the *Cinnamoyl CoA Reductase* gene and for protein-DNA complex formation. *Plant Journal*, **23**, 663–676.

Lauvergeat, V., Rech, P., Jauneau, A., Guez, C., Coutos-Thevenot, P., and Grima-Pettenati, J. (2002) The vascular expression pattern directed by the *Eucalyptous gunnii* cinnamyl alcohol dehydrogenase *EgCAD2* promoter is conserved among woody and herbaceous plant species. *Plant Molecular Biology*, **50**, 497–509.

Lee, C., O'Neill, M.A., Tsumuraya, Y., Darvill, A.G., and Ye, Z.H. (2007a) The *irregular xylem9* mutant is deficient in xylan xylosyltransferase activity. *Plant Cell Physiology*, **48**, 1624–1634.

Lee, C., Zhong, R., Richardson, E.A., Himmelsbach, D.S., McPhail, B.T., and Ye, Z.H. (2007b) The *PARVUS* gene is expressed in cells undergoing secondary wall thickening and is essential for glucuronoxylan biosynthesis. *Plant Cell Physiology*, **48**, 1659–1672.

Lee, C., Teng, Q., Huang, W., Zhong, R., and Ye, Z.H. (2009) Down-regulation of PoGT47C expression in poplar results in a reduced glucuronoxylan content and an increased wood digestibility by cellulase. *Plant Cell Physiology*, **50**, 1075–1089.

Lee, C., Teng, Q., Huang, W., Zhong, R., and Ye, Z.H. (2010) The Arabidopsis family GT43 glycosyltransferases form two functionally nonredundant groups essential for the elongation of glucuronoxylan backbone. *Plant Physiology*, **153**, 526–541.

Lee, C., Teng, Q., Zhong, R., and Ye, Z.H. (2011) Molecular dissection of xylan biosynthesis during wood formation in poplar. *Molecular Plant*, **4**, 730–747.

Lee, C., Teng, Q., Zhong, R., and Ye, Z.H. (2012a) Arabidopsis GUX proteins are glucuronyltransferases responsible for the addition of glucuronic acid side chains onto xylan. *Plant Cell Physiology*, **53**, 1204–1216

Lee, C., Teng, Q., Zhong, R., Yuan, Y., Haghighat, M., and Ye, Z.H. (2012b) Three Arabidopsis DUF579 domain-containing GXM proteins are methyltransferases catalyzing 4-*O*-methylation of glucuronic acid on xylan. *Plant Cell Physiology*, **53**, 1934–1949.

Lee, C., Zhong, R., and Ye, Z.H. (2012c) Arabidopsis family GT43 members are xylan xylosyltransferases required for the elongation of the xylan backbone. *Plant Cell Physiology*, **53**, 135–143.

Lee, C., Zhong, R., and Ye, Z.H. (2012d) Biochemical characterization of xylan xylosyltransferases involved in wood formation in poplar. *Plant Signaling and Behavior*, **7**, 332–337.

Legay, S., Lacombe, E., Goicoechea, M., Briere, C., Seguin, A., Mackay, J., and Grima-Pettenati, J. (2007) Molecular characterization of *EgMYB1*, a putative transcriptional repressor of the lignin biosynthetic pathway. *Plant Science*, **173**, 542–549.

Legay, S., Sivadon, P., Blervacq, A.S., Pavy, N., Baghdady, A., Tremblay, L., Levasseur, C., Ladouce, N., Lapierre, C., Séguin, A., Hawkins, S., Mackay, J., and Grima-Pettenati, J. (2010) EgMYB1, an R2R3 MYB transcription factor from eucalyptus negatively regulates secondary cell wall formation in Arabidopsis and poplar. *New Phytologist*, **188**, 774–786.

Leyva, A., Liang, X., Pintor-Toro, J.A., Dixon, R.A., and Lamb, C.J. (1992) cis-element combinations determine phenylalanine ammonia-lyase gene tissue-specific expression patterns. *Plant Cell*, **4**, 263–271.

Li, E., Wang, S., Liu, Y., Chen, J.G., and Douglas, C.J. (2011) OVATE FAMILY PROTEIN4 (OFP4) interaction with KNAT7 regulates secondary cell wall formation in *Arabidopsis thaliana*. *Plant Journal*, **67**, 328–341.

Li, E., Bhargava, A., Qiang, W., Friedmann, M.C., Forneris, N., Savidge, R.A., Johnson, L.A., Mansfield, S.D., Ellis, B.E., and Douglas, C.J. (2012a) The Class II KNOX gene KNAT7 negatively regulates secondary wall formation in Arabidopsis and is functionally conserved in *Populus*. *New Phytologist*, **194**, 102–115.

Li, Q., Lin, Y.C., Sun, Y.H., Song, J., Chen, H., Zhang, X.H., Sederoff, R.R., and Chiang, V.L. (2012b) Splice variant of the SND1 transcription factor is a dominant negative of SND1 members and their regulation in Populus trichocarpa. *Proceedings of the National Academy of Science, USA*, **109**, 14699–14704.

Liepman, A.H., Nairn, C.J., Willats, W.G., Sørensen, I., Roberts, A.W., and Keegstra, K. (2007) Functional genomic analysis supports conservation of function among cellulose synthase-like A gene family members and suggests diverse roles of mannans in plants. *Plant Physiology*, **143**, 1881–1893.

McCann, M.C., Wells, B., and Roberts, K. (1990) Direct visualization of cross-links in the primary plant cell wall. *Journal of Cell Science*, **96**, 323–334.

McCarthy, R.L., Zhong, R., and Ye, Z.H. (2009) MYB83 is a direct target of SND1 and acts redundantly with MYB46 in the regulation of secondary cell wall biosynthesis in *Arabidopsis*. *Plant Cell Physiology*, **50**, 1950–1964.

McCarthy, R.L., Zhong, R., Fowler, S., Lyskowski, D., Piyasena, H., Carleton, K., Spicer, C., and Ye, Z.H. (2010) The poplar MYB transcription factors, PtrMYB3 and PtrMYB20, are involved in the regulation of secondary wall biosynthesis. *Plant Cell Physiology*, **51**, 1084–1090.

McCarthy, R.L., Zhong, R., and Ye, Z.H. (2011) Secondary wall NAC binding element (SNBE), a key cis-acting element required for target gene activation by secondary wall NAC master switches. *Plant Signaling and Behavior*, **6**, 1282–1285.

Mellerowicz, E.J., Baucher, M., Sundberg, B., and Boerjan, W. (2001) Unravelling cell wall formation in the woody dicot stem. *Plant Molecular Biology*, **47**, 239–274.

Miao, Y.C. and Liu, C.J. (2010) ATP-binding cassette-like transporters are involved in the transport of lignin precursors across plasma and vacuolar membranes. *Proceedings of the National Academy of Science, USA*, **107**, 22728–22733.

Mitsuda, N., Iwase, A., Yamamoto, H., Yoshida, M., Seki, M., Shinozaki, K., and Ohme-Takagi, M. (2007) NAC transcription factors, NST1 and NST3, are key regulators of the formation of secondary walls in woody tissues of *Arabidopsis*. *Plant Cell*, **19**, 270–280.

Mortimer, J.C., Miles, G.P., Brown, D.M., Zhang, Z., Segura, M.P., Weimar, T., Yu, X., Seffen, K.A., Stephens, E., Turner, S.R., and Dupree, P. (2010) Absence of branches from xylan in Arabidopsis gux mutants reveals potential for simplification of lignocellulosic biomass. *Proceedings of the National Academy of Science, USA*, **107**, 17409–17414.

Ohashi-Ito, K. and Fukuda, H. (2010) Transcriptional regulation of vascular cell fates. *Current Opinion in Plant Biology*, **13**, 670–676.

Ohashi-Ito, K., Oda, Y. and Fukuda, H. (2010) Arabidopsis VASCULAR-RELATED NAC-DOMAIN6 directly regulates the genes that govern programmed cell death and secondary wall formation during xylem differentiation. *Plant Cell*, **22**, 3461–3473.

Ohman, D., Demedts, B., Kumar, M., Gerber, L., Gorzsás, A., Goeminne, G., Hedenström, M., Ellis, B., Boerjan, W., and Sundberg, B. (2013) MYB103 is required for FERULATE-5-HYDROXYLASE expression and syringyl lignin biosynthesis in Arabidopsis stems. *Plant Journal*, **73**, 63–76.

Ohtani, M., Nishikubo, N., Xu, B., Yamaguchi, M., Mitsuda, N., Goué, N., Shi, F., Ohme-Takagi, M., and Demura, T. (2011) A NAC domain protein family contributing to the regulation of wood formation in poplar. *Plant Journal*, **67**, 499–512.

Patzlaff, A., McInnis, S., Courtenay, A., Surman, C., Newman, L.J., Smith, C., Bevan, M.W., Mansfield, S., Whetten, R.W., Sederoff, R.R., and Campbell, M.M. (2003a) Characterisation of a pine MYB that regulates lignification. *Plant Journal*, **36**, 743–754.

Patzlaff, A., Newman, L.J., Dubos, C., Whetten, R.W., Smith, C., McInnis, S., Bevan, M.W., Sederoff, R.R., and Campbell, M.M. (2003b) Characterization of PtMYB1, an R2R3-MYB from pine xylem. *Plant Molecular Biology*, **53**, 597–608.

Pena, M.J., Zhong, R., Zhou, G.K., Richardson, E.A., O'Neill, M.A., Darvill, A.G., York, W.S., and Ye, Z.H. (2007) *Arabidopsis irregular xylem8* and *irregular xylem9*, implications for the complexity of glucuronoxylan biosynthesis. *Plant Cell*, **19**, 549–563.

Plomion, C., Leprovost, G., and Stokes, A. (2001) Wood formation in trees. *Plant Physiology*, **127**, 1513–1523.

Preston, J., Wheeler, J., Heazlewood, J., Li, S.F., and Parish, R.W. (2004) AtMYB32 is required for normal pollen development in *Arabidopsis thaliana. Plant Journal*, **40**, 979–995.

Raes, J., Rohde, A., Christensen, J.H., Peer, Y.V., and Boerjan, W. (2003) Genome-wide characterization of the lignification toolbox in *Arabidopsis. Plant Physiology*, **133**, 1051–1071.

Rahantamalala, A., Rech, P., Martinez, Y., Chaubet-Gigot, N., Grima-Pettenati, J., and Pacquit, V. (2010) Coordinated transcriptional regulation of two key genes in the lignin branch pathway–CAD and CCR–is mediated through MYB- binding sites. *BMC Plant Biology*, **10**, 130.

Raven, P.H., Evert, R.F., Eichhorn, S.E. (1999) *Biology of Plants*, 6th edition. W.H. Freeman and Company, New York.

Rennie, E.A., Hansen, S.F., Baidoo, E.E., Hadi, M.Z., Keasling, J.D., and Scheller, H.V. (2012) Three members of the Arabidopsis glycosyltransferase family 8 are xylan glucuronosyltransferases. *Plant Physiology*, **159**, 1408–1417.

Rensing, S.A., Lang, D., Zimmer, A.D., Terry, A., Salamov, A. *et al.* (2008) The *Physcomitrella* genome reveals evolutionary insights into the conquest of land by plants. *Science*, **319**, 64–69.

Roberts, A.W. and Bushoven, J.T. (2007) The cellulose synthase (CESA) gene superfamily of the moss *Physcomitrella patens. Plant Molecular Biology*, **63**, 207–219.

Scheller, H.V. and Ulvskov, P. (2010) Hemicelluloses. *Annual Review of Plant Biology*, **61**, 263–289.

Schuetz, M., Smith, R., and Ellis, B. (2013) Xylem tissue specification, patterning, and differentiation mechanisms. *Journal of Experimental Botany*, **64**, 11–31.

Séguin, A., Laible, G., Leyva, A., Dixon, R.A., and Lamb, C.J. (1997) Characterization of a gene encoding a DNA-binding protein that interacts in vitro with vascular specific cis elements of the phenylalanine ammonia-lyase promoter. *Plant Molecular Biology*, **35**, 281–291.

Shen, H., He, X., Poovaiah, C.R., Wuddineh, W.A., Ma, J., Mann, D.G., Wang, H., Jackson, L., Tang, Y., Stewart, C.N. Jr,, Chen, F., and Dixon, R.A. (2012) Functional characterization of the switchgrass (*Panicum virgatum*) R2R3-MYB transcription factor PvMYB4 for improvement of lignocellulosic feedstocks. *New Phytologist*, **193**, 121–136.

Shi, R., Sun, Y.H., Li, Q., Heber, S., Sederoff, R., and Chiang, V.L. (2010) Towards a systems approach for lignin biosynthesis in *Populus trichocarpa*, transcript abundance and specificity of the monolignol biosynthetic genes. *Plant Cell Physiology*, **51**, 144–163.

Shimizu, K., Ishihara, M., and Ishihara, T. (1976) Hemicellulases of Brown Rotting Fungus, *Tyromyces palustris*. II. The oligosaccharides from the hydrolysate of a hardwood xylan by the intracellular xylanase. *Mokuzai Gaikkashi*, **22**, 618–625.

Somerville, C. (2006) Cellulose synthesis in higher plants. *Annual Review of Cell and Developmental Biology*, **22**, 53–78.

Sonbol, F.M., Fornalé, S., Capellades, M., Encina, A., Touriño, S., Torres, J.L., Rovira, P., Ruel, K., Puigdomènech, P., Rigau, J., and Caparrós-Ruiz, D. (2009) The maize ZmMYB42 represses the phenylpropanoid pathway and affects the cell wall structure, composition and degradability in *Arabidopsis thaliana. Plant Molecular Biology*, **70**, 283–296.

Soyano, T., Thitamadee, S., Machida, Y., and Chua, N.H. (2008) *ASYMMETRIC LEAVES2-LIKE19/LATERAL ORGAN BOUNDARIES DOMAIN30* and *ASL20/LBD18* regulate tracheary element differentiation in Arabidopsis. *Plant Cell*, **20**, 3359–3373.

Suzuki, S., Li, L., Sun, Y.H., and Chiang, V.L. (2006) The cellulose synthase gene superfamily and biochemical functions of xylem-specific cellulose synthase-like genes in *Populus trichocarpa. Plant Physiology*, **142**, 1233–1245.

Taylor, N.G., Gardiner, J.C., Whiteman, R., and Turner, S.R. (2004) Cellulose synthesis in the Arabidopsis secondary cell wall. *Cellulose*, **11**, 329–338.

Timell, T.E. (1967) Recent progress in the chemistry of wood hemicelluloses. *Wood Science and Technology*, **1**, 45–70.

Urbanowicz, B.R., Peña, M.J., Ratnaparkhe, S., Avci, U., Backe, J., Steet, H.F., Foston, M., Li, H., O'Neill, M.A., Ragauskas, A.J., Darvill, A.G., Wyman, C., Gilbert, H.J., and York, W.S. (2012) 4-*O*-methylation of glucuronic acid in Arabidopsis glucuronoxylan is catalyzed by a domain of unknown function family 579 protein. *Proceedings of the National Academy of Science, USA*, **109**, 14253–14258.

Valdivia, E.R., Herrera, M.T., Gianzo, C., Fidalgo, J., Revilla, G., Zarra, I., and Sampedro, J. (2013) Regulation of secondary wall synthesis and cell death by NAC transcription factors in the monocot *Brachypodium distachyon*. *Journal of Experimental Botany*, **64**, 1333–1343.

Wang, H.Z. and Dixon, R.A. (2012) On-off switches for secondary cell wall biosynthesis. *Molecular Plant*, **5**, 297–303.

Wang, H., Zhao, Q., Chen, F., Wang, M., and Dixon, R.A. (2011) NAC domain function and transcriptional control of a secondary cell wall master switch. *Plant Journal*, **68**, 1104–1114.

Wu, A.M., Rihouey, C., Seveno, M., Hörnblad, E., Singh, S.K., Matsunaga, T., Ishii, T., Lerouge, P., and Marchant, A. (2009) The *Arabidopsis* IRX10 and IRX10-LIKE glycosyltransferases are critical for glucuronoxylan biosynthesis during secondary cell wall formation. *Plant Journal*, **57**, 718–731.

Wu, A.M., Hörnblad, E., Voxeur, A., Gerber, L., Rihouey, C., Lerouge, P., and Marchant, A. (2010) Analysis of the Arabidopsis IRX9/IRX9-L and IRX14/IRX14-L pairs of glycosyltransferase genes reveals critical contributions to biosynthesis of the hemicellulose glucuronoxylan. *Plant Physiology*, **153**, 542–554.

Wu, L., Joshi, C.P., and Chiang, V.L. (2000) A xylem-specific cellulose synthase gene from aspen (*Populus tremuloides*) is responsive to mechanical stress. *Plant Journal*, **22**, 495–502.

Xiong, G., Cheng, K., and Pauly, M. (2013) Xylan O-acetylation impacts xylem development and enzymatic recalcitrance as indicated by the Arabidopsis mutant tbl29. *Molecular Plant*, **6**, 1373–1375.

Xu, Z., Zhang, D., Hu, J., Zhou, X., Ye, X., Reichel, K.L., Stewart, N.R., Syrenne, R.D., Yang, X., Gao, P., Shi, W., Doeppke, C., Sykes, R.W., Burris, J.N., Bozell, J.J., Cheng, M.Z., Hayes, D.G., Labbe, N., Davis, M., Stewart, C.N. Jr,, and Yuan, J.S. (2009) Comparative genome analysis of lignin biosynthesis gene families across the plant kingdom. *BMC Bioinformatics*, **10**, S3.

Yamaguchi, M., Kubo, M., Fukuda, H., and Demura, T. (2008) Vascular-related NAC-DOMAIN7 is involved in the differentiation of all types of xylem vessels in *Arabidopsis* roots and shoots. *Plant Journal*, **55**, 652–664.

Yamaguchi, M., Goué, N., Igarashi, H., Ohtani, M., Nakano, Y., Mortimer, J.C., Nishikubo, N., Kubo, M., Katayama, Y., Kakegawa, K., Dupree, P., and Demura, T. (2010a) VASCULAR-RELATED NAC-DOMAIN6 and VASCULAR-RELATED NAC-DOMAIN7 effectively induce transdifferentiation into xylem vessel elements under control of an induction system. *Plant Physiology*, **153**, 906–914.

Yamaguchi, M., Ohtani, M., Mitsuda, N., Kubo, M., Ohme-Takagi, M., Fukuda, H., and Demura, T. (2010b) VND-INTERACTING2, a NAC domain transcription factor, negatively regulates xylem vessel formation in Arabidopsis. *Plant Cell*, **22**, 1249–1263.

Yamaguchi, M., Mitsuda, N., Ohtani, M., Ohme-Takagi, M., Kato, K., and Demura, T. (2011) VASCULAR-RELATED NAC-DOMAIN7 directly regulates the expression of a broad range of genes for xylem vessel formation. *Plant Journal*, **66**, 579–590.

Yuan, Y., Teng, Q., Zhong, R., and Ye, Z.H. (2013) The Arabidopsis DUF231 domain-containing protein ESK1 mediates 2-*O*- and 3-*O*-acetylation of xylosyl residues in xylan. *Plant Cell Physiology*, **54**, 1186–1199.

Zhang, J., Elo, A., and Helariutta, Y. (2011) Arabidopsis as a model for wood formation. *Current Opinion in Biotechnology*, **22**, 293–299.

Zhao, C., Craig, J.C., Petzold, H.E., Dickerman, A.W., and Beers, E.P. (2005) The xylem and phloem transcriptomes from secondary tissues of the Arabidopsis root-hypocotyl. *Plant Physiology*, **138**, 803–818.

Zhao, C., Avci, U., Grant, E.H., Haigler, C.H., and Beers, E.P. (2008) XND1, a member of the NAC domain family in *Arabidopsis thaliana*, negatively regulates lignocellulose synthesis and programmed cell death in xylem. *Plant Journal*, **53**, 425–436.

Zhao, Q., Gallego-Giraldo, L., Wang, H., Zeng, Y., Ding, S.Y., Chen, F., and Dixon, R.A. (2010a) An NAC transcription factor orchestrates multiple features of cell wall development in *Medicago truncatula*. *Plant Journal*, **63**, 100–114.

Zhao, Q., Wang, H., Yin, Y., Xu, Y., Chen, F., and Dixon, R.A. (2010b) Syringyl lignin biosynthesis is directly regulated by a secondary cell wall master switch. *Proceedings of the National Academy of Science, USA*, **107**, 14496–14501.

Zhong, R. and Ye, Z.H. (2007) Regulation of cell wall biosynthesis. *Current Opinion in Plant Biology*, **10**, 564–572.

Zhong, R. and Ye, Z.H. (2009) Transcriptional regulation of lignin biosynthesis. *Plant Signaling and Behavior*, **4**, 1028–1034.

Zhong, R. and Ye, Z.H. (2010) The poplar PtrWNDs are transcriptional activators of secondary cell wall biosynthesis. *Plant Signaling and Behavior*, **5**, 469–472.

Zhong, R. and Ye, Z.H. (2012) MYB46 and MYB83 bind to the SMRE sites and directly activate a suite of transcription factors and secondary wall biosynthetic genes. *Plant Cell Physiology*, **53**, 368–380.

Zhong, R., Burk, D.H., and Ye, Z.H. (2001) Fibers. A model for studying cell differentiation, cell elongation, and cell wall biosynthesis. *Plant Physiology*, **126**, 477–479.

Zhong, R., Pena, M.J., Zhou, G.K., Nairn, C.J., Wood-Jones, A., Richardson, E.A., Morrison, W.H., Darvill, A.G., York, W.S., and Ye, Z.H. (2005) Arabidopsis *Fragile Fiber8*, which encodes a putative glucuronyltransferase, is essential for normal secondary wall synthesis. *Plant Cell*, **17**, 3390–3408.

Zhong, R., Demura, T., and Ye, Z.H. (2006) SND1, a NAC domain transcription factor, is a key regulator of secondary wall synthesis in fibers of *Arabidopsis*. *Plant Cell*, **18**, 3158–3170.

Zhong, R., Richardson, E.A., and Ye, Z.H. (2007a) The MYB46 transcription factor is a direct target of SND1 and regulates secondary wall biosynthesis in Arabidopsis. *Plant Cell*, **19**, 2776–2792.

Zhong, R., Richardson, E.A., and Ye, Z.H. (2007b) Two NAC domain transcription factors, SND1 and NST1, function redundantly in regulation of secondary wall synthesis in fibers of *Arabidopsis*. *Planta*, **225**, 1603–1611.

Zhong, R., Lee, C., Zhou, J., McCarthy, R.L., and Ye, Z.H. (2008) A battery of transcription factors involved in the regulation of secondary cell wall biosynthesis in *Arabidopsis*. *Plant Cell*, **20**, 2763–2782.

Zhong, R., Lee, C., and Ye, Z.H. (2010a) Evolutionary conservation of the transcriptional network regulating secondary cell wall biosynthesis. *Trends in Plant Science*, **15**, 625–631.

Zhong, R., Lee, C., and Ye, Z.H. (2010b) Functional characterization of poplar wood-associated NAC domain transcription factors. *Plant Physiology*, **152**, 1044–1055.

Zhong, R., Lee, C., and Ye, Z.H. (2010c) Global analysis of direct targets of secondary wall NAC master switches in *Arabidopsis*. *Molecular Plant*, **3**, 1087–1103.

Zhong, R., Lee, C., McCarthy, R.L., Reeves, C.K., Jones, E.G., and Ye, Z.H. (2011a) Transcriptional activation of secondary wall biosynthesis by rice and maize NAC and MYB transcription factors. *Plant Cell Physiology*, **52**, 1856–1871.

Zhong, R., McCarthy, R.L., Lee, C., and Ye, Z.H. (2011b) Dissection of the Transcriptional Program Regulating Secondary Wall Biosynthesis during Wood Formation in Poplar. *Plant Physiology*, **157**, 1452–1468.

Zhong, R., McCarthy, R.L., Haghighat, M., and Ye, Z.H. (2013) The poplar MYB master switches bind to the SMRE site and activate the secondary wall biosynthetic program during wood formation. *PLoS One*, **8**, e69219.

Zhou, J., Lee, C., Zhong, R., and Ye, Z.H. (2009) MYB58 and MYB63 are transcriptional activators of the lignin biosynthetic pathway during secondary cell wall formation in Arabidopsis. *Plant Cell*, **21**, 248–266.

14 Phloem Cell Development

Raffael Lichtenberger, Kaori Furuta-Miyashima, Eva Hellmann and Ykä Helariutta

Institute of Biotechnology/Department of Biology and Environmental Sciences, University of Helsinki, FIN, Finland

Introduction

A defining feature of higher plants such as *Arabidopsis thaliana* is the sophisticated vascular transport system throughout the entire plant organism. This vascular system consists of two conducting tissues, xylem and phloem. Xylem is mostly responsible for water and mineral transport from the root system to the aerial organs and for the structural stability of the entire plant, while phloem is the main system for transporting photoassimilates and long-distance molecular signals. These compounds are transported from the photosynthetic organs to photosynthetically inactive parts of the plant, the sink tissues.

The phloem consists primarily of two highly specialized cell types: sieve elements (SE) and companion cells (CC). Furthermore, additional cell types with unique morphology and physiology can be present in the phloem such as sclereids, fibers, resin ducts, or laticifers. Sieve elements are enucleate, but still living, cells which form continuous transport channels throughout the plant. The sieve elements constitute the main conducting component of the phloem. Individual sieve elements are connected through sieve plates, specialized cell walls which are heavily perforated to enable the flow of phloem sap.

Closely associated with the sieve elements are the companion cells, which derive from the same precursor cells. Companion cells are fully functional cells, although they are small in comparison to sieve elements. They are symplastically connected to the adjacent sieve element cells through plasmodesmata. Companion cells retain an active nucleus and are responsible for the functional regulation of the enucleate sieve element cells (Oparka and Turgeon, 1999).

In recent years, several novel regulatory mechanisms that control the specification of vascular patterning and differentiation have been uncovered. Through the use of new genomic and molecular techniques in multiple plant model systems such as Arabidopsis, Populus, and Zinnia, new insights have become available regarding the regulation of vascular development. The

Plant Cell Wall Patterning and Cell Shape, First Edition. Edited by Hiroo Fukuda.

importance of signaling in the control of vascular morphogenesis has become increasingly apparent. The primary agents involved include well-known phytohormones such as auxin, cytokinin, and various brassinosteroids, as well as other small regulatory molecules. This level of understanding also involves the transporters and receptors of these factors. There is increasing evidence for the notion that xylem and phloem development are highly coordinated. While many of the factors which will be described in this review act cell-autonomously, the emerging importance of mobile regulatory factors will also be highlighted.

Cell-to-cell communication has to be strictly controlled during development. Significant advances have been made recently in the understanding of such control mechanisms. The new key studies in this field are discussed here, and we attempt to construct as complete a picture as possible of phloem development and differentiation.

Primary Development of the Vasculature

Overview

The vasculature of higher plants – phloem, xylem, and procambium – derives from a discrete population of stem cells, the vascular meristem. Generally, these are organized in discrete tissues according to their origin: provasculature, procambium, cambium, and preprocambium. The provasculature consists of still undifferentiated procambium, first established by periclinal cell divisions during embryogenesis. Eventually, it develops into the mature vasculature consisting of phloem, xylem, and procambium.

What is known as primary growth describes the generation of new vascular bundles, where the procambium derives from apical meristems. In the stem this is the shoot apical meristem, whereas in the root it is the root apical meristem. In leaf venation, preprocambium is induced during the early stages of leaf development and eventually develops into the complete vascular bundle. In contrast, secondary growth describes the process of new vascular tissue deriving from a new lateral meristem, the cambium.

Provascular Development in the Embryo

During embryogenesis, the progenitor cells that will eventually become the vascular tissues are first established as provascular cells. Surrounded by the epidermal and ground tissue layers, this provascular tissue forms the innermost domain of the plant embryo.

By the late globular embryonic stage, the four provascular cells have undergone periclinal divisions to generate the future pericycle and the vascular primordium. The complete root promeristem with all initials and derived cell types is already contained in the early-torpedo-stage embryo (Scheres *et al.*, 1995). Asymmetric cell divisions within the vascular primordium go on to establish the number of vascular initials present in the seedling root meristem.

Shoot Vascular Development

In the aerial parts of the seedling, the provascular elements are first specified and start to differentiate in the cotyledons, and only subsequently in the axis (Bauby *et al.*, 2007). Protophloem differentiation can first be observed in the leaf midvein by typical cell elongation, followed by extension to the distal loops and cotyledon node.

Continuous procambial strands can be observed (Fig. 14.1) prior to provascular differentiation. The main agent in the establishment of this pattern is the phytohormone auxin. The accumulation of auxin through

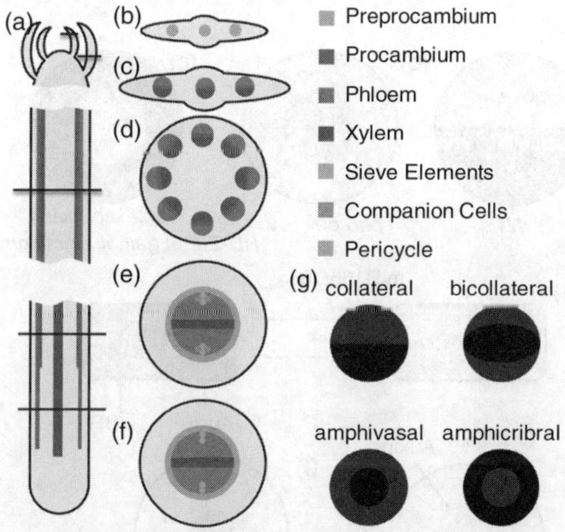

Figure 14.1 Primary phloem organization in Arabidopsis shoot and root. (a) Longitudinal section through the shoot apex, shoot, and root. (b) Cross-section of an early developing leaf showing the pre-procambial bundles which precede the vascular bundles. (c) Cross-section of a leaf showing established vascular bundles where primary phloem and xylem differentiate asymmetrically from a separating layer of procambium. (d) Cross-section of the stem showing primary vascular bundles. (e) Cross-section of the root showing the primary vascular patterning with two phloem poles consisting of sieve elements and companion cells flanking the xylem axis. (f) Cross-section of the root tip showing two poles of protophloem sieve elements flanking the xylem axis. (g) In different species, the radial patterning of the vasculature falls into distinct categories. In the collateral pattern, abaxial phloem is separated from adaxial xylem. In the bicollateral pattern, a central xylem pole is flanked by adaxial and abaxial phloem poles. In the amphivasal pattern, a xylem core is completely surrounded by phloem. In the amphicribal pattern, a phloem core is completely surrounded by xylem.

polar transport mechanisms, as shown by the synthetic auxin reporter DR5 and the auxin-induced pre-procambial marker AtHB8, precedes the formation of vascular strands in leaves (Scarpella *et al.*, 2004). Facilitating this highly localized auxin accumulation is the auxin transporter PIN1, which channels auxin to the provascular regions. PIN1 is already expressed before procambium formation.

After the initial differentiation, the vasculature develops in bundles (Esau, 1969). These bundles have a very distinct radial pattern of abaxial phloem and adaxial xylem divided by an active procambium. The radial patterning of the vascular bundles is already established during embryogenesis by the main factors of radial patterning, KANADI (KAN) and the class III HD-ZIP – PHABULOSA (PHB), PHAVOLUTA (PHV), REVOLUTA (REV), and CORONA (CNA) – transcription factors (Fig. 14.2; McConnell *et al.*, 2001; Emery *et al.*, 2003). Auxin also plays a role in this determination of the radial vascular patterning (Izhaki and Bowman, 2007). PIN1 localization is affected in *kan* mutants, showing the integration of the auxin transport pathway and KAN signaling.

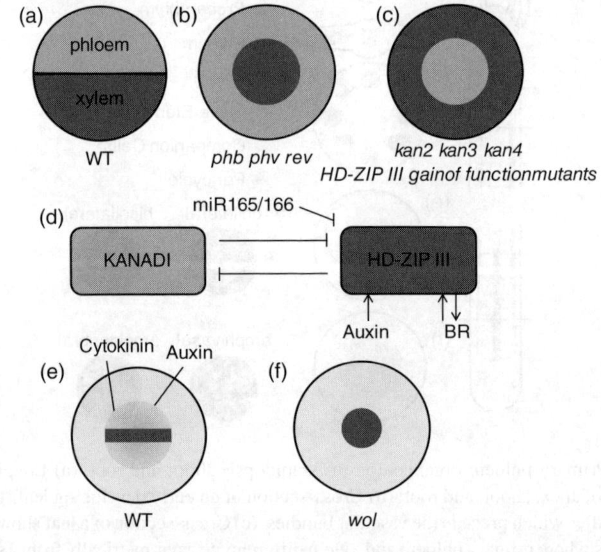

Figure 14.2 Vascular patterning is regulated by KANADI and HD-ZIP II genes and the phytohormones auxin and cytokinin. (a) In the shoot vascular bundles of Arabidopsis, the default radial pattern has phloem located abaxially and xylem adaxially. (b) In the *phb phv rev* mutant, phloem surrounds xylem. (c) In HD-ZIP III gain-of-function mutants and *kan1 kan2 kan3*, xylem surrounds phloem. (d) HD-ZIP III genes are regulated by miR165/166 and interact with both auxin and brassinosteroids. (e) In the root, auxin is restricted to the xylem axis by the presence of cytokinin. (f) In mutants with deficient cytokinin signalling such as *wol*, auxin is abundant throughout the stele which results in ubiquitous protoxylem differentiation and loss of phloem identity.

The triple mutant *phb phv rev* has radialized as well as abaxialized leaves and vascular bundles. In contrast, gain-of-function HD-ZIP III mutants with faulty microRNA (miRNA) regulation and *kan1 kan2 kan3* have radialized and adaxialized leaves and bundles. In addition, the HD-ZIP III genes also appear to regulate vascular tissue proliferation.

As described in more detail in 'Transport', the phloem translocation stream, or phloem sap, contains not only photosynthates but also a wide array of macromolecules such as mRNA, small RNAs, and proteins. In both animals and plants, small RNAs have already been identified as important regulatory factors controlling cell fate. A bidirectional cell-to-cell communication network involving the mobile transcription factor SHORTROOT (SHR) and microRNA165/166 species specifies the radial position of two types of xylem vessels in Arabidopsis roots (Carlsbecker *et al.*, 2010; Miyashima *et al.*, 2011). The microRNAs provide short-range non-cell-autonomous developmental signals that are transported through plasmodesmatal connections via a symplastic pathway (Vatén *et al.*, 2011). Since microRNA165/166 is a factor restricting PHB activity, it also regulates phloem development.

Root Vascular Development

In the root, protophloem initially differentiates from an independent differentiation locus in the upper hypocotyls; only later does protophloem differentiation from the root apical meristem start.

As roots grow, the cellular pattern is established and maintained by the self-renewal of pluripotent root meristem cells. Different cell identities are initiated from the stem cells around the quiescent center (QC): the provascular initials of stele, the cortex/endodermal initials, the epidermal/lateral root initials, and columella initials. In Arabidopsis roots, a central vascular cylinder (consisting of xylem, phloem, and procambium) is surrounded by radially symmetric layers of pericycle, endodermis, cortex, and endodermis cells. In Arabidopsis roots, the vascular tissue comprises a central axis of water-conductive xylem tissue that is flanked by two poles of photoassimilate-conductive phloem tissue.

In the root apical meristem, the phloem cell lineages arise from two domains of initials through asymmetric cell divisions (Mähönen *et al.*, 2000). Periclinal divisions establish companion cells and tangential divisions establish sieve elements. The asymmetry allows these initials to give rise to multiple cell lineages with different fates; in addition to the phloem lineage, they also precede undifferentiated procambial cell lineages. As opposed to the invariant pattern of cell lineages in the endodermis and outer layers, the number and exact pattern of these procambial divisions vary between individual seedlings.

Mähönen *et al.* (2000) have described these initial asymmetric cell divisions in great detail through sequential cross-sections of the region immediately above the quiescent centre, allowing the precise determination of the first true phloem domains. Although cell divisions and early xylem specification can be observed at first (but not yet the complete xylem axis), phloem identity is not observable directly above the QC. The first newly formed cell walls associated with phloem development are only visible 27 μm above the QC. At 69 μm above the QC, the first protophloem sieve elements are clearly present.

Cytokinin, an essential phytohormone for development in the root, is required for vascular patterning and the differentiation of all cell types except protoxylem. It has recently been shown that the root vascular pattern is defined by a mutually inhibitory interaction between cytokinin and auxin (Bishopp *et al.*, 2011a).

If cytokinin signaling is disturbed, as in the mutant *wol* (WOODEN LEG) or the triple cytokinin receptor mutant *ahk2 ahk3 ahk4* (ARABIDOPSIS HISTIDINE KINASE), or if cytokinin levels are reduced, as is found in transgenic plants overexpressing CKX (CYTOKININ OXIDASE), the effect is always an increased number of protoxylem cell files and the loss of other cell types in the root vasculature. The domain of cytokinin activity is restricted by the action of the cytokinin signaling inhibitor ARABIDOPSIS HISTIDINE PHOSPHOTRANSFER PROTEIN 6 (AHP6). Only through this mechanism does protoxylem differentiation occur in a spatially specific manner, allowing for the proper development of the phloem cell types.

While AHP6 infers the protoxylem identity, other genes have been described to be specifically involved in the specification of xylem and phloem identity. Most of these genes encode transcription factors, but cell death regulators and putative signaling proteins have also been implicated in this scheme. Currently, almost no data exist on the regulation of the specific expression of these genes.

Differentiation and Characteristics of Phloem

Phloem Differentiation

Sieve elements (SE), also known as sieve tube elements, form the conductive part of the phloem. As for companion cells (CC), they originate from phloem precursor cells in the procambium. They undergo severe changes in their morphology (Fig. 14.3) very early during primary phloem development. During maturation, the SEs experience extensive degradation of their organelles. The nucleus, vacuoles, rough endoplasmic reticulum (ER), and Golgi are all broken down in a process which has not yet been described in exhaustive detail. This reduction in cellular contents creates an effective transport route through the sieve tubes. In contrast to the tracheary elements of the xylem for example,

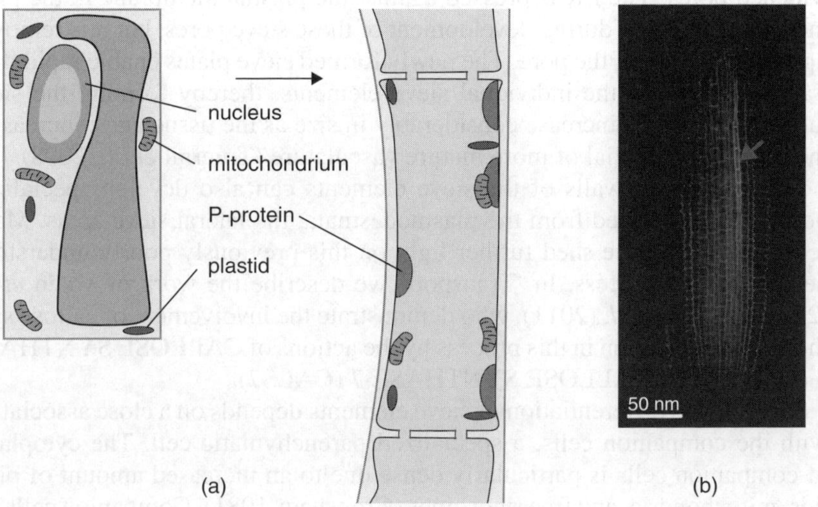

Figure 14.3 Maturation of sieve elements. (a) Sieve element differentiation (from left to right) involves sieve plate formation with callose-rich sieve pore development, cell wall thickening, organelle (nucleus, Golgi, rough ER, vacuole) degradation, cytoplasmic autolysis, phloem plastid differentiation, and P-protein development. (b) Protophloem imaging by mPS-PI staining in the Arabidopsis root. Two protophloem cell files are visualized in this root. The arrow highlights the already visible cell wall thickening.

sieve elements still remain living as they retain a plasma membrane and a reduced number of other organelles such as the smooth endoplasmic reticulum (ER), plastids, mitochondria, and P-proteins (Knoblauch and Peters, 2010). P-proteins were first identified as 'phloem slime' (Hartig, 1854; Esau, 1969). In some species, they are known to be involved in forisomes (Pélissier *et al.*, 2008). Since obstructions in the sieve elements can be used for flow control, the contractile P-proteins are suggested to reversibly block sieve tubes in case of injury (Knoblauch *et al.*, 2001, 2003).

Concurrently with the changes in cytosolic composition, the cell wall of the sieve elements also undergoes drastic changes. The first observable process is a vast increase in callose deposition. Callose is a β-1,3-glucan which is synthesized by callose synthases (Verma and Hong, 2001) and is deposited as platelets around the plasmodesmata of sieve elements, replacing cellulose which accumulated there previously. The cell walls which form the interface to the adjacent sieve elements are heavily enriched with these callose-rich plasmodesmata.

As the sieve elements mature, the callose in this region is dissolved and the middle lamella cell wall material is removed, forming a plate structure with widened pores. The ER is pressed against the plasma membrane in the plasmodesmatal region during development of these sieve pores, but it is removed upon completion of the pore. The newly formed sieve plates enable rapid fluid transport between the individual sieve elements, thereby forming the sieve tubes. These pores increase considerably in size as the tissue ages, increasing the transport potential of more mature vasculature (Truernit *et al.*, 2008).

The lateral cell walls of the sieve elements can also develop specialized porous zones derived from the plasmodesmata, the lateral sieve areas. Many recent findings have shed further light on this previously poorly understood developmental process. In 'Transport' we describe the work of Vatén *et al.* (2011) and Xie *et al.* (2011), who demonstrate the involvement of callose synthesis and deposition in this process by the actions of CALLOSE SYNTHASE 3 (CALS3) and CALLOSE SYNTHASE 7 (CALS7).

Survival and differentiation of sieve elements depends on a close association with the companion cells, a specialized parenchymatic cell. The cytoplasm of companion cells is particularly dense due to an increased amount of plastids, mitochondria, and free ribosomes (Cronshaw, 1981). Companion cells are connected to the neighboring sieve elements by multiple branched plasmodesmata. Through these connections, the enucleate sieve elements are supplied with energy, assimilates, and macromolecular compounds such as various proteins (Raven, 1991). The size exclusion limit of these connections lies between 10 and 40 kD, much higher than in plasmodesmata outside the phloem (Kempers and van Bel, 1997).

The morphological and physiological distinctiveness of the phloem cells as described above is the result of unique gene expression patterns, as shown by recent transcriptome studies (Lee *et al.*, 2006; Brady *et al.*, 2007, 2011). These transcriptional programs are tightly controlled in a spatio-temporal manner. The most comprehensive and high-resolution tissue-specific gene expression maps today have been supplied by extensive microarray experiments.

Many distinct expression patterns have been identified in these analyses; some are specific to sieve elements or to sieve elements and companion cells together. More than a thousand genes have been identified as having phloem-specific expression, highlighting phloem as a highly specialized and unique tissue within the stele. The data from these microarray studies have been made publically available in the AREX database, providing a huge boon to researchers looking for new phloem functionality.

Bauby *et al.* (2007) have identified several phloem markers, which they named Phloem Differentiation 1–5 (PD1–PD5) by screening the Versailles collection of gene trap mutants for plant lines expressing the uidA reporter gene in immature vascular tissues. PD1–PD4 were restricted to protophloem cells as determined by their cell shape. However, PD5 was expressed in both protophloem and metaphloem cells. Using these markers, the authors could track the onset of phloem development directly after embryogenesis. PD4 is expressed in the tip of leaf primordia 3 days after germination. Spreading from there, this marker has already traced the entire future leaf vasculature by 7 days after germination. The first expression of PD1 and PD3 was detected 3 days after germination in the proximal protophloem of leaf primordia. These results show that PD1 and PD3 are expressed during the differentiation of protophloem. PD4 and PD5 reporter gene expression was also detected in the location of midveins and higher-order veins before the procambium differentiation, thereby defining the pre-patterning of the future veins.

It has been confirmed that the phloem proteome is indeed very abundant in unique expression patterns; Lin *et al.* (2009) identified over 1000 unique proteins based on their analysis of pumpkin phloem sap. The phloem signaling network likely constitutes an integral part of the whole plant communication network. If particular importance is the abundance of RNA-binding proteins within the phloem, emphasizing the concept of a developmentally integrated signaling network acting through the phloem.

Molecular Mechanisms

Only two genes are known of at this point which specifically regulate phloem identity: *APL* (ALTERED PHLOEM DEVELOPMENT) and *OPS* (OCTOPUS).

APL (Bonke *et al.*, 2003), a MYB-coiled-coil transcription factor, is essential for the proper differentiation of both sieve elements and companion cells. Furthermore, it also contributes to the spatial limitation of xylem differentiation. It is expressed both in sieve elements and companion cells and has been shown to be nuclear localized.

As can be observed in the loss-of-function mutant *apl1* (Fig. 14.4), APL is of the utmost importance to proper phloem development as its loss has extraordinary effects: plants become short-rooted and die at the seedling stage. In cross-sections, neither sieve elements nor companion cells can be detected in the presumptive vasculature. When utilizing molecular phloem markers such as *SUC2* (SUCROSE TRANSPORTER 2) or the proto-SE reporter J0701, no trace of expression can be found. In the same vein, callose deposition in the sieve elements cannot be detected. This phenotype is however restricted entirely to the phloem, and all other tissues appear to develop normally.

As mentioned in 'Root Vascular Development', it is asymmetrical cell divisions that establish the phloem poles in wild type: periclinal divisions establish companion cells and tangential divisions establish sieve elements. In the *apl* mutant, these cell divisions are usually delayed. They still take place however, so APL does not appear to be required for the asymmetric cell divisions as such (only for their initiation). However, since the subsequent differentiation of sieve elements cannot be observed, it can be concluded that APL is

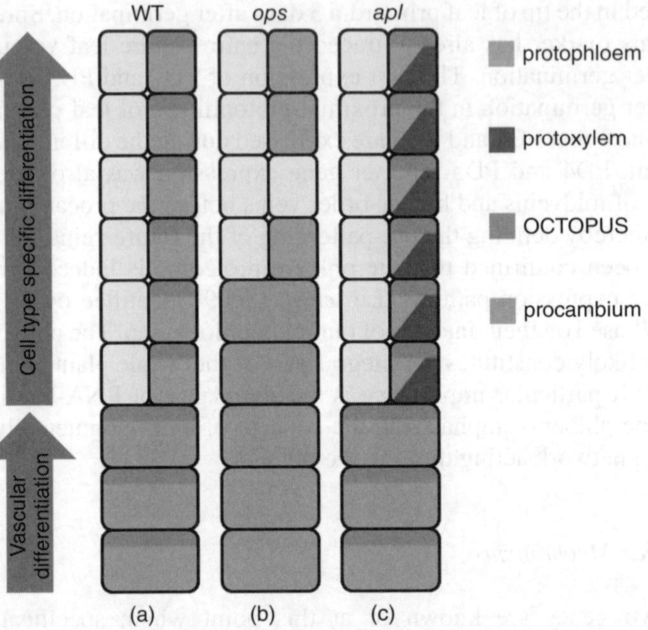

Figure 14.4 Model of OPS and APL action. (a) OPS is located at the apical plasmamembrane in the procambium and phloem lineage. OPS interprets vascular signals for phloem differentiation such as APL. (b) In the *ops* mutant, phloem differentiation is delayed. The procambial cell number is increased and gaps of undifferentiated cells are visible in the protophloem strand. (c) In the *apl* mutant, the initiation of phloem differentiation is largely unperturbed and proper protophloem fails to emerge. In its place, protophloem/protoxylem hybrid cells appear. (See plate section for the color version of this figure.)

responsible for phloem differentiation rather than establishment of the phloem cell lineage.

Furthermore, APL has also been proposed as an inhibitor to xylem differentiation. In the *apl* mutant, ectopic xylem strands are seen in place of the phloem poles. When APL is expressed ectopically in the vascular bundles, xylem formation is inhibited.

In more recent experiments, the *apl* mutant has been analyzed in more detail by using novel imaging techniques (Truernit *et al.*, 2012). With this higher-resolution data, it was discovered that protophloem differentiation actually proceeds regularly in this mutant until two days after germination. The cells in the protophloem position display the characteristic shape and cell wall thickening. After this time however, the previously described acquisition of xylem characteristics is observed. Strangely, the cells in place of the sieve elements do form sieve plates. These cells can therefore be described as hybrid cell forms between phloem and xylem.

In conclusion, APL is absolutely required for the later stages of phloem development, although it is not the earliest factor in this process. The identity of such a factor or factors has yet to be determined. This line of thought is further reinforced by the research of Lee *et al.* (2006), who point out that there are numerous phloem markers with earlier sieve element expression than APL.

OPS (OCTOPUS) is, after APL, the second currently known gene which specifically regulates phloem differentiation. OPS is required for phloem continuity during phloem development (Truernit *et al.*, 2012). It was first reported as a gene with vascular expression (Nagawa *et al.*, 2006); further detailed analysis revealed that OPS is initially expressed in provascular cells at the heart stage embryo and later becomes restricted to the phloem lineage cells when the phloem cell types are specified (Truernit *et al.*, 2012; Bauby *et al.*, 2007). Unlike APL, OPS expression is already observed in the phloem and procambium initials near the QC.

In the *ops* loss-of-function mutant, the vascular patterning of cotyledons in the mature embryo is delayed and the number of completed vascular loops of developing cotyledon is lowered (Truernit *et al.*, 2012). In contrast, the progression of vascular patterning is accelerated by OPS overexpression in the cotyledons of mature embryos and the number of completes vascular loops of developing cotyledon is increased. This suggests that OPS promotes the progression of vascular patterning. In *ops* hypocotyls and roots, the phloem sieve element cell files are interrupted by undifferentiated sieve element cells which failed to undergo the increasing of cell wall thickness, callose deposition, and nuclear degradation. SE-specific markers such as PD1 are not expressed either (Truernit *et al.*, 2012). These cell differentiation defects are the cause of inefficient phloem transport in root as demonstrated by the reduced transport of the AtSUC2::GFP marker, which is produced in companion cells in the differentiation zone and transported to root tip via sieve elements. These observations indicate that OPS function is required for continuous phloem development.

OPS encodes a membrane-associated protein (Benschop *et al.*, 2007) specific to higher plants (Nagawa *et al.*, 2006). The molecular function and mechanism of OPS remains unknown; no functional domain has been identified in the OPS protein so far. The functionally complementing OPS-GFP protein was located at the apical end of phloem SE cells (Truernit *et al.*, 2012). Inductive cell-to-cell communication from differentiated to an undifferentiated neighboring cell is known to occur during xylem differentiation; XYLOGEN is a polar localized proteoglycan-like factor and required to direct continuous xylem differentiation in *Zinnia elegans* L. and *Arabidopsis thaliana* (Motose *et al.*, 2004). It is assumed that OPS contributes to longitudinal signaling in a similar manner, thus inducing SE differentiation in undifferentiated SE precursor cells. Further study of OPS function and finding the related factors of OPS function will be important

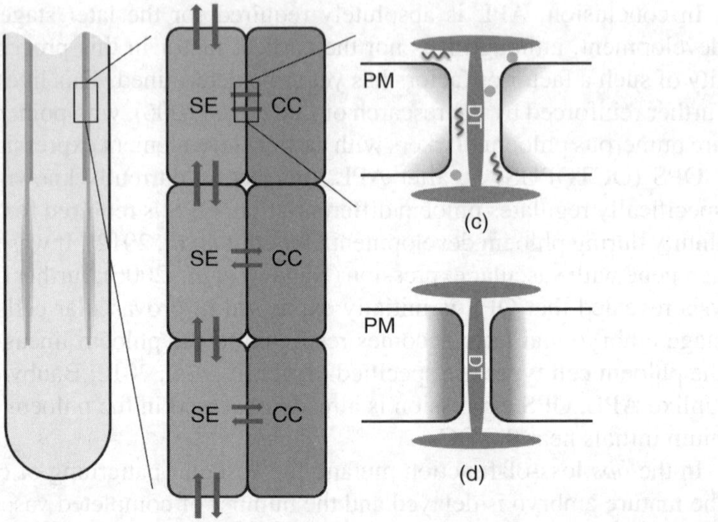

Figure 14.5 Schematic overview of phloem transport. (a) Phloem unloading occurs in protophloem sieve element strands (light blue). (b) Sieve elements are connected to each other through the sieve plates, enabling high-throughput long-distance transport in the phloem. They also share abundant symplastic connections with companion cells through plasmodesmata. (c) Schematic close-up of a plasmodesmatal connection with the desmotubule (DT). This connection enables the symplastic movement of compounds such as small RNAs or proteins. Callose (light blue) regulates the aperture size. (d) An increase in callose accumulation, e.g., through the action of the modified CalS3m, reduces aperture size to block all movement. (See plate section for the color version of this figure.)

for understanding how the phloem continuity is organized and how phloem develops.

Transport

The driving force for long-distance transport of sap through the phloem is phloem loading (Turgeon and Hepler, 1989; van Bel, 1993). This process describes the accumulation of photosynthates in minor veins, leading to sap flow towards the photosynthate-depleted sink tissues. In Arabidopsis, phloem loading occurs symplastically through the plasmodesmata (Fig. 14.5).

However, the function of phloem transport is not merely to supply the sink tissues with photosynthates, but also to transport signaling compounds throughout the plant. Analyses of phloem sap have found an impressive number of non-nutrient compounds such as large amounts of small RNAs, especially in response to nutrient starvation (Buhtz *et al.*, 2010).

These studies have already hinted at the operation of a sophisticated long-distance communication network utilizing the vascular system of higher plants. From an evolutionary perspective, the development of such a network reflects the need to communicate, and react to, environmental inputs which can be sensed by mature organs but may require a response from the meristematic regions of the plant. One consequence of this long-distance signaling system is that the newly formed organs can develop properly optimized for the environment into which they will all eventually emerge.

Researchers have identified a broad array of small RNAs and proteins which are being transported through the phloem translocation stream (Lee *et al.*, 2009). The authenticity of these macromolecules as *bona fide* components of the phloem sap offers support for the hypothesis that they participate in an inter-organ signaling network. The plethora of compounds found in the phloem sap has led some authors to refer to this conducting tissue as the 'information superhighway' of plants (Jorgensen *et al.*, 1998). One of the earliest demonstrations of long-distance phloem transport of diverse compounds was the observation that the dye CF-diacetate (CFDA) could be transported throughout the phloem (Oparka *et al.*, 1994). This work was followed by the discovery of even more phloem-mobile dyes such as the fluorescent probe HPTS (Birschwilks *et al.*, 2006). The regulatory protein FLOWERING LOCUS T (FT) is also believed to be transported through phloem and exported by companion cells (Mathieu *et al.*, 2009). Even phytohormones vital to vascular development have been shown to be transported plant-wide through the phloem (Bishopp *et al.*, 2011b).

Plant cells are connected through plasmodesmata (PD), membrane-lined channels that allow symplastic movement of molecules between cells. During plant morphogenesis, PDs can mediate the signaling network. One of the key factors in this process is CALS3 (CALLOSE SYNTHASE 3, Vatén *et al.*, 2011).

Callose synthases are membrane-bound complexes which synthesize the β-1,3-glucan callose (Colombani *et al.*, 2004), which in turn accumulates at plasmodesmata and regulates their aperture size. Conversely, the action of β-1,3-glucanases degrades callose (Levy *et al.*, 2007).

CALS3 is expressed in the phloem and meristematic tissues. Gain-of-function *cals3* mutants lead to the increased deposition of callose at the plasmodesmata, a decrease in the plasmodesmatal aperture size, multiple defects in root development, and the reduction of intercellular trafficking. An increase of CALS3 expression during phloem development suppresses loss-of-function mutations in another phloem-abundant callose synthase CALS7 (CALLOSE SYNTHASE 7), indicating that CALS3 is indeed a bona fide callose synthase.

Inducible expression of a modified CALS3 version provides a novel tool to spatially and temporally control the aperture size of plasmodesmata between plant tissues. Using this tool, it can be observed that visibly increased callose deposition inhibits the movement of the transcription factor SHORT-ROOT and microRNA165 between the stele and the endodermis, proving that this process is indeed dependent on symplastic transport through the plasmodesmata. The regulated biosynthesis of callose at the plasmodesmata is therefore essential for cell-to-cell communication.

As mentioned earlier, CALS7 is another phloem-localized callose synthase. CALS7 is specifically expressed in the phloem (Xie *et al.*, 2011). In *cals7* loss-of-fuction mutants, callose deposition in the phloem is severely reduced. Ultrastructure analysis of the developing sieve elements reveal that callose fails to accumulate in the plasmodesmata of incipient sieve plates at the early perforation stage of phloem development, resulting in the formation of sieve plates with fewer pores. While callose deposition can normally be induced by wounding as well, this process is also disrupted by the loss of CALS7 function. However, the mutant plants only exhibit a moderate reduction in seedling height and, while producing some aberrant pollen grains and short siliques with aborted embryos, remain fertile. This suggests that while CALS7 may play a role in plant growth and reproduction, it is not absolutely essential for plant development.

ARABIDOPSIS LATERAL ROOT DEVELOPMENT 3 (*LRD3*) is another important gene that regulates the early phloem development and controls the transport function of phloem (Ingram *et al.*, 2011). The *LRD3* gene encodes a LIM-domain protein which is specifically expressed in the companion cells of phloem. The normal function of LRD3 is to maintain a balance between primary and lateral root growth and phloem-mediated resource allocation among these roots. The *lrd3* loss-of-function mutant has decreased primary root growth and increased lateral root growth and density without having a significant effect on sucrose uptake. Aniline blue staining of the *lrd3* primary root shows a reduction of callose deposition in root meristems, developing phloem sieve elements and plasmodesmata pore units. This suggests a non-cell-autonomous role for LRD3 in early phloem development. A detailed long-distance transport analysis using different transport assays, including 14C-sucrose, a fluorescent dye carboxyfluorescein diacetate (CFDA), and companion cell-driven GFP (AtSUC2::GFP), demonstrates that delivery of phloem to the primary root tip is severely limited in young plants while phloem loading and export from the shoot occur normally. These phloem defects are subsequently rescued spontaneously in older plants, along with the subsequent increase in phloem delivery rates and primary root growth. An exogenous continuous treatment with auxin rescues the early phloem developmental defects and transport function in the primary root of *lrd3*, suggesting that auxin functions downstream of LRD3 or that it may have an independent role in the early development of phloem. In conclusion, there might exist some tightly regulated mechanisms which selectively support a biased phloem-mediated resource allocation in the lateral roots when the primary root is compromised.

The earlier mentioned P-proteins are encoded by the *SEO* gene family (*SIEVE ELEMENT OCCLUSION*, or *SEO-F* for *SIEVE ELEMENT OCCLUSION BY FORISOMES*; Pélissier *et al.*, 2008; Ruping *et al.*, 2010; Froelich *et al.*, 2011). Live imaging of the YFP-tagged version in Arabidopsis showed the accumulation of AtSEOR1 in sieve elements. AtSEOR meshworks are usually observed in the peripheral region and slot-like lumen structures in the SEs. However, the accumulation of AtSEOR alone does not significantly alter the flow capacity, suggesting further factors are involved.

FT-INTERACTING PROTEIN 1 (FTIP1) has recently been reported as a factor mediating phloem export from CCs into SEs (Liu *et al.*, 2012); CC-specific FTIP1 physically interacts with FT during export for long-distance transport in the flowering mechanism.

Evolutionary and Divergence Aspects

The development of a dedicated vasculature allowed plants to colonize the landmasses. While non-vascular plants mainly include green algae (chrolophytes), brown algae, red algae, and parts of bryophytes, vascular plants (tracheophytes) include lycophytes, ferns, gymnosperms including conifers, and angiosperms including dicots and monocots. Some bryophytes (mosses) develop primitive vascular systems with water-conducting cells (WCCs) and food-conducting cells (FCCs). The evolution of vasculature presumably occurred simultaneously and independently across the plant kingdom.

Shapes and Forms of Vascular Bundles

As mentioned above, vascular bundles are composed of procambium/cambium, xylem, and phloem. Most dicots and gymnosperms develop cambium in the vascular bundles to allow subsequent secondary growth. In monocots and ferns however, lack of cambium prevents any kind of secondary growth. The predominant bundle layout in these latter plants is the collateral placement, where an inner xylem tissue is set against an outer phloem tissue. Several less-common vascular tissue patterns also occur. Some monocot plants, such as Acorus, develop amphivasal vascular bundles where the phloem is surrounded by the xylem. In contrast to the amphivasal bundles, some angiosperms and ferns develop amphicribral vascular bundles where the xylem is surrounded by the phloem. Some species, such as Cucurbitaceae (cucurbits) and Solanaceae, arrange bicollateral bundles where xylem is located between external phloem (EP) and internal phloem (IP) in fascicular phloem (FP) (Clark and Schiefelbein, 1997). In particular, cucurbits also have spatially distinguishable extrafascicular phloem (EFP) veins (Crafts, 1932; Esau, 1969). The EFP is distributed in a scattered pattern in which EFP is located in the periphery of the FP, inside (entocyclic) and outside (ectocyclic) the sclerenchyma cell ring in the cortex, and commissural sieve tubes between FPs (Kempers and van Bel, 1997).

As described in 'Molecular Mechanisms', some mutants show alteration of the vascular cell patterns within vascular bundles in Arabidopsis. The loss-of-function triple mutant of HD-ZIP III genes, *phb phv rev*, shows amphicribal vascular bundles (Emery *et al.*, 2003). The gain-of-function mutations of *PHB*, *PHV*, or *REV/IFL1* and the loss-of-function double or triple mutants of *KANADI* genes, *kan1 kan2* mutant, or *kan1 kan2 kan3* mutant show amphivasal vascular bundles in the leaf (McConnell and Barton, 1998; McConnell *et al.*, 2001; Emery *et al.*, 2003; Eshed and Bowman, 2004). The loss-of-function mutant of *PHAN* in *Antirrhium majus* shows amphicribral vascular bundles (Waites and Hudson, 2001). In a cross-species approach, transcriptome analyses were performed to compare the divergent vascular tissues of monocots and dicots. These experiments are beginning to identify the conserved vascular genes and biological processes fundamental to vascular development (Xu *et al.*, 2013).

Cytological Evolution of Phloem

Early in the evolution of phloem, extant bryophytes developed food-conducting cells (FCCs). Specialized FCCs, which are morphologically similar to a tracheophyte sieve cell, are formed in some bryophytes (polytrichales) around hydroids and associated with specialized parenchyma cells. These specialized FCCs are also known as leptoids (Tansley and Chick, 1901). This developmental strategy is similar to the association between sieve elements and companion cells in tracheophytes. FCCs and leptoids in bryophytes are similar to the sieve elements of tracheophytes, as they also lose a central vacuole and develop secondary cell walls with callose-associated pores. However, in the FCCs, leptoids, and the sieve elements of lycophytes, ferns, and conifers, the desmotubules remain in the pores and P-proteins are not accumulated. The nuclear degradation aspect is observed in some species of bryophytes (Ligrone *et al.*, 2000; Pressel *et al.*, 2006; Ligrone *et al.*, 2012).

Ongoing Efforts

During the last decade, comparative genomics and transcriptome analyses of various species were performed by investigating the alterations of genome structure, conserved genes, and regulation of gene expression in order to address plant evolution (Feuilet and Keller, 2002; Kirst *et al.*, 2003; Brenner *et al.*, 2005; Pavy *et al.*, 2005; Ujino-Ihara *et al.*, 2005; Bowman *et al.*, 2007; Ralph *et al.*, 2008; Xu *et al.*, 2013). Individual genes which are known to be involved in vascular tissue formation were also analyzed in various species. For example, HD-ZIP III genes are a highly conserved gene family in land plants (Floyd *et al.*, 2006). No close homolog of the MYB domain transcription factor *APL*, a master regulator of phloem specification, is found in non-vascular chlorophytes however, while the MYB domain genes are found in chlorophytes (Prouse and Campbell, 2012). This suggests a different role in ancestral species and later participation in vascular development machinery.

In phloem evolution and divergence, the regulatory upstream sequences of Arabidopsis phloem genes (originally identified by Ruiz-Medrano *et al.*, 2011) were compared in the species from chlorophytes to angiosperms (Martinez-Navarro *et al.*, 2013). The regulatory upstream sequences of Arabidopsis are conserved in bryophytes and lycophytes, but not in non-vascular chlorophytes.

Conclusions

The importance of phloem to the plant is not to be underestimated; it is indeed vital for more than just the well-established function of photoassimilate transport from the photosynthetic organs to the sink tissues. Newly discovered transport functions continue such as the phloem-based transport of phytohormones, small RNAs, and even larger proteins. These transport mechanisms help facilitate the coordinated developmental programs in meristematic regions. In addition to these specialized transport functions, phloem is also noteworthy because of the highly distinctive morphology of its cell types. The interaction of enucleated sieve elements and still functional companion cells is highly integrated. Recently, callose has been identified as being vital to the regulation of symplastic trafficking in phloem and from phloem to adjacent tissues. Callose deposition is regulated by the activity of callose synthases, such as CALS3 and CALS7, and the betaglucanases. Tissue-specific callose deposition by a modified CalS3m is a novel and powerful tool in regulation of transport between cells.

Acknowledgements

We would like to thank Timo 'Tinde' Päivärinta for his help with the figures.

References

Bauby, H., Divol, F., Truernit, E., Grandjean, O., and Palauqui, J.C. (2007) Protophloem differentiation in early Arabidopsis thaliana development. *Plant Cell Physiology*, **48**(1), 97–109.

Benschop, J.J., Mohammed, S., O'Flaherty, M., Heck, A.J., Slijper, M., and Menke, F.L. (2007) Quantitative phosphoproteomics of early elicitor signaling in Arabidopsis. *Molecular and Cellular Proteomics*, **6**(7), 1198–1214.

Birschwilks, M., Haupt, S., Hofius, D., and Neumann, S. (2006) Transfer of phloem-mobile substances from the host plants to the holoparasite Cuscuta sp. *Journal of Experimental Botany*, **57**(4), 911–921.

Bishopp, A., Help, H., El-Showk, S., Weijers, D., Scheres, B., Friml, J., Benková, E., Mähönen, A.P., and Helariutta, Y. (2011a) A mutually inhibitory interaction between auxin and cytokinin specifies vascular pattern in roots. *Current Biology*, **21**(11), 917–926.

Bishopp, A., Lehesranta, S., Vatén, A., Help, H., El-Showk, S., Scheres, B., Helariutta, K., Mähönen, A.P., Sakakibara, H., and Helariutta, Y. (2011b) Phloem-transported cytokinin regulates polar auxin transport and maintains vascular pattern in the root meristem. *Current Biology*, **21**(11), 927–932.

Bonke, M., Thitamadee, S., Mähönen, A.P., Hauser, M.T., and Helariutta, Y. (2003) APL regulates vascular tissue identity in Arabidopsis. *Nature*, **426**(6963), 181–186.

Bowman, J.L., Floyd, S.K., and Sakakibara, K. (2007) Green genes-comparative genomics of the green branch of life. *Cell*, **129**(2), 229–234.

Brady, S.M., Orlando, D.A., Lee, J.Y., Wang, J.Y., Koch, J., Dinneny, J.R., Mace, D., Ohler, U., and Benfey, P.N. (2007) A high-resolution root spatiotemporal map reveals dominant expression patterns. *Science*, **318**(5851), 801–806.

Brady, S.M., Zhang, L., Megraw, M., Martinez, N.J., Jiang, E., Yi, C.S., Liu, W., Zeng, A., Taylor-Teeples, M., Kim, D., Ahnert, S., Ohler, U., Ware, D., Walhout, A.J., and Benfey, P.N. (2011) A stele-enriched gene regulatory network in the Arabidopsis root. *Molecular Systems Biology*, **7**, 459.

Brenner, E.D., Katari, M.S., Stevenson, D.W., Rudd, S.A., Douglas, A.W., Moss, W.N., Twigg, R.W., Runko, S.J., Stellari, G.M., McCombie, W.R., and Coruzzi, G.M. (2005) EST analysis in Ginkgo biloba, an assessment of conserved developmental regulators and gymnosperm specific genes. *BMC Genomics*, **6**, 143.

Buhtz, A., Pieritz, J., Springer, F., and Kehr, J. (2010) Phloem small RNAs, nutrient stress responses, and systemic mobility. *BMC Plant Biology*, **10**, 64.

Carlsbecker, A., Lee, J.Y., Roberts, C.J., Dettmer, J., Lehesranta, S., Zhou, J., Lindgren, O., Moreno-Risueno, M.A., Vatén, A., Thitamadee, S., Campilho, A., Sebastian, J., Bowman, J.L., Helariutta, Y., and Benfey, P.N. (2010) Cell signalling by microRNA165/6 directs gene dose-dependent root cell fate. *Nature*, **465**(7296), 316–321.

Clark, S.E. and Schiefelbein, J.W. (1997) Expanding insights into the role of cell proliferation in plant development. *Trends in Cell Biology*, **7**(11), 454–458.

Colombani, A., Djerbi, S., Bessueille, L., Blomqvist, K., Ohlsson, A., Berglund, T., Teeri, T.T., Bulone, V. (2004) In vitro synthesis of (1→3)-β-D-glucan (callose) and cellulose by detergent extracts of membranes from cell suspension cultures of hybrid aspen. *Cellulose*, **1**(3–4), 313–327.

Crafts, A.S. (1932) Phloem anatomy, exudation, and transport of organic nutrients in cucurbits. *Plant Physiology*, **7**(2), 4–22.

Cronshaw, J. (1981) Phloem structure and function. *Annual Review of Plant Physiology*, **32**, 465–484.

Emery, J.F., Floyd, S.K., Alvarez, J., Eshed, Y., Hawker, N.P., Izhaki, A., Baum, S.F., and Bowman, J.L. (2003) Radial patterning of Arabidopsis shoots by class III HD-ZIP and KANADI genes. *Current Biology*, **13**, 1768–1774.

Esau, K. (1969) *The Phloem. Encyclopedia of Plant Anatomy*. Borntaeger, Berlin.

Eshed, Y. and Bowman, J.L. (2004) MicroRNAs guide asymmetric DNA modifications guiding asymmetric organs. *Developmental Cell*, **7**(5), 629–630.

Feuillet, C. and Keller, B. (2002) Comparative genomics in the grass family, molecular characterization of grass genome structure and evolution. *Annals of Botany*, **89**(1), 3–10.

Floyd, S.K., Zalewski, C.S., and Bowman, J.L. (2006) Evolution of class III homeodomain-leucine zipper genes in streptophytes. *Genetics*, **173**(1), 373–388.

Froelich, D.R., Mullendore, D.L., Jensen, K.H., Ross-Elliott, T.J., Anstead, J.A., Thompson, G.A., Pélissier, H.C., and Knoblauch, M. (2011) Phloem ultrastructure and pressure flow: Sieve-Element-Occlusion-Related agglomerations do not affect translocation. *Plant Cell*, **23**(12), 4428–4445.

Hartig, T. (1854) Über die Querscheidewände zwischen den einzelnen Gliedern der Siebröhren in Cucurbita pepo. *Botanische Zeitung (Berlin)*, **12**, 51–54.

Ingram, P., Dettmer, J., Helariutta, Y., and Malamy, J.E. (2011) Arabidopsis Lateral Root Development 3 is essential for early phloem development and function, and hence for normal root system development. *Plant Journal*, **68**(3), 455–467.

Izhaki, A. and Bowman, J. (2007) KANADI and Class III HD-Zip Gene Families Regulate Embryo Patterning and Modulate Auxin Flow during Embryogenesis in Arabidopsis. *Plant Cell*, **19**, 495–508.

Jorgensen, R.A., Atkinson, R.G., Forster, R.L., and Lucas, W.J. (1998) An RNA-based information superhighway in plants. *Science*, **279**(5356), 1486–1487.

Kempers, R. and van Bel, A.J.E. (1997) Symplasmic connections between sieve element and companion cell in the stem of Vicia fava L. have a molecular exclusion limit of at least 10 kD. *Planta*, **201**, 195–201.

Kirst, M., Johnson, A.F., Baucom, C., Ulrich, E., Hubbard, K., Staggs, R., Paule, C., Retzel, E., Whetten, R. and Sederoff, R. (2003) Apparent homology of expressed genes from wood-forming tissues of loblolly pine (Pinus taeda L.) with Arabidopsis thaliana. *Proceedings of the National Academy of Science, USA*, **100**(12), 7383–7388.

Knoblauch, M. and Peters, W.S. (2010) Münch, morphology, microfluidics - our structural problem with the phloem. *Plant Cell Environment*, **33**(9), 1439–1452.

Knoblauch, M., Peters, W.S., Ehlers, K., and van Bel, A.J. (2001) Reversible calcium-regulated stopcocks in legume sieve tubes. *Plant Cell*, **13**(5), 1221–1230.

Knoblauch, M., Noll, G.A., Müller, T., Prüfer, D., Schneider-Hüther, I., Scharner, D., Van Bel, A.J., and Peters, W.S. (2003) ATP-independent contractile proteins from plants. *Nature Materials*, **2**(9), 600–603.

Lee, B.R., Jin, Y.L., Avice, J.C., Cliquet, J.B., Ourry, A., and Kim, T.H. (2009) Increased proline loading to phloem and its effects on nitrogen uptake and assimilation in water-stressed white clover (Trifolium repens*). New Phytologist*, **182**(3), 654–663.

Lee, J.Y., Colinas, J., Wang, J.Y., Mace, D., Ohler, U., and Benfey, P.N. (2006) Transcriptional and post-transcriptional regulation of transcription factor expression in Arabidopsis roots. *Proceedings of the National Academy of Science, USA*, **103**(15), 6055–6060.

Levy, A., Erlanger, M., Rosenthal, M., and Epel, B.L. (2007) A plasmodesmata-associated beta-1,3-glucanase in Arabidopsis. *Plant Journal*, **49**(4), 669–682.

Ligrone, R., Ducket, J.G., and Renzaglia, K.S. (2000) Conducting tissues and phyletic relationships of bryophytes. *Philosophical Transactions of the Royal Society of London, B: Biological Sciences*, **355**(1398), 795–813.

Ligrone, R., Duckett, J.G., and Renzaglia, K.S. (2012) The origin of the sporophyte shoot in land plants, a bryological perspective. *Annals of Botany*, **110**(5), 935–941.

Lin, M.K., Lee, Y.J., Lough, T.J., Phinney, B.S., and Lucas, W.J. (2009) Analysis of the pumpkin phloem proteome provides insights into angiosperm sieve tube function. *Molecular and Cellular Proteomics*, **8**(2), 343–356.

Liu, L., Liu, C., Hou, X., Xi, W., Shen, L., Tao, Z., Wang, Y., and Yu, H. (2012) FTIP1 is an essential regulator required for florigen transport. *PLoS Biology*, **10**(4), e1001313.

Mähönen, A.P., Bonke, M., Kauppinen, L., Riikonen, M., Benfey, P.N., and Helariutta, Y. (2000) A novel two-component hybrid molecule regulates vascular morphogenesis of the Arabidopsis root. *Genes and Development*, **14**(23), 2938–2943.

Martínez-Navarro, A.C., Galván-Gordillo, S.V., Xoconostle-Cázares, B., and Ruiz-Medrano, R. (2013) Vascular gene expression, a hypothesis. *Frontiers in Plant Science*, **4**, 261.

Mathieu, J., Warthmann, N., Küttner, F., and Schmid, M. (2009) Export of FT protein from phloem companion cells is sufficient for floral induction in Arabidopsis. *Current Biology*, **17**(12), 1055–1060.

McConnell, J.R. and Barton, M.K. (1998) Leaf polarity and meristem formation in Arabidopsis. *Development*, **125**(15), 2935–2942.

McConnell, J.R., Emery, J., Eshed, Y., Bao, N., Bowman, J., and Barton, M.K. (2001) Role of PHABULOSA and PHAVOLUTA in determining radial patterning in shoots. *Nature*, **411**, 709–713.

Miyashima, S., Koi, S., Hashimoto, T., and Nakajima, K. (2011) Non-cell-autonomous microRNA165 acts in a dose-dependent manner to regulate multiple differentiation status in the Arabidopsis root. *Development*, **138**(11), 2303–2313.

Motose, H., Sugiyama, M., and Fukuda, H. (2004) A proteoglycan mediates inductive interaction during plant vascular development. *Nature*, **429**(6994), 873–878.

Nagawa, S., Sawa, S., Sato, S., Kato, T., Tabata, S., and Fukuda, H. (2006) Gene trapping in Arabidopsis reveals genes involved in vascular development. *Plant Cell Physiology*, **47**(10), 1394–1405.

Oparka, K.J. and Turgeon, R. (1999) Sieve elements and companion cells-traffic control centers of the phloem. *Plant Cell*, **11**(4), 739–750.

Oparka, K.J., Duckett, C.M., Prior, D.A.M., and Fisher, D.B. (1994) Realtime imaging of phloem unloading in the root tip of Arabidopsis. *The Plant Journal*, **6**, 759–766.

Pavy, N., Paule, C., Parsons, L., Crow, J.A., Morency, M.J., Cooke, J., Johnson, J.E., Noumen, E., Guillet-Claude, C., Butterfield, Y., Barber, S., Yang, G., Liu, J., Stott, J., Kirkpatrick, R., Siddiqui, A., Holt, R., Marra, M., Seguin, A., Retzel, E., Bousquet, J., and MacKay, J. (2005) Generation, annotation, analysis and database integration of 16,500 white spruce EST clusters. *BMC Genomics*, **6**, 144.

Pélissier, H.C., Peters, W.S., Collier, R., van Bel, A.J., and Knoblauch, M. (2008) GFP tagging of sieve element occlusion (SEO) proteins results in green fluorescent forisomes. *Plant Cell Physiology*, **49**(11), 1699–1710.

Pressel, S., Ligrone, R., and Duckett, J.G. (2006) Effects of de- and rehydration on food-conducting cells in the moss Polytrichum formosum, a cytological study. *Annals of Botany*, **98**(1), 67–76.

Prouse, M.B. and Campbell, M.M. (2012) The interaction between MYB proteins and their target DNA binding sites. *Biochimica et Biophysica Acta*, **1819**(1), 67–77.

Ralph, S.G., Chun, H.J., Kolosova, N., Cooper, D., Oddy, C., Ritland, C.E., Kirkpatrick, R., Moore, R., Barber, S., Holt, R.A., Jones, S.J., Marra, M.A., Douglas, C.J., Ritland, K., and Bohlmann, J. (2008) A conifer genomics resource of 200,000 spruce (Picea spp.) ESTs and 6,464 high-quality, sequence-finished full-length cDNAs for Sitka spruce (Picea sitchensis). *BMC Genomics*, **9**, 484.

Raven, J.A. (1991) Long-term functioning of enucleate sieve elements: possible mechanisms of damage avoidance and damage repair. *Plant, Cell and Environment*, **14**(2), 139–146.

Ruiz-Medrano, R., Xoconostle-Cázares, B., Ham, B.K., Li, G., and Lucas, W.J. (2011) Vascular expression in Arabidopsis is predicted by the frequency of CT/GA-rich repeats in gene promoters. *Plant Journal*, **67**(1), 130–144.

Rüping, B., Ernst, A.M., Jekat, S.B., Nordzieke, S., Reineke, A.R., Müller, B., Bornberg-Bauer, E., Prüfer, D., and Noll, G.A. (2010) Molecular and phylogenetic characterization of the sieve element occlusion gene family in Fabaceae and non-Fabaceae plants. *BMC Plant Biology*, **10**, 21.

Scarpella, E., Francis, P., and Berleth, T. (2004) Stage-specific markers define early steps of procambium development in Arabidopsis leaves and correlate termination of vein formation with mesophyll differentiation. *Development*, **131**(14), 3445–3455.

Scheres, B., Di Laurenzio, L., Willemsen, V., Hauser, M.-T., Janmaat, K., Weisbeek, P., and Benfey, P.N. (1995) Mutations affecting the radial organisation of the Arabidopsis root display specific defects throughout the embryonic axis. *Development*, **121**, 53–62.

Tansley, A.G. and Chick, E. (1901) Notes on the conducting tissue-system in Bryophyta. *Annals of Botany*, **15**, 1–38.

Truernit, E., Bauby, H., Dubreucq, B., Grandjean, O., Runions, J., Barthélémy, J., and Palauqui, J.C. (2008) High-resolution whole-mount imaging of three-dimensional tissue organization and gene expression enables the study of Phloem development and structure in Arabidopsis. *Plant Cell*, **20**(6), 1494–1503.

Truernit, E., Bauby, H., Belcram, K., Barthélémy, J., and Palauqui, J.C. (2012) OCTOPUS, a polarly localised membrane-associated protein, regulates phloem differentiation entry in Arabidopsis thaliana. *Development*, **139**(7), 1306–1315.

Turgeon, R. and Hepler, P.K. (1989) Symplastic continuity between mesophyll and companion cells in minor veins of mature Cucurbita pepo L. leaves. *Planta*, **179**(1), 24–31.

Ujino-Ihara, T., Kanamori, H., Yamane, H., Taguchi, Y., Namiki, N., Mukai, Y., Yoshimura, K., and Tsumura, Y. (2005) Comparative analysis of expressed sequence tags of conifers and angiosperms reveals sequences specifically conserved in conifers. *Plant Molecular Biology*, **59**(6), 895–907.

Vatén, A., Dettmer, J., Wu, S., Stierhof, Y.D., Miyashima, S., Yadav, S.R., Roberts, C.J., Campilho, A., Bulone, V., Lichtenberger, R., Lehesranta, S., Mähönen, A.P., Kim, J.Y., Jokitalo, E., Sauer, N., Scheres, B., Nakajima, K., Carlsbecker, A., Gallagher, K.L., and Helariutta, Y. (2011) Callose biosynthesis regulates symplastic trafficking during root development. *Developmental Cell*, **21**(6), 1144–1155.

Verma, D.P. and Hong, Z. (2001) Plant callose synthase complexes. *Plant Molecular Biology*, **47**(6), 693–701.

Waites, R. and Hudson, A. (2001) The Handlebars gene is required with Phantastica for dorsoventral asymmetry of organs and for stem cell activity in Antirrhinum. *Development*, **128**(11), 1923–1931.

Xie, B., Wang, X., Zhu, M., Zhang, Z., and Hong, Z. (2011) CalS7 encodes a callose synthase responsible for callose deposition in the phloem. *Plant Journal*, **65**(1), 1–14.

Xu, P., Kong, Y., Li, X., and Li, L. (2013) Identification of molecular processes needed for vascular formation through transcriptome analysis of different vascular systems. *BMC Genomics*, **14**, 217.

Index

Plant Cell Wall Patterning and Cell Shape, First Edition. Edited by Hiroo Fukuda.

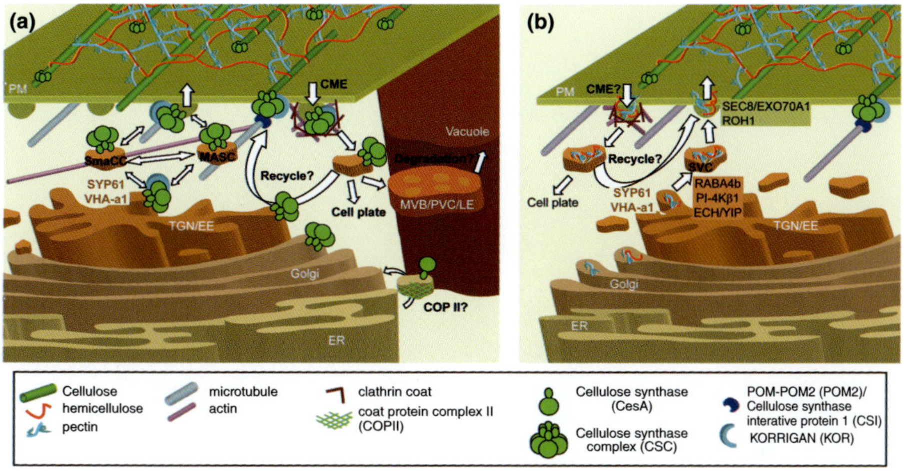

Figure 2.3 Membrane trafficking in cell wall polysaccharide metabolism. (a) Membrane trafficking during cellulose synthesis. Cellulose synthase (CesA) subunits are assembled into CesA complexes (CSCs) at the endoplasmic reticulum (ER) or the Golgi apparatus (in the figure the last option is represented). Individual CesAs, or assembled CSCs, should reach the Golgi via coat protein complex II (COP-II). From the trans-Golgi network/early endosome (TGN/EE), the CSCs travel to the plasma membrane (PM) probably via SmaCCs/MASCs (small CESA compartments/microtubule-associated CSCs), which move over the actin and microtubule cytoskeleton. SmaCCs/MASCs partially co-localize with the TGN-markers SYP61 and VHA-a1. Once at the PM, the CSCs follow linear paths that are typically guided by cortical microtubules, due to the CSC connection to the microtubule-binding protein POM2/CSI1 (POM-POM2/cellulose synthase interactive protein 1). The CSCs are internalized, probably via clathrin-mediated endocytosis (CME). The CSCs might then be recycled back to the PM via TGN/EE or sent for degradation to the vacuole. In dividing cells, the CSCs are recruited to the cell plate through the CME machinery. The CSC interacting protein KORRIGAN (KOR) is also necessary for cellulose synthesis and traffics via similar paths as that of CSCs. (b) Membrane trafficking of matrix polysaccharides. Both pectins and hemicelluloses are synthesized in the Golgi apparatus and delivered to the PM via secretory vesicle clusters (SVCs). The SVCs are formed at the TGN/EE by the coordinated activity of various proteins such as the small-GTPase RABA4b, the phosphatidylinositol-4-kinase (PI-4Kβ1), and ECH/YIP complex (ECHIDNIA/YPT/RAB GTPase interacting proteins). The SVCs and some of the proteins required for their formation co-localize with the TGN/EE-markers SYP61 and VHA-a1. During seed coat formation, pectin-containing vesicles tether to the PM through the exocyst subunits SEC8 and EXO70A1 and their interactor ROH1. In dividing root cells xyloglucans and pectins are internalized and relocated to the cell plate, probably via CME. After endocytosis, part of the matrix polysaccharides might be recycled back to the PM.

Plant Cell Wall Patterning and Cell Shape, First Edition. Edited by Hiroo Fukuda.

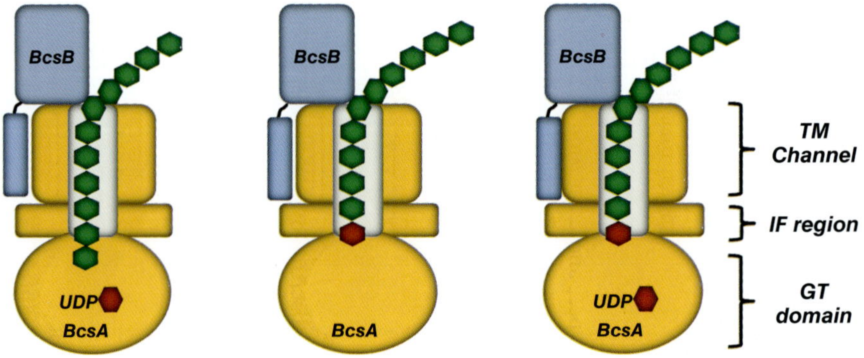

Figure 3.1 Features and proposed mechanism of the bacterial cellulose synthase complex. The BcsA-B dimer from Rhodobacter sphaeroides was crystallized as a 1:1 heterodimer (Morgan *et al.*, 2013). BcsA (yellow) consists of distinct transmembrane, interfacial, and glycosyltransferase domain regions. The single transmembrane domain of BcsB (blue) interacts with the transmembrane helices of BcsA to form a stable complex in the bacterial plasma membrane. BcsA was crystallized with a nascent glucan chain (green hexagons) that begins in the BcsA glycosyltransferase domain, transits a pore in the BcsA transmembrane region, and interacts with BcsB in the extracellular space. The glycosyltransferase domain binds a single UDP-glucose molecule (red hexagon), which serves as a sugar nucleotide donor that elongates the nascent glucan chain by a single glucose residue in each catalytic cycle. As a new glucose residue is added to the glucan chain (red hexagon), the polysaccharide is processively pushed further into the extracellular matrix.

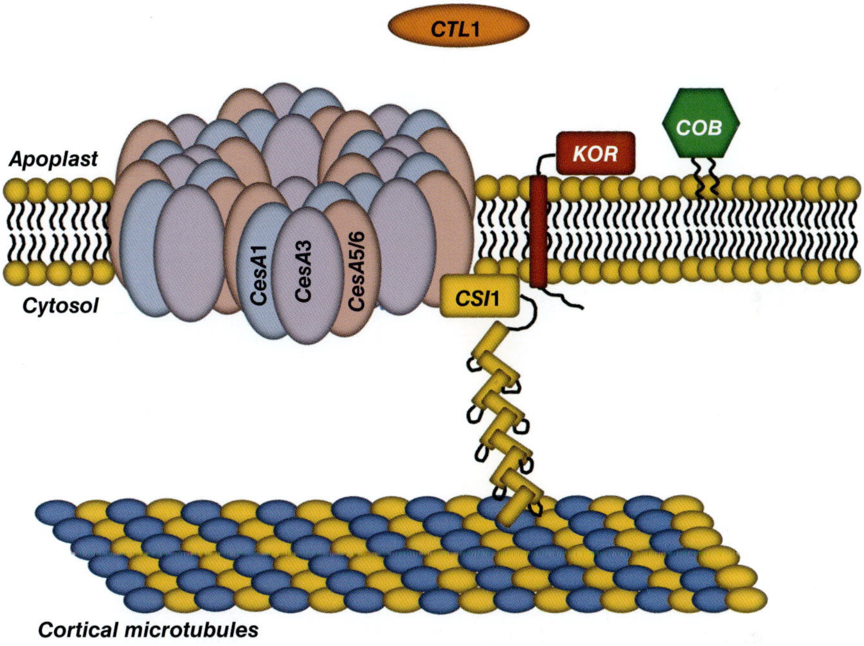

Figure 3.2 Features of the higher plant cellulose synthase complex and accessory proteins involved in cellulose biosynthesis. The active cellulose synthase complex (CSC) is composed of three unique Cellulose Synthase A (CesA) subunits. The complex that produces cellulose in the primary cell wall contains CesA1 (blue), CesA3 (purple), and a CesA6-like subunit (pink). Multiple copies of these proteins are present in the complex, and the most likely multimeric organization consists of up to 36 CesA subunits arranged in a hexamer of hexamers to form a structure that resembles the rosette terminal complexes observed in freeze-fracture electron microscopy. The active CSC forms discrete complexes in the plasma membrane, and these complexes are tethered to cortical microtubules through the influence of Cellulose Synthase Interacting protein 1 (CSI1). The complex moves along cortical microtubules as it is driven by synthesis of new crystalline cellulose in the apoplast. Other proteins have been implicated as positive regulators of cellulose biosynthesis. The KORRIGAN endoglucanase (red) localizes to the plasma membrane and contains a single membrane spanning region as well as an extracellularly localized endoglucanase domain. COBRA (green) is implicated in cellulose biosynthesis, is localized to the plasma membrane by a GPI anchor, and potentially facilitates the organization of nascent microfibrils. CTL1 (orange) is a secreted glucan binding protein that may also play a role in organizing glucan chains once they have been extruded into the apoplast.

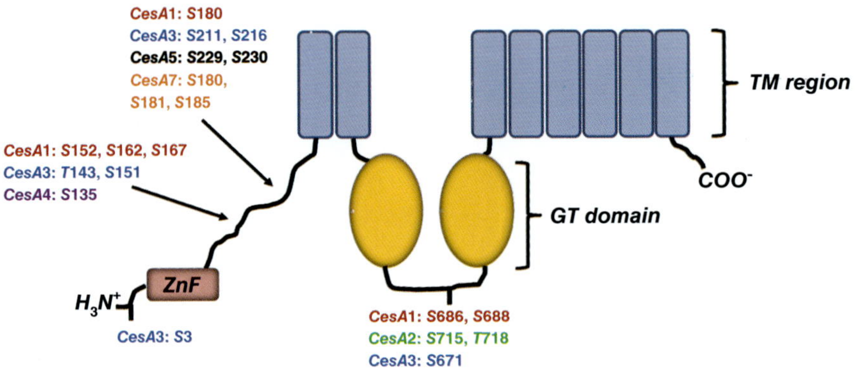

Figure 3.4 Summary of proteomically supported phosphorylation events on CesA subunits. CesAs exhibit a conserved domain architecture consisting of an N-terminal zinc finger (pink), eight transmembrane domains (blue), and a large central loop containing sequence motifs that are typical of processive glycosyltransferases (yellow). Phosphoproteomic surveys indicate that Arabidopsis CesA1, 2, 3, 4, 5, and 7 are phosphorylated at multiple positions within the cytosolic N-terminus and the large central catalytic loop. The relative position of each phosphorylation site is shown on a generalized model of CesA structure, with the residue indices and CesA isoform indicated. The color code for phosphorylation sites is: CesA1 (red), CesA2 (green), CesA3 (blue), CesA4 (purple), CesA5 (black), and CesA7 (orange).

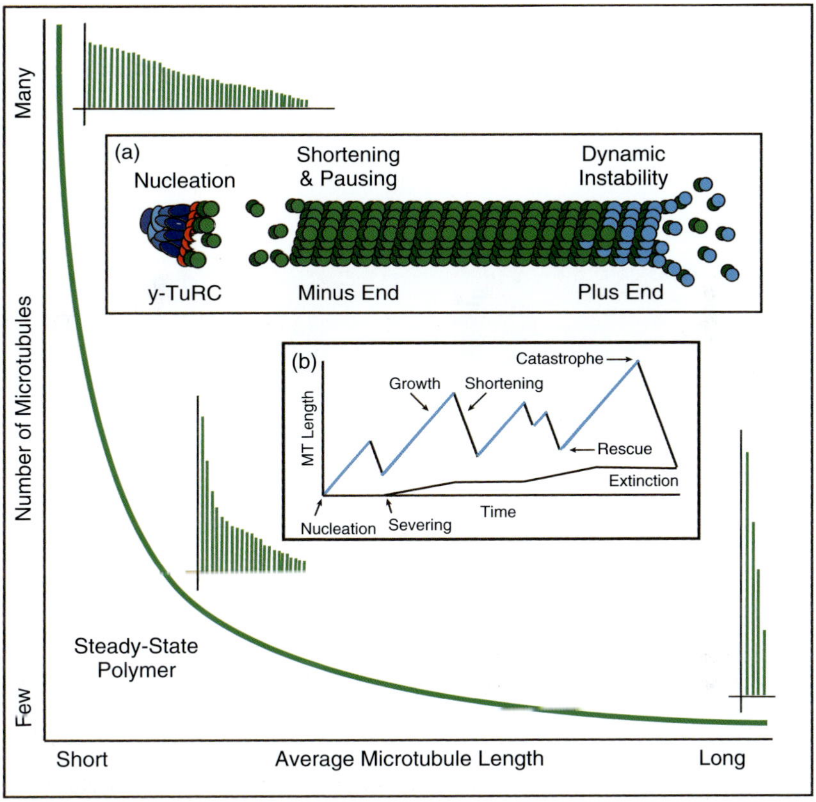

Figure 4.1 Microtubule structure, dynamics, and steady-state length distribution. (a) Microtubules in cells nucleate from templates containing a ring of γ-tubulin and associated proteins. Dimeric tubulin subunits line up head-to-tail in 13 offset rows, termed protofilaments, forming a tube. Microtubules in flowering plants are severed from the γ-TuRC and undergo polymer treadmilling at the cell cortex. (b) The microtubule minus end undergoes switching between slow depolymerization and pausing. The plus end switches between states of more rapid growth, accumulating GTP-bound tubulin dimers (light blue), and very rapid shortening when GTP hydrolysis catches up to the microtubule end. The combined dynamic instability at the plus end and depolymerization at the minus end will produce a roughly constant total polymer length in the cell. The microtubule dynamics, coupled with additional severing, will partition the polymer into a number of microtubules with an average length that form a length distribution within the cell.

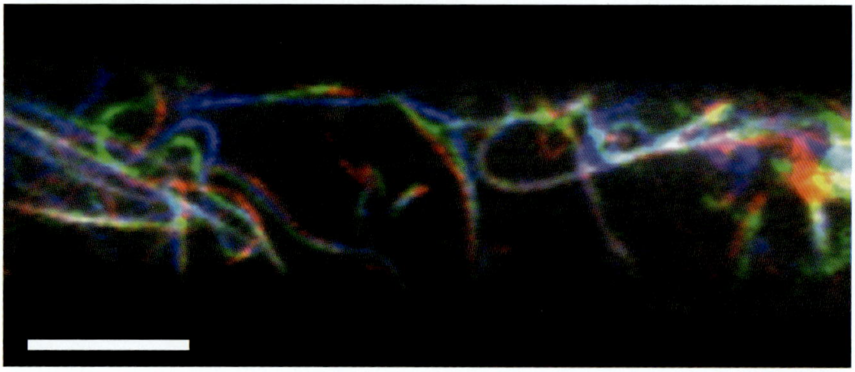

Figure 5.1 Actin organization and remodeling in epidermal cells is complex and fast. A time-series projection for an epidermal cell from the dark-grown hypocotyl of Arabidopsis expressing GFP-fABD2 is shown. Time-lapse images were collected by variable-angle epifluorescence microscopy (VAEM). Three consecutive images, separated by 3 s each, were colored red, green and blue and the three images superimposed. A white color indicates stable actin structures, such as actin filament bundles. Colored filaments are fainter and more ephemeral structures; they grow, disappear and translocate over the elapsed time of 6 s. These latter structures are presumed to be individual actin filaments based on their fluorescence intensity and dynamic properties (Staiger *et al.*, 2009). Scale bar: 10 μm.

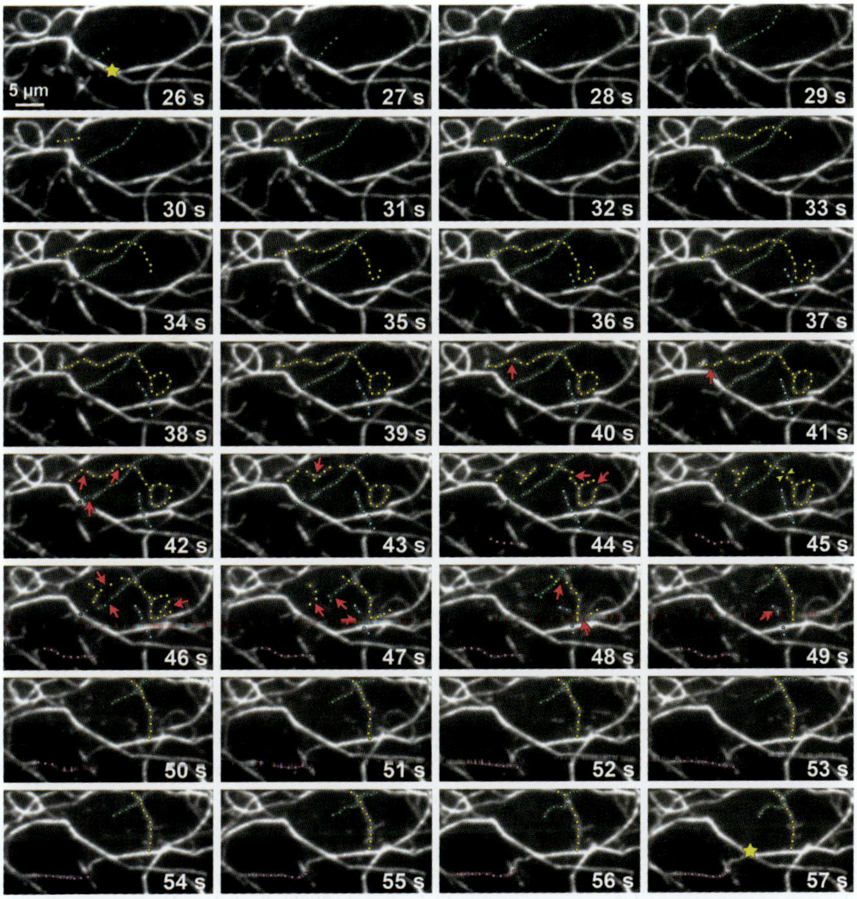

Figure 5.2 Dynamic behavior of individual actin filaments in the cortical array of Arabidopsis. Time-lapse VAEM was used to image the cortical actin cytoskeleton of dark-grown Arabidopsis hypocotyl epidermal cells expressing GFP-fABD2 (Staiger *et al.*, 2009). Two populations of actin filaments exist in the cortical array: single actin filaments and actin filament bundles. Representative single actin filaments originate from three locations: the side of a pre-existing filament or bundle (frame 26, green dots and frame 44, purple dots); *de novo* in the cytoplasm (frame 29, yellow dots) or the end of a pre-existing fragment (frame 36, blue dots). These filaments elongate rapidly and are disassembled by severing events (red arrows). Newly created ends occasionally anneal with another filament (yellow arrowheads). By contrast, a representative actin bundle (yellow star) remains relatively stationary throughout the elapsed time of approximately 30 s. Images were recorded from wild-type epidermal cells at 1 s intervals. Scale bar: 5 μm.

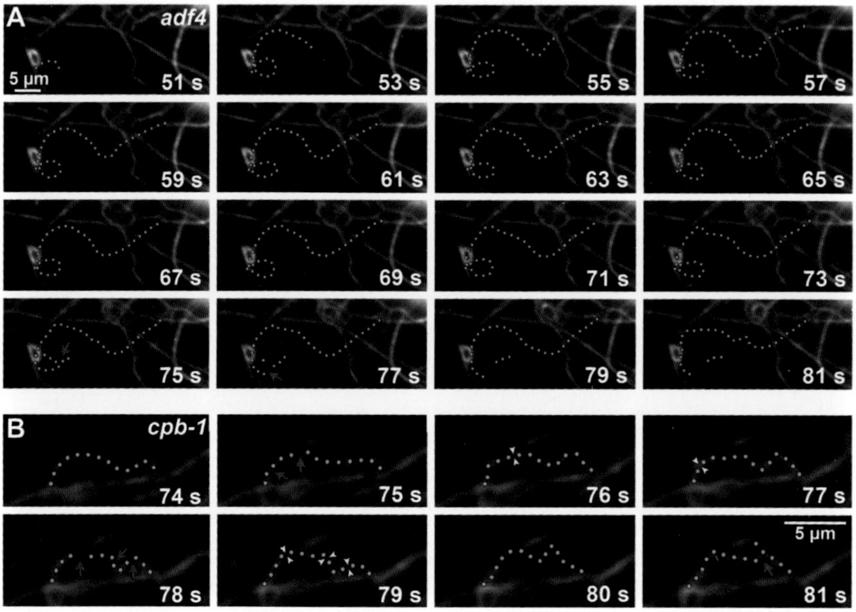

Figure 5.3 Individual actin filament dynamic properties are altered in actin-binding protein mutants. (a) Time-lapse VAEM series from an epidermal cell of *adf4* knockout mutant shows alteration of single filament dynamics. The frequency of filament severing is reduced in *adf4* epidermal cells. The representative filament (green dots) elongates rapidly and grows up to 42 μm in length (frame 73). It also persists for approximately 20 s before a few instances of severing occur (red arrows). Images were taken from a non-elongating epidermal cell of an 11-day-old dark-grown *adf4* mutant hypocotyl. Scale bar: 5 μm. (b) The frequency of filament–filament annealing is enhanced in *cpb-1* mutant cells. The highlighted filament suffers many severing events; however, several newly created ends join together within approximately 1 s. Five annealing events (yellow arrowheads) occur in the sequential time-lapse frames. Images were recorded from an epidermal cell in the non-growing region of 5-day-old *cpb-1* mutant hypocotyl by time-lapse VAEM. Scale bar: 5 μm.

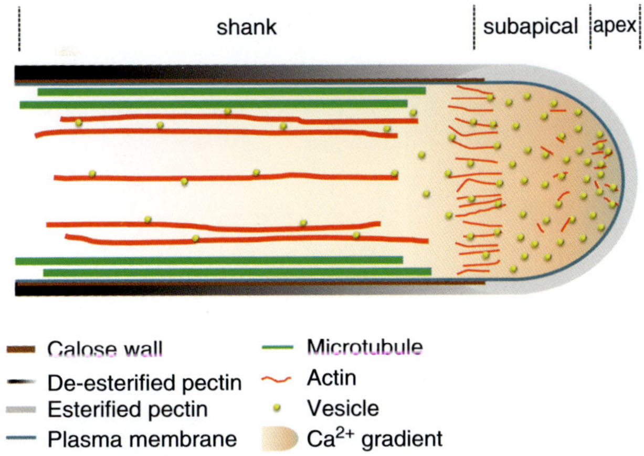

Figure 8.1 Polarized structural elements in pollen tubes. The orange shading represents the tip-focused (Ca2+) gradient. Exocytic vesicles are accumulated to a inverted cone-shaped domain at the apex of pollen tubes. Highly dynamic short F-actin is presented in the apical zone, and organized into a ring/collar structure in the subapical region. Longitudinal thick F-actin cables (red lines) and microtubule cables (green lines) exist in the shank region. The cell wall structure also varies in distinct zones. The apex contains a single soft esterified pectin apical wall, whereas the shank wall contain an outer cellulosic layer coating of de-esterified hard pectin and an inner callose layer.

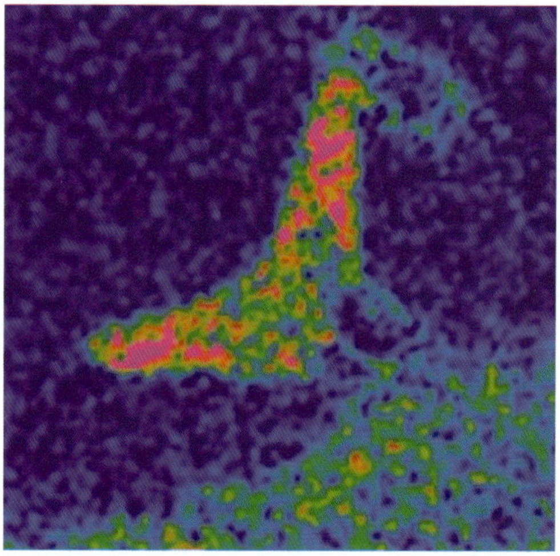

Figure 10.4 FRET analysis of stage 4 trichome expressing cameleon 3.6. The gradient of Ca^{++} from high to low is represented by the changes in color from pink to red to yellow to green to blue. Image captured and analyzed by Kevin Dorn (University of Minnesota) on a Leica TCS SP2 AOBS confocal microscope with the assistance of Elison B. Blancaflor at the Samuel Roberts Noble Foundation, Ardmore OK.

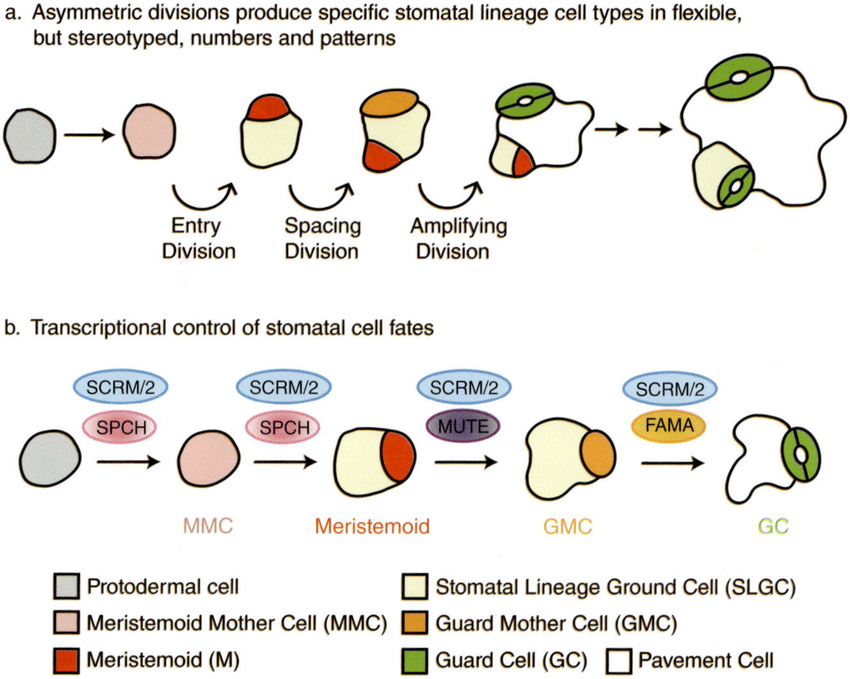

Figure 12.1 Diagram of stomatal lineage cell divisions and fates. (a) In the typical progression through differentiation, a protodermal cell (gray) becomes a meristemoid mother cell (MMC; pink) when it acquires the ability to divide asymmetrically. An MMC's first asymmetric cell division, also called an entry division, creates a smaller meristemoid (red) and larger stomatal lineage ground cell (SLGC; beige). Both meristemoids and SLGCs can divide again asymmetrically. SLGCs can either differentiate into a pavement cell or undergo another asymmetric 'spacing' cell division to form another meristemoid/SLGC pair with the meristemoid spaced away from committed stomatal precursor cells. 'Amplifying division' refers to the divisions made by meristemoids to produce another meristemoid/SLGC pair. After amplification, a meristemoid transitions into a guard mother cell (GMC; orange) and eventually differentiates into paired guard cells (GCs; green). (b) Stomatal lineage progression is mediated by heterodimers of three closely related and cell-type specifically expressed bHLHs, SPEECHLESS(SPCH), MUTE, and FAMA with partners SCRM and/or SCRM2. These transcription factor complexes are necessary and sufficient to determinate the meristemoid, GMC, and GC fates during stomatal development.

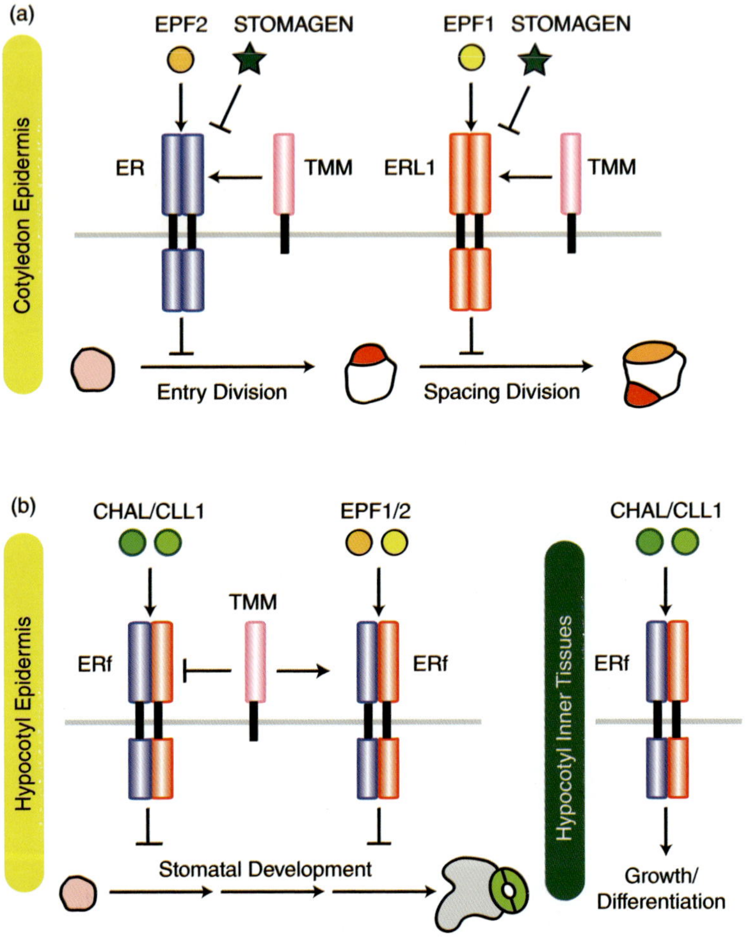

Figure 12.2 Diagram of ligand-receptor interactions and function of TMM during stomatal development. (a) In the cotyledon or leaf epidermis, EPIDERMAL PATTERNING FACTOR 2 (EPF2; orange) secreted from MMCs is perceived by the receptor-like kinase ERECTA (ER; blue) present in protodermal cells. The activity of ER is enhanced here by TOO MANY MOUTHS (TMM; pink) resulting in a signal that inhibits entry division. EPF1 (yellow) produced in meristemoids, GMCs, and GCs interacts with ERECTA-LIKE 1 (ERL1; red) which, in association with TMM, transduces the signal to inhibit spacing divisions. STOMAGEN (green) made by internal mesophyll cells opposes EPF1/2 activity, possibly by competing for receptor binding, and promotes divisions. (b) In hypocotyls or stems, EPF-Like (EPFL) ligand and ER family (ERf) receptor are required for both stomatal pattern and stem growth. CHALLAH (CHAL; dark green) and its family members (e.g. CHALLAH-Like 1, CLL1; light green), whose normal roles are in stem growth, could 'escape' from internal tissues and negatively regulate stomatal development. TMM is only expressed in stomatal lineage cells. When TMM is present, CHAL/CLL1 signaling through ERf receptors is dampened while EPF1/2 signaling is permitted. In the absence of TMM however, a CHAL/CLL1-ERf module transmits information to prevent stomatal production; consequently, a 'no stomata' phenotype is found in *tmm* hypocotyls. Arrows indicate positive influences in signaling process and T-bars represent negative influences.

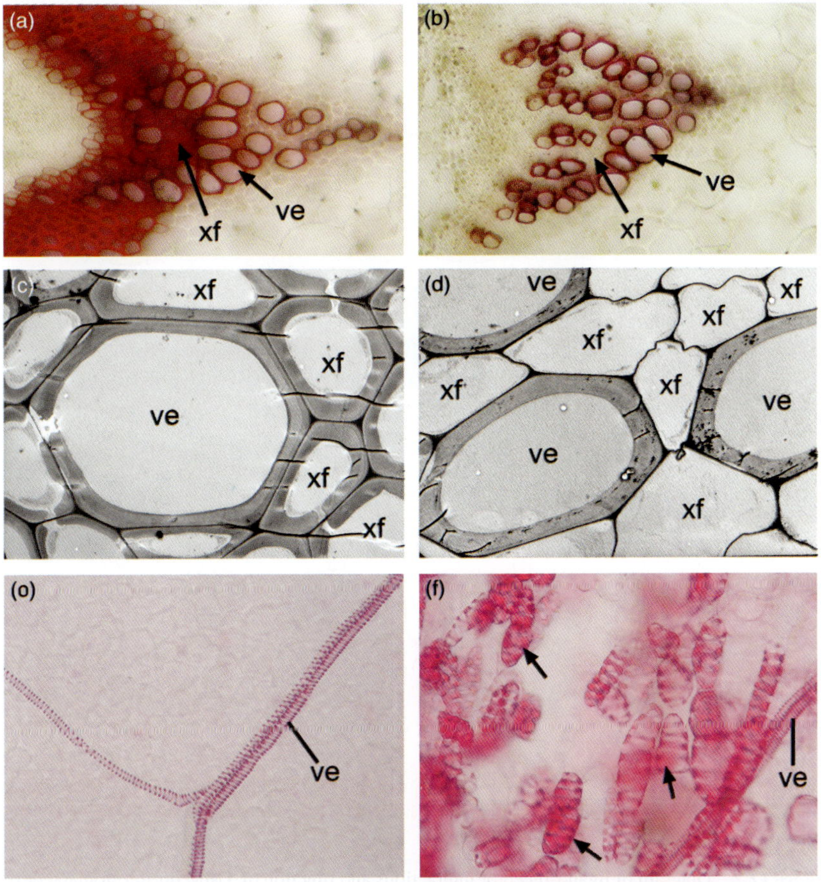

Figure 13.1 The Arabidopsis secondary wall NAC gene SND1 is a master switch for activating the secondary wall biosynthetic program in xylary fibers. (a, b) Cross-sections of stem xylem bundles showing that simultaneous mutations of the *SND1* and *NST1* genes (b) cause a specific loss of lignified secondary walls in xylary fibers but not in vessels compared with the wild type (a) (stained red with phloroglucinol-HCl). (c, d) Transmission electron micrographs of stem xylem cells showing that the *snd1 nst1* double mutant (d) lacks thickened secondary walls in xylary fibers but not in vessels compared with the wild type (c). (e, f) SND1 overexpression induces helical secondary wall thickening (arrows) in mesophyll cells of Arabidopsis leaves (f) compared to the wild type in which only vessels in leaf veins have helical secondary wall thickening (e).

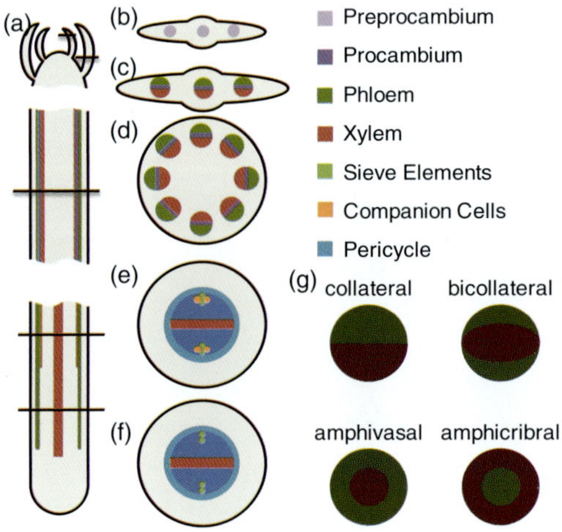

Figure 14.1 Primary phloem organization in Arabidopsis shoot and root. (a) Longitudinal section through the shoot apex, shoot, and root. (b) Cross-section of an early developing leaf showing the pre-procambial bundles which precede the vascular bundles. (c) Cross-section of a leaf showing established vascular bundles where primary phloem and xylem differentiate asymmetrically from a separating layer of procambium. (d) Cross-section of the stem showing primary vascular bundles. (e) Cross-section of the root showing the primary vascular patterning with two phloem poles consisting of sieve elements and companion cells flanking the xylem axis. (f) Cross-section of the root tip showing two poles of protophloem sieve elements flanking the xylem axis. (g) In different species, the radial patterning of the vasculature falls into distinct categories. In the collateral pattern, abaxial phloem is separated from adaxial xylem. In the bicollateral pattern, a central xylem pole is flanked by adaxial and abaxial phloem poles. In the amphivasal pattern, a xylem core is completely surrounded by phloem. In the amphicribal pattern, a phloem core is completely surrounded by xylem.

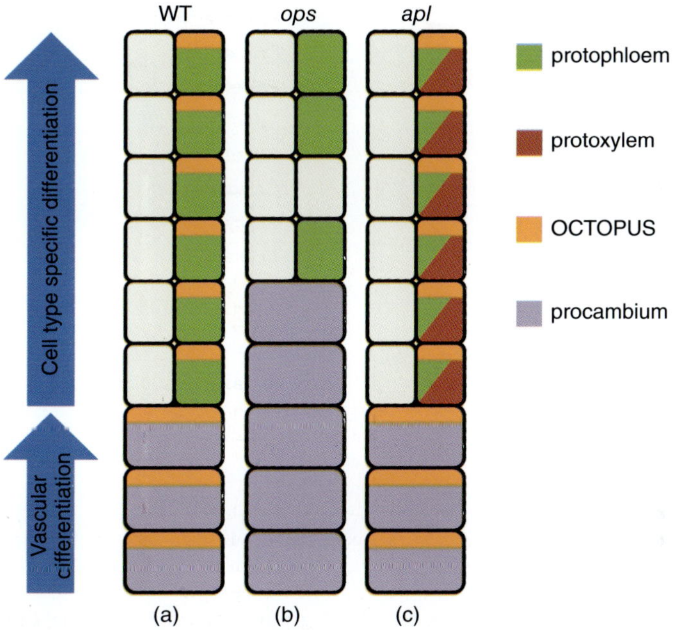

Figure 14.4 Model of OPS and APL action. (a) OPS is located at the apical plasmamembrane in the procambium and phloem lineage. OPS interprets vascular signals for phloem differentiation such as APL. (b) In the *ops* mutant, phloem differentiation is delayed. The procambial cell number is increased and gaps of undifferentiated cells are visible in the protophloem strand. (c) In the *apl* mutant, the initiation of phloem differentiation is largely unperturbed and proper protophloem fails to emerge. In its place, protophloem/protoxylem hybrid cells appear.

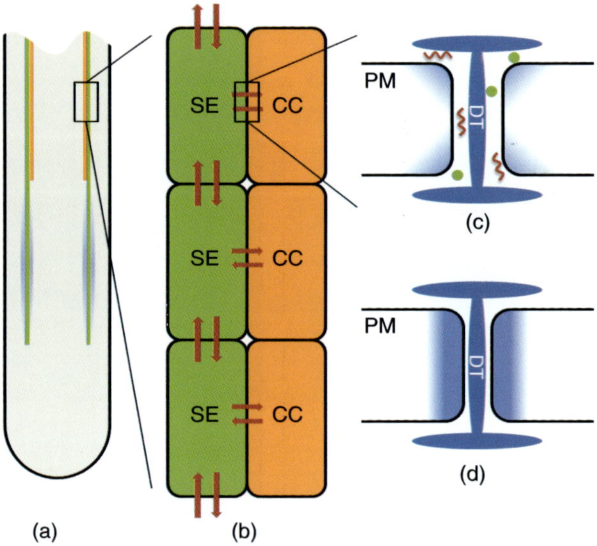

Figure 14.5 Schematic overview of phloem transport. (a) Phloem unloading occurs in protophloem sieve element strands (light blue). (b) Sieve elements are connected to each other through the sieve plates, enabling high-throughput long-distance transport in the phloem. They also share abundant symplastic connections with companion cells through plasmodesmata. (c) Schematic close-up of a plasmodesmatal connection with the desmotubule (DT). This connection enables the symplastic movement of compounds such as small RNAs or proteins. Callose (light blue) regulates the aperture size. (d) An increase in callose accumulation, e.g., through the action of the modified CalS3m, reduces aperture size to block all movement.